COMPSTAT 1984

Proceedings in Computational Statistics

6th Symposium held at Prague 1984

Edited by
T. Havránek
Z. Šidák
M. Novák

Springer-Verlag Berlin Heidelberg GmbH 1984

CIP-Kurztitelaufnahme der Deutschen Bibliothek

COMPSTAT: 06, 1984, Prag
COMPSTAT: proceedings in computational statistics;
6. symposium.

held at Prague 1984/ed by T. Havranek; Z. Sidák;
M. Novák.
ISBN 978-3-7051-0007-7 ISBN 978-3-642-51883-6 (eBook)
DOI 10.1007/978-3-642-51883-6

Originally published by Physica-Verlag Ges.m.b.H., Vienna/Austria in 1984

for IASC (International Association for Statistical Computing) 1984

ISBN 978-3-7051-0007-7

Preface

Looking into the recent history of computational statistics, major changes in this field as well as in attitudes of workers to it may be noticed over the past decade. The preface of the COMPSTAT 1974 conference volume started by the following words: "Perhaps neither mathematicians specialized in probability theory or statistics nor experts in electronic data processing will look on computational statistics as a serious scientific subject." We feel now, ten years later, that this field has changed to be considered a serious scientific subject. This leads, on the one hand, to an increasing number of scientific papers devoted to it, but, on the other hand, it is much more complicated task to recognize substantial scientific contributions. In the present volume we have collected papers that we believe to give some new insight into the problems of computational statistics.

It was a difficult task to select 65 contributed papers, presented here, from 301 submitted abstracts. The main criterion, as in preceding COMPSTAT conferences, was that the papers should show some novelty of their contents both in statistics and in computing. The selection was done by the Scientific Programme Committee consisting of J. Anděl, A. Bartkowiak, J. Chambers, A. Czurgay, A. de Falguerolles, T. Havránek (Secretary), N. Lauro, J. de Leeuw, G. Mélard, S. Mustonen, G.J.S. Ross, Z. Šidák (Chairman), O.V. Staroverov, R. Struck, and N. Victor.

The meeting was arranged by the Central Computing Center of the Czechoslovak Academy of Sciences in cooperation with the Department of Biomathematics of the Institute of Physiology of the Czecjoslovak Academy of Sciences, Mathematical Institute of the Czechoslovak Academy of Sciences, and the Faculty of Mathematics and Physics of the Charles University, Prague.

The conference was organized by the Local Organizing Committee consisting of M.K. Chytil, Z. Fabian, T. Havránek (Secretary), D. Húsek, P. Jirků, O. Kufudaki, B. Louvar, M. Novák (Chairman) P. Pačes, M. Pravdová, S. Řízek, G. Vítková, and Z. Závorková. We wish to express our gratitude to the members of both committees and to other colleagues of ours from the above mentioned institutions whose efforts enabled that the conference could be held in Prague.

Editors

Contents

Sint, P.: Roots of computational statistics (invited paper) 9

Computational statistics in random processes

Ansley, C.F., R. Kohn: New algorithmic developments for estimation problems in time series (invited paper) 23
Mélard, G.: On fast algorithms for several problems in time series models. 41
Krzyśko, M., Smoczyński: Parameter estimation and order determination of a multivariate autoregressive process 35

Computational aspects of robustness

Michálek, Jiří, Vajda, I., Víšek: New topic in robust statistics with applications (invited paper). 73
Antoch, J., Collomb, G., Hassani, S.: Robustness in parametric and non-parametric regression estimation: an investigation by computer simulations 49
Glasbey, C.A.: CEVOPE – a program for finding conservative estimates of the variances of regression parameter estimators when errors are serially correlated 55
Gomes, M.I.: Robustness of Gumbel statistic for distribution functions in the domain of attraction of a type I distribution of largest values. 61
White, A.P., Still, A.W.: Monte-Carlo analysis of variance 90
Müller, H.-G.: Boundary effects in nonparametric curve estimation models. 83
Lužar, V.: Asymptotic distribution of the roots of a correlation matrix. 67

Computational graphics in statistical data analysis

Gordesch, J.: Graphical methods for the manipulation and solution of systems of linear equations. 99
Puranen, J.: DISTEP – dynamic interactive statistical teaching package. 105

Various statistical and data analytic methods

Zagorujko, N.: On discriminant analysis (invited paper) 172
Van der Sluis, D.M., Schaafsma, W.: POSCON – a decision-support system in diagnosis and prognosis based on a statistical approach. 160
Tracy, D.S., Tajuddin, I.H.: Simulated powers under two different approximations to the exact distribution of MRPP statistics. 166

Perez, A.: "Barycenter" of a set of probability measures and its application in statistical decision 154
Payne, R.W., Dixon, T.J.: A study of selection criteria for constructing identification keys. 148
Läuter, J.: Algorithms of discriminant analysis using parameter restrictions for diminishing the error rate 131
Diday, E., Moreau, J.V.: Hierarchical inference 119
Murtagh, F.: A review of fast techniques for nearest neighbour searching 143
Meulman, J., Heiser, W.J.: Constrained multidimensional scaling: more directions than dimensions. 137
Korhonen, P.J.: An interactive multiple criteria approach to finding subjective principal components. 125
Aluja Banet, T., Lebart, L.: Local and partial principal component analysis (PCA) and correspondence analysis (CA) 113

Numerical aspects and complexity of statistical algorithms

Křivánek, M., Morávek, J.: On NP-hardness in hierarchical clustering. 189
Jöckel, K.-H.: Computational aspects of Monte-Carlo tests 183

Linear Models

Kubáček, L.: Reliability of calculations in the linear estimation and some problems of the analysis of a multistage regression experiment (invited paper). . . . 214
Gilchrist, R., Scallan, A.: Parametric link functions in generalized linear models . 203
Kleffe, J.: Statistical analysis of mixed linear models with inhomogeneous variances. 209
Bethlehem, J.G.: Linweight: a program for weithting sample survey data 197

Intelligent software in statistical data analysis

Gale, W.A., Pregibon, D.: Constructing an expert system for data analysis by working examples (invited paper). 227
Pregibon, D., Gale, W.A.: REX: an expert system for regression analysis 242
Kowalski, A., Schütt, A.: Analysis of suppositions. 237
Pukkila, T., Puntanen, S., Stenman, O.: On the possibilities to automate the study of statistical dependence between two variables 249

Integration of statistical data base management and data analysis

Neumann, K.: Metainformation as a tool for integration. 273
Rauch, L.: Interface problems between software packages used in statistical information system 285

Borzemski, L., Koszalka, L.: Statistical data analysis using a data base for experiment design and systems identification 257
Kurzyński, M.W., Lebiediewa, S.: Processing of statistical data for multistage recognition system using data base 268
Rafanelli, M., Ricci, F.L.: Statistical database: an interactive language for logical schema definition by means of a model based on graphs 279
Jomier, G., Kezouit, O., Ralambondrainy, H.: SICLA-PEPIN: A system integrating data analysis and relational data base management system. 263

Software for parallel processing in statistical data analysis

Lafaye de Micheaux, D.: Parallelization of algorithms in the practice of statistical data analysis 293
Ososkov, G.A., Vajteršic, M.: A fast parallel algorithm for the recognition of ionized particle tracks 301

Evaluation and enhancements of statistical software

Anderson, A.J.B.: The enhancement of Fortran for statistical analysis. 311
Nedoma, P.: Design of experimental statistical packages 347
De Jong, V.J.: A multilevel approach to improve the flexibility of statistical software 323
Molenaar, I.W., Broersma, H.: Interpretation of statistical software output: some behavioral studies 341
Dekker, A.L.: Comparing statistical software packages on tabulation efficiency. . 317
Kuik, D.J.: QUESTOR, a conversational preprocessor to BMDP. 329
Michálek, Jaroslav, Němcová, M., Popelínský, Rebíčková, M.: The system for multivariate statistical data processing. 335

Software in categorical data analysis

Whitakker, J.: Fitting all possible decomposable and graphical models to multiway contingency tables. 401
Edwards, D.: A computer intensive approach to the analysis of sparse multidimensional contigency tables 355
Murphy, B.P.: Similarities and differences between linear, logistic and log-linear models for survival analysis; some practical observations. 378
Lane, P.W.: Methods for summarizing the analysis of categorical data 366
Rudas, T.: Stepwise discriminant analysis procedure for categorical variable. . . . 389

Morineau, A.: Computational and statistical methods for exploratory analysis of textual data 372
Scherb, H., Welzl, G.: A recursive algorithm for the determination of the uniformly most powerful unbiased test in fourfold tables 395
Pokorný, D.: Optimal collapsing of two-way contingency tables including structural zeroes 383
Hájek, P.: The new version of the GUHA procedure ASSOC (generating hypotheses on associations) – mathematical foundations. 360

Optimization techniques in statistics

Narula, S.C., Wellington, J.F.: On special purpose algorithms for the MSAE and the MMAE regression. 440
Prékopa, A.: Programming under probabilistic constraints and its applications in statistics (invited paper) 446
Meester, I.: Maximum likelihood estimation for mixtures of distributions 427
Lejeune, M.: Optimization in nonparametric estimation 421
Ross, G.J.S.: Parallel model analysis: fitting non-linear models to several sets of data 458
Messean, A., Nicole, O., Vila, J.P.: HAUS 82: a new tool for nonlinear fitting and the study of non-linear models. 434
Kutas, T.: A new method for solving nonlinear least squares problem 415
Kaishev, V.K.: A computer program package for solving spline-regression problems 409

Software for data preprocessing (checking, handling of missing values reduction)

Chytil, M.K.: Is data preprocessing a computational process only? 467
Liski, E.: Missing data problems in growth curves 479
Dickson, J.M.: Data capture and validation using portable terminals 473

Error free and formal computations in statistics

Collombier, D.: Error-free computation in design of experiments (invited paper). 487
Vuchkov, I.N., Damgaliev, D.L., Donev, A.N.: Comparison of some algorithms for discrete D-optimal design generation 503
Tiit, E.-M.: Formal computation of regression parameters. 497

Computational problems in spatial data analysis

Košmelj, M.: Means for analysis of space-time relationship 509

Roots of Computational Statistics

P. Sint, Wien

This paper describes some of the nodes in history where the traditions of computing with machinery and the develooment of statistical methods met. It will give also some ideas about the driving forces which brought these developments about and why there exist fundamental connections between them.

The preparation of this paper was partly sponsored by IBM Austria

***

A base for a report of this kind should be a collection of monographs on the topics concerned and an overwiew of the whole development which I can't pretend to possess. However as there has not been too much work going on in this direction it may well be worthwile to point out some factors in the development. The paper as it stands scratches, however, scarcley the surface, especially concerning the analysis of the facts. As historical work is cumulative in the sense that the information on relevant influences, e.g. on the persons acting, is scattered in varying places, and may be found sometimes only after long study of the specific topic under consideration and often only accidentally, it is likely that most readers will be able to point out factors overlooked or given a wrong emphasis. This paper should stimulate this kind of interest in historical developments, especially as we are still living in a period where many of those who have contributed to a quantum leap in the development of statistical computation are still alive, and where many of you can contribute to the preservation of historical records.

Already most of the developers of mechanical calculating machines are, sometimes in roundabout ways, connected to statistics:

Least of them Wilhelm Schickard (1592-1635) who built the first mechanical calculator which could already perform multiplications and divisions, following an algorithmic idea of Kepler in 1623.

The builder of the next calculating machine was Blaise Pascal (1623-1662) who constructed a machine for addition and subtraction for accounting purposes in 1642-44, and studied well known early problems in the theory of games of chance, which is now considered to be the beginning of the modern theory of probability. Thus Pascal combined activities related to two early and important roots of computational statistics.

With Gottfried Wilhelm Leibnitz (1646-1716) the mathematical - computational aspect, the original "old" statistics and the political arithmetic - the forerunner of the "modern" statistics - meet for the first and probably only time:

The word statistics is derived from the italian/latin ragione di stato/disciplina status meant to describe the knowledge of statesmanship and/or state affairs. The "statist" in Shakespears Hamlet and Cymbeline is used that way. Gabriel Naude (1639) was the first to use the adjective statisticus in print (Pearson 1978 p.4). It was the older contemporary of Leibnitz, Hermann Conring (1606 - 1681), personal physician to a Swedish queen, and a real polyhistor, who founded, based on these Italian and also on French, Dutch and German forerunners, the later German and continental school of "statistics" as an art of collecting data relevant for decision making by the government. Conring was advisor to Baron Boineburg, a political figure of European importance, who used later on Leibnitz's services. Conring - jealousy may have played a part - disapproved of Leibnitz's mathematical studies and warned him not to waste his time. A fuller picture of their interesting relationship may be found in Lazarsfeld (1961). Conring, who is well known in the history of law, approaches the "questions of state" by concepts based on Aristotelian thinking. Leibnitz contributed to the discussion about the "old statistics" as an active diplomat e.g. by the memoir on the "Entwurf gewisser Staatstafeln" (design of certain state tables, cf. Rassem 1980). We would describe his and Conrings efforts as a structured approach to data collection, and as far as Leibnitz is concerned as a systematic effort of filing this information in a readily accessible data base, the entries of the table being of non numerical nature. While Leibnitz used the philosophical and practical vocabulary of Conring and his contemporaries the whole idea was related to his "Arte Combinatoria" (1666) even though he did not mention those connections in the "Staatstafeln". The combinatory art has again to do with combinatorics and probability, following Pascal, on the one hand, and with the art of logical inference and calculation on the other. This last aspect led Leibnitz to the construction of a four species machine, incorporating the Leibnitz wheel, from now on nearly always present in such machines. It was this machine which was of principle interest for the members of the Academie des Sciences in Paris, and of the Royal Society in London, where he became a Fellow in 1673.

This last fact shows that Leibnitz could have known the "political arithmeticians" John Graunt (1620 - 1674, mortality tables 1662) and William Petty (1623-1687) the founders of quantitative statistics in Britain, but he seems never to have written on this topic.
Nevertheless he brought the data based on registration figures of christenings and funerals in Breslau to the attention of the Royal Society who in turn asked Halley for advice and this resulted in Halleys famous paper on mortality in 1693 which contained the first real life table (John 1884, Pearson 1978).

Conring avoided the word statistics which had in his time assumed a derogatory meaning, statisticus, "statistisch" being a mode of using ones knowledge in a dubious way, used at best in an ironical sense, a statist even being a criminal or at least a "pseudopoliticus". But already around 1700 Leibniz proposed the planned "Academy" in Vienna to include "statistics" in its agenda (Adam 1973 p.116).

The word was made respectable again by jurists like Martin Schmeitzel (+1747) who called his lectures "Collegio politico-statisticum". He and the once famous Gottfried Achenwall (1719-1772), recognised as the "founder" or "father" of statistics by some even late in the 19th century, saw their "Statistik" as a further development of Conring's ideas. Contrary to his pupils he knew and appreciated the mathematical methods, especially those of the political arithmeticians, however he considered himself a collector of data on the "constitution" of countries and regions under an "historical" point of view. Numerical data were not in the centre of his interest. His pupils, especially August L.v.Schlözer (1735 - 1809) fought the "table statisticians" who took out the "real" flesh and bone of statistics and replaced it by the analysis of a "skeleton" of pure numbers by "table serfs" (John 1884, p.128). This kind of reasoning has remained with us ever since. At that time the Scotsman John Sinclair had adopted the term statistics for the collection of numerical data in 1790, and this new meaning replaced very fast the term "political arithmetic" in Britain (Pearson 1973 p.8).

The next chapter is probably one of the most vivid and at the same time most tragic in the development of computing: Charles Babbage (1791-1871) started from work typical for his period. He wanted to improve the calculation of tables which were the main tools for calculations for still another century to come. By calculation on reasonably adjusted differences many tables could be calculated or at least filled starting from marginal values. The idea of such a machine seems first to have been proposed by J.H.Müller in 1786 (d'Ocagne), but he did not build a prototype. Babbage planned his first calculating machine in 1812 and finished the first model in 1822. Babbage's first report on his difference engine was accepted in 1823. While a number of working parts were constructed and demonstrated he met construction problems with others. The net result of his efforts was that work on the machine was stopped in 1833. However, because of the technological difficulties met, Babbage carefully studied the technological processes and economic conditions under which a technology would become viable. This was related to the "statistics" of his period. Babbage had already published on statistical topics e.g. on games of chance (1823, read before the Royal Society of Edinburgh 1820) and on the "Comparative View of various Institutions for the Assurance of Life" (London 1826).

All this seems to be the reason why Babbage became a driving force in the establishment, first of the section F - Statistics - in the British Association for the Advancement of Science (of which he was a cofounder) and later under the influence of Quetelet (1796 -1874) of the "Statistical Society of London". Indeed another famous 19th century statistician William Farr (1807 - 1883) stated in his presidential adress before the Statistical Society of London 1871 that "Babbage was, in reality more than any other man its founder" (JStSL, vol.34, p.411). Later during the "Congress period of statistics" (1853 - 1878), he was an active participant and speaker in those congresses.

Perhaps the main reason that the difference engine was never finished was that Babbage went further to construct the analytical machine, a real programmable mechanical computer which consisted of 1000 mechanical registers and "mill", a kind of CPU, to do the calculations directed by punched cards of the Jacquard loom type (invented by Jacquard 1805).

Even though Babbage did not finish his machines, Peter Georg Scheutz (1785-1873) and Edward Scheutz from Sweden built a difference engine during his lifetime and displayed it in London with Babbage's active help. A copy of this machine, the only one in practical use, was built for the English Government and installed at the Registrar General's Department at Somerset house. This was a machine for statistical purposes and the life tables used for several years by insurance companies were calculated on this machine. An improved version of this machine was constructed by Martin Wiberg (1826-1905), also from Stockholm about 1874. Torres Quevedo (1852-1936) from Spain proposed an electromechanical solution to Babbage's ideas in 1893 (La Nature 1914, p.56), and Babbage's son finished his analytical engine 1905. Meanwhile the industrial production of calculating machines had been started 1824 by the insurance banker Fr.X.Thomas in Colmar.

While natural scientists had already been prominent in the development of statistical methods, those statistical methods now penetrated the natural sciences to change them permanently. On the one hand we have Francis Galton (1822 - 1911), a cousin of Charles Darwin, introducing statistics in a fundamental way into biology which in its turn triggered of the important development in British biometry, and laid the base for the British prominence in the field. On the other hand Ludwig Boltzmann (1844 - 1906) founded statistical thermodynamics, based on preliminary work by Maxwell, by giving a statistical explanation of entropy and describing the dynamical behavior of thermic processes. The development of statistical theory in physics was parallel and partly independent of the development in vital and biometric statistics. However, this development based on Laplace, the French mathematicians from Condorcet to Poisson and the Russian school of probability and statistics (Maistrov 1974) transformed, statistics to an abstract science separated from its roots in "political" arithmetic and state sciences.

The next and most crucial advance was happening in the US Bureau of Census. John Shaw Billings (1839-1892), a surgeon in charge of the work on vital statistics both for the 1880 and 1890 census, was interested in getting the work of hand tallying the statistical items done in some mechanical way. Thus he proposed to Herman Hollerith (1860-1929) a young engineer "something on the principle of the Jacquard loom" to do the work. Hollerith tells in his recollections 1919: "After studying the problem I went back to Dr.Billings and said that I thought I could work out a solution to the problem and asked if he would go in with me. The Doctor said he was not interested any further than to see some solution of the problem worked out." Thus, concernig the result the <u>punched card statistical machine</u>, the biostatistician Raymond Pearl could state: "Billings was the originator, the <u>discoverer</u>, who contributed that which lies at the core of every scientific discovery, namely an original idea that proved in the trial to be sound and good; Hollerith built a machine that implemented the idea in practical performance, the accomplishment of the successful <u>inventor</u>."

Holleriths machines were used successfully in the 1880 census in the US and in the same year a version improved by the German born Viennese Otto Schäffler in the Austro - Hungarian Empire (Schäfflers machine was, different to Holleriths machines, which could be used only for one type of counting, a machine "programmable" by a telephone switchboard cf. Zemanek 1970).

Now Samuel W.Stratton (1861-1931) enterd the picture who had built a analog harmonic analyzer for determining Fourier coefficients by using the forces of spiral springs together with the physicist Michelson. He became the founder of the National Bureau of Standards (1903) and was responsible for setting up the Census Machine shop, commissioned to continue the developments of Hollerith's machine.
The laboratory developed several refinements to the machine after some disagreement on rental charges with Hollerith. Printing count-

ers were developed 1906/7 and James Powers developed in this laboratory an automatic card punching machine. The Powers Tabulating Machine Co. founded by him 1911 was for many years the main competitor of Hollerith (CTR, later IBM).

It is in this context that Goldstine states: It is a noteworthy feature of our American system that much of the computer field owes its existence to the generosity of our government in giving to its employees and university contractors the rights to inventions made with government funds (p.71).

Starting as late as 1928 the Hollerith tabulating machines were extensively used in scientific calculations (Leslie John Comrie 1893-1950, tables of th moon positions). In 1929 Thomas J.Watson chief executive officer of IBM set up the Columbia University Statistical Bureau with such success that he authorized the construction of a special "Difference Tabulator" a modern version of the Babbage/Scheutz machine. The astronomer Wallace J. Eckert (1902-1971) expanded the bureau to an astronomical computing bureau. The whole enterprise brought IBM into contact with scientific computing.

Computation was done at that time by the method of progressive digiting. Roughly this worked the following way: Let us assume we have to calculate (X' X). The cards, containing each a matrix row, were sorted by a one digit column with which to multiply, receiving 10 card packs corresponding to the digits 9,8...1,0. Now the card contents were added, running through first the 9 deck, afterwards "progressively" the deck with the 9 and the 8 digits, and so on each time adding a new deck to the cards taken out of the machine. Adding up the partial results gave the correct multiplied values. (for a somewhat more complete description see Hartley in: Owen 1976, for a real example Comrie et al. 1937, see also below).

Karl Pearson (1857-1936) was not only a famous statistician, calculating many important statistical tables, but as a computational specialist he introduced the method of finite differences into ballistic computation in England (Goldstine p.76). In the United States this was done by the astronomer Forest Ray Moulton heading one of the two important ballistic groups during the first world war. The second group was headed by Oswald Veblen (1880-1960) brother of the famous sociologist Thorstein Veblen. After many years a a professor in Priceton he became later, together with Einstein, one of the first two members of the famous Institute of Advanced Studies in Princeton. The physicist Ehrenfest proposed to Oswald Veblen to bring over Eugene P.Wigner later Nobel Laureate in Physics and John von Neumann (1903-1957), and this resulted in joint appointments in mathematics and physics, starting from January 1930. Neumann had given the right formulation of the entropy concept in quantum physics and as a modern physicist statistical problems were for him a natural environment. Thus he had developed the minimax principle to study probabilistic strategies for games, applying them to economic behavior.

Before those events, however, the computer had been further developed: electromechanically by Aiken at Harvard, Stibitz at Bell Telephone Laboratories and, before and during the war, by Konrad Zuse in Germany. A first electronic calculator has been built by John V.Atanasoff (Iowa State College) together with Clifford Berry "early in 1940". At about the same time Alan Mathison Turing

(1912-1954) who had already shortly after graduation in Cambridge, UK, written on "Computable Numbers" and invented the theoretical "Turing Machine", up to now not surpassed in its generality, participated in the construction of an electronic automaton, able to decipher the German military code produced by the ENIGMA machine, in several generations (ULTRA with Good and Michie, COLOSSUS using 2000 electronic tubes).

Atanasoff's machine influenced John W.Mauchly who visited him in 1941 (then at Ursinus College, Philadelphia and soon after at the Moore School of Electrical Engineering, University of Pennsylvania). He was interested in computing because he had worked with Vannevar Bush's analogue "differential analyzer" and because of discussions on the application of statistics to diverse fields, including weather prediction, with Sam Wilks (Princeton). He interested his graduate student Presper Eckert Jr., an able electronic engineer, in the problems of electronic computation and came into contact with the mathematician Hermann Goldstine who was responsible for preparing shooting tables at the Ballistic Research Laboratory at Aberdeen Proving Grounds during the war. Due to Veblen's influence this was a center of highest scientific excellence organized on wartime basis (even all regular army officers trained at MIT and one assistant director, Leslie E.Simon a specialist in statistical quality control of ammunition). This resulted in the joint development of the first electronic computer for practical applications. The ENIAC was dedicated and running 15 Feb. 1945. The machine was finished mainly without the help of von Neumann. However Goldstine came into contact with Neumann, who was at that time (1944)at Los Alamos working on the development of the atomic bomb, and soon von Neumann came to influence the development.

The next machine, the EDVAC, the concept of which was elaborated at the Moore School is described especially in the "First Draft of a Report on the EDVAC" by J. v.Neumann (Philadelphia 1945, June) which shows von Neumann's immense influence and was a blueprint of developments in Europe. In fact the EDSAC in GB was finished before the EDVAC (Wilkes 1956).

Eckert and Mauchly set up a partnership in 1946 to build the BINAC the first American Machine after the ENIAC for the Northcraft Air Co. Afterwards they formed the Eckert Mauchly Computer Corporation which was successful in obtaining a contract with the National Bureau of Standards to build the UNIVAC for the Bureau of the Census. This machine, the first to perform primarily work in statistics, became the base of a very powerful industrial development.

The machine built after the war at the Institute of Advanced Studies in Priceton bears von Neumann's mark, as does the whole computer development influenced by this "von Neumann" machine. Goldstine (p253) states that "the logical design was articulated by Burks and himself in collaboration with von Neumann and with assists from John Tukey". The statistician Tukey had been on the advisary board to von Neumann nearly from the beginning of the project (thus he became e.g. the inventor of the word bit). The first input-output organ for the institute computer was developed 1949, by the National Bureau of Standards which was eager to achieve its proper role in the computer field by modifying some teletype equipment. The i/o was however too slow and had to be replaced by an IBM punched card equipment.
Nevertheless NBS built later a number of successful machines. The IAS machine was publically announced 1952.

Now we come into the region of real electronic statistical computing. In some sense much of the early computing has influence on statistics. Naturally the problem of inverting matrices or calculating eigenvalues has implications to statistics. Thus a heuristic estimate of Hotelling seemed to show that Gauss' method for solving linear equations by elimination was unstable. This view has soon been corrected by von Neumann, Goldstine and Turing. Developments of this kind will not be our topic.

Electronic computational statistics in its narrower sense starts with a bang: The Monte Carlo Method. This is a method that really needs the computer to be applicable practically. It seems that it was first proposed by the physicist Ulam to von Neumann (Letter to Richtmeyer 1947). In any case we can already read in von Neumanns proposal for the IAS machine (memo 8 Nov.1945, see Goldstine p.254): "It will certainly open up a new approach in mathematical statistics: the approach by computed statistical experiments...". And it seems therefore to be rather likely that one of the first - originally classified - Los Alamos problems may have been a Monte Carlo problem. In any case Metropolis and Ulam describe in the very first article on electronic computing that I found in the Journal of the American Statistical Association (1949) that the problems around the Boltzmann equations governing fission gave rise to the Monte Carlo method. This is also the content of an abstract of von Neumann and Ulam in 1945 (I have not seen the Los Alamos Declassified Document by D.Hawkins and S. Ulam 1944, see Puri in: Owen 1976). S.Ulam from Poland was already active much earlier in the field of probability: together with Z.Lomnicki he found the association of the concept of independence with product measures. The principal results were presented at the 1932 International Congress of Mathematicians in Zürich, one year before the appearance of Kolmogorov's book (published only 1934; cf. Kac in:Gani 1982)

The basic method to study random phenomena by mathematical experiments is much older, and the need of random variables for that purpose was already encouraged by the tables of random numbers produced by Leonard H.C.Tippet in 1927 under the close guidance of Karl Pearson. What was really new was the possibility of large scale use of pseudo random numbers and, because of its importance, the new name.

The further development of the Monte Carlo methods was very fast, visible by conferences in the 40ies. For its early history and for the forerunners one should refer to a classical monograph in the field by Hammersley and Handscomb (1964) or NBS (1951).

The next important computational step in this direction was the emergence of simulation systems. Geoffrey Gordon started his work on simulation 1954 at the Telecommunication Research Establishment in England (for this and the following cf. Wexelblatt p.403ff). Later he arrived via Bell Telephone Laboratories at an IBM division, where he developed, based on internal experience, GPSS (General Purpose Simulation System) the program which made simulation easily accessible for a wide class of users (available 1961).

In Norway a different development took place: Kristen Nygaard was working at the Norwegian Defence Research Establishment where Monte Carlo simulation by hand, related to the construction of Norways first nuclear reactor, was introduced 1949/50. Ole-Johan Dahl joined this institution a little later (1952) and they worked together on operation research problems for which Monte Carlo simulation turned out to be an appropriate tool. 1961 Nygaard started in the newly founded Norwegian Computing Center (NCC), a semigovernmental research institute, the development of a simulation language. In the first document on this language SIMULA dated 1962 he states: "The status of the Simulation Language (Monte Carlo Compiler) is that I have rather clear ideas on how to describe the queuing systems,...I believe that these results have some interest even isolated from the compiler". This language, based on ALGOL, had indeed a far reaching influence on the development of computer languages. It stood at the cradle of many concepts which are present in the most modern programming languages. Obviously the need to serve the practical purpose of the simulation of stochastic systems led Nygaard and Dahl who developed this language in close cooperation (and in furious fights) to the development of programming tools fitting the needs of many other applications.

At the same time and shortly afterwards there appeared a number of further simulation languages (for an overview of the state of the art Bradley et al. 1983). Today representatives of different languages (SLAM and SIMAN by C.D.Pegden and/or A.Allen Pritsker) are fighting each other in the law courts, not just academically any more, a certain sign that simulation technology has come of age.

In another domain, the domain where it originally came from, statistical physics, the Monte Carlo Method has become a tool to analyze the fundamental behaviour of elementary particles, of matter and of our world. One may ask whether those questions are still questions of statistics. But one may answer, that it is just the question of what is statistics which may be answered by such investigations. Indeed, development in this field is intrinsically related to the questions of the reversability of time, and with it to the questions on the fundamental meaning of empirical probability. Relevant results are discussed in the two volumes edited by K.Binder (1979,1984). An elementary treatment of some aspects is given in an article by Brian Hayes in the Scientific American.

In his introduction Binder states clearly that "the development leads to : i)use of array or vector processors specifically suited for a wide class of simulation purposes, ii)special processors ... for a more specific class of problems or in the extreme case for the simulation of one single model system. In this way, Monte Carlo simulations make a considerable impact on the further develpment of scientific computing including hardware".

The Monte Carlo approach has also led to important advances in "pure" mathematics. It was used as the only practical method to compute integrals over high dimensional spaces. Statistically the sequences calculated converge with probability one to the value of the integral. The new "numerical analysis" shows, that with the proper choice of the pseudo random numbers one may achieve a guaranteed convergence (for an overview: Hlawka 1981).

It is quite clear that during the introduction of the new electronic computer, work on the more traditional "low cost" business machines and the work on the new machines went in parallel. The continuity may be seen e.g. from an article by Laderman and Abramowitz (JASA 1949) who described the application of accounting machines for the calculation of tables, following Babbage using 7th differences.

Milton Abramowitz was already at the Mathematical Tables Project of the US Bureau of Standards. He was later to become one of the driving forces, and the over-all editor, of the Handbook of Mathematical Functions (edited by him and Irene Stegun, produced 1952-1964). The tables were often taken over from other sources, the probability functions e.g. from the Biometrika tables of Pearson and Hartley, and from the Kelley Statistical Tables (pre electronic).

In statistics, as in all other branches with elaborate mathematical models in applied fields, important efforts of computing in the times before the advent of the electronic computer, went into the production of tables, because more complicated calculations had to be done, so to say, in advance. And most of the people applying statistical methods relied upon those tables, just as many today rely upon the finished programs and statistical packages. The tables by Fisher and Yates contain electronically computed additions only from its fifth edition (finished 1956) computed mostly at the computer at Rothamstead one of the early locations of statistical computation world-wide.

Nevertheless Healy and Dyke from this institution described in 1953 a Hollerith technique for the solution of normal equations using only a sorter and a tabulator following an idea of F.Yates.

In the same volume appears the article on electronic Computation in Economic Statistics advocating strongly the use of the new instrument to a group outside the electronic engineer - mathematician users group. The authors Brown, Houthakker and Prais had access to the EDSAC finished by Wilkes and Hartree for the Cavendish Laboratories in Cambridge England in 1949 and saw already clearly the importance to input output analysis, of Monte Carlo methods and the possibility to solve "full maximum likelihood" models in econometrics, though no programs of that kind existed yet.

The situation concerning tables in 1972 is characterized by the introduction to the second volume of the Biometrica tables (E.S. Pearson, H.O.Hartley): "It is noteworthy that about three quarters of the 69 tables were computed in North America, sometimes at Biometrikas suggestion, a fact which shows the much greater extent, for which computer facilities have been made available in recent years in that continent compared with what has been the case in the United Kingdom". This volume is dedicated rightly "to K.Pearson and L.J.Comrie who, without the aid of electronic computers contributed so much to to the production of mathematical tables".

We have looked extensively at the development of Monte Carlo because it seems to be the first, one of the most typical and perhaps one of the most spectacular developments in statistical computing. Every branch of statistical computing has its own history. Nearly no subject of statistics survived its transformation to the computer unchanged. Nearly everywhere new opportunities opened up. There are, however, a number of applications which are again specific to the advent of the electronic computer. One of them is the wide spread use of non parametric statistics, and especially the jackknife method and its offspring.

Whilst it was Quenouille who first proposed the sample reuse method of leaving one out for the reduction of bias, it was John Wilder Tukey (1915-) who brought the method to computing and invented a name for it as belonging to a class of simple instruments well adapted to a variety of jobs. Bradley Efron who advocates this approach, developed further by him, gives an overview in several papers (1979 general on the influence of the computer on statistical theory). Tukey himself, during the war "weaned away from topology to statistics" (B.Harshbarger in: Owen 1976 p.139) in the statistical Fire Control Research group, remained later in Princeton, a location so well suited for the development of statistical computing. We spoke already of his contributions to the development of the first machines. But he pioneered as well many applications in statistics, especially in time series analysis. David Brillinger (in Owen 1976 p.272) points out the many techniques and terms which have become standard in the data analysis of the field: "prewhitening", "alias", "smoothing and decimation", "taper", "bispectrum", "complex demodulation" and "cepstrum". In elementary data analysis and in data exploration he has set new standards for the intelligent and interactive use of the computer, and his methods and the lines of approach started by him will become more important with the spreading use of micros in statistics.

Perhaps this is the right place to mention another early bird: Willy Feller. The well known statistician was interested in electronic computation from the early 1950s, and served as a consultant to Goldstine, while at IBM, and followed him there as a leader of a mathematical science department of first quality.

The factor which contributed most to the adoption of statistical methods in the computer age is the development of ready to use computer packages. One complains about the easy misuse of such packages but in the overall evaluation they have done more to educate non statisticians and statisticians alike on the proper use of statistics. The high turn-around time they enable is a way to speed up the process of learning.

I will not go into details of the history of the packages, there will be even more knowledgable persons on this in this conference than on any other topic. I am not even sure on the order in which the major general purpose packages appeared: W.J.Dixon worked in Princeton during the war and his project on stand alone BMD programs was only combined in a coherent package with a unique English-based control language starting 1968. The development of SPSS began 1965 at Stanford by a group frustrated by incompatible, badly documented programs. Roald Buhler another Princeton man started PSTAT in 1963 as a one (or two) man effort, with a first user manual in 1964. SAS was conceived 1966 in Raleigh, North Carolina, where R.L.Anderson cooperated both with them, and with BMD (Gani 1962). A first comprehensive user guide appeared 1972 Other early packages were OSIRIS and DATA-TEXT(+). In Britain GLIM was an effort by R.J.Baker, M.R.B.Clarke and J.A.Nelder under the sponsorship of the Royal Statistical society, originally designed for the fitting of linear models. Similar to SAS matrix operations, GLIM's offspring GENSTAT handles matrices and allows a flexible formulation of ones own models.

Speaking of the roots of computational statistics, one is also tempted to look into the future. We have not yet mentioned the ever growing data bases on a huge variety of topics. With the help of telecommunications they are accessible to an increasingly larger group of people. This will give the opportunity and the necessity to apply statistical methods to a larger class of problems, by this larger group of people, using intelligent terminals and ever mightier personal computers. But it may be necessary to hide the statistical technology and give the user only the answer to the question he poses. This, the pressure to process increasingly non-numeric information, and other demands for more flexibility could lead to the progressive use of artificial intelligence tools. Statistics will, on the other hand, supply such tools to the field of AI. The effect may be a closely interwoven system of brainware and hardware for the handling of the important resource "information" to tackle the human problems of the future.

There are those who consider this development dangerous, and a certain doubt concerning the controllability of the new technologies can hardly be avoided. Perhaps we should regain some of the interests of the "old" statistics, to study the welfare of the state and to use the results of our scientific and technological efforts more purposefully for the welfare of the people or, as Graunt poses it: "to preserve the Subject in Peace and Plenty". This will mean that we should plan our solutions not only for our immediate needs, but to see also their long term and far - reaching implications.
This is especially true for "computational" statistics, because if we want to isolate the most important driving factor which led to it, and which is still important today, we will find, that it is the fact that it forms very often the link between statistical theory and the problems of the "real" world.

References

Adam Adolf (1973), Vom Himmlischen Uhrwerk zur Statistischen Fabrik, Munk, Wien/Vienna 1973 (circulated at the meeting of the ISI 1973)

Archibald R.C. (1947), "P.G.Scheutz, Publicist, Author, Scientific Mechanician, and Edward Scheutz " Mathematical Tables and other Aids to Computation, vol.2, pp.238-245

Beauclair W.de (1968), Rechnen mit Maschinen, Eine Bildgeschichte der Rechentechnik, Braunschweig

Binder K. (ed. 1979,1982), Application of Monte Carlo Methods in Statistical Physics, Springer, Berlin, New York, Tokyo 2 vol.

Bradley Paul, L.Fox Bennet, Linus E.Schrage (1983) A Guide to Simulation, Springer, Berlin, New York, Tokyo 1983

Brown J.A.C., H.S.Houthakker S.J.Prais (1953), Electronic Computation in Economic Statistics, JASA 43,414-428, 1953

Cave Brown Anthony (1976), A Bodyguard of Lies, New York, 1976

Comrie L.J., G.B.Hey, H.G.Hudson (1937), Application of Hollerith equipment to an agricultural investigation, J.Roy.Statist. Soc. Suppl. 4, 210-224

d'Ocagne Philbert Maurice (1905), Le Calcul Simplifie, Paris 1905, p.74

Eckert Wallace J. (1940), Punched Card Methods in Scientific Computation, New York 1940

Efron Bradley (1979), Computers and the theory of Statistics: Thinking the Unthinkable. SIAM Review, 21(4) 1979 p.460-480

Everett C.J., S.Ulam (1948), Multiplicative systems in several variables III, Los Alamos Declassified Document 707, 1948 (cf. LADC-533 and 534)

---(1948), Multiplicative systems. I. Proc.Nat.Acad.Sci.(USA) 34,403-405, 1948

Gani J., The Making of Statisticians, New York, Heidelberg 1982

Hawkins D., S.Ulam, Theory of multiplicative processes. Los Alamos Declassified Document 265, 1944

Goldstine Herman H. (1972), The Computer from Pascal to von Neumann, Princeton, New Jersey 1972

Hammersley J.M., D.C.Handscomb (1964), Monte Carlo Methods, London 1964

Hayes Brian (1984), Computer Recreations, Scientific American

Healy J.R., G.V.Dyke (1953), A Hollerith Technique for the Solution of Normal equations, JASA 48,p.809-815
John Vinzenz (1884), Geschichte der Statistik , Stuttgart
Kelley (1948), The Kelley Statistical Tables, Harvard U.Press, Cambridge, Mass. 1948
Kendall Maurice G., R.L.Plackett (ed. 1977), Studies in the History of Statistics and Probability, London vol.II
Laderman Jack, Milton Abramowitz (1949), Application of Machines to Differencing of Tables, JASA 41, 1949 p.233-237
Lazarsfeld Paul F. (1977), Notes on the History of Quantification in Sociology - Trends, Sources and Problems, in: Kendall,Plackett 1977 Plackett 1977, pp.277-333
Maistrov Leonid E. (1974), Probability Theory, a Historical Sketch. Academic Pr., New York, London 1974
Metropolis Nicholas (1949), S.Ulam, The Monte Carlo method, JASA 44,p.335 1949 (I have not seen yet the "History of Computing in the 20th Century" edited by Metropolis, New York)
NBS, National Bureau of Standards (1951), Monte Carlo Method (Applied Math.Ser.12), Washington, US Gov.Printing O., 1951
Owen D.B. (1976), On the History of Statistics and Probability, New York
Pearson E.S., H.O.Hartley (ed.1954), Biometrika Tables for Statisticians, vol I. Cambridge U.Press, Cambridge, England 1954
Pearson E.S., M.G.Kendall (ed., 1970), Studies in the History of Statistics and Probability, London vol.I 1970
Pearson Karl (1978), The History of Statistics in the 17th and 18th Centuries against the changing background of intellectual, scientific and religious thought, Ed.E.S.Pearson, London 1978
Quenouille M.H. (1957), Notes on Bias in estimation, Biometrika 43, p.353-360, 1957
Rassem Mohammed (1980), Stichproben aus dem Wortfeld der alten Statistik, in:Rassem M., J.Stagl, Statistik und Staatsbeschreibung in der Neuzeit vornehmlich im 16-18Jh., Paderborn, Wien 1980 p.17-36
Truesdell Leon E. (1965), The Development of Punch Card Tabulation in the Bureau of Census, 1890-1940, US Government Printing Office, 1965
Tukey J.W.(1964), Bias and confidence in not-quite large samples, Abstract in: Ann.Math.Stat. 29,614 1964
Vallee Jacques (1982), The Network Revolution. Berkeley, 1982
von Neumann John, Letter to Richtmeyer, 11 March 1947; collected works, vol. V, pp.751-762
--- (1951), Various Techniques Used in Connection with Random Digits, J.of Research of the Nat.Bur. of Standards, Appl.Math.Ser., 3 1951, pp.36-38; Coll.Works, V, pp.768-770
--- , S.Ulam (1945), Bulletin A.M.S., Abstract 51-9-165, 1945
Westergaard Harald (1932), Contributions to the History of Statistics, London 1932
Wexelblatt Richard L.(ed.1981), History of Programming Languages, Academic Press, New York, London 1981
Wilberg Ernst-Eberhard (1977), Die Leibniz'sche Rechenmaschine und die Julius-Universität in Helmstedt, Beiträge zur Geschichte der Carolo-Wilhelmina, Braunschweig 1977
Wilkes M.V. (1949), Progress in High Speed Calculating Machine Design, Nature 164,341-343, 1949
Wilkes M.V. (1956), Automatic Digital Computers, London 1956
Wilkes Maurice Vincent, D.G.Wheeler, S.Gill (1951), The Preparation of Programs for an Electronic Digital Computer, With Special Reference to the EDSAC and the Use of Library Subroutines, Camb.Mass.1951
Zemanek H.(1970), Otto Schäffler, Wiener Pionier der Lochkartentechnik, Elektr.Rechenanl.12 p.133-134

Computational Statistics in Random Processes

New Algorithmic Developments for Estimation Problems in Time Series

C.F. Ansley, and R. Kohn, Chicago

SUMMARY

New filtering results are described for a nonstationary ARIMA model having any pattern of missing data. These results are based on a modification of the Kalman filter which allows the initial state vector to partially diffuse. The problem of defining a likelihood for a nonstationary ARIMA process is discussed, and several alternative approaches are shown to be equivalent. The modified Kalman filter is then used to compute the likelihood. Computational savings arising from the special structure of the Kalman filter conditional covariance matrices are discussed, and different state space representations are compared.

KEYWORDS: ARIMA process, state space, Kalman filter, diffuse initial state, maximum likelihood estimation.

1. Introduction

In this paper we outline general filtering results for a nonstationary integrated autoregressive moving average (ARIMA) time series model having any pattern of missing data. The general form of the model we consider is:

$$\phi(B)\nabla^d y = \theta(B)e(t) \qquad (1.1)$$

where (i) B is the backshift operator defined by Box and Jenkins (1976), and

$$\phi(B) = 1 - \sum_{j=1}^{p} \phi_j B^j \ , \quad \theta(B) = 1 - \sum_{j=1}^{q} \theta_j B^j$$

with $\phi(B)$ having all its zeros outside the unit circle; (ii) $\{e(t)\}$ is a sequence of independent normal random variables having mean zero and variance σ^2 ; and (iii) $\nabla \equiv 1 - B$ is the backward difference operator. For convenience we write $\nabla^d \equiv 1 - \delta_1 B - \dots - \delta_d B^d$.

Our development also applies to the multiplicative seasonal ARIMA (p,d,q).(P,D,Q) model

$$\Phi(B^s)\phi(B)\nabla_s^D\nabla^d y(t) = \Theta(B^s)\theta(B)e(t) \ ; \qquad (1.2)$$

$$\Phi(B^s) = 1 - \sum_{j=1}^{P} \Phi_j B^{js}, \quad \Theta(B^s) = 1 - \sum_{j=1}^{Q} \Theta_j B^{js} \ ,$$

with $\nabla_s \equiv 1 - B^s$. We do not consider the seasonal model explicitly, but treat it as a special case of the nonseasonal model.

Our results are based on the Kalman filter (see e.g. Anderson and Moore, 1979, Chapter 5). However, because the initial state vector is partially diffuse, we cannot apply the Kalman filter directly, but instead we use the modified Kalman filter obtained in Ansley and Kohn (1983a). This paper will be referred to below as AKI.

In Section 2 we write the ARIMA model in state space form, and indicate how the initial state vector is to be given a diffuse prior distribution. Justification for this is given later. Section 3 sets out the modified Kalman filter for the ARIMA model, and in Section 4 we show how it can be applied to compute the likelihood for an ARIMA process. In Section 5 we explore sources of computational efficiency and briefly compare the computational advantages of different possible state space representations.

State space representations for ARMA models were pioneered by Akaike (1974, 1978) and their use in computing the likelihood has been discussed by Gardner et. al. (1979) and Jones (1980). Jones also shows how to handle missing observations for stationary series. Harvey and Pierse (1984) extend these techniques to the case of missing observations in nonstationary ARIMA models, but they do not give a complete solution to the problem of missing starting values. The methods set forth in this paper apply to all patterns of missing data.

2. A State Space Representation for an ARIMA Process

We will write

$$u(t) = \nabla^d y(t) \qquad (2.1)$$

so that u(t) is a stationary ARMA(p,q) process, and

$$\upsilon(B) = \nabla^d \phi(B) = 1 - \upsilon_1 B - \ldots - \upsilon_r B^r$$

where $r = d + p$.

The ARIMA(p,d,q) process described by (1.1) can be written in state space form as follows:

$$y(t) = h'x(t) \qquad (2.2)$$

$$x(t+1) = Fx(t) + ge(t+1) \qquad (2.3)$$

x(t) is the (r+q+1) x 1 state vector with elements given by

$$x_j(t) = y(t-j+1) \qquad j = 1,\ldots,r$$

$$x_{r+j+1}(t) = -\sum_{i=j}^{q} \theta_j e(t+j-i) \qquad j = 0,\ldots,q$$

with $\theta_0 = -1$. The (r+q+1) x 1 vectors h and g are given by $h = (1,0,\ldots,0)'$ and $g = (1,0,\ldots,0,1,-\theta_1,\ldots,-\theta_q)$. F is an (r+q+1) x (r+q+1) matrix given by

$$F = \begin{bmatrix} F_{11} & F_{12} \\ 0 & F_{22} \end{bmatrix} \qquad (2.4)$$

with F_{11}, F_{12}, and F_{22} matrices of dimensions $r \times r$, $r \times (q+1)$ and $(q+1) \times (q+1)$ respectively, with elements given as follows:

$$F_{11} = \begin{bmatrix} \nu_1 & \cdot & \cdot & \nu_r \\ I_{r-1} & & & 0 \end{bmatrix} \quad F_{12} = \begin{bmatrix} 0 & 1 & 0 & \cdots & 0 \\ & & 0 & & \end{bmatrix} \quad F_{22} = \begin{bmatrix} 0 & I_q \\ 0 & 0 \end{bmatrix}$$

For $d > 0$, let $\eta = [y(0),\ldots,y(1-d)]'$, and assume that $\eta \sim N(0, kI_d)$ and that η is independent of the $e(t)$'s for all t. In the development below we will let $k \to \infty$, i.e. we will assume that η has a diffuse prior distribution. Justification for this assumption will be given in later sections.

We now define the $(p+q+1) \times 1$ vector ξ as

$$\xi = [u(0),\ldots,u(1-p),x_{r+1}(0),\ldots,x_{r+q+1}(0)]' .$$

Then ξ is a zero mean random vector that is independent of η, and whose covariance matrix we can compute given the parameters of the process. The following lemma shows the connection between the initial state vector $x(0)$ and the vectors η and ξ.

Lemma 2.1

Define the $(r+q+1) \times (r+q+1)$ matrix M by

$$M = \left[\begin{array}{c|c} \begin{matrix} I_d \\ M_{21} \\ 0 \end{matrix} & \begin{matrix} 0 \\ 0 \\ I_{q+1} \end{matrix} \end{array}\right]; \quad M_{21} = \begin{bmatrix} 1 & -\delta_1 & \cdot & \cdot & -\delta_d & & 0 \\ & 1 & -\delta_1 & \cdot & \cdot & -\delta_d & \\ & & \cdot & \cdot & \cdot & \cdot & \cdot \\ 0 & & & 1 & -\delta_1 & \cdot & \cdot\ -\delta_d \end{bmatrix}$$

Then M is a nonsingular matrix because $\delta_d \neq 0$ and

$$E[x(0)] = 0 \ ; \quad \mathrm{Var}[x(0)] = M^{-1} \begin{bmatrix} kI_d & 0 \\ 0 & V \end{bmatrix} (M')^{-1} \qquad (2.5)$$

The following lemma gives a method for constructing V.

Lemma 2.2

Let $\gamma(j) = \mathrm{cov}\{u(t),y(t-j)\}$ and write $u(t)$ as the infinite moving average

$$u(t) = \sum_0^\infty \psi_j\, e(t-j)$$

where $\psi(B) = \phi(B)^{-1}\theta(B) = \sum_{j=0}^\infty \psi_j B^j$. Define the $q \times q$ matrix A and for $p > q$ the $p \times q$ matrix B as

$$A = \begin{bmatrix} 1 & -\theta_1 & \cdot & \cdot & \cdot & -\theta_q \\ -\theta_1 & -\theta_2 & \cdot & \cdot & -\theta_q & \\ \cdot & \cdot & \cdot & & & \\ -\theta_{q-1} & -\theta_q & & & & 0 \\ -\theta_q & 0 & & & & 0 \end{bmatrix} \qquad B = \begin{bmatrix} 1 & \psi_1 & \cdot & \cdot & \psi_q \\ \psi_1 & \cdot & \cdot & \psi_q & \\ \cdot & \cdot & \cdot & & \\ \psi_q & 0 & & & \\ 0 & 0 & \cdot & \cdot & 0 \end{bmatrix}$$

For $p < q+1$ define B as the top p rows of the matrix B above. Then

$$\mathrm{Var}(\zeta) = V = \begin{bmatrix} V_{11} & V_{12} \\ V_{12}' & V_{22} \end{bmatrix}$$

where V_{11} is the $p \times p$ Toeplitz matrix with (i,j)th element $\gamma(|i-j|)$, $V_{12} = \sigma^2 BA'$ and $V_{22} = \sigma^2 AA'$.

Lemma 2.2 follows immediately from the definition of ξ . Algorithms for computing the γ's and ψ's are given by McLeod (1975) and Ansley and Kohn (1982).

3. The Modified Kalman Filter

Suppose we observe $y(t_1),\ldots,y(t_T)$ with $1 = t_1 < \ldots < t_T = n$. For each $\tau > 0$ let

$$x(t|0) = E\{x(t)\} = 0, \quad S(t|0) = \mathrm{Var}\{x(t)\}/\sigma^2 .$$

Using (2.5) we have

$$x(0|0) = 0 \tag{3.1}$$

$$S(0|0) = kS^{(1)}(0|0) + S^{(0)}(0|0) \tag{3.2}$$

with

$$S^{(1)}(0|0) = M^{-1} \begin{bmatrix} I_d & 0 \\ 0 & 0 \end{bmatrix} (M')^{-1} \quad S^{(0)}(0|0) = M^{-1} \begin{bmatrix} 0 & 0 \\ 0 & V \end{bmatrix} (M')^{-1}$$

For each $\tau > 1$, let ℓ be the greatest integer such that $t_\ell < \tau$, and define

$$x(t|\tau) = E\{x(t)|y(t_1),\ldots,y(t_\ell)\}$$

$$S(t|\tau) = E\{[x(t) - x(t|\tau)][x(t) - x(t|\tau)]'\}/\sigma^2 .$$

It is shown in AKI Section 3 that for all t, $x(t|t)$ and $S(t|t)$ have decompositions of the form

$$x(t|t) = x^{(0)}(t|t) + O(1/k)$$

$$S(t|t) = kS^{(1)}(t|t) + S^{(0)}(t|t) + O(1/k)$$

with similar decompositions holding for $x(t|t-1)$ and $S(t|t-1)$. Also, writing $\varepsilon(t) = y(t) - y(t|t-1)$ and $\lambda_t = \mathrm{Var}(e(t))$, it is shown in AKI that

$$\varepsilon(t) = \varepsilon^{(0)}(t) + O(1/k)$$

and

$$\lambda_t = k\lambda_t^{(1)} + \lambda_t^{(0)} + O(1/k) .$$

The modified Kalman filter obtains these quantities recursively as follows:

Step 1

$$x^{(0)}(t+1|t) = Fx^{(0)}(t|t) \tag{3.3}$$

$$S^{(1)}(t+1|t) = FS^{(1)}(t|t)F' \tag{3.4}$$

$$S^{(0)}(t+1|t) = FS^{(0)}(t|t)F' + gg' \tag{3.5}$$

If y(t+1) is missing:

Step 2

$$x^{(0)}(t|1|t+1) = x^{(0)}(t+1|t) \tag{3.6}$$

$$S^{(1)}(t+1|t+1) = S^{(1)}(t+1|t) \tag{3.7}$$

$$S^{(0)}(t+1|t+1) = S^{(0)}(t+1|t) \tag{3.8}$$

Return to Step 1

If y(t+1) is not missing

Step 3

$$\varepsilon^{(0)}(t+1) = y(t+1) - h'x^{(0)}(t+1|t) \tag{3.9}$$

$$\lambda_{t+1}{}^{(1)} = h'S^{(1)}(t+1|t)h \tag{3.10}$$

$$\lambda_{t+1}{}^{(0)} = h'S^{(0)}(t+1|t)h + 1 \tag{3.11}$$

To simplify the notation, for the remaining steps we write $S^{(1)} = S^{(1)}(t+1|t)$, $S^{(0)} = S^{(0)}(t+1|t)$, $x^{(0)} = x^{(0)}(t+1|t)$ and omit the time index (t+1) from $\lambda^{(1)}$, $\lambda^{(0)}$ and $\varepsilon^{(0)}$.

Step 4 If $\lambda^{(1)} > 0$ then

$$x^{(0)}(t+1|t+1) = x(0) + S^{(1)}h\varepsilon^{(0)}\lambda^{(1)} \tag{3.12}$$

$$S^{(1)}(t+1|t+1) = S^{(1)} - S^{(1)}hh'S^{(1)}/\lambda^{(1)} \tag{3.13}$$

$$S^{(0)}(t+1|t+1) = S^{(0)} + S^{(1)}hh'S^{(1)}(\lambda^{(0)}/[\lambda^{(1)}]^2) - S^{(1)}hh'S^{(0)}/\lambda^{(1)} - S^{(0)}hh'S^{(1)}/\lambda^{(1)} \tag{3.14}$$

In this case,

$$\text{rank } S^{(1)}(t+1|t+1) = \text{rank } S^{(1)}(t|t) - 1 . \tag{3.15}$$

Return to Step 1

Step 5 If $\lambda^{(1)} = 0$ then

$$x^{(0)}(t+1|t+1) = x^{(0)} + S^{(0)}h'\varepsilon^{(0)}/\lambda^{(0)} \tag{3.16}$$

$$S^{(1)}(t+1|t+1) = S^{(1)} \tag{3.17}$$

$$S^{(0)}(t+1|t+1) = S^{(0)} - S^{(0)}hh'S^{(0)}/\lambda^{(0)} \tag{3.18}$$

The above algorithm represents a simple modification of the usual Kalman filter. Noting that rank $S^{(1)}(0|0) = d$, it is clear from (3.15) that $\lambda^{(1)} > 0$ for only d values of t, after which $S^{(1)}(t|t)$ becomes identically zero, and the modified algorithm degenerates to the usual Kalman filter.

We note that Step 2 above automatically handles the problem of missing data.

4. Defining the Likelihood

The likelihood is not defined in the usual sense because the density does not exist except conditionally. However, we have been able to show that the following three possible approaches are equivalent.

1. The posterior density (suitably normalized) of the observation vector y given a diffuse prior distribution on the starting values for the nonstationary difference equation;
2. The conditional density of the observations y given observations on d elements (our result is in fact more general than this);
3. The density for an $(n-d) \times n$ singular transformation of y which is independent of the starting values η.

The equivalence between 1 and 2 is pointed out without proof by Harvey (1981, p. 205).

Before considering these three approaches, we rewrite the state space representation of Section 2 in a canonical form. Partition M from Lemma 2.1 as $[M_1, M_2]$ with M_1 consisting of the first r columns of M. Write

$$x^{(0)} = x^{(1)}(0) + x^{(0)}(0) = M_1\eta + M_2\xi \qquad (4.1)$$

and break (2.3) into two parts:

$$x^{(0)}(t+1) = Fx^{(1)}(t) \qquad (4.2)$$

$$x^{(1)}(t+1) = Fx^{(0)}(t) + ge(t+1) \qquad (4.3)$$

Now define the $T \times 1$ vectors y and ω and the $T \times d$ matrix A by setting rows j of y, ω and A equal to $y(t_j)$, $H(t_j)x^{(0)}(t_j) + e(t_j)$ and $H(t_j) \prod_{\ell=0}^{t_j-1} F(\ell)M_1$ respectively, for $j = 1,\dots,T$. It is now easily checked that

$$y = A\eta + \omega \qquad (4.4)$$

Note that the density of ω is well-defined and that the starting values η affect the observations y only through the matrix A, which is independent of the parameters $\beta = (\phi_1,\dots,\phi_p, \phi_1,\dots,\phi_q,\sigma)'$ of the process. Our results depend on the following two theorems from AKI.

Theorem 4.1

Assume that y is generated by the model (4.4) and that $\eta \sim N(0,kI)$, where k is an arbitrary positive constant.

(i) Let $f(y;k)$ be the density function of y. Then the limit

$$\lim_{k\to\infty} k^{d/2} f(y;k)$$

exists and is nonzero.

(ii) Suppose $H(k)$ is an $T \times T$ matrix function of k, independent of the parameters of the ARMA process, such that $\det H(k) = 1$ and $\lim H(k)$ exists. Now let $w(k) = H(k)y$ and suppose $H_1(k)$ consists of d rows of $H(k)$ so that $H_1(k)A$ has rank d for all k. Put $w_1(k) = H_1(k)y$ and let $w_2(k)$ consist of the other $T-d$ rows of $w(k)$. Then

$$\lim_{k\to\infty} f(w_2|w_1;k) = c \lim_{k\to\infty} k^{d/2} f(w;k)$$

where c is a constant, independent of k and β.

(iii) Suppose $\hat{H}(k)$ is another $T \times T$ matrix having similar properties to $H(k)$. Define $\hat{H}_1(k)$, $\hat{w}_1(k)$ and $\hat{w}_2(k)$ with respect to $\hat{H}(k)$ in the same way as $H_1(k)$, $w_1(k)$ and $w_2(k)$ are defined with respect to $H(k)$. Then

$$\lim_{k\to\infty} f(\hat{w}_2|\hat{w}_1;k) = c_1 \lim_{k\to\infty} f(w_2|w_1;k)$$

where c_1 is a constant, independent of k and β.

Theorem 4.2

Suppose J is an $T \times T$ matrix such that

(i) $\det J = 1$;

(ii) the elements of J do not depend on $\Omega^{(0)}$;

(iii) the $T \times d$ matrix JA has exactly d nonzero rows.

Let $w = Jy$ and take w_2 to be the subvector of w corresponding to the zero rows of JA; let w_1 consist of the other elements of w. Now if $f(w_2)$ is the density of w_2 and $f(w_2|w_1;k)$ is the density of w_2 conditional on w_1, then

$$f(w_2) = \lim_{k\to\infty} f(w_2|w_1;k) = c \lim_{k\to\infty} k^{d/2} f(w;k)$$

where c is a positive constant, independent of k and β.

Theorem 4.1 states that with normalization factor $k^{d/2}$ the limit of the posterior distribution $f(y;k)$ exists and is proportional to the limit of the conditional distribution $f(w_2|w_1;k)$ where (w_1,w_2) is obtained by a linear transformation of y. The choice of linear transformation is arbitrary, providing that w_1 is related to the starting values η through a nonsingular matrix H_1A. Often, w_1 may consist of the first d elements of y; this is the way in which Harvey and Pierse (1984) start their Kalman filter recursions for nonstationary ARIMA processes. However, if there are missing values amongst the first d positions a simple choice of w_1 cannot be made.

Theorem 4.2 can be interpreted as follows. Suppose we can find a nonsingular transformation.

$$J = \begin{bmatrix} J_1 \\ J_2 \end{bmatrix}$$

where J_2 is an (T-d) x r matrix such that $J_2A = 0$. Then $w_2 = J_2y$ is independent of the starting values η, and we can find its density explicitly. This density is exactly that given by the normalized posterior and conditional distributions introduced in Theorem 4.1. The importance of this result is that the transformation matrix J_2 is a generalization of the differencing transformation used by Box and Jenkins (1976), Harvey and Pierse (1984) and others, so that Theorem 4.2 gives a precise interpretation to their informal definition of a likelihood. The result is general in that the choice of J_1 and J_2 is arbitrary, and the partition of J need not be by consecutive rows. Given the above results, the choice of method for constructing the likelihood can be made purely on the grounds of computational efficiency. We contend that the first method, based on the limit of the renormalized posterior density of y, is most efficient because it can be calculated recursively via the modified Kalman filter.

To see how this can be done, we need the following generalization of the Cholesky factorization. The proof is given in AKI.

Theorem 4.3

Suppose that $\Omega(k)$ is a T x T matrix such that

$$\Omega(k) = k\Omega^{(1)} + \Omega^{(0)} \qquad (4.5)$$

where $\Omega^{(1)}$ and $\Omega^{(0)}$ are positive semidefinite matrices independent of k. Then for $k > 0$, $\Omega(k)$ is a positive semidefinite matrix and can be factorized as

$$\Omega(k) = L(k)\Lambda(k)L(k)' \qquad (4.6)$$

where:

(i) $\Lambda(k)$ is a diagonal matrix and $L(k)$ a lower triangular matrix with 1's on the diagonal.

(ii) We can write

$$L(k) = L^{(0)} + 1/k\, L^{(-1)}(k) \qquad (4.7)$$

$$\Lambda(k) = k\Lambda^{(1)} + \Lambda^{(0)} + 1/k\, \Lambda^{(-1)}(k) \qquad (4.8)$$

where $L^{(0)}$, $\Lambda^{(1)}$ and $\Lambda^{(0)}$ do not depend on k, and the elements of $L^{(-1)}(k)$ and $\Lambda^{(-1)}(k)$ are uniformly bounded in absolute value.

(iii) Let the i-th diagonal elements of $\Lambda^{(1)}, \Lambda^{(0)}$ and $\Lambda^{(-1)}(k)$ be respectively $\lambda_i^{(1)}$, $\lambda_i^{(0)}$ and $\lambda_i^{(-1)}(k)$. Then $\lambda_i^{(1)} \geq 0$ and $\lambda_i^{(0)} \geq 0$, and $\lambda_i^{(-1)}(k) = 0$ whenever both $\lambda_i^{(1)} = 0$

and $\lambda_i^{(0)} = 0$. Further, the number of nonzero values of $\lambda_i^{(1)}$ is equal to the rank of $\Omega^{(1)}$.

Theorem 4.3 allows us to define the likelihood via the following theorem.

Theorem 4.4

Suppose y is generated by (4.4). Then $\mathrm{Var}(y) = \Omega(k) = kAA' + \Omega$ where $\Omega = \mathrm{Var}(w)$, and by Theorem 4.3 we can write $\Omega(k) = L(k)\Lambda(k)L(k)'$ with $L(k)$ and $\Lambda(k)$ as described in Theorem 4.3. Write $\varepsilon = (L^{(0)})^{-1}y$ and denote the T-d rows for which $\lambda_j^{(1)} = 0$ by $t_1,\ldots,t_{T-d}$. Then

$$\lim_{k\to\infty} k^{-d/2} f(y;k) = c\left(\prod_{j=1}^{T-d} \lambda_{t_j}^{(0)}\right)^{-1/2} \exp -1/2 \sum_{j=1}^{T-d} \varepsilon_{t_j}^2 / \lambda_{t_j}^{(0)}$$

where c is a positive constant, independent of k and the parameters β .

Because both the Kalman filter and Cholesky factorization are equivalent to the Gram-Schmidt orthogonalization, it can be shown that the ε's in Theorem 4.4 coincide with the $\varepsilon^{(0)}$'s given by (3.9), and the $\lambda^{(0)}$'s and $\lambda^{(1)}$'s coincide with those given by (3.10) and (3.11). Thus all the quantities appearing in Theorem 4.4 can be computed recursively with the modified Kalman filter of Section 3.

5. Some Computational Efficiencies

The form of the state vector x(t) in (2.3) leads to important computational efficiences. Because the recursions on the $S^{(1)}$ components in the modified Kalman filter are required for only d steps, we consider only the standard Kalman filter recursion for the $S^{(0)}$ terms assuming $\lambda^{(1)} = 0$.

When there are no missing values, we see that (i) the first d rows of x(t), i.e. (y(t),...,y(t-d+1))', are observed, and thus have zero variance, and (ii) the state vector element $x_{r+1}(t) = \phi(B)\nabla^d y(t)$ is observed and so it also has zero variance. Thus the conditional covariance matrix $S(t|t)$ has the form shown in Figure 1, where a shaded area indicates nonzero elements.

The pattern of zero elements can be exploited in a computer program, and it can be shown that the number of operations (multiplications/divisions) required per Kalman filter cycle is p+d+1+q(q+5)/2 together with one square root. This is exactly the same as for the Cholesky method of Ansley (1979), although considerably less bookkeeping and working space is required in the Cholesky method.

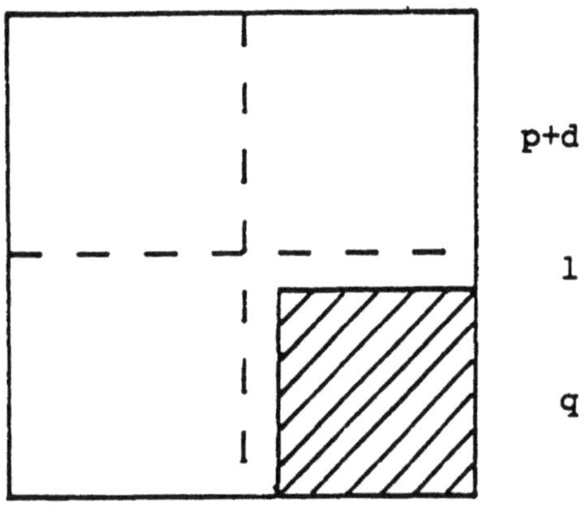

Figure 1

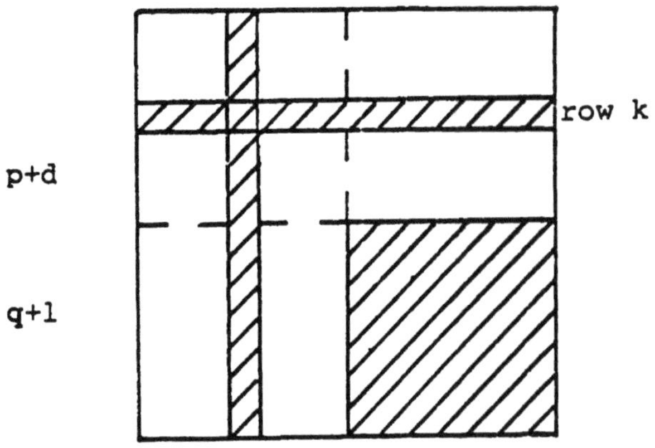

Figure 2

Suppose now there is a missing value at row t, but there are no other missing values. Then none of the elements $x_{r+j}(t)$, $j > 0$, are observable for steps $t+1,\ldots,t+p+d$, after which time the situation for no missing value is restored. In this case, for row $t+k$ where $1 < k < d$, $S(t+k|t+k)$ has the form shown in Figure 2.

Notice also that the matrix F in (2.4) consists largely of a subdiagonal of 1's, so that the multiplications in (3.4) and (3.5) consist mainly of shifts of indexes. This feature, combined with the sparse form of the matrices in Figures 1 and 2, allow us to realize considerable gains in computing time. It is very simple to construct and update an indexing vector to keep track of the zero rows and columns for the representation (2.2-2.3) .

There are at least two other state space representations which are candidates for estimation of ARIMA models with missing values. The first is that used by Harvey and Pierse (1984), and is a generalization of the representation for stationary ARMA processes used by Gardner et al. (1979) and Jones (1980). Let $m=\max(p,q+1)$. Then $y(t)$ has the state space representation (2.2-2.3) where the state vector $x(t)$ is now $(m+d) \times 1$ with elements:

$$x_j(t) = y(t-j+1) \qquad j = 1,\ldots,d$$

$$x_{d+j}(t) = \sum_{i=j}^{m} \phi_i u(t+j-1-i) - \sum_{\ell=j+1}^{m} \theta_{\ell-1} e(t+j+1-\ell) \qquad j = 1,\ldots,m$$

where we take $\theta_0 = -1$ and define $\phi_i = 0$ for $i > p$ and $\theta_\ell = 0$ for $\ell > q$. h and g are $(m+d) \times 1$ and are defined similarly to their definitions in Section 2. F is now an $(m+d) \times (m+d)$ matrix given by:

$$F = \begin{bmatrix} F_{11} & F_{12} \\ 0 & F_{22} \end{bmatrix} \quad F_{11} = \begin{bmatrix} \delta_1 & \cdot & \cdot & \delta_d \\ I_{d-1} & & & 0 \end{bmatrix} \quad F_{22} = \begin{bmatrix} \phi_1 & I_{m-1} \\ \vdots & \\ \phi_m & 0 \end{bmatrix}$$

F_{12} has first row given by $(\phi_1, 1, 0, \ldots, 0)$, with all other elements zero.

For this representation, it can be seen that if there are missing values, $S(t|t)$ has zeros in rows and columns 1 through d+1. When there are missing values, the situation is more complicated, but it is still possible to realize substantial computational savings by exploiting the structure carefully.

A third representation is obtained by defining $x(t)$ to be a $m' \times 1$ state vector, $m'=\max(p+d, q+1)$, with elements given by writing

$$x_j(t) = \sum_{i=j}^{m'} \nu_i u(t+j-1-i) - \sum_{\ell=j+1}^{m'} \theta_{\ell-1} e(t+j+1-\ell) \qquad j = 1, \ldots, m'$$

In this case, we have

$$F = \begin{bmatrix} \nu_1 & I_{m'-1} \\ \vdots & \\ \nu_{m'} & 0 \end{bmatrix}$$

with $\nu_j = 0$ for $j > p+d$. We have $g = [1, 0, \ldots, 0]'$ and h given as before. In this case, the first row and final $m'-q-1$ rows of $S(t|t)$ are zero if there are now missing data, and similar, though more complicated, patterns of zeros can be found when there are missing values. As before, there are substantial computational savings available by exploiting these features.

The two additional representations given here require their own versions of Lemmas 2.1 and 2.2, but these are not difficult. Note that the last representation has dimension $m'=\max(p+d, q+1)$, that of Harvey and Pierse (1984) discussed earlier has dimension $m+d=\max(p, q+1)+d$, and that in Section 2 has dimension $p+d+q+1$. Thus the last representation, which has not as yet been proposed in the literature, has the smallest dimension, and may prove to be the most efficient in practice, although the computational costs of the various bookkeeping requirements must be investigated before any final assessment can be made. It is interesting that in all three cases, when there are no missing values the operation count (multiplications and divisions) is identical to that for the Cholesky method of Ansley (1979).

6. Concluding Remarks

Using the methods of Ansley and Kohn (1983b), it is possible to extend the diffuse initial condition approach to provide modified versions of the fixed interval, fixed point and fixed lag smoothing algorithms (see Anderson, 1979, Chapter 7). These algorithms enable us to interpolate missing values and to provide appropriate standard errors.

The approach has been generalized to multidimensional state space processes in Ansley and Kohn (1983a,b), and these methods can be applied to multivariate ARIMA models also. In this case, only part of the observation vector y(t) may be missing, and we must adapt our method using the techniques of Ansley and Kohn (1983c).

References

Akaike H. (1974), Markovian representation of a stochastic process and its application to the analysis of autoregressive moving average processes, Ann. Inst. Stat. Math. 26, 363-387.

Akaike H. (1978), Covariance matrix computation of the state variable of a stationary Gaussian process, Ann. Inst. Stat. Math. 30, 499-504.

Anderson B.D.O. and Moore J.B. (1979), Optimal Filtering, Englewood Cliffs, New Jersey, Prentice Hall.

Ansley C.F. (1979), An algorithm for the exact likelihood of a mixed autoregressive moving average process, Biometrika 66, 59-65.

Ansley C.F. and Kohn R. (1982), On the computation of the theoretical autocovariance function for a vector ARMA process, J. Statist. Comput. Simul. 15, 272-283.

Ansley C.F. and Kohn R. (1983a), State space models with diffuse initial conditions I: Filtering and likelihood, Technical Report #13, Statistics Research Center, Univ. Chicago.

Ansley C.F. and Kohn R. (1983b), State space models with diffuse initial conditions II: Smoothing and continuous time processes, Technical report #15, Statistics Research Center, Univ. Chicago.

Ansley C.F. and Kohn R. (1983c), Obtaining the exact likelihood of a vector ARMA process when some of the data are missing or aggregated, Biometrika 70, 275-278.

Box G.E.P. and Jenkins G.M. (1976), Time Series Analysis Forecasting and Control (rev. ed.), San Francisco, Holden Day.

Gardner G., Harvey A.C. and Phillips G.D.A. (1979), An algorithm for exact maximum likelihood estimation of autoregressive-moving average models by means of Kalman filtering, Appl. Stat. 29, 311-322.

Harvey A.C. (1981), Time Series Models, New York, Halsted.

Harvey A.C. and Pierse R.G. (1984), Estimating missing observations in economic time series, J. Am. Statist. Ass. 79, 125-131.

Jones R.H. (1980), Maximum likelihood fitting of ARMA models to time series with missing observations, Technometrics 22, 389-395.

McLeod A.I. (1977), Derivation of the theoretical autocovariances of autoregressive moving average time series, Appl. Stat. 26, 194.

Parameter Estimation and Order Determination of a Multivariate Autoregressive Process

M. Krzyśko, and D. Smoczyński, Poznań

Summary: The least squares estimators of the parameters of the multivariate autoregressive process and the asymptotic properties of these estimators are investigated. The determination of the true order of the multivariate autoregressive equation is also described; the estimator of the true order is obtained by minimizing the posterior risk.

Keywords: Multivariate Autoregression, Least Squares Estimators, Autoregressive Order, Bayesian Method of Estimation.

1. Estimating the parameters of the multivariate autoregressive process

Let $\{\underline{X}(t),\ t = 0, \pm 1, \ldots\}$ be a p-dimensional stochastic process with expectation $E(X(t)) = \underline{0}$ and covariance matrix $E(\underline{X}(m)\underline{X}'(n)) = \underline{G}(n-m) = \underline{G}'(m-n)$ described by an autoregressive equation of finite order q

$$\sum_{u=0}^{q} \underline{B}(u)\underline{X}(t-u) = \underline{e}(t), \quad \text{where} \quad \underline{B}(0) = \underline{I}_p \tag{1}$$

and $\{\underline{e}(t),\ t = 0, \pm 1, \pm 2, \ldots\}$ is a sequence of independent $p \times 1$ random variables with common normal distribution $N(\underline{0}, \underline{V})$, $\underline{V} > 0$. The process $\{\underline{X}(t)\}$ can be expressed in the unique infinite moving average representation $\underline{X}(t) = \sum_{r=0}^{\infty} \underline{U}(r)\underline{e}(t-r)$.

It is clear that the model (1) can also be expressed as a first-order vector autoregressive model i.e. $\underline{Y}(t) = \underline{A}\,\underline{Y}(t-1) + \underline{\varepsilon}(t)$ where $\underline{Y}(t) = (\underline{X}'(t), \underline{X}'(t-1), \ldots, \underline{X}'(t-q+1))'$, $\underline{\varepsilon}(t) = (\underline{e}'(t), \underline{0}, \ldots, \underline{0})'$ and

$$\underline{A} = \begin{bmatrix} -\underline{B}(1), & -\underline{B}(2), & \ldots, & -\underline{B}(q) \\ & \underline{I}_{p(q-1)} & & \underline{0} \end{bmatrix}.$$

Note that $\underline{X}(t) = \underline{W}'\underline{Y}(t)$, $\underline{e}(t) = \underline{W}'\underline{\varepsilon}(t)$ and $\underline{\varepsilon}(t) = \underline{W}\,\underline{e}(t)$ where $\underline{W}' = (\underline{I}_p, \underline{0}, \ldots, \underline{0})$ is a $(p \times pq)$ matrix.

Also, $\underline{Y}(t)$ has the representation $\underline{Y}(t) = \sum_{r=0}^{\infty} \underline{A}^r\, \underline{\varepsilon}(t-r)$ so that $\underline{X}(t) = \sum_{r=0}^{\infty} \underline{W}'\underline{A}^r\underline{W}\ \underline{e}(t-r)$.

Thus, from the unique moving average representation of $\underline{X}(t)$ we have that $\underline{U}(r) = \underline{W}'\underline{A}^r\underline{W}$, $r = 0,1,\ldots$, i.e. $\underline{U}(r)$ is the upper $(p \times p)$ block of the matrix $\underline{A}^r$. Hence

$$\underline{G}(n-m) = \sum_{r=0}^{\infty} \underline{W}'\underline{A}^r\underline{W}\ \underline{V}\ \underline{W}'\underline{A}^{r+n-m}\underline{W} \ .$$

The spectral decomposition of a matrix $\underline{A}$ has the form $\underline{A}=\underline{P}\ \underline{\Lambda}\ \underline{P}^{-1}$. Hence $\underline{A}^r = \underline{P}\ \underline{\Lambda}^r\underline{P}^{-1}$ and

$$(2) \qquad \underline{G}(n-m) = \underline{W}'\underline{P}\ \underline{H}^*\underline{P}\ \underline{W} \text{ , where}$$

$$\underline{H}^* = (h^*_{ij}) = \sum_{r=0}^{\infty} \underline{\Lambda}^r\underline{H}\ \underline{\Lambda}^{r+n-m} \ , \qquad \underline{H} = (h_{ij}) = \underline{P}^{-1}\underline{W}\ \underline{V}(\underline{P}^{-1}\underline{W})' \ ,$$

$$h^*_{ij} = h_{ij}/(1-\lambda_i\lambda_j).$$

We first assume that the order q of (1) is known and that the parameters $\underline{B}(1),\ldots,\underline{B}(q)$, $\underline{V}$ are unknown. We will now consider the estimation of the parameters $\underline{B}(1),\ldots,\underline{B}(q),\underline{V}$ on the basis of N independent realizations $\underline{X}_n(1),\ldots,\underline{X}_n(T)$, $n=1,\ldots,N$ of the process $\{\underline{X}(t)\}$.

If we introduce the following notation

$$\underline{B} = (\underline{B}(1),\ldots,\underline{B}(q)), \quad \underline{Y}_n(t-1) = (\underline{X}_n'(t-1),\ldots,\underline{X}_n'(t-q))', \quad n=1,\ldots,N,$$

$$\underline{X} = (\underline{X}_1(q+1),\ldots,\underline{X}_1(T),\underline{X}_2(q+1),\ldots,\underline{X}_2(T),\ldots,\underline{X}_N(q+1),\ldots,\underline{X}_n(T)),$$

$$\underline{Y} = (\underline{Y}_1(q),\ldots,\underline{Y}_1(T-1),\underline{Y}_2(q),\ldots,\underline{Y}_2(T-1),\ldots,\underline{Y}_N(q),\ldots,\underline{Y}_N(T-1)),$$

$$\underline{E} = (\underline{e}_1(q+1),\ldots,\underline{e}_1(T),\underline{e}_2(q+1),\ldots,\underline{e}_2(T),\ldots,\underline{e}_N(q+1),\ldots,\underline{e}_N(T)),$$

then model (1) for $t = q+1,\ldots,T$, $n = 1,\ldots,N$ may be rewritten as

$$(3) \qquad \underline{X} = -\underline{B}\ \underline{Y} + \underline{E} \ .$$

<u>LEMMA 1</u>. If $N(T-q) \geq pq$, then

$$\underline{D} = \sum_{n=1}^{N}\sum_{t=q+1}^{T} \underline{Y}_n(t-1)\underline{Y}_n'(t-1)$$

is a symmetric and positive definite matrix with probability 1.

Proof. It is clear that $\underline{D}$ is symmetric and nonnegative definite. To show positive definiteness, let $\underline{c} = (\underline{c}_1',\underline{c}_2',\ldots,\underline{c}_q')$ be an arbitrary $pq \times 1$ vector and consider

$$\underline{c}'\underline{D}\ \underline{c} = \sum_{n=1}^{N}\sum_{t=1}^{T-q}\Big(\sum_{j=1}^{q}\underline{c}_j'\ \underline{X}_n(t+q-j)\Big)^2 \ .$$

If $\underline{c}'\underline{D}\ \underline{c} = 0$, then

$$\sum_{j=1}^{q}\underline{c}_j'\ \underline{X}_n(t+q-j) = 0, \quad t=1,2,\ldots,T-q, \quad n=1,2,\ldots,N,$$

which, since $N(T-q) \geq pq$, is a system of linear equations of pq unknowns in at least pq equations. In particular, consider

$$\sum_{j=1}^{q}\underline{c}_j'\ \underline{X}\ (t+q-j) = 0 \quad \text{for} \quad t = 1,2,\ldots,q, \quad n = 1,2,\ldots,p.$$

Since $\{\underline{X}(t)\}$ is continuous and not deterministic, we have that $\det(\underline{Z}) \neq 0$ with probability 1 where $\underline{Z}$ is a $pq \times pq$ matrix whose elements are the coefficients of a system of the linear equations. This implies that $\underline{c} = \underline{0}$ and the conclusion follows. □

The ordinary least squares estimate of $\underline{B}$ is

(4) $$\hat{\underline{B}} = -\underline{X}\,\underline{Y}'(\underline{Y}\,\underline{Y}')^{-1} = -\Big(\sum_{n=1}^{N}\sum_{t=q+1}^{T}\underline{X}_n(t)\underline{Y}_n'(t-1)\Big)\underline{D}^{-1}$$

and Lemma 1 implies that $\hat{\underline{B}}$ exists for $N(T-q) \geq pq$. Having an estimator for $\underline{B}$, we may estimate $\underline{E}$ from (3) by $\hat{\underline{E}} = \underline{X} + \hat{\underline{B}}\,\underline{Y}$ and thus matrix $\underline{V}$ by

(5) $$\hat{\underline{V}} = (N(T-q))^{-1}\hat{\underline{E}}\,\hat{\underline{E}}' = (N(T-q))^{-1}(\underline{X}\,\underline{X}' + \underline{X}\,\underline{Y}'\hat{\underline{B}}').$$

The model (3) may be rewritten as

$$\text{vec}\,\underline{X} = -(\underline{Y}'\otimes\underline{I}_p)\text{vec}\,\underline{B} + \text{vec}\,\underline{E}$$

and

(6) $$\text{vec}\,\hat{\underline{B}} = -\Big[\Big(\sum_{n=1}^{N}\sum_{t=q+1}^{T}\underline{Y}_n(t-1)\underline{Y}_n'(t-1)\Big)^{-1}\otimes\underline{I}_p\Big]\Big[\sum_{n=1}^{N}\sum_{t=q+1}^{T}(\underline{Y}_n(t-1)\otimes\underline{I}_p)\underline{X}_n(t)\Big].$$

THEOREM 1. Consider the process $\{\underline{X}(t)\}$ which satisfies (1). If $E[X_{ni}^2(t)] < \infty$, for $n=1,\dots,N$, $i=1,\dots,p$, where $\underline{X}_n(t) = (X_{n1}(t),\dots,X_{np}(t))'$, then $\text{vec}\,\hat{\underline{B}}$ defined by (6) is a strongly consistent estimate of $\text{vec}\,\underline{B}$.

Furthermore, if $E[X_m^4(t)] < \infty$ then $\sqrt{N(T-q)}\,\text{vec}\,(\hat{\underline{B}}-\underline{B})$ has a distribution which converges to that of a normally distributed random vector with zero mean and covariance matrix $\underline{G}_q^{-1}\otimes\underline{V}$, where $\underline{G}_q$ is a block matrix of q^2 blocks, where block (u,v) contains matrix $\underline{G}(v-u)$.

Proof. We have

$$\text{vec}(\hat{\underline{B}}-\underline{B}) = -\Big[\Big(\sum_{n=1}^{N}\sum_{t=q+1}^{T}\underline{Y}_n(t-1)\underline{Y}_n'(t-1)\Big)^{-1}\otimes\underline{I}_p\Big]\Big[\sum_{n=1}^{N}\sum_{t=q+1}^{T}(\underline{Y}_n(t-1)\otimes\underline{I}_p)\underline{e}_n(t)\Big].$$

From the ergodic theorem (cf. Hannan 1970, Chapter IV.2, Theorem 2) we have

$$(T-q)^{-1}\sum_{T=q+1}^{T}\underline{Y}_n(t-1)\underline{Y}_n'(t-1)\otimes\underline{I}_p \xrightarrow{a.s.} \{E[\underline{Y}_n(t-1)\underline{Y}_n'(t-1)]\}\otimes\underline{I}_p =$$

$= \underline{G}_q\otimes\underline{I}_p$ so that $\text{vec}(\hat{\underline{B}}-\underline{B})$ has the same asymptotic properties as

$$(\underline{G}_q^{-1}\otimes\underline{I}_p)(N(T-q))^{-1}\sum_{n=1}^{N}\sum_{t=q+1}^{T}[\underline{Y}_n(t-1)\otimes\underline{I}_p]\,\underline{e}_n(t).$$

It can easily be seen that

$$E\{[\underline{Y}_n(t-1)\otimes\underline{I}_p]\,\underline{e}_n(t)\} = \underline{0}$$

since $\underline{X}_n(t-q)$ can only contain $\underline{e}_n(u)$ variates to the moment $t-q$, which are not correlated with $\underline{e}_n(t)$, and since $\underline{Y}_n(t)$ and $\underline{e}_n(t)$ are both strictly stationary and ergodic, then so also is $\{\underline{Y}_n(t-1)\otimes \underline{I}_p\}\ \underline{e}_n(t)$. Hence, once more from the ergodic theorem, since $E(X_{ni}^2(t))<\infty$, $i=1,2,\dots,p$, $n=1,2,\dots,N$, we have

$$(N(T-q))^{-1}\sum_{n=1}^{N}\sum_{t=q+1}^{T}\left[\underline{Y}_n(t-1)\otimes \underline{I}_p\right]\underline{e}_n(t) \xrightarrow{a.s.} E\{[\underline{Y}_n(t-1)\otimes \underline{I}_p]\underline{e}_n(t)\} = \underline{0}$$

and from this we find that vec $(\hat{\underline{B}} - \underline{B}) \xrightarrow{a.s.} \underline{0}$, so the estimator vec $\hat{\underline{B}}$ defined by (6) is a strongly consistent estimator of vec $\underline{B}$. Let us further observe that if $\underline{c}$ is a $p^2q\times 1$ constant vector, then, from the Lindeberg-Lévy theorem for martingales (cf. Billingsley 1961), the expression

$$(N(T-q))^{-1}\sum_{n=1}^{N}\sum_{t=q+1}^{T}\underline{c}'\left[\underline{Y}_n(t-1)\otimes \underline{I}_p\right]\ \underline{e}_n(t)$$

has a distribution which converges to the normal distribution with zero mean and variance

$$E\{[\underline{c}'(\underline{Y}_n(t-1)\otimes \underline{I}_p)\underline{e}_n(t)]^2\} =$$

$$= E\{\ \underline{c}'[\underline{Y}_n(t-1)\otimes \underline{I}_p]\ E[\underline{e}_n(t)\underline{e}_n'(t)][\underline{Y}_n'(t-1)\otimes \underline{I}_p]\ \underline{c}\} =$$

$$= \underline{c}'\,E\{[\underline{Y}_n(t-1)\otimes \underline{I}_p]\ \underline{V}\ [\underline{Y}_n'(t-1)\otimes \underline{I}_p]\}\,\underline{c} =$$

$$= \underline{c}'\{E[\ \underline{Y}_n(t-1)\underline{Y}_n'(t-1)\otimes \underline{V}\}\,\underline{c}\ = \underline{c}'(\underline{G}_q\otimes \underline{V})\underline{c}\ ,$$

assuming that $E(X_{ni}^4(t))<\infty$, for $i=1,2,\dots,p$, $n=1,2,\dots,N$.

Thus $\sqrt{N(T-q)}\,\text{vec}(\hat{\underline{B}}-\underline{B})$ converges in distribution to the normal distribution with zero mean and covariance matrix $\underline{G}_q^{-1}\otimes \underline{V}$, as required. □

2. Determining the order of the multivariate autoregressive process

In practice, it is not only the parameters of the model (1) that are unknown, but so is the order of the model. We will now consider the estimation of the true order q^* of the autoregressive equation (1) on the basis of N independent realizations $\underline{X}_n(1),\dots,\underline{X}_n(T)$, $n = 1,\dots,N$ of process $\{\ \underline{X}(t)\}$.

Suppose that the order q of the autoregression equation is a random variable with a known prior density function $h(q)$. If we have no information as to the choice of $h(q)$, we assume $h(q) = Q^{-1}$, where Q is an arbitrarily chosen number larger than the true order q^*. An incorrect decision in choosing the order of the model (1) results in a loss of the form

(7) $$S(q,r) = c\,|q-r|,$$

where c is an a constant chosen beforehand, which satisfies the

following conditions: $c = 1$ for $r \leq q$ and $c > 1$ for $r > q$, where r is the assumed order the model (1). The loss function of the form (7) has the following properties. The cost of choosing an order larger than the true one is proportional to the error. The cost of choosing an order smaller than the true one is smaller than cost of a higher order because lowering the order of the model progressively lowers the computing cost involved in the data analysis.
The estimator q^* of the true order of the model (1) is obtained by minimizing the posterior risk

$$(8) \qquad R(r) = \sum_{q=1}^{Q} S(q,r)\, z(q|\underline{X}_n(1),\ldots,\underline{X}_n(T),\ n=1,\ldots,N)$$

over $r \in \{1,2,\ldots,Q\}$, where $z(q|\underline{X}_n(1),\ldots,\underline{X}_n(T),\ n=1,\ldots,N)$ is the posterior density function of the random variable q:

$$z(q|\ \underline{X}_n(1),\ldots,\underline{X}_n(T),\ n=1,\ldots,N) = \frac{h(q)\prod_{n=1}^{N} f(\underline{X}_n(1),\ldots,\underline{X}_n(T)|\,q)}{\sum_{q=1}^{Q} h(q)\prod_{n=1}^{N} f(\underline{X}_n(1),\ldots,\underline{X}_n(T)|\,q)}.$$

The function $f(\underline{X}_n(1),\ldots,\underline{X}_n(T)|\,q)$, $n=1,\ldots,N$ is a density function of a pT-dimensional normal distribution and may be written as follows (Krzyśko, 1983):

$$(9) \quad f(\underline{X}_n(1),\ldots,\underline{X}_n(T)|\,q) =$$

$$= (2\pi)^{-\frac{pT}{2}} |\hat{\underline{G}}_q|^{-\frac{1}{2}} |\hat{\underline{V}}|^{-\frac{T-q}{2}} \exp\Big\{ -\frac{1}{2}\Big[\mathrm{vec}'\,\underline{X}_{nq}\hat{\underline{G}}_q^{-1}\,\mathrm{vec}\,\underline{X}_{nq} +$$

$$+ \sum_{t=q+1}^{T} \Big(\sum_{u=0}^{q} \hat{\underline{B}}(u)\underline{X}_n(t-u)\Big)'\hat{\underline{V}}^{-1}\Big(\sum_{u=0}^{q} \hat{\underline{B}}(u)\underline{X}_n(t-u)\Big)\Big]\Big\},$$

where

$$\underline{X}_{nq} = (\underline{X}_n(1),\ldots,\underline{X}_n(q)), \qquad n=1,\ldots,N.$$

From (7) and (8) we obtain that the estimator $\hat{q}^*$ of the true order of the model (1) can be espressed as

$$\hat{q}^* = \arg\Big\{\min_{r} \sum_{q=1}^{Q} c|q-r|\,h(q) \prod_{n=1}^{N} f(\underline{X}_n(1),\ldots,\underline{X}_n(T)|\,q)\Big\},$$

i.e. $\hat{q}^*$ is that values of the parameter r for which the risk function $R(r)$ takes the minimal value.
As the true order of the model (1) is not known, we can assume that $q \in \{1,2,\ldots,Q\}$. For each assumed order we can obtain the estimators of the remaining parameters of the model by (4) and (5). In this manner we can compute the values of the function (9) for

each $q \in \{1,2,\dots,Q\}$, and we can find the estimator of the order of the model by minimizing the posterior risk.
The computer program which gives the least squares estimators of the unknown parameters and the Bayesian estimator of the true order of the autoregression equation is written in Fortran. The program is part of the Statistical Programs Package which is being prepared in the Mickiewicz University Computing Center in Poznań. The package is implemented on R-32 computer with an OS/360-MVT operating system.

References

Billingsley P. (1961), The Lindeberg-Lévy theorem for martingales, Proc. Am. Math. Soc., 12, 788-792.

Hannan E.J. (1970), Multiple time series, Wiley, New York.

Krzyśko M. (1983), Asymptotic distribution of the discriminant function, Statistics and Probability Letters, 1, 243-250.

On Fast Algorithms for Several Problems in Time Series Models

G. Mélard, Bruxelles

SUMMARY

In a recent paper the author has given a Fortran program for the computation of the likelihood function of an ARMA process. The algorithm is extended here to handle exactly and efficiently four related problems :

(a) the computation of forecasts
(b) the generation of artificial time series according to a given ARMA model
(c) the computation of the likelihood function of a transfer function model or a regression model with ARMA disturbances
(d) the computation of the first-order derivatives of the log-likelihood of an ARMA(p,q) process with respect to the coefficients.

Keywords and phrases : ARMA model, transfer function model, regression model with ARMA disturbances, fast algorithms, likelihood function, first-order derivatives of the log-likelihood, computation of forecasts, generation of artificial time series.

1. INTRODUCTION

In a recent paper, the author (Mélard, 1984) has given a Fortran program for the computation of the likelihood function of a stationary ARMA(p,q) process, defined by the equation

$$w_t = \phi_1 w_{t-1} + \dots + \phi_p w_{t-p} + a_t - \theta_1 a_{t-1} - \dots - \theta_q a_{t-q} \quad (1)$$

where the a_t are n.i.d. random variables with mean zero and variance σ^2. Using the AR and MA polynomials in the backshift operator B, (1) is written as $\phi(B)\, w_t = \theta(B)\, a_t$. Let $\underline{\phi} = (\phi_1, \dots, \phi_p)'$ and $\underline{\theta} = (\theta_1, \dots, \theta_q)'$, the parameters of the model. Given a time series $\underline{w} = (w_1, \dots, w_n)$ the likelihood function is obtained by

$$L(\underline{\phi}, \underline{\theta}, \sigma^2; \underline{w}) = (2\pi)^{-n/2} \left(\prod_{t=1}^{n} \tilde{\sigma}_t\right)^{-1} \exp\left\{-\frac{1}{2} \sum_{t=1}^{n} \left(\frac{\hat{a}_t}{\tilde{\sigma}_t}\right)^2\right\}$$

where $\hat{a}_t = w_t - \hat{w}_t$, $\hat{w}_t$ is the orthogonal projection of w_t into the subspace spanned by $(w_1, \dots, w_{t-1})$ and $\tilde{\sigma}_t$ is the norm of $\hat{a}_t$. For maximizing the likelihood function with respect to $\underline{\phi}$ and $\underline{\theta}$ all what is required is the sum of squares

$$S(\underline{\phi}, \underline{\theta}) = \left(\prod_{t=1}^{n} h_t^2\right)^{1/n} \sum_{t=1}^{n} \left(\frac{\hat{a}_t}{h_t}\right)^2, \quad (2)$$

where $h_t = \tilde{\sigma}_t/\sigma$.

Let $r = \max\{p, q+1\}$ and define the rx1 vectors $\underline{\varphi} = (\phi_1, \dots, \phi_r)'$,

$\underline{G} = -(-1, \theta_1, \ldots, \theta_{r-1})'$ and $\underline{H}' = (1, 0, \ldots, 0)'$ where $\phi_i = 0$ for $i > p$ and $\phi_j = 0$ for $j > q$. We consider the state space representation of (1)

$$w_t = \underline{H}\,\underline{W}_t$$

$$\underline{W}_t = \underline{F}\,\underline{W}_{t-1} + \underline{G}\,a_t \quad \text{with } \underline{F} = \begin{pmatrix} \underline{\varphi} & \begin{array}{c} \underline{I}_{r-1} \\ \hline \underline{0} \end{array} \end{pmatrix},$$

where $\underline{W}_t$ is the rx1 state vector and $\underline{I}_{r-1}$ is the identity matrix of order r-1. According to Pearlman (1980), the $\hat{a}_t$ can be computed by the following recursions

$$\hat{a}_t = w_t - \underline{H}\,\hat{\underline{W}}_t \tag{3}$$

$$\hat{\underline{W}}_{t+1} = \underline{F}\,\hat{\underline{W}}_t + \underline{K}_t\,(\hat{a}_t/h_t^2) \tag{4}$$

$$\underline{K}_{t+1} = \underline{K}_t - \alpha_t\,\underline{F}\,\underline{L}_t \quad \text{with } \alpha_t = \underline{H}\,\underline{L}_t/h_t^2 \tag{5}$$

$$\underline{L}_{t+1} = \underline{F}\,\underline{L}_t - \alpha_t\,\underline{K}_t \tag{6}$$

$$h_{t+1}^2 = h_t^2\,(1 - \alpha_t^2) \tag{7}$$

for $t = 1, \ldots, n$. Using improvements proposed by Mélard (1984) no more than $p + 3q + \min\{p,q\}$ multiplications or divisions are needed for each t and the total storage size is limited to three vectors of length r+1. The starting values for these recursions are $\hat{\underline{W}}_1 = \underline{0}$, $\underline{K}_1 = \underline{L}_1 = \underline{F}\,\underline{\mu}$ and $h_1^2 = \underline{H}\,\underline{\mu}$ where the i-th component of $\underline{\mu}$ is given by

$$\mu_i = \sum_{j=i}^{r} \left(\phi_j\,\gamma_{j-i+1} - \theta_{j-1}\,\lambda_{j-i}\right), \tag{8}$$

where the $\gamma_j = \mathrm{cov}\,(w_t, w_{t-j})/\sigma^2$ and $\lambda_j = \mathrm{cov}\,(w_t, a_{t-j})/\sigma^2$ can be obtained by the algorithm of Wilson (1979) with a number of operations bounded by a quadratic function of p and q. Memory space is also limited to three vectors of length r+1. Despite its apparent complexity, this new algorithm for computing the likelihood function of the ARMA(p,q) process is presently the most efficient in terms of computer time and storage. In practice the method is even faster than the traditional unconditional least squares method which can give only approximate (and sometimes bad) results.

In this paper, we extend the algorithm in order to handle four related problems.

2. COMPUTATION OF THE FORECASTS

If the a_t were known for $t = 1, \ldots, n$, (1) could be used to compute forecasts at times n+1, ..., n+h (Box and Jenkins, 1976). Since we can only compute exactly the $\hat{a}_t$, the forecasts should be derived from them (Ansley and Newbold, 1980). After the algorithm of Section 1 has been called by a nonlinear least-squares routine up to convergence, giving the maximum likelihood estimates of $\underline{\phi}$ and $\underline{\theta}$, the following relations are used for t = n+1, ..., n+h :

$$w_t = \underline{H}\, \hat{\underline{W}}_t$$
$$\hat{\underline{W}}_{t+1} = \underline{F}\, \hat{\underline{W}}_t. \tag{9}$$

The w_t are the forecasts.

3. GENERATION OF ARTIFICIAL TIME SERIES USING ARMA MODELS

Generally, (1) is used for $t > p$ and $a_{p-1} = a_{p-2} = \ldots = 0$. To avoid the transient caused by these starting values, a certain number of data points are neglected at the beginning of the series. Ansley and Newbold (1980) and Wilson (1979) have recommended the use of a more exact method. It appears that a very efficient method is given by the algorithm of Section 1, with the following modification : (3) is replaced by

$$w_t = \hat{a}_t + \underline{H}\, \hat{\underline{W}}_t \tag{10}$$

where the $(\hat{a}_t/h_t)$, t = 1, ..., n, are generated by pseudo-random variables with a given arbitrary distribution.

4. THE LIKELIHOOD OF A TRANSFER FUNCTION MODEL

The reader is referred to Liu (1983) for methods of estimation of transfer function models. Let the model be

$$w_t = \sum_{i=1}^{k} \frac{\omega_i(B)}{\delta_i(B)} X_{i,t} + \frac{\theta(B)}{\phi(B)} a_t \tag{11}$$

where $\omega_i(B)$, $\delta_i(B)$, $\theta(B)$, $\phi(B)$ are polynomials of respective degrees s_i, r_i, q, p, such that $\delta_i(0) = \theta(0) = \phi(0) = 1$ (i = 1, ..., k). It is assumed that data for w_t are available for $t > 0$ and that $X_{i,t}$ is known for $t > t_i$. For each i = 1, ..., k, two cases are possible :

1° $X_{i,t}$ is non-stochastic and $t_i = -\infty$. This occurs especially if X_i is an intervention variable, with the convention that $X_{i,t} = 0$ before the beginning of the series. Then, let $X'_{i,t} = \delta_i^{-1}(B)\, \omega_i(B)\, X_{i,t}$, $\omega'_i(B) = \delta'_i(B) = 1$.

2° otherwise we let $X'_{i,t} = X_{i,t}$, $\omega'_i(B) = \omega_i(B)$ and $\delta'_i(B) = \delta_i(B)$.

Denote by $\delta^*(B)$ the smallest common multiple of $\delta'_i(B)$, i = 1, ..., k, and r^* its degree. Then, the polynomials $\delta_i^*(B) = \delta^*(B)/\delta'_i(B)$ are computed and their degrees denoted by r_i^*. With these notations (11) is written under the form

$$\delta^*(B)\ w_t = \sum_{i=1}^{k} \omega_i'(B)\ \delta_i^*(B)\ X_{i,t}' + \frac{\delta^*(B)\ \theta(B)}{\phi(B)}\ a_t \tag{12}$$

for $t > t_o = \max \{r;\ \max_{i=1,\dots,k} (t_i + s_i' + r_i^*)\} \geq 0$, where s_i' is the degree of $\omega_i'(B)$. By cancellation of common factors $\delta^*(B)\ \theta(B)/\phi(B)$ is written as $\theta^*(B)/\phi^*(B)$. An algorithm for computing the likelihood function of a transfer function model is now :

1° for $t > t_o$ evaluate

$$w_t^* = \delta^*(B)\ w_t - \sum_{i=1}^{k} \omega_i'(B)\ \delta_i^*(B)\ X_{i,t}'; \tag{13}$$

2° use the algorithm of Section 1 with two modifications :

- t varies from t_o+1 to n instead of from 1 to n;
- with the present notations the model is written

$$\phi^*(B)\ w_t^* = \theta^*(B)\ a_t.$$

The regression model with autoregressive-moving average disturbances of Harvey and Phillips (1979) is a special case of (11) with $\delta_i(B) = 1$, $\omega_i(B) = \omega_i$. Consequently the algorithm of Section 1 can be used directly on

$$w_t^* = w_t - \sum_{i=1}^{k} \omega_i\ X_{i,t}$$

with a significant improvement in efficiency.

5. THE EXACT FIRST-ORDER DERIVATIVES OF THE LIKELIHOOD

Minimization of (2) can be made easier if the exact derivatives with respect to the ϕ_i and θ_j are known. Box and Jenkins (1976) have given recursions for the derivatives of the conditional log-likelihood, i.e. when starting values of the w_t and a_t are provided. In that case $S(\underline{\phi}, \underline{\theta}) = \Sigma_{t=1}^{n} \hat{a}_t^2$, where the $\hat{a}_t$ are linear with respect to the ϕ_i and θ_j. Khabie-Zeitoune (1980) has described a method for the derivatives of (2) where the first factor is omitted. Furthermore, the algorithm is based on the method of Levinson (1949) which requires a number of operations of order n^2. Given that the recursions used by Mélard (1984) are simple, it is easy to obtain similar recursions for the first-order derivatives. For example, the i-th relation of (6) is

$$(\underline{L}_{t+1})_i = \phi_i (\underline{L}_t)_1 + (\underline{L}_t)_{i+1} - \alpha_t (\underline{K}_t)_i,$$

and differentiation with respect to ϕ_k gives the recursion

$$\frac{\partial(\underline{L}_{t+1})_i}{\partial\phi_k} = \delta_{ik} (\underline{L}_t)_1 + \phi_i \frac{\partial(\underline{L}_t)_1}{\partial\phi_k} + \frac{\partial(\underline{L}_t)_{i+1}}{\partial\phi_k} - \frac{\partial\alpha_t}{\partial\phi_k} (\underline{K}_t)_i - \alpha_t \frac{\partial(\underline{K}_t)_i}{\partial\phi_k} .$$

Similarly $(\partial\mu_i/\partial\theta_k)$ requires knowledge of the first-order derivatives of the γ_i and the λ_i which can be obtained by differentiating all the recursions in the algorithm of Wilson (1979). The number of multiplications and divisions is multiplied by a factor less than 2(p+q) and the storage size is multiplied by (p+q). Thanks to the inherent efficiency of the algorithm for computing (2), the resources needed are theoretically still quite reasonable. In practice, however, since the recursions operate on two-dimensional arrays instead of vectors, data management is heavier and the computing time is significantly larger than 2(p+q) times what is required by the evaluation of (1).

ACNOWLEDGMENT

This paper was written while the author was at the Université de Montréal with a grant from the co-operation between the Province of Quebec and the French Community of Belgium.

REFERENCES

Ansley, C.F. and Newbold, P. (1980), Finite sample properties of estimators for autoregressive moving average models, *J. Econometrics* *13*, 159-183.
Box, G.E.P. and Jenkins, G.M. (1976), *Time Series Analysis Forecasting and Control*, Holden-Day, San Francisco (revised edition).
Harvey, A.C. and Phillips, G.D.A. (1979), Maximum likelihood estimation of regression models with autoregressive-moving average disturbances, *Biometrika 66*, 49-58.
Khabie-Zeitoune, E. (1980), Closed form likelihood in multivariate weakly stationary time series. In O.D. Anderson (ed.) : *Time Series*, North-Holland, Amsterdam, 311-318.
Levinson, N. (1979), A heuristic exposition of Wiener's mathematical theory of prediction and filtering. In N. Wiener (ed.) : *Extrapolation, Interpolation and Smoothing of Stationary Time Series, with Engineering Applications*, Wiley, New York.
Liu, L.M. (1983), Estimation of rational transfer function models, Unpublished paper, University of Chicago, Illinois.
Mélard, G. (1984), Algorithm AS 197 - A fast algorithm for the exact likelihood of autoregressive-moving average models, *J. Roy. Statist. Soc. Ser. C Appl. Statist. 33*, n°1 (to appear).
Pearlman, J.G. (1980), An algorithm for the exact likelihood of a high-order autoregressive-moving average process, *Biometrika 67*, 232-233.
Wilson, G.T. (1979), Some efficient computational procedures for high order ARMA models, *J. Statist. Comp. Simul. 8*, 301-309.

Computational Aspects of Robustness

Robustness in Parametric and Non-Parametric Regression Estimation: An Investigation by Computer Simulations

J. Antoch, Prague, and G. Collomb, and S. Hassani, Toulouse

Summary
Let $\{(X_i,Y_i),\ i=1,\dots,n\}$ be a sequence of n independent observations of a bivariate random variable (X,Y). The problem of the estimation of the regression $r(.)=E(Y/X=.)$ from these observations is investigated here by means of computer simulations for various parametric (§2) and non-parametric (§3) estimates. The main aim of these simulations (results given in §4) is the comparison of these estimates from the point of view of their robustness against contaminations relative to the law of $\varepsilon = Y-r(X)$ assumed independent on X and Y and the regression function assumed to be a polynomial.

Key words : Polynomial regression, regression quantile, least-squares estimate, α-trimmed and α-winsorized least-squares estimates, kernel estimate.

AMS classification : 62E25, 62J05, 62G05, 62F35.

I - INTRODUCTION

In this contribution we shall investigate the behaviour of parametric and non parametric estimates associated with the model[1]

$$(M1)\quad r(x) = \sum_{j=0}^{q} \theta_j x^j, \quad \mathcal{L}(\varepsilon) \sim (1-\beta)N+\beta C, \quad \mathcal{L}(X) \sim N, \quad q > 1, \ \beta\in[0,1].$$

To illustrate their behaviour from the point of view of the distributional robustness, we use $\beta=0$ and $\beta=0.1$ in our simulations. In addition to this problem we are also interested in the design robustness of our estimates. More precisely, if $\{(X_i,Y_i),\ i=1,\dots,n\}$ corresponds to the "true model" (M1) with q fixed, besides this true model we estimate also the model

$$(M2)\quad r(x) = r_1(x) = \sum_{j=0}^{q-1} \theta_j x^j, \quad \mathcal{L}(\varepsilon) \sim (1-\beta)N+\beta C, \quad \mathcal{L}(X) \sim N, \ \beta\in[0,1].$$

In our simulations we use (M1) with q=2,3,4 and (M2) with q=3,4.

In all upper situations the estimates are compared by means of the statistics

$$\max_{x\in A} |r_n(x)-r(x)|, \quad \int_A |r_n(x)-r(x)|dx \quad \text{and} \quad \int_A (r_n(x) - r(x))^2 dx,$$

where $r_n(.)$ is a generic estimate of r(.) and A is a fixed interval.

(1) Note : We shall use the following convention : N is the standard normal law, C is the standard Cauchy law and $\mathcal{L}(Z) \sim L$ means that the probability law of the r.r.v. Z is L. In the present work we do not consider robustness against dependence, heteroscedaticity of the law of Y given X and contamination of the law of X.

II - PARAMETRIC ESTIMATES.

Let q be fixed. In the problem of the regression estimation (in our models) from (X_i, Y_i), $i=1,\dots,n$ we consider besides the classical least-squares estimate three robust estimates : the α-trimmed and α-winsorized least-squares estimates and the ℓ_1-norm (least absolute error) estimate. Here we give only definitions, for a review of their mathematical properties we refer to the papers of Koenker-Basset (1978), Ruppert-Carrol (1980) and Jurečková (1983).

The common basis of all considered robust estimates is the notion of γ-regression quantile $r_n^{\gamma RQ}(x)$, introduced and studied by Koenker-Basset (1978). We can define it for our purpose as

$$r_n^{\gamma RQ}(x) = \sum_{j=o}^{q} \hat{t}_j x^j , \tag{2.1}$$

where $(\hat{t}_o,\dots,\hat{t}_q)$ minimizes $\sum_{i=1}^{n} \rho_\gamma(Y_i - \sum_{j=o}^{q} t_j X_i^j)$, $\gamma \in [0,1]$ and

$\rho_\gamma(x) = x.(\gamma - I[x<0])$.

It is evident that the ℓ_1-norm estimate $r_n^{LAE}(x)$ is nothing else than the regression median, i.e. $r_n^{\gamma RQ}(x)$ for $\gamma = 1/2$.

The α-trimmed least squares estimator $r_n^{\alpha T}(x)$ is the least-squares estimate associated with the sample

$$\{(X_i,Y_i) : r_n^{\alpha RQ}(X_i) \leqslant Y_i \leqslant r_n^{(1-\alpha)RQ}(X_i) , \quad i=1,\dots,n \} . \tag{2.2}$$

This estimate, proposed by Koenker-Basset (1978) and later studied by Rupert-Caroll (1980), provides the natural extension of the α-trimmed mean in the location case to the linear regression model.

The α-winsorized least-squares estimate $r_n^{\alpha W}(x)$ is defined by

$$r_n^{\alpha W}(x) = n^{-1}.\{ [n\alpha](r_n^{\alpha RQ}(x)+r_n^{(1-\alpha)RQ}(x))+(n-2\,[n\alpha]).r_n^{\alpha T}(x)\}, \tag{2.3}$$

where $[u] = \max\{v : v \in \mathbb{N}, v \leq u\}$, $\forall\, u \in \mathbb{R}$. This estimate, which is an extension of the α-winsorized mean, was proposed and studied by Jurečková (1983).

The least-squares estimate $r_n^{LSE}(x)$ associated with our sample is here the function $r_n^{LSE}(x) = \sum_{i=o}^{q} \hat{\theta}_j x^j$, where $(\hat{\theta}_o,\dots,\hat{\theta}_q)$ minimizes $\sum_{i=1}^{n} (Y_i - \sum_{j=o}^{q} \theta_j X_i^j)^2$.

While for computing the least squares estimates we use the QR-method as described e.g. in Hanson-Lawson (1974), for computing the regression quantiles we use modified method of Barrodale-Roberts (1974).

III - NON-PARAMETRIC ESTIMATE.

We consider here the kernel estimate, defined by

$$r_n(x) = \frac{\sum_{i=1}^{n} Y_i\, K((x-X_i)/h_n)}{\sum_{i=1}^{n} K((x-X_i)/h_n}\ , \quad (\tfrac{0}{0} = 0) , \tag{3.1}$$

with $h_n > 0$ satisfying $h_n \to 0$ for $n \to \infty$, $nh_n \to \infty$ for $n \to \infty$ and where we choose for the kernel K the function $K(u) = 1$ if $u \in [-.5, .5]$ and $K(u) = 0$ otherwise. This estimate is selected in the class of all the available non-parametric estimates (see Collomb, 1984) because of its simplicity and good asymptotic properties considered from an applied statistics point of view by Collomb (1982). In our application the number h_n was chosen so that the associated curve r_n is rather smooth ; the cross validation method, see Collomb (1982, p.177) , was not employed.

IV - RESULTS.

For every combination "MODEL[(M1), resp.(M2)] and DISTRIBUTION OF ε [normal, contaminated normal]", 1000 normally distributed pseudorandom variables representing X_i's and independently 1000 i.i.d. errors were generated. The values of the corresponding Y_i's were computed according (M1), resp. (M2) with fixed θ_i's [we used $\theta_0 = 1$, $\theta_1 = .25$, $\theta_2 = .1$, $\theta_3 = .04$ and $\theta_4 = .01$ in our study]. For each of 10 obtained situations we computed all upper mentioned estimates. The proportion of trimming (resp. winsorizing) which is chosen in all cases is $\alpha = .05$.

Results are summarized in tables 1-5. and figures 1-6. In all figures line 1 corresponds to the true model, line 2 to the $r_n^{LSE}(x)$, line 3 to the $r_n^{LAE}(x)$, line 4 to the $r_n^{.05T}(x)$ and line 5 to the $r_n^{.05W}(x)$, the signs "-" represent $r_n(x)$.

ESTIMATION OF THE REGRESSION $r(x) = 1 + \frac{x}{4} + \frac{x^2}{10}$
FOR VARIOUS ESTIMATES AND/OR LAWS OF $\varepsilon = Y - r(x)$

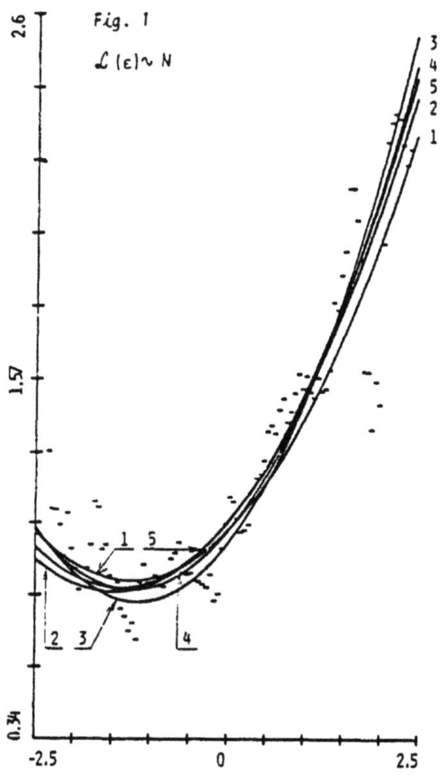

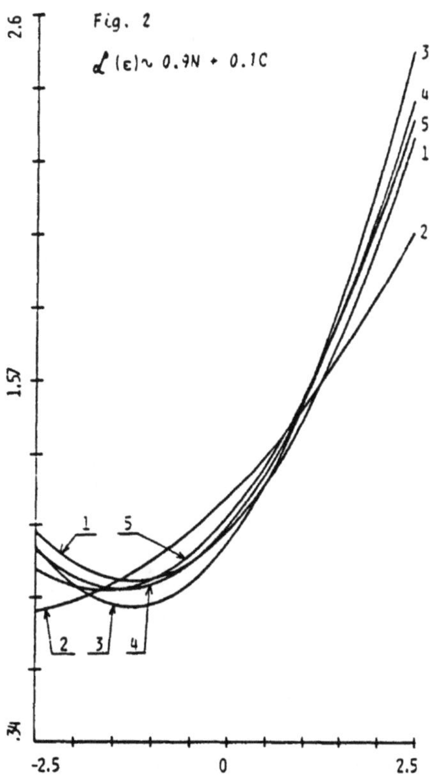

ESTIMATION OF THE REGRESSION $r(x) = 1 + \frac{x}{4} + \frac{x^2}{10} + \frac{4x^3}{25} + \frac{x^4}{100}$
FOR VARIOUS ESTIMATES AND/OR LAWS OF $\varepsilon = Y - r(x)$

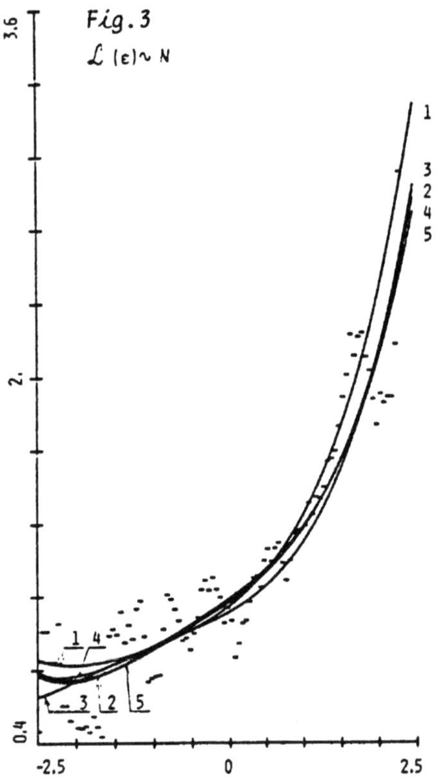

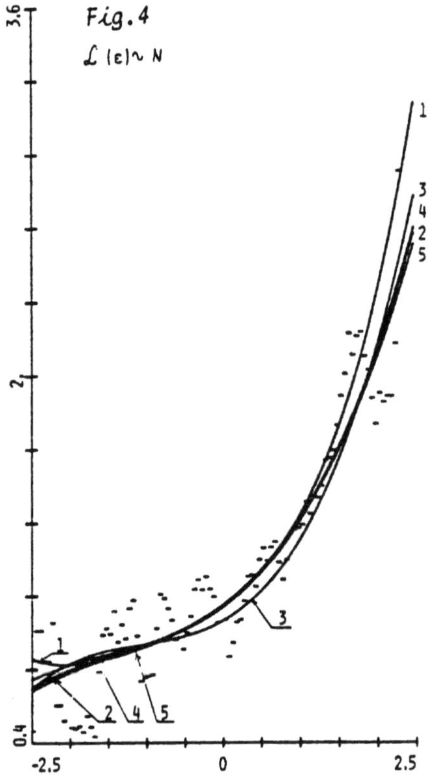

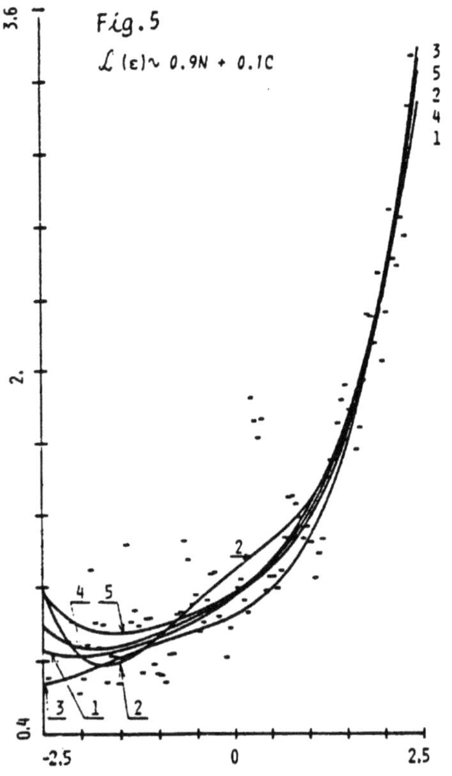

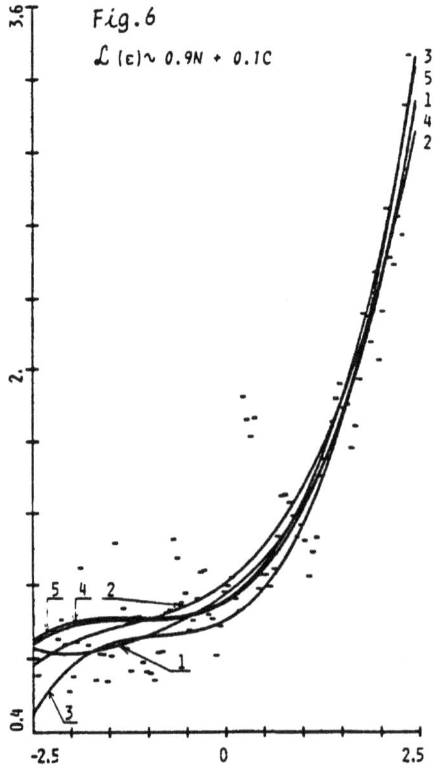

SOME NUMERICAL RESULTS FOR PARAMETRIC AND NONPARAMETRIC ESTIMATORS

	Real value	LSE	5% TLSE	5% WLSE	L1 NORM	LSE	5% TLSE	5% WLSE	L1 NORM	KERNEL
TABLE 1		$\mathcal{L}(\varepsilon) \sim N$, see Fig.1				$\mathcal{L}(\varepsilon) \sim 0.9N+0.1C$, see Fig.2				
θ_0	1.	1.024	0.997	1.028	0.959	1.108	1.015	1.046	0.962	-
θ_1	0.25	0.291	0.293	0.296	0.310	0.239	0.284	0.284	0.314	-
θ_2	0.1	0.098	0.118	0.106	0.133	0.038	0.102	0.090	0.124	-
$\int\|\hat{f}-f\|$	-	0.271	0.271	0.304	0.403	0.491	0.222	0.265	0.424	0.721
$\int\|\hat{f}-f\|^2$	-	0.020	0.032	0.030	0.060	0.070	0.014	0.017	0.054	0.257
$\max\|\hat{f}-f\|$	-	0.117	0.220	0.180	0.315	0.307	0.116	0.119	0.273	1.156
TABLE 2		$\mathcal{L}(\varepsilon) \sim N$				$\mathcal{L}(\varepsilon) \sim N$				
θ_0	1.	1.020	1.029	1.016	1.034	1.021	1.033	1.022	1.035	-
θ_1	0.25	0.211	0.195	0.206	0.253	0.355	0.354	0.359	0.353	-
θ_2	0.1	0.104	0.095	0.104	0.108	0.105	0.094	0.104	0.109	-
θ_3	0.04	0.053	0.059	0.057	0.048	NOT ESTIMATED				-
$\int\|\hat{f}-f\|$	-	0.166	0.213	0.190	0.278	0.419	0.418	0.415	0.454	0.856
$\int\|\hat{f}-f\|^2$	-	0.010	0.016	0.014	0.027	0.077	0.069	0.067	0.084	0.312
$\max\|\hat{f}-f\|$	-	0.149	0.168	0.190	0.214	0.412	0.369	0.400	0.461	1.181
TABLE 3		$\mathcal{L}(\varepsilon) \sim 0.9N + 0.1C$				$\mathcal{L}(\varepsilon) \sim 0.9N + 0.1C$				
θ_0	1.	1.357	1.004	1.001	1.006	1.358	1.014	1.004	1.002	-
θ_1	0.25	0.236	0.293	0.324	0.271	0.355	0.339	0.344	0.358	-
θ_2	0.1	-0.018	0.132	0.145	0.122	-0.017	0.119	0.133	0.139	-
θ_3	0.04	0.044	0.016	0.009	0.038	NOT ESTIMATED				-
$\int\|\hat{f}-f\|$	-	1.106	0.382	0.502	0.257	1.112	0.466	0.493	0.516	2.247
$\int\|f-f\|^2$	-	0.305	0.084	0.125	0.024	0.369	0.104	0.123	0.420	1.950
$\max\|\hat{f}-f\|$	-	0.408	0.477	0.585	0.169	0.739	0.537	0.600	0.602	2.173
TABLE 4		$\mathcal{L}(\varepsilon) \sim N$, see Fig.3				$\mathcal{L}(\varepsilon) \sim N$, see Fig.4				
θ_0	1.	1.034	1.017	1.025	0.974	1.002	0.991	1.003	0.932	-
θ_1	0.25	0.232	0.226	0.243	0.185	0.251	0.246	0.252	0.189	-
θ_2	0.1	0.035	0.056	0.042	0.070	0.104	0.111	0.099	0.129	-
θ_3	0.04	0.033	0.032	0.027	0.044	0.024	0.025	0.023	0.040	-
θ_4	0.01	0.014	0.010	0.012	0.009	NOT ESTIMATED				-
$\int\|\hat{f}-f\|$	-	0.437	0.405	0.458	0.500	0.369	0.364	0.408	0.490	0.906
$\int\|f-f\|^2$	-	0.086	0.088	0.111	0.099	0.110	0.101	0.132	0.089	0.363
$\max\|f-f\|$	-	0.387	0.426	0.488	0.362	0.602	0.575	0.649	0.427	0.786
TABLE 5		$\mathcal{L}(\varepsilon) \sim 0.9N + 0.1C$, see Fig.5				$\mathcal{L}(\varepsilon) \sim 0.9N + 0.1C$, see Fig.6				
θ_0	1.	1.115	1.002	1.017	0.917	1.048	0.964	0.975	0.886	-
θ_1	0.25	0.259	0.201	0.205	0.163	0.259	0.200	0.213	0.179	-
θ_2	0.1	-0.025	0.070	0.085	0.036	0.137	0.167	0.181	0.174	-
θ_3	0.04	0.036	0.049	0.046	0.067	0.036	0.047	0.050	0.070	-
θ_4	0.01	0.033	0.018	0.018	0.014	NOT ESTIMATED				-
$\int\|\hat{f}-f\|$	-	0.319	0.186	0.297	0.416	0.305	0.234	0.372	0.400	1.843
$\int\|\hat{f}-f\|^2$	-	0.031	0.010	0.033	0.046	0.021	0.018	0.041	0.049	1.270
$\max\|\hat{f}-f\|$	-	0.274	0.149	0.253	0.262	0.152	0.127	0.153	0.288	0.354

* We note $\int|\hat{f}-f|^i = \int_{-2.5}^{2.5} |\hat{f}(x)-f(x)|^i \, dx$, $i=1,2$ and $\max|\hat{f}-f| = \max_{x\in[-2.5,2.5]} |\hat{f}(x)-f(x)|$

V - CONCLUSIONS

The content of the tables 1 - 5 and figures 1 - 6 indicates that the behavior of all studied robust estimators is very good from the point of view of distributional robustness. If the distribution of ε is normal, than all studied parametric estimates are almost equivalent, see. tables 1,2 and 4 and figures 1 and 3. But when some grave errors are present, robust estimates, and especially α-trimmed and α-winsorized least-squares estimates, are preferable, see tables 1,3 and 5 and figures 2 and 5. The choice of the trimming proportion α will depend of course on the confidence in the quality of available data. The experience of Antoch (1984 a,b) and Ruppert-Carroll (1980) shows that $\alpha\epsilon[.05,.1]$ is sufficient for most of the situations. Nevertheless, for data of very dubious quality, ℓ_1-norm estimate seems to be preferable.

Against our expectation, behaviour of studied robust estimates is good even in the case when we underestimate the model, i.e. we use (M1) when (M2) is true model, see. tables 2-5 and figures 4,6. For more results and broadly varied situations see Antoch (1984 a,b).

Kernel estimation leads to curves which are very irregular. The important number of jumps is inherent to the definition of this method and comes from the rather large dispersion of the law of ε. In many practical applications this conditional variance is less important so that the obtained curves are smoother even in the case of heteroscedaticicity and regressions which are not polynomial, see e.g. the application in biology of Collomb (1982) and the simulations of Hassani (1982).

However, in spite of these irregularities we can observe that the kernel method gives an information on the true regression line in the uncontamined case, see figures 1,3 and 4. On the other hand, curves 5-6 and tables 1,3 and 5 show that the kernel estimate is not robust against outliers in the Y_i, $i=1,\dots,n$.

It is clear that the non-parametric estimate is a data analysis tool very useful in the problem of the choice of the model. When a parametric model is available then non-parametric estimate will be worse than any reasonable parametric estimate associated with the model. This intuitive statement is confirmed by our simulations as well as by mathematical results on the rate of convergence of all studies estimates.

REFERENCES.

ANTOCH J. (1984a). L-estimates in linear regression model. Preprint, Univ. Paul Sabatier, Toulouse, France.
ANTOCH J. (1984b). Numerical behaviour of some L-estimates in linear regression model. Submitted.
BARRODALE I.-ROBERTS F.D.K. (1974). Solution of an overdetermined system of equations in ℓ_1-norm. CACM, June 1974, 319-320.
COLLOMB G. (1982). COMPSTAT 1982, Physica-Verlag, Wien.
COLLOMB G. (1984). Non-parametric regression : an up to date bibliography. To appear in Math. Operationsforsch. Statis.
HASSANI S. (1982). Etudes par simulation de la méthode du noyau en estimation de la régression. Publication interne, Université Paul Sabatier, Toulouse.
JUREČKOVÁ J. (1983). Winsorized least squares estimator. Essays in Honour of N.B. Johnson, P.K. Sen ed., 237-245, North-Holland P.C.
KOENKER R., BASSET G. (1978). Econometrica 46, 33-50.
LAWSON C.L., HANSON R.J. (1974). Solving least squares problems. Prentice Hall Inc., Englewood Cliffs., New-Jersey.
MARAZZI A. (1980), COMPSTAT 1980, Physica-Verlag, Wien.
RUPPERT D., CARROLL R.J. (1980). J.A.S.A. 75, 828-838.

CEVOPE – a Program for Finding Conservative Estimates of the Variances of Regression Parameter Estimators when Errors are Serially Correlated

C.A. Glasbey, Edinburgh

Summary

The paper discusses computational problems associated with implementing a large algorithm in Fortran 77 which involves alternately solving a linear program with of order 10 000 constraints and 30 variables, and finding a vector such that a quadratic form approximately 200 by 200 in size is negative. The end-products are estimates of the variances of regression parameter estimators which are valid when errors are serially correlated in an unknown manner. Therefore the method makes only light use of statistical models but at the cost of a heavy computing burden.

Keywords

Linear regression, correlated errors, conservative estimates, robust estimates of variances, linear program, quadratic form, Fortran.

1. Introduction

A standard assumption in linear regression is that the errors are independent of one another, in which case ordinary least-squares regression parameter estimators are efficient and the usual estimators of the variances of these estimators are unbiased. However, if the errors are correlated, then in general neither of these properties holds true. Malinvaud (chapter 12, 1980), for example, discusses the case when errors are serially correlated and seeks to overcome the deficiencies in the standard analysis by assuming that the correlation coefficients are known except for a few parameters. But there are situations where very little is known about the form of correlations in a data set and certainly not enough to reduce $n(n+1)/2$ unknown terms in an n by n variance-covariance matrix to a few parameters.

Least-squares regression parameter estimators have an intuitive appeal, irrespective of the assumption of independent errors, because they give a simply understood summary of a set of data. However, when errors are correlated, valid estimates of the variances of these estimators are lacking. What are required are estimators of the variances which are unbiased for all possible types of correlation among the errors. These do not usually exist, so the second best alternatives are estimators which are slightly upward biased, that is conservative, for all possible correlation structures. These will be robust because they do not depend upon assumptions being made about the error correlations.

A vector of observations $\underset{\sim}{y}$ of length n is assumed to have an expectation denoted by

$$E(\underset{\sim}{y}) = \underset{\sim}{X}\,\underset{\sim}{\beta}\ ,$$

and an n by n variance matrix denoted by $\underset{\sim}{V}$, where $\underset{\sim}{X}$ is the known regression design matrix of size n by r and $\underset{\sim}{\beta}$ is the r-vector of regression parameters. The least-squares estimator of $\underset{\sim}{\beta}$ is

$$\hat{\underset{\sim}{\beta}} = (\underset{\sim}{X}^T\underset{\sim}{X})^{-1}\underset{\sim}{X}^T\underset{\sim}{y}\ ,$$

provided that $(\underset{\sim}{X}^T\underset{\sim}{X})$ is non-singular, where the superscript T denotes a matrix transpose and the superscript -1 denotes a matrix inverse. An estimator of the variance of $\hat{\beta}_k$, a function of $\underset{\sim}{y}$ denoted by $\hat{\text{var}}(\hat{\beta}_k)$, is sought for each value of k between 1 and r such that

$$E(\hat{\text{var}}(\hat{\beta}_k)) \geq \text{var}(\hat{\beta}_k) \quad \text{for all } \underset{\sim}{V} \text{ in } \Omega\ ,$$

where Ω is a set of n by n matrices which includes all possible variance matrices of $\underset{\sim}{y}$ in a particular application. More particularly, the estimator required is the one with smallest relative bias, that is which minimizes

$$(E(\hat{\text{var}}(\hat{\beta}_k)) - \text{var}(\hat{\beta}_k))/\text{var}(\hat{\beta}_k)\ ,$$

for either a particular choice of $\underset{\sim}{V}$, or an average value over Ω . This is a constrained optimization problem, the mathematical and computational tractability of which depends critically upon the choices of Ω and the functional form of $\hat{\text{var}}(\hat{\beta}_k)$. These are major considerations which will be discussed in detail in a more mathematical paper (Glasbey, 1985). For the purposes of this paper it is simply stated that when Ω consists of band matrices whose elements decrease monotonically away from the diagonal, and $\hat{\text{var}}(\hat{\beta}_k)$ is a positive semi-definite quadratic form of the least-squares residuals, the problem reduces to alternately (a) solving a large linear program (LP) and (b) finding a vector such that a large quadratic form is negative.

(a) The LP is

$$\text{minimize } \underset{\sim}{c}^T\underset{\sim}{x} \text{ with respect to } \underset{\sim}{x}\ ,$$

$$\text{subject to } \underset{\sim}{A}\,\underset{\sim}{x} \leq \underset{\sim}{b} \text{ and } \underset{\sim}{x} \geq 0\ ,$$

where $\underset{\sim}{x}$ and $\underset{\sim}{c}$ are vectors of length t (where t takes values between 1 and 30), $\underset{\sim}{b}$ is a vector of length m (where m depends on the choice of Ω but lies between n and n(n+1)/2) , and $\underset{\sim}{A}$ is an m by t matrix. So, for example, if n is 200, then m may typically take a value of 10 000 and the LP has 10 000 constraints and up to 30 variables.

(b) Once the LP has been solved any vector $\underset{\sim}{z}$ of length n is sought such that

$$\underset{\sim}{z}^T\underset{\sim}{P}\,\underset{\sim}{Q}\,\underset{\sim}{P}\,\underset{\sim}{z} < 0\ ,$$

where $\underset{\sim}{P}$ is the n by n idempotent matrix $(\underset{\sim}{I} - \underset{\sim}{X}(\underset{\sim}{X}^T\underset{\sim}{X})^{-1}\underset{\sim}{X}^T)$ with $\underset{\sim}{I}$ the n by n identity matrix, and $\underset{\sim}{Q}$ is an n by n symmetric matrix each of whose elements is a function of the Lagrangian multipliers in the solution of the LP. If, as above, n is equal to 200, then this is a large problem with $\underset{\sim}{P}$ and $\underset{\sim}{Q}$ each having 40 000 elements. Once $\underset{\sim}{z}$ has been found it is used to add a (t+1)st variable into the LP with c_{t+1} and the elements in the (t+1)st column of $\underset{\sim}{A}$ derived as functions of $\underset{\sim}{z}$. At the same time, if an element in $\underset{\sim}{x}$ was zero in the solution of the preceding LP then that variable is removed before solving the new LP.

The algorithm has been implemented as a Fortran 77 computer program called CEVOPE (Conservative Estimates of the Variances Of Regression Parameter Estimators). In the sections that follow, the computational problems associated with implementing such a large algorithm are considered, and the time taken to run CEVOPE in a particular application is examined. The effort put into the writing of CEVOPE has been sufficient to ensure that it can run in reasonable time on the AFRC Unit of Statistics' Prime 550 mini-computer, but there are many ways in which the algorithm could be made even more efficient. For example, the payoff in costs between computer processing time and array sizes has not been fully explored. The statistical technique which CEVOPE implements is novel, and only when this technique becomes accepted will it become worthwhile to spend more time streamlining the algorithm. In the final section, some general points are made about the relationship between statistical models and the use of computer time.

2. Linear Program

The revised simplex algorithm (see for example Garvin, chapter 13, 1960) has been used to solve the dual of the above LP problem in order to avoid having to use the enormous m by m matrix corresponding to the m auxiliary variables required by the inequality constraints. Because elements in the large matrix $\underset{\sim}{A}$ are accessed by the algorithm row by row whereas Fortran 77 stores array elements by columns, it is more efficient to store the transpose of $\underset{\sim}{A}$. Also, if the first dimension of the array holding $\underset{\sim}{A}$ is of fixed size 30, that is the maximum value of t, and m is equal to 10 000, then with double precision variables the array is of size 2.4 Mbytes. However, the central processor unit (CPU) space available on the Prime 550 computer is only 1.7 Mbytes. Therefore, in paged virtual memory parts of the array have to be moved in and out of the CPU as required. This incurs disc I/O time and dramatically increases the total run time for the program. If instead, the first dimension of the array is set close to t then, when t is less than 30, elements in $\underset{\sim}{A}$ may be placed in closer proximity in computer store and if no other job is running concurrently on the computer, disc I/O may be unnecessary. In order to achieve this, the first dimension of the array has to be variable so that it can be changed as t is changed between LPs. Another technique that is being used is to place all variables used in the LP algorithm into a single common block. This increases the chances of variables required consecutively being on the same page of virtual memory, which reduces access time. Another critical consideration in programming, because of the size of $\underset{\sim}{A}$, is the method used to calculate its coefficients. It is sufficient here to note that a recursive algorithm is used to calculate the coefficients and that when $\underset{\sim}{A}$ is updated after solving the quadratic form problem the strategy involves making the minimum number of alterations.

An LP routine in the NAG library, H01ADF (Numerical Algorithms Group, 1983), has been used in which any degeneracy encountered is resolved by the perturbed problem technique described for example by Garvin (chapter 14, 1960). It is almost certainly possible to improve upon the general routine for this particular problem, for example by using as a starting value the solution of the preceding LP rather than allowing H01ADF to set all the variables to zero initially. Also, until the final few iterations, the time-consuming search performed by the simplex algorithm could be restricted to a subset only of the large number of constraints.

3. <u>Quadratic Form Problem</u>

The quadratic form problem may be treated as an eigenvector problem because the vectors $\underset{\sim}{z}$ corresponding to stationary values of the quadratic form, subject to the constraint that $\underset{\sim}{z}^T\underset{\sim}{z}$ is equal to unity, are the eigenvectors of $\underset{\sim}{P}\,\underset{\sim}{Q}\,\underset{\sim}{P}$. In particular, the eigenvector which minimizes the quadratic form is the one corresponding to the largest negative eigenvalue.

A modified version of an iterative algorithm for finding the eigenvector corresponding to the eigenvalue of greatest magnitude has been used. (See for example Wilkinson (1954).) From an initial random vector $\underset{\sim}{z}_0$ of length n orthogonal to $\underset{\sim}{X}$, the transformation

$$e_i\,\underset{\sim}{z}_i = \underset{\sim}{P}\,\underset{\sim}{Q}\,\underset{\sim}{P}\,\underset{\sim}{z}_{i-1}$$

is applied repeatedly for $i=1, 2, \ldots,$ where e_i is a normalizing constant selected so that the largest coefficient in $\underset{\sim}{z}_i$ is unity. However, $\underset{\sim}{z}_{i-1}$ is orthogonal to $\underset{\sim}{X}$ for $i=1, 2, \ldots$ by its definition, so

$$\underset{\sim}{P}\,\underset{\sim}{z}_{i-1} = \underset{\sim}{z}_{i-1}$$

and the transformation may be simplified to

$$\underset{\sim}{e}_i\,\underset{\sim}{z}_i = \underset{\sim}{P}\,\underset{\sim}{Q}\,\underset{\sim}{z}_{i-1} \,.$$

If $\underset{\sim}{P}\,\underset{\sim}{Q}\,\underset{\sim}{P}$ has a unique eigenvalue of largest magnitude, say e, then e_i converges to it and $\underset{\sim}{z}_i$ converges to $\underset{\sim}{z}$, the corresponding eigenvector. If e is negative then the quadratic form problem is solved. Otherwise, a new transformation can be used

$$(f_i - e)\,\underset{\sim}{z}_i = \underset{\sim}{P}\,(\underset{\sim}{Q} - e\underset{\sim}{I})\,\underset{\sim}{z}_{i-1} \,,$$

to ensure that the eigenvalue of largest negative magnitude (say f) is found, where

$(f_i - e)$ is the normalizing constant corresponding to e_i above and f_i converges to f. If f is negative then the quadratic form problem is solved because

$$\underset{\sim}{z}^T \underset{\sim}{P}\, \underset{\sim}{Q}\, \underset{\sim}{P}\, \underset{\sim}{z} = f\, \underset{\sim}{z}^T \underset{\sim}{z} < 0 ,$$

and otherwise $\underset{\sim}{P}\, \underset{\sim}{Q}\, \underset{\sim}{P}$ is positive semi-definite and the problem has no solution.

One exception is that when the largest negative eigenvalue is not unique the iterative procedure does not converge. However, if at the ith iteration $\underset{\sim}{z}_i$ is such that

$$\underset{\sim}{z}_i^T \underset{\sim}{P}\, \underset{\sim}{Q}\, \underset{\sim}{P}\, \underset{\sim}{z}_i < 0 ,$$

then this is all that is required and the algorithm may stop without an eigenvector being found.

The above algorithm has the advantage of only finding one eigenvector, not the full set generated by some other eigenvector algorithms. Also, it avoids the explicit evaluation of $\underset{\sim}{P}\, \underset{\sim}{Q}\, \underset{\sim}{P}$, which could be time-consuming with the sizes of arrays involved, because the transformation $\underset{\sim}{P}\, \underset{\sim}{Q}\, \underset{\sim}{z}$ can be achieved by first multiplying by $\underset{\sim}{Q}$ and then by $\underset{\sim}{P}$, using its decomposition in terms of $\underset{\sim}{X}$. As in the LP algorithm, all variables are grouped together within computer store by placing them in a single common block, and an efficient algorithm is used to calculate the large number of elements in $\underset{\sim}{Q}$ using a recursive algorithm which in some senses is the inverse of that used to evaluate $\underset{\sim}{A}$.

Possibilities exist for improving the algorithm, although they have not yet been implemented. These include not choosing $\underset{\sim}{z}_0$ at random but in some systematic manner, and using an accelerated convergence procedure to find the eigenvector.

4. Example

In a particular example with n equal to 200 and m equal to 5 735, the number of iterations and computer time required every fifth step by the two algorithms are tabulated below:

t	LP			Quadratic Form		
	iterations	CPU time	I/O time	iterations	CPU time	I/O time
1	6	10 sec	0 sec	3	9 sec	0 sec
6	18	77	0	40	74	0
10	33	213	0	96	177	1
14	95	822	0	31	60	0
18	61	673	0	188	332	0
23	60	832	5	89	162	11
27	54	889	147	42	85	9

The CPU time used by the LP algorithm becomes much larger than that used by the quadratic form problem once t approaches 30. For this value of m the disc I/O time moves from zero only when t exceeds 25. However, if m were larger this would occur for a lower value of t, because it is the size of $\underset{\sim}{A}$, that is mt, which is critical. The program stopped after 34 iterations because t exceeded its maximum permitted value of 30, with a total CPU time of 23 400 sec and an elapsed time of 27 100 sec. Because the disc I/O time has been kept low the difference between CPU time and elapsed time is small.

5. Discussion

CEVOPE makes heavy use of computer time, but the results are valuable because they provide valid estimates of the variances of regression parameter estimators whilst making minimal assumptions about the form of the error correlations. This is an illustration of a general principle: current developments in computers provide an opportunity for many of the assumptions in classical statistical models to be dropped, even if a lot of computer time has to be used to analyse data. Other examples are jackknifing and bootstrapping. Efron (1982) comments that they involve "the substitution of computational power for theoretical analysis. The payoff is freedom from the constraints of traditional parametric theory with its overreliance on a small set of standard models for which theoretical solutions are available."

In the future it ought to become a general rule that, if possible, the required information should be extracted from a data set without making arbitrary model assumptions, even if the procedure uses a lot of computer time. If precision is lost, even though the new technique is efficient within the new set of assumptions, then this is a measure of the information being added into the analysis by use of a specific model, and if that model is purely conjectural the extra precision it generates is entirely spurious.

REFERENCES

Efron B. (1982), The jackknife, the bootstrap and other resampling plans, Society for Industrial and Applied Mathematics, Philadelphia.
Garvin N.W. (1960), Introduction to linear programming, McGraw-Hill, New York.
Glasbey C.A. (1985), Conservative estimates of the variances of regression parameter estimators for classes of serially correlated errors (to be submitted).
Malinvaud E. (1980), Statistical methods of econometrics (3rd edition), North Holland, Amsterdam.
Numerical Algorithms Group. (1983), Library manual mark 10, NAG Central Office, Oxford.
Wilkinson J.H. (1954), The calculation of the latent roots and vectors of matrices on the pilot model of the A.C.E., Proc. Cambridge Phil. Soc. 50, 536-566.

Robustness of Gumbel Statistic for Distribution Functions in the Domain of Attraction of a Type I Distribution of Largest Values

M.I. Gomes, Lisboa

Abstract: Robustness of Gumbel's statistic W_n, weighting the balance between the right and left handsides of a distribution, and a powerful test statistic for statistical choice between extreme value models, is investigated for a class of distribution functions Ω in the domain of attraction of Gumbel's Λ extreme value distribution. While $W_n|\Lambda$ is close to Λ, and the same happens to W_n for an underlying exponential distribution, for other members F of Ω the rate of convergence of $W_n|F^m$ towards Λ is quite slow, matching the "penultimate" behaviour of rate of convergence of $F^m(a_m x+b_m)$ towards the "ultimate" limiting form.

Keywords: stochastic simulation, robustness, extreme value theory, rates of convergence.

1. Let $(X_1,X_2,\dots,X_n)$ denote a random sample of size n from a population of maximum values. In practical situations we generally assume that each of the X_i's has a generalized extreme value (G.E.V.) or a von Misès-Jenkinson distribution function (d.f.) $G(\delta^{-1}(x-\lambda);k)$, whith support A, where

$$G(y;k) = \exp(-(1+ky)^{-1/k}) \quad , \quad A = \{y:\ 1+ky>0\},\ k\in \mathbb{R}-\{0\}, \tag{1.1}$$

with the usual continuation when $k\to 0$; indeed $G(y;k) \underset{k\to 0}{\to} \Lambda(y)=\exp(-\exp(-y))$, the Gumbel distribution. For k>0, G(y;k) is a Fréchet distribution (of index k^{-1}) and with k<0, G(y;k) is a Weibull distribution (of index $|k|^{-1}$).

The Gumbel d.f. plays a proeminent role in theory and practice of statistical extremes — it is referred to sometimes as the "extreme value distribution" — mainly because of mathematical convenience (and, to a certain extent, because the normal d.f. lies in its domain of attraction). Statistical choice between extremal models, testing k=0 in the G.E.V. model versus suitable one-sided or two-sided alternatives, turns out to be an interesting theoretical and practical problem, and has received some attention from statisticians in the last few years, mainly by van Montfort, Otten, Galambos, Tiago de Oliveira and Gomes; cf. Tiago de Oliveira and Gomes, 1983; van Montfort and Gomes, 1984, and references therein.

Among suitable statistics for testing H_0:k=0 versus k>0 (k<0) or k≠0, the one presented by Gumbel (1965)

$$W_n = \{Q_{1/2} - X_{1:n}\}^{-1} \{X_{n:n} - Q_{1/2}\} \qquad (1.2)$$

where $X_{i:n}$ denotes the i-th ascending order statistic associated to the sample $(X_1, X_2,\ldots,X_n)$ and $Q_{1/2}= \{X_{[n/2]:n} + X_{[n/2+1]:n}\}/2$ seems to be the most simple, although powerful, test statistic (Gomes, 1982).Properties and asymptotic behaviour of W_n have been studied by Tiago de Oliveira and Gomes (1983), who have shown that

$$\log(\log n) \{W_n - [\log(n)+\log(\log 2)]/[\log(\log n)-\log(\log 2)]\} \underset{w}{\to} Z \qquad (1.3)$$

Z a standard Gumbel random variable (r.v.) when the underlying d.f. of X is Gumbel.

In the present paper we study the robustness of W_n, by investigating the effects of replacing a Gumbel r.v. X in the definition of W_n by Y with d.f. F^m, $m\geq 1$, where F belongs to the domain of attraction of the Gumbel distribution (fact that we shall denote by $F\varepsilon\mathcal{D}(\Lambda)$), i.e.,

$$P\{M_m \leq a_m x+b_m\} = F^m(a_m x+b_m) \underset{m\to\infty}{\to} \Lambda(x),\ x\varepsilon \mathbb{R} \qquad (1.4)$$

for suitably chosen attraction coefficients $\{(a_m,b_m)\}, m\geq 1$, $a_m>0$, $b_m \varepsilon \mathbb{R}$, where $M_m = \max(Y_1,\ldots,Y_m)$, the Y_i's replica of Y. In other words, $F^m(a_m x+b_m) \approx \Lambda(x)$, and in practical situations we often identify $F^m(x)$, F usually unknown, with $\Lambda((x-b)/a)$ for some estimated $a>0$ and b, even for moderate values of m.

2. The d.f. of W_n defined as in (1.2) could not yet be obtained analytically, even for the easy situation X an exponential r.v., and our study had to be conducted in the light of stochastic simulation. The class of d.f.'s covered in this paper are the exponential, the normal, and certain d.f.'s with right-tail $1-F(x)= A x^r \exp(-x^q)$, $x\geq c$, $r\varepsilon \mathbb{R}$, $q>0$, $A>0$, c trivially obtained. Notice that all these d.f.'s belong to the family

$$\Omega = \{F(x;\alpha,\beta;A;r,q;\eta(x))\} \qquad (2.1)$$

where

$$F(x;\alpha,\beta;A;r,q;\eta(x)) = \{1-A(\alpha x+\beta)^r \exp(-(\alpha x+\beta)^q)(1+\eta(x))\}\, I_{[c,\infty)} \qquad (2.2)$$

of absolutely continuous d.f.'s in the domain of attraction of Λ, thoroughly investigated by Gomes(1978, 1981, 1984) in connection to penultimate behaviour: if $F\varepsilon\Omega$, $(r,q)\neq(0,1)$, even for large m, F^m is better approximated by a penultimate $G(.;k)$, $k<0$, if $q>1$, or $G(.;k)$, $k>0$, if $q<1$, than by the ultimate $\Lambda(.)=G(.;0)$. Denoting by Φ, as usual, the normal d.f., observe that $\Phi(x)=F(x;1/\sqrt{2},0;1/\sqrt{2\pi};-1,2;-x^{-2}+O(x^{-4}))$; the penultimate phenomenon in normal populations has first been pointed out by Fisher and Tippett (1928). Observe, further, that unless $q=1$ the rate of convergence of $F^m(.)$ towards $\Lambda(.)$ is rather slow, though in certain cases F(x) itself is close to $\Lambda(.)$.

To a certain extent, the present study may thus be viewed as a further in-investigation on the closeness of $F^m(a_m x+b_m)$ to $\Lambda(x)$.

3. First of all a stochastic simulation of the behaviour of the statistic W_n for underlying Gumbel lead us to the conclusion that such a statistic is reasonably well approximated by a Gumbel r.v., even for moderate values of n. A second order Kolmogorov-Smirnov test was undertaken and even for n=20 we were not lead to the rejection of such a model at the significance level .05. More than that, for all the 20 samples of size 500 generated we were far from the critical point .06 of Kolmogorov-Smirnov statistic. The mean value of those 20 distances is, for instance for n=20, equal to .04.

In order to evaluate the robustness of W_n, when allowing the underlying distribution to be in the domain of attraction of the Gumbel distribution, we perform a stochastic simulation with replicas and the use of antithetic variables of $W_n|F_i^m$; $W_n|F_i^m$ denotes the statistic W_n corresponding to an underlying population with d.f. $F_i^m(.)$. We present here results for m=1, 10, 50 and 200, $F_1(x)=\{1-\exp(-x^{.75})\}I_{[0,\infty)}$, $F_2(x)=\{1-\exp(-x)\}\ I_{[0,\infty)}$, $F_3(x)\equiv\Phi(x)$, $\Phi(.)$ the standard normal d.f., $F_4(x)=\{1-\exp(-x^4)\}\ I_{[0,\infty)}$, which are representatives of the patterns of behaviour of d.f.'s in the class Ω; details on other parent distributions are available from the author. Notice that, since $\Lambda^m(x)=\Lambda(x-\log m)$ and W_n is invariant under location and scale transformations, $W_n|\Lambda^m$ is obviously invariant with m.

Implementation of suitable algorithms that enable the efficient generation of pseudo random numbers with d.f. $F\in\mathcal{D}(\Lambda)$, as well as F^m, m>1, is undertaken. Usually, different methods are used for different ranges of the real line. Among them: von Neumann's rejection method, Newman and Forsythe's odd-even method (Knuth, 1981) as well as an assymetrization method (Sibuya, 1983).

In table 1 below, and for the particular value n=20, some estimates of population characteristics — quantiles, mean value, variance, skewness and kurtosis — are exhibited for $W_n|F_i^m$, i=1,2,3,4 and m=10,50,200. Similar estimates for $W_n|\Lambda$, n=20 as well, appear at the bottom of the table. Since the results are based on 20 independent replications of 500 runs each, and $t_{19;.975}=2.093$, $t_{n;\xi}$ the ξ-quantile of a t-distribution with n degrees of freedom, a 95% confidence interval for a characteristic θ, to which is associated an error $e_{\theta*}$, placed between brackets close to the corresponding estimate $\theta*$ is $(\theta*-2.093\ e_{\theta*},\theta*+2.093\ e_{\theta*})$. The closeness of corresponding estimates for the exponential (F_2) and for the Gumbel itself is striking, even for m as moderate as 10.

In table 2 we present, for sample sizes n=10 and 20 of maxima from samples of size m=10,50,200 from populations with d.f. F_i, i=1,2,3,4 as above, three measures of robustness that we now describe:

Table 1 – Estimates of population characteristics of $W_n|F_i^m$ for i=1,2,3,4 and for n=20.

Underlying d.f.		F_1	F_2	F_3	F_4
m=10	$x_{1/4}(.)$	1.84 (.02)	1.46 (.02)	.86 (.03)	.85 (.004)
	$x_{1/2}(.)$	2.63 (.05)	2.03 (.03)	1.16 (.02)	1.14 (.02)
	$x_{3/4}(.)$	3.70 (.08)	2.83 (.07)	1.51 (.01)	1.57 (.04)
	$\mu(.)$	2.98 (.04)	2.28 (.03)	1.25 (.01)	1.26 (.01)
	$\sigma^2(.)$	2.71 (.18)	1.96 (.08)	.30 (.006)	.35 (.02)
	$\beta_1(.)$	1.67 (.23)	1.49 (.24)	1.34 (.11)	1.35 (.26)
	$\beta_2(.)$	7.58(1.55)	6.61(1.64)	6.17(1.22)	6.00(1.70)
m=50	$x_{1/4}(.)$	1.61 (.02)	1.39 (.01)	.97 (.03)	.97 (.008)
	$x_{1/2}(.)$	2.26 (.04)	1.92 (.13)	1.31 (.03)	1.31 (.02)
	$x_{3/4}(.)$	3.20 (.09)	2.67 (.08)	1.70 (.02)	1.80 (.05)
	$\mu(.)$	2.57 (.03)	2.15 (.03)	1.41 (.01)	1.45 (.02)
	$\sigma^2(.)$	2.02 (.13)	1.90 (.07)	.38 (.01)	.47 (.03)
	$\beta_1(.)$	1.63 (.25)	1.50 (.25)	1.33 (.11)	1.35 (.25)
	$\beta_2(.)$	7.39(1.80)	6.65(1.76)	6.14(1.13)	5.97(1.68)
m=200	$x_{1/4}(.)$	1.54 (.02)	1.37(.01)	1.06 (.03)	1.05 (.01)
	$x_{1/2}(.)$	2.15 (.04)	1.89 (.03)	1.43 (.03)	1.41 (.02)
	$x_{3/4}(.)$	3.02 (.09)	2.64 (.08)	1.86 (.02)	1.95 (.04)
	$\mu(.)$	2.44 (.03)	2.12 (.03)	1.55 (.01)	1.57 (.02)
	$\sigma^2(.)$	1.89 (.12)	1.87 (.08)	.46 (.01)	.56 (.03)
	$\beta_1(.)$	1.61 (.25)	1.50 (.25)	1.34 (.12)	1.36 (.25)
	$\beta_2(.)$	7.29(1.87)	6.66(1.79)	6.19(1.18)	5.98(1.67)

Gumbel parent				
	$x_{1/4}(.)$	1.37 (.009)	$\mu(.)$	2.01 (.003)
	$x_{1/2}(.)$	1.88 (.03)	$\sigma^2(.)$	1.89 (.01)
	$x_{3/4}(.)$	2.63 (.08)	$\beta_1(.)$	1.50 (.25)
			$\beta_2(.)$	6.67(1.80)

i) $K^{(n)}_{F^m}$ is a location indicator, that measures the Kolmogorov- Smirnov distance between the estimated d.f. of $W_n|F^m$ and the estimated d.f. of $W_n|\Lambda$..

ii) $Q^{(n)}_{F^m}$ is a location measure, expressed in terms of a quadratic discrepancy between the estimated quantiles Q_p of $W_n|F^m$ and Q^*_p of $W_n|\Lambda$,

$$Q^{(n)}_{F^m} = \frac{1}{9} \sum_{p \in S} (Q_p - Q^*_p)^2 \qquad (3.1)$$

where $S = \{.025, .05, .10, .25, .50, .75, .90, .95, .975\}$.

iii) $R^{(n)}_{F^m}$ is the Kolmogorov-Smirnov distance between the estimated d.f. of $W_n|F^m$ and a fitted Gumbel d.f.; large values of $R^{(n)}_{F^m}$ obviously indicate departure of a Gumbel behaviour for W_n.

These computations have also been based on 20 replicas of 500 runs each.

Table 2 — Estimates of the three measures of robustness specified, for sample sizes n=10,20

		$F \equiv F_1$			$F \equiv F_2$			$F \equiv F_3$			$F \equiv F_4$		
		$K^{(n)}_{F^m}$	$Q^{(n)}_{F^m}$	$R^{(n)}_{F^m}$	$K^{(n)}_{F^m}$	$Q^{(n)}_{F^m}$	$R^{(n)}_{F^m}$	$K^{(n)}_{F^m}$	$Q^{(n)}_{F^m}$	$R^{(n)}_{F^m}$	$K^{(n)}_{F^m}$	$Q^{(n)}_{F^m}$	$R^{(n)}_{F^m}$
n=10	m=10	.18	2.13	.08	.07	.37	.06	.30	1.07	.06	.32	.95	.06
		(.02)	(.52)	(.01)	(.02)	(.21)	(.01)	(.02)	(.36)	(.01)	(.02)	(.39)	(.01)
	m=50	.10	.77	.07	.06	.001	.05	.26	.77	.06	.24	.55	.06
		(.03)	(.43)	(.01)	(.01)	(.00)	(.01)	(.01)	(.24)	(.01)	(.02)	(.24)	(.01)
	m=200	.08	.38	.06	.05	.000	.05	.22	.62	.05	.18	.35	.06
		(.01)	(.12)	(.01)	(.01)	(.00)	(.01)	(.01)	(.18)	(.01)	(.01)	(.11)	(.01)
n=20	m=10	.27	1.63	.05	.07	.05	.04	.46	1.49	.04	.43	1.39	.04
		(.01)	(.12)	(.00)	(.01)	(.09)	(.01)	(.01)	(.16)	(.01)	(.01)	(.09)	(.01)
	m=50	.15	.45	.04	.04	.002	.04	.37	1.06	.04	.33	.86	.04
		(.01)	(.05)	(.00)	(.01)	(.00)	(.01)	(.01)	(.14)	(.01)	(.01)	(.05)	(.01)
	m=200	.11	.24	.04	.04	.001	.04	.29	.74	.04	.27	.59	.04
		(.01)	(.03)	(.00)	(.01)	(.00)	(.01)	(.01)	(.12)	(.01)	(.01)	(.04)	(.01)

4. Simulation results for greater values of n and m, and for d.f.'s $F \in \Omega$ with different values of r and q only emphasize what is clear from the tables: for q=1 closeness between $W_n|F^m$ and $W_n|\Lambda$ is striking even for moderate values of m and n, and there are indications that the rate of convergence of $W_n|F^m$ towards $W_n|\Lambda$, as $m \to \infty$, is fast; on the other hand, we may observe departure of $W_n|F^m$ from $W_n|\Lambda$ when $F \in \Omega$ but $|q-1| >> 0$ even for large values of m. Notice also that if the parameter q in $F \in \Omega$ is greater than or equal to 1 it were always possible to fit a Gumbel d.f. to any of

the 20 samples of 500 observations of W_n generated. In this sense we may consider that there is a certain robustness of W_n. In what regards the location and scale parameters of such Gumbel d.f.'s they differ significantly from the limiting ones when we consider small values of m and values of $q \neq 1$ in $F \in \Omega$.

This pattern of behaviour could be expected from what is known about the rate of convergence of F^m, $F \in \Omega$ towards the limiting Gumbel d.f. (Gomes, 1978, 1984), and is clearly related to standard (or normal) and non-standard (or non-normal) domains of attraction (Gomes and Pestana, 1984).

References

Fisher, R.A. & L.H.C. Tippett (1928). Limiting forms of the frequency of the largest or smallest member of a sample. Proc. Camb. Phil. Soc. 24, 180-190.

Gomes, M.I. (1978). Some Probabilistic and Statistical Problems in Extreme Value Theory. Ph.D. Thesis, Sheffield, England.

Gomes, M.I. (1981). Closeness of penultimate approximations in extreme value theory. Abs. 14th European Meeting of Statisticians, 149-150, Wroclaw.

Gomes, M.I. (1982). A note on statistical choice of extremal models. Actas IX Jornadas Matemáticas Hispano-Lusas, II, 653-656, Salamanca.

Gomes, M.I. (1984). Penultimate limiting forms in extreme value theory. Ann. Inst. Statist. Math. (Tokyo) 36A.

Gomes, M.I. and D.D. Pestana (1984). Domains of attraction and penultimate behaviour. Submitted to 16th European Meeting of Statisticians, Marburg.

Gumbel, E.J. (1965). A quick estimation of parameters in Fréchet's distribution. Rev. Internat. Statist. Inst. 33, 349-363.

Knuth, D.E. (1981). The Art of Computer Programming, vol. II, 2nd ed., Addison-Wesley

Sibuya, M. (1983). Doubly exponential random number generators. To appear in J. Tiago de Oliveira, ed., Statistical Extremes and Applications, D. Reidel, Dordrecht.

Tiago de Oliveira, J. and M.I. Gomes (1983). Two test statistics for choice of univa-extreme models. To appear in J. Tiago de Oliveira, ed., Statistical Extremes and Applications, D. Reidel, Doedrecht.

van Montfort, M.A.J. and M.I. Gomes (1984). Statistical choice of extremal models for complete and censored data. To appear in J. Hydrology.

Aknowledgements: The research project of the author is currently supported by Centro de Estatística e Aplicações (INIC). The author is thankful to Calouste Gulbenkian Foundation for willingness to support her participation in successive editions of Compstat.

Asymptotic Distribution of the Roots of a Correlation Matrix

V. Lužar, Zagreb

SUMMARY:

Asymptotic expansion of the distribution of the latent root of a correlation matrix is obtained, when the corresponding population root is assumed to be simple. The distribution is derived by applying perturbation method yielding the expansion of characteristic root and by expanding the characteristic function.

KEY WORDS: asymptotic expansion; correlation matrix; latent roots; perturbation expansion.

1. INTRODUCTION

Depending on the nature of the problem and on the aim of the study, component analysis can be applied to different types of data matrices.

In order to faciliate interpretation, raw score data matrix is most frequently transformed to standardized form prior to performing component analysis.

Using correlation matrix as a basis for component analysis creates a problem of determining the number of principal components to retain and to rotate. Although a number of techniques and criteria has been proposed for determining the numer of non-trivial components, there are neither exact nor asymptotic tests for testing significancy of components. In last two or three decades much attention has been given to the derivation of the distribution of the latent roots of covariance matrices but very little is known about the distribution of the eigenvalues of other symmetric matrices (like correlation, Harris or Guttman's matrices) appearing in component analysis. Asymptotic expansions of the distributions of the latent roots of the covariance matrices, based on the properties of hypergeometric function, were derived for the case of simple roots by G.A. Anderson (1965) and Sugiura (1973) and in general case when the population roots have arbitrary multiplicity by T.W. Anderson (1963), James (1969), Chattopadhyay and Pillai (1973), Chikuse (1976), Sugiura (1976) and Contantine and Muirhead (1976). Exact joint density function was given by James (1964), Krishnaiah and Waikar (1971) in terms of zonal polynomials.

For the case of correlation matrix, some asymptotic results for likelihood ratio test statistics were obtained by Anderson (1963) and Lawley

(1956).

The purpose of this paper is to study the asymptotic expansion of the distribution of the latent roots of the sample correlation matrix drawn from multivariate normal population. This paper is devoted to the case when the popalation roots are simple.

2. PERTURBATION EXPANSION OF THE LATENT ROOTS OF THE CORRELATION MATRIX

Let $U = U_i; \{i = 1, n\}$ be the random sample drawn from the population Π and let each entity $U_i \in U$ be described on random sample $V = \{V_j; j = 1, m,\}$ of m random variables. Suppose that the variables in V are multivariatelly normaly distributed with the expected valùe μ and population covariance matrix Σ.

Let

$$Z = (z_{ij}) \quad \begin{matrix} i = 1,n \\ j = 1,m \end{matrix}$$

denote description of the set U over V. Then the estimate of covariance matrix on the sample U is

$$S_{/n} = (Z - 11^T Z)^T (Z - 11^T Z)/n,$$

where $1^T = (1,1,...1)$ is a vector of unities.

Putting

$$D^2 = \text{diag } S/n$$

and $\quad \tilde{D}^2 = \text{diag } \Sigma$

we can write sample and population correlation matrix as

$$R = D^{-1} S/n \, D^{-1} \quad ; \quad R = (r_{ij})$$

and

$$\tilde{R} = \tilde{D}^{-1} \Sigma \, \tilde{D}^{-1} \quad ; \quad \tilde{R} = (\tilde{r}_{ij})$$

respectivelly.

<u>LEMA</u>

Let λ_i be the i-th largest characteristic root of R and let $\tilde{\Lambda} = \text{diag}\,(\tilde{\lambda}_1, \tilde{\lambda}_2, \dots \tilde{\lambda}_m)$ $\tilde{\lambda}_1 \geq \tilde{\lambda}_2 \geq \dots \geq \tilde{\lambda}_m$ and $\tilde{X} = (\tilde{X}_1, \tilde{X}_2, \dots \tilde{X}_m)$ be the matrices of latent roots and vectors of $\tilde{R}$. If $\tilde{\lambda}_i$ is a simple root of $\tilde{R}$, then λ_i can be expanded as:

$$\lambda_i = \tilde{\lambda}_i + n^{-1/2}\, \tilde{x}_i^T [V - \tfrac{1}{2} (\tilde{R} V_o + V_o \tilde{R})] \tilde{x}_i +$$

$$+ n^{-1} \{-\frac{1}{2} \tilde{x}_i^T(VV_o+V_oV)\tilde{x}_i + \sum_{\substack{j=1\\ j\neq i}}^{m} \frac{\tilde{x}_j^T W\tilde{x}_i \tilde{x}_i^T W\tilde{x}_j}{\tilde{\lambda}_i - \tilde{\lambda}_j}\} + o(n^{-3/2}) \quad , \qquad (1)$$

where

$$V = \tilde{D}^{-1}(n^{-1/2}\, S - n^{1/2}\Sigma)\tilde{D}^{-1}.$$

$$V_o = \text{diag } V \qquad (2)$$

$$W = V - \frac{1}{2}\,(\tilde{R}\,V_o + V_o\tilde{R}\,) - \frac{n^{-1/2}}{2}\,(VV_o + V_oV)$$

Proof

Put

$$T = n^{-1/2}S - n^{1/2}\,\Sigma$$

$$V = n^{-1/2}\tilde{D}^{-1}S\tilde{D}^{-1} - n^{1/2}\tilde{R} = \tilde{D}^{-1}T\tilde{D}^{-1} \qquad (3)$$

and

V_o = diag V.

Anderson (1963) has shown that V is asymptotically normally distributed with expected value zero and covariances given by

$$E\,(v_{ij},\, v_{kl}) = \tilde{r}_{ik}\,\tilde{r}_{jl} + \tilde{r}_{il}\,\tilde{r}_{jk} \qquad (4)$$

Further, it is easy to show that

$$\lim_{n\to\infty} P\,(|n^{-1/2}\,v_{ii}| \leq 1) = 1 \qquad (5)$$

By expanding a diagonal matrix $B_o^{-1/2}$, in terms of V_o, where B_o is given by

$$B_o = \text{diag } \tilde{D}^{-1}S\tilde{D}^{-1},$$

and after some rearrangement, we get the following expression:

$$R-\tilde{R} = n^{-1/2}\,[V - \frac{1}{2}(\tilde{R}V_o+V_o\tilde{R})] - \frac{n^{-1}}{2}\,(VV_o+V_oV)+o(n^{-3/2}) =$$

$$= n^{-1/2}W + o(n^{-3/2}) \qquad (6)$$

The above expression was obtained by Anderson (1963) using different arguments.

By applying perturbation method to the matrices R and $\tilde{R}$ as for example in Sugiura (1976) in deriving the asymptotic joint distribution of multivariate F matrix, we get the expansion (1).

3. ASYMTOTIC EXPANSION OF THE DISTRIBUTION OF THE LATENT ROOT OF CORRELATION MATRIX

The above expression for the i-th largest latent root of R yields:

THEOREM:

Let λ_i be the i-th largest characteristic root of R. Let S has Wishart distribution $W_m(n, \Sigma)$. Let $\Phi(x)$ denotes the standard normal distribution function. If $\tilde{\lambda}_i$ is a simple root of $\tilde{R}$, the distribution function of $\sqrt{n}(\lambda_i - \tilde{\lambda}_i)/\tau$ can be expanded for large n as

$$\Phi(x)+\alpha_2\Phi^{(2)}(x)/\tau^2+n^{-1/2}\sum_{k=1}^{3}\beta_{2k-1}\Phi^{(2k-1)}(x)/\tau^{2k-1}+ \tag{7}$$

$$+\,n^{-1}\sum_{k=1}^{4}\gamma_{2k}\Phi^{(2k)}(x)/\tau^{2k}+o(n^{-3/2})$$

where $\tau = \sqrt{2}\,\lambda_i$ and the coefficients are given by

$$\alpha_2 = -2\tilde{\lambda}_i^3\xi_{ii}$$

$$\beta_1 = -2\tilde{\lambda}_i + 2\,\Sigma'(g_{ij}-\tilde{\lambda}_i f_{ij}\xi_{jj})$$

$$\beta_3 = \frac{4}{3}\tilde{\lambda}_i^3-6\tilde{\lambda}_i^4\xi_{ii}+2\tilde{\lambda}_i^4\,\Sigma' f_{ij}\xi_{ji}\xi_{ij}, \qquad \beta_5 = \frac{4}{3}\tilde{\lambda}_i^6\xi_{ii} \tag{8}$$

$$\gamma_2 = 4\tilde{\lambda}_i^2\,\Sigma\, f_{ij}\rho_{jj} - 4\tilde{\lambda}_i^2(1+\tilde{\lambda}_i\xi_{ii})$$

$$\gamma_4 = 2\tilde{\lambda}_i^4 - 24\tilde{\lambda}_i^5\xi_{ii} + 8\tilde{\lambda}_i^5\,\Sigma' f_{ij}\xi_{ji}\xi_{ij}$$

$$\gamma_6 = \frac{8}{9}\tilde{\lambda}_i^6 - 12\tilde{\lambda}_i^7\xi_{ii} + \frac{8}{3}\tilde{\lambda}_i^7\,\Sigma' f_{ij}\xi_{ji}\xi_{ij}, \qquad \gamma_8 = -\frac{19}{9}\tilde{\lambda}_i^9\xi_{ii}$$

Coefficients ξ_{ij}, ρ_{ij}, f_{ij} and g_{ij} are defined by

$$\xi_{ij} = \tilde{x}_i^T\,\mathrm{diag}(\tilde{x}_i\tilde{x}_i^T)\tilde{x}_j$$

$$\rho_{ij} = \tilde{x}_i^T\,\mathrm{diag}(\tilde{x}_i\tilde{x}_i^T)\tilde{R}\,\mathrm{diag}(\tilde{x}_i\tilde{x}_i^T)\tilde{x}_j \tag{9}$$

$$f_{ij} = \frac{\tilde{\lambda}_i(\tilde{\lambda}_i+\tilde{\lambda}_j)}{\tilde{\lambda}_i-\tilde{\lambda}_j}$$

$$g_{ij} = \frac{\tilde{\lambda}_i\tilde{\lambda}_j}{\tilde{\lambda}_i-\tilde{\lambda}_j} \qquad \Sigma' \equiv \sum_{\substack{j=1\\ j\neq 1}}^{m}$$

Proof

Applying the above lemma and using the fact that

$$x_i^T V x_i = tr\,(KT)$$

where

$$K = D^{-1} x_i x_i^T D^{-1}$$

and T is given in (3), the characteristic function ϕ of $\sqrt{n}\,(\lambda_i - \tilde{\lambda}_i)$ can be expressed as

$$E(e^{(it)tr(KT)}\{1-(it)\tilde{\lambda}_i tr(KT_o)+n^{-1/2}(it)[-\tfrac{1}{2}\tilde{x}_i^T(VV_o+V_oV)\tilde{x}_i + \sum_{\substack{j=1\\ j\neq i}}^{m} \frac{\tilde{x}_j^T W\tilde{x}_i \tilde{x}_i^T W\tilde{x}_j}{\tilde{\lambda}_i - \tilde{\lambda}_j}] + o(n^{-1})\} \tag{10}$$

By applying lemma 5.1 of Sugiura (1973) to each term in the above expanssion, expectations can be computed, yielding after differentiations and some rearrangements

$$\phi(\sqrt{n}(\lambda_i-\tilde{\lambda}_i)) = e^{-t^2\tilde{\lambda}_i^2}\{1+\alpha_2(it)^2+n^{-1/2}\sum_{k=1}^{3}\beta_{2k-1}(it)^{2k-1} + n^{-1}\sum_{k=1}^{4}\gamma_{2k}(it)^{2k}\}+o(n^{-3/2}) \tag{11}$$

where the coefficients α, β, γ are given in (8). Inversion of the characteristic function completes the proof.

This paper is based on author's doctoral thesis at University of Zagreb done under the supervision of professor K. Momirović.

REFERENCES:

(1) Anderson, T.W. (1963). Asymptotic theory for principal component analysis. Ann. Math. Statist. 34, 122-148.

(2) Anderson, G.A. (1965). An asymptotic expansion for the distribution of the latent roots of the estimated covariance matrix. Ann. Math. Statist. 36, 1153-1173.

(3) Chattopadhyay, A.K. and Pillai, K.C.S. (1973). Asymptotic expansions for the distributions of characteristic roots when the parameter matrix has several multiple roots. In Multivariate Analysis - III (P.R. Krishnaiah, ed.), Academic Press, New York.

(4) Chikuse (1976). Asymptotic distributions of the latent roots of the covariance matrix with multiple population roots. J. Multivariate Anal. 6, 237--249.

(5) James, A.T. (1964). Distributions of matrix variates and latent roots derived from normal samples. Ann. Math. Statist. 35, 475-501.

(6) Krishnaiah, P.R. and Waikar, V.B. (1971). Exact distributions of any few ordered roots of a class of random matrices. J. Multivariate Anal. 1, 308-315.

(7) Lawley, D.N. (1956). Tests of significance for the latent roots of covariance and correlation matrices. Biometrika 43, 128-136.

(8) Lužar, V. (1984). Approximations to the distributions of the latent roots of the matrices in standardized, Guttman's and Harris spaces. Ph. D. thesis, University of Zagreb.

(9) Sugiura, N. (1973). Derivates of the characteristic root of a symmetric or a hermitian matrix with two applications in multivariate analysis. Comm. in Statist. 1, 393-417.

(10) Sugiura, N. (1976). Asymptotic expansions of the distributions of the latent roots and the latent vector of the Wishart and multivariate F matrices. J. Multivariate Anal. 6, 500-525.

New Topics in Robust Statistics with Applications

J. Michalek, I. Vajda, and J.A. Víšek, Praha

Abstract: General statistical procedures robust in a broad or minimax sense are introduced and their sensitivity with respect to a contamination level is studied. Broad-sense-robust directed D-estimators of abstract statistical parameters are introduced and their sensitivity and asymptotic characteristics are evaluated. For minimax -sense-robust tests the sensitivity of size and power is evaluated. Numerical results, computational algorithms and practical experiences are emphasized throughout the paper.

Key words: Robust estimation and testing, sensitivity to contamination, directed D-estimators.

1. Robustness and sensitivity of statistical procedures. Let $\mathcal{N}$ be the set of all positive integers and R the real line and let $\mathcal{P}$ be the class of all probability distributions on a Borel space $(\mathcal{X},\mathcal{B})$. A general statistical procedure can be described as a mapping d: $\mathcal{X}^n \to D$, $n \in \mathcal{N}$, in the framework of a statistical decision problem with sample space $(\mathcal{X},\mathcal{B})$, decision space D, and sample family $\mathcal{Q} \subset \mathcal{P}$. If the risk $r(d,\mathcal{Q})$ of d (e. g. a minimax or a Bayesian mean of the average losses $r(d,Q) \triangleq$ $\triangleq E_{Q^n} L(Q,d(\underline{x}))$, $Q \in \mathcal{Q}$, where $\underline{x} = (x_1,\dots,x_n) \in \mathcal{X}^n$ is a data-vector, Q^n is the product of Q and L: $\mathcal{P}\times D \to R$ is a loss function) minimizes $r(\tilde{d},\mathcal{Q})$ over a suitable subclass of procedures $\tilde{d}$: $\mathcal{X} \to D$, then d is optimum for $\mathcal{Q}$ and it is denoted by $d_{\mathcal{Q}}$.

An important endeavour of many statisticians which has started from very early times of data processing was a try to cope with a bad behaviour of statistical procedures $d_{\mathcal{Q}}$ in cases when assumptions concerning the family $\mathcal{Q}$ are violated (see Stiegler (1973)). In two last decades a great deal of attention was paid to this problem undoubtedly also due to the fact that statistical procedures are quite frequently used in computers of various data-transmission networks, where one can hardly imagine any permanent supervision of data by a statistician. Initiated with papers of Huber (1964, 1972), Huber and Strassen (1973), and Hampel (1974), many attempts occured to suggest statistical estimators and tests with limited sensitivity to modifications $\mathcal{Q} \to \mathcal{Q}_*$ of assumed sample families and to suggest tools for evaluation of the sensitivity itself (see e. g. Andrews et al (1972), Portnoy (1977), Millar (1981), Boos (1981), Hampel (1981), Vajda (1982-1984e) or Ronchetti and Rousseeuw (1979), Lambert (1981), Ronchetti (1982), Víšek (1983b-1984c)).

In term of a general statistical procedure d the conceptual approaches can be summarized and completed as follows.

Let for a fixed $s \in N$ and every $\varepsilon \in (0,1)^s$, $U_\varepsilon : P \to \exp P$, where $P \in U_\varepsilon(P)$ for all $P \in P$ (a model of contamination neighborhoods). Further let J be a bijection $P \to P$ and $\mathcal{Q}_\varepsilon(\mathcal{Q}) = \{J(\mathcal{Q}) : \forall (J : J(Q) \in U_\varepsilon(Q), \forall Q \in \mathcal{Q})\}$. A procedure d is considered *robust in a broad sense* for a class $\mathcal{Q}$ of families $\mathcal{Q} \subset P$ if at each $\mathcal{Q} \in \mathcal{Q}$ it is stable in the sense

$$r_\varepsilon(d,\mathcal{Q}) \triangleq \sup_{\mathcal{Y} \in \mathcal{Q}_\varepsilon(\mathcal{Q})} r(d,\mathcal{Y}) \to r(d,\mathcal{Q}) \text{ as } \|\varepsilon\| \to 0$$

and if the risk increase $r(d,\mathcal{Q}) - r(d_{\mathcal{Q}},\mathcal{Q})$ is reasonably limited on $\mathcal{Q}$ (as to the inference concerning location parameter a class $\mathcal{Q}$ of reasonable alternatives to the normal sample family $\mathcal{Q}$ can be found in Andrews et al (1972)). A procedure d_ε is considered *robust in a minimax sense* for $\mathcal{Q}$ if the least favourable risk $r_\varepsilon(d_\varepsilon,\mathcal{Q})$ minimizes the least favourable risks $r_\varepsilon(d,\mathcal{Q})$ over a suitable subclass of procedures d: $X \to D$. Further we are interested in derivatives

$$D_\varepsilon(d,\mathcal{Q}) \triangleq \lim_{\tau \downarrow 0} \frac{r_{\varepsilon+\xi(\tau)}(d,\mathcal{Q}) - r_\varepsilon(d,\mathcal{Q})}{\tau}$$

for coordinatewise nondecreasing mappings $\xi: [0,1] \to [0,1]^s$ having right-hand derivatives $\xi'(\tau)$ at zero. If $D_\varepsilon(d,\mathcal{Q})$ exists, then it characterizes the *dependency* of the least favourable risk $r_\varepsilon(d,\mathcal{Q})$ on slight deviations of the assumed contamination level ε. Thus the derivatives are extremely interesting at procedures $d=d_\varepsilon$ fitted to a given $\mathcal{Q}$ at a particular level ε.

We shall apply the concepts of robustness and dependency to the fundamental statistical procedures: the point estimation and testing of hypotheses.

Following Vajda (1984a) let us consider an arbitrary non-empty parameter space Θ and a parametrized sample family $\mathcal{Q}_\Theta = \{Q_\theta : \theta \in \Theta\} \subset P$.

Definition 1. An estimator of a parameter from Θ is a mapping $T: P(T) \to \Theta$ where $P(T)$ is supposed to be a non-empty subset of P. The T is said *(continuously) well-defined* if

(1) $$P_{emp} \triangleq \{P_n \triangleq \frac{1}{n}\sum_{i=1}^{n} 1_{\{x_i\}} : \underline{x}=(x_1,\dots,x_n) \in X^n,\ n=1,2,\dots\} \subset P(T)$$

and if the estimates $T(P_n)$ as functions of sample vectors $\underline{x}$ (cf. (1)) are measurable (continuous) on X^n, $n \in N$. The estimator T is said *strongly consistent* for a sample family $\mathcal{Q}_\Theta$ if it is well-defined and $T(P_n) \to \theta$ as $n \to \infty$ a. s. Q_θ^∞ for all $\theta \in \Theta$. If $\Theta \subset R^m$ then we say that T is *asymptotically normal* for a sample family $\mathcal{Q}_\Theta$ with an asymptotic mean $M_\theta(T)$ and asymptotic variance $V_\theta(T)$ if it is well-defined and $n^{\frac{1}{2}}(T(P_n)-\theta) \to$

$\to N(M_\theta(T), V_\theta(T))$ as $n\to\infty$ Q_θ^∞-weakly for all $\theta\in\Theta$. If $\Theta\subset R^m$ and, for some $Q\in P$ and all $x\in X$, $Q_{\varepsilon,x} \triangleq (1-\varepsilon)Q+\varepsilon.1_{\{x\}} \in P(T)$ for all sufficiently small $\varepsilon>0$ and $\Omega_Q(x) \triangleq \lim_{\varepsilon\downarrow 0} (T(Q_{\varepsilon,x})-T(Q))/\varepsilon$ exists then $\Omega_Q: X\to R^m$ is called an *influence curve* of T at Q.

For each estimator T of a real valued parameter $\theta\in\Theta\subset R$ we can define average losses $r(T,Q_\theta) = \infty$ if the asymptotic variance $V_\theta(T)$ is non-existing and $r(T,Q_\theta) = V_\theta(T)$ in the opposite case. The dependency of estimators of real valued parameters on the deviation of contamination level is considered in this paper under the above specified risk $r(T,\mathcal{Q}_\theta)$ and under Huber's contamination neighborhoods $U_\varepsilon(Q) = \{(1-\varepsilon)Q+\varepsilon Q_*: Q_*\in P\}$, $\varepsilon\in[0,1)$. In estimating location sample families $\mathcal{Q} = \mathcal{Q}_\theta$ are defined by one parent $Q\in P$ and each $\tilde{\mathcal{Q}}\in\mathfrak{Q}_\varepsilon(\mathcal{Q})$ by one parent $\tilde{Q}\in U_\varepsilon(Q)$ so that the class of families $\mathfrak{Q}_\varepsilon(\mathcal{Q})$ can in this case be replaced by a family $\mathcal{Q}_\varepsilon = \{\tilde{Q}\in U_\varepsilon(Q)\}\subset P$. This simplification is taken into account in the next chapter.

In testing a simple hypothesis $H \triangleq P_0\in P$ against an alternative $A \triangleq P_1\in P$ we consider tests Φ_T based on statistics $T(\underline{x}) = \Sigma_1^n \Omega(x_i)$, $\underline{x}\in X^n$, with T_n written instead of T when convenient, and the so-called general contamination neighborhoods (cf. Rieder (1977)) given for $\varepsilon = (\varepsilon_1,\varepsilon_2)\in[0,1)^2$, $0<\varepsilon_1+\varepsilon_2<1$, by

$$U_\varepsilon(Q) = \{P\in P: P(A)\leq(1-\varepsilon_1)Q(A)+\varepsilon_1+\varepsilon_2, \forall A\in B\}.$$

Here the sample family $\mathcal{Q}$ is a two-distributions set $\{P_0,P_1\}$ and $\mathfrak{Q}_\varepsilon = \mathcal{Q}_{0,\varepsilon^0}\cup\mathcal{Q}_{1,\varepsilon^1}$ where $\mathcal{Q}_{i,\varepsilon^i}=\{\tilde{Q}\in P: \tilde{Q}\in U_{\varepsilon^i}(P_i)\}$ for i=0,1. In what follows we consider $\xi^i(\tau) = (\xi^i(\tau),\xi^i(\tau))$ and fixed ε^i and we put $P_{i,\tau} = Q_{i,\varepsilon^i+\xi^i(\tau)}$, $\tau\in[0,1]$, i=0,1. For all sufficiently small ε_1^i and ε_2^i it holds $\mathcal{Q}_{0,\varepsilon^0}\cap\mathcal{Q}_{1,\varepsilon^1} = \emptyset$ (see Rieder (1977)). This implies (cf. Remark 1 in Víšek (1983c)) that there is $\tau_0\in[0,1]$ such that $P_{0\tau}\cap P_{1\tau} = \emptyset$ for all $\tau\in[0,\tau_0]$. In what follows we restrict ourselves to $\{P_{i\tau}: \tau\in[0,\tau_0]\}$. The following definition is a special version of the general definition of the dependency above which takes into account the fact that the risk $r_\varepsilon(\Phi_T,\mathcal{Q})$ of a test Φ_T may be represented by a pair of size and power s and p, where $s = s(t,n,P_{0\tau}) \triangleq \sup\{Q^n(T_n>t): Q\in P_{0\tau}\}$, $p = p(t,n,P_{1\tau}) \triangleq \inf\{Q^n(T_n>t): Q\in P_{1\tau}\}$, $\tau\in[0,\tau_0]$ (cf. Huber and Strassen (1973)).

Definition 2. If the limits

$$\lim_{\tau\downarrow 0} \tau^{-1}.[s(t,n,P_{0\tau})-s(t,n,P_{00})] \tag{2}$$

$$\lim_{\tau\downarrow 0} \tau^{-1}.[p(t,n,P_{1\tau})-p(t,n,P_{10})]$$

exist, we call them *dependency of the size and power of test* Φ_T (*or size and power dependency, for short*) *on the deviation of probabi-*

lity model and denote them by SD(t,n) and PD(t,n), respectively.

2. Robustness and sensitivity of directed D-estimators. In this chapter we consider along with sample families $\mathcal{Q}_\Theta \subset \mathcal{P}$ also projection families $\mathcal{P}_\Theta = \{P_\theta : \theta \in \Theta\} \subset \mathcal{P}$, $P_\theta \neq P_{\theta'}$ for $\theta \neq \theta' \in \Theta$.

Definition 3. Directed D-estimators T^α, $\alpha \in [0,1]$, with a projection family $\mathcal{P}_\Theta$ dominated by a directing measure W are defined on domains $\mathcal{P}(T^\alpha) = \{Q \in \mathcal{P}: T^\alpha(Q) \neq \emptyset\}$ by $T^\alpha(Q) = \tau(\mathbf{T}^\alpha(Q))$, where $\mathbf{T}^\alpha(Q)$ denotes a subset of Θ at which the directed D^α-divergence $D^\alpha(P_\theta,Q) = E_Q f^\alpha(p_\theta)$, $p_\theta = dP_\theta/dW$, $f^\alpha(p) = (1-p^\alpha)/\alpha$, between theoretical and empirical probability P_θ and Q is minimum and where τ is a fixed rule of choice on $\exp\Theta-\emptyset$.

Obviously T^0 is a general maximum likelihood estimator (MLE) with projection family $\mathcal{P}_\Theta$. As shown in Vajda (1984c,f), all T^α with $\alpha>0$ and reasonably smooth $\mathcal{P}_\Theta$ are well-defined. Moreover, for a wide variety of sample families $\mathcal{Q}_\Theta$, they are strongly consistent and asymptotically normal with the asymptotic mean zero and asymptotic variance $V_\theta(T^\alpha) = \mathrm{var}_{Q_\theta}\Omega^\alpha_{Q_\theta}$, where the influence curves $\Omega^\alpha_{Q_\theta}$ are explicitely evaluated as well. Contrary to wider classes of estimators considered by Huber (1964) or Pfanzagl (1969), all estimators T^α are well motivated in a sense specified and proved in Vajda (1984a). A computationally important difference is that our estimates $T^\alpha(Q)$ are chosen from sets $\mathbf{T}^\alpha(Q)$ by simpler rules of choice depending on Q through $\mathbf{T}^\alpha(Q)$ only.

Next we consider *directed D-estimators of location* T^α, i. e. $\Theta = X = R$, W is Lebesgue measure and a projection parent density p is supposed to be unimodal, even, twice continuously differentiable. The next assertion follows from Sec. 4 of Vajda (1984f).

Theorem 1. All estimators of location T^α, $\alpha \in (0,1)$, are (a) continuously well-defined and location equivariant, (b) strongly consistent for any sample family with even, unimodal sample parent density q, (c) asymptotically normal for sample families sub (b) with asymptotic mean zero and asymptotic variance $\mathrm{var}_Q\Omega^\alpha_Q$ where

$$\Omega^\alpha_Q = p'p^{\alpha-1} \,/\, E_Q p^{\alpha-2}\,(p''p-(1-\alpha)p'^2).$$

Restrict ourselves to the standard normal p. Then the MLE T^0 is the well-known sample mean with $T^0(Q) = E_Q X$ for $Q \in \mathcal{P}(T^0) = \{Q \in \mathcal{P}: E_Q X^2 < \infty\}$. A difference in robustness of estimators T^α for $\alpha>0$ and $\alpha=0$ is indicated already by (b): the former are consistent for all symmetric unimodal sample parent densities q while the latter is not (see e. g. the C auchy parent). The broad-sense-robustness of the former estimators for the above mentioned class of sample parents follows from Hampel's (1974) quantitative criteria of robustness applied to their influence curves $\Omega_Q^\alpha(x) = c_Q^\alpha\, x \exp(-\alpha x^2/2)$, where $c_Q^\alpha = (E_Q(1-\alpha x^2)\exp(-\alpha x^2/2))^{-1}$.

To see a practical difference between broad-sense and minimax-sense robust estimators, in the next table we compare in percents relative deviations of variances $V(T^{0.2})$, $V(T_{0.01})$ and $V(T_{0.1})$ from the least one of them. Here T_ε denotes the estimator of Huber (1964) which is minimax-sense-robust for the standard normal Q at a contamination level ε. The sample parents $\tilde{Q}$ are subsequently standard normal, standard Cauchy, standard uniform on $(-\sqrt{3},\sqrt{3}) \subset R$, and Q_ε or $Q^\varepsilon \triangleq (1-\varepsilon)Q + \varepsilon P_\varepsilon$ or $+ \varepsilon P^\varepsilon$ where P_ε or P^ε is least favourable for T_ε or $T^{0.2}$ respectively.

	N(0,1)	Ca(0,1)	$U(-\sqrt{3},\sqrt{3})$	$Q_{0.01}$	$Q_{.01}$	$Q^{0.01}$	$Q^{0.1}$
$T^{0.2}$	3	3	0	0	0	1	5
$T_{0.01}$	0	51	140	0	12	0	6
$T_{0.1}$	3	0	227	1	5	0	0

This table indicates that practical rules for selection of good robust estimators cannot be based on the somewhat too pesimistic minimax point of view only. Since the asymptotic inefficiences of T^α for all common sample parents are reasonably small provided $\alpha \in (0.1, 0.3)$ and since, at the same time, the influence curves $(1+\alpha)^{3/2}\, x \exp(-\alpha x^2/2)$ of these estimators quite well approximate the curves of broad-sense robust estimators A12-A25 found experimentally by Andrews et al (1972), T^α with $\alpha \in (0.1, 0.3)$ reccommended can be for practical use. Simulated performances of $T^{0.1}$ carried out by Vošvrda (1984) are justifying this step as well.

The dependency of T^α for the standard normal Q at a contamination level ε can be found by differentiating the asymptotic variance function

$$v(x) = \frac{(1-\varepsilon)(1+2\alpha)^{-3/2} + \varepsilon x^2 \exp(-\alpha x^2)}{[(1-\varepsilon)(1+\alpha)^{-3/2} + (1-\alpha x^2)\exp(-\alpha x^2/2)]^2}$$

with respect to ε at the unique $x = x(\alpha,\varepsilon) > 1/\sqrt{\alpha}$ which is a root of the equation $dv(x)/dx = 0$.

As to the computational algorithms for estimators T^α under consideration, the most suitable one can be simply described by considering the data as beads. Starting with the beads at positions $x_1, \ldots, x_n$ on the horizontal x-axis, let them roll down the curve $\Psi(x) = -\Sigma_1^n\, p(x-x_i)^d$. At the end register all local minima of the function Ψ occupied by at least one bead. Just one local minimum is mostly desired since it means that the whole data file $(x_1, \ldots, x_n) \in X^n$ is homogeneous. Two or more minima automatically yield a classification of the file into clusters of data, each one homogeneous in the above described sense (in this case one "main" cluster yielding a global minimum is desirable). Thus the "roll data" algorithm provides an automatic testing of the homogenity of the data file and an automatic clustering of the file into homogeneous subfiles. This algorithm has been studied in detail by Grim (1982).

For *directed D-estimators of autoregression coefficients* we refer to Feistauerová (1984). Note that the estimators T^{α} have successfully been tested in discrete families as well. E. g. $T^{0.1}$ was used to estimate an instant ratio of data packets transmitted to a distant receiver with detected errors. The usual binary statistics sent back to the sender via real telecommunication channels was considered. This estimator was found to be much more noise-resistant than the classical frequency - count MLE.

We note that in Vajda (1984a) another class of robust estimators called *weak D-estimators* has been introduced as well. One of these estimators has been applied in Vajda (1984e) to results of measurements on several radio and telephone data transmission links. In this manner several surprisingly satisfactory Markov chain models of noise have been obtained. In case of location this class of estimators minimizes functions $E_W|F(x)-G(x+\theta)|^{\alpha}$, $\alpha \in [1,\infty)$, where F is a projection and G an empirical distribution function and W an arbitrary fixed "weight" on R. The estimators of Millar (1981) and Boos (1981) belong to this class.

3. Power and size sensitivity of robust tests. In this chapter we consider the tests Φ_T introduced in Chap. 1. Denote by (Q_0, Q_1) a *least favourable pair* (LFP) from $(\mathcal{P}_{00}, \mathcal{P}_{10})$ defined by the property that the test Φ_T with $\Omega = \ln(dQ_1/dQ_0)$, i. e. with

$$T_n(\underline{x}) = \sum_{i=1}^{n} \ln(dQ_1/dQ_0)(x_i), \tag{3}$$

is robust in a minimax sense. The last property means that Φ_T minimizes the least favourable risk or, in some sense equivalently, the least favourable power and size. For the existence of the LFP and for the property $-\infty<d^0 \le d^1<\infty$ where $d^0 = \text{ess min } \Omega(x)$, $d^1 = \text{ess max } \Omega(x)$ for Q_0+Q_1, we refer to Rieder (1977).

Theorem 3. $SD(t,n) = -n[(1-\varepsilon^0)^1 \xi_1^{\alpha} Q_0^n(T_n(\underline{x})>t) + (\omega^0 \xi_1^{\alpha}+\xi_2^{\alpha}) Q_0^{n-1}(T_{n-1}(\underline{x}) >t-d^0) - ((1+\nu^0) \xi_1^{0\prime}+\xi_2^{0\prime}) Q_0^{n-1}(T_{n-1}(\underline{x})>t-d^1)$ where $\nu^i=(\varepsilon_1^i+\varepsilon_2^i)/(1-\varepsilon_1^i)$,

$\omega^i=\varepsilon_2^i/(1-\varepsilon_1^i)$ and $\xi_j^{i\prime}$ denote $\frac{d\xi_j^i}{d\tau}$, and for PD(t,n) the upper indexes 0 and 1 must be interchanges and Q_1 must be considered instead of Q_0.

For the proof see Víšek (1983c), Theorem 1.

Choosing for an $\alpha\in(0,1)$ a sequence $\{t_\alpha(n)\}_{n=1}^{\infty}\subset R$ so that

$$\lim_{n\to\infty} Q_0^n(T_n(\underline{x})>t_\alpha(n)) = \alpha,$$

we may give the following theorem.

Theorem 4. Let

$$\lim_{v\to\infty} \sup|E_Q \exp\{iv\ \Omega(x)\}| < 1.$$

Then

(4) $$SD(t_\alpha,n) = n^{\frac{1}{2}}\{\xi_1^{0\prime}.(1-\varepsilon^0)^1(d^1-E_{Q_0}\Omega(x)) + (\omega^0\xi_1^{0\prime}+\xi_2^{0\prime})(d^1-d^0)\}\{\phi(u_\alpha).\mathrm{var}_{Q_0}^{\frac{1}{2}}\Omega(x)+o(1)\},$$

where ϕ denotes the density of the standard normal law and u_α its upper α-quantile. Moreover

$$\lim_{n\to\infty}\frac{1}{n}\log|PD(t_\alpha,n)| = \inf_{v<0} E_{Q_1}\exp\{v[\Omega(x)-E_{Q_0}\Omega(x)]\}.$$

For the proof see Lemma 3 of Víšek (1983b) together with Corollary 1 of Víšek (1983c).

Remark 1. It is easy to see that the results might be given in a little more general form assuming dependence of P_i, ε^i and ξ^i on n (see Víšek (1983c and d)). Then it is possible under a usual collection of regularity conditions to prove analogous asymptotic results in local alternative setting (see Víšek (1983c)). But it seems preferable not to overcrowd the paper by a large number of notations and formulae and instead of it to offer the reader a fraction of results which were computed in numerical study of this topic.

From (4) it may be deduced that in (3) the expression in parentheses is of order $n^{-\frac{1}{2}}$ and hence to evaluate SD(t,n) by means of (3) in high precision needs to compute Q^n and Q^{n-1} very precisely. It considerably burdens the computational time even when a very fast (e. g. small sample asymptotic) method is used. Fortunately we need usually only an estimate of size of changes of the first and second kind error probabilities and for this purpose the asymptotic approximation given by (4) serveswell. It was confirmed numerically by a large study a small fraction of which is present below.

The above described model of general contamination neighbourhoods was used. Corresponding P_0 and P_1 were specified on the upper margine of tables, sample size was taken 30. For all probability models the values $\varepsilon_1^0 = \varepsilon_1^1 = .050$ and $\varepsilon_2^0 = \varepsilon_2^1 = .025$ were considered (for a discussion see Víšek (1983a)). Having used the results of Rieder (1977) LFP for (P_{00}, P_{10}) was found and the distribution of T_n was approximated by means of small sample asymptotic method (see Field and Hampel (1982)). A family of "deviated" models $P_{i\tau}$, $\tau \in [0, \tau_0]$ was given by $\xi_1^0(\tau) = \xi_1^1(\tau) = \xi_2^0(\tau) = \xi_2^1(\tau) = \tau$, the values .0125, .0032, .0008 and .0004 for τ were assumed. Then, using repeatedly method of small sample asymptotic together with results of Víšek (1983c) $s(t_\alpha, n, P_{0\tau})$ were found for the just mentioned values of τ. The following tables contain in the first column the first kind error probabilities α (= $s(t_\alpha, n, P_{00})$). Starting from the second one to the fifth the values of expression behind the sign "lim" in (2) for above given τ's accompanied by $s(t_\alpha, n, P_{0\tau})$ (placed in parentheses) are displayed. Last but one column includes the values of $SD(t_\alpha, n)$ given by (3) and the last one gathers the values of its approximation evaluated by means of (4).

Normal model: $p(x) = (2\pi\sigma^2)^{-\frac{1}{2}} \exp\{-(x-\mu)^2(2\sigma^2)^{-1}\}$

P_0: $\mu = 0$, $\sigma^2 = 1$ $\quad$ P_1: $\mu = 1$, $\sigma^2 = 1$

α	$\tau = .0125$	$\tau = .0032$	$\tau = .0008$	$\tau = .0004$		
.045	2.91 (.081)	2.45 (.053)	2.37 (.047)	2.33 (.046)	2.26	2.18
.024	1.84 (.047)	1.50 (.029)	1.42 (.025)	1.41 (.024)	1.41	1.31
.012	1.08 (.026)	.846 (.015)	.797 (.012)	.786 (.012)	.815	.726

Gamma model: $p(x) = a^p \Gamma^{-1}(p) \exp\{-ax\} x^{p-1}$

P_0: $a = 1$, $p = 3$ $\quad$ P_1: $a = 2$, $p = 3$

.056	4.23 (.109)	3.54 (.067)	3.39 (.059)	3.23 (.057)	3.11	3.06
.024	2.33 (.053)	1.84 (.030)	1.74 (.025)	1.66 (.025)	1.63	1.52
.009	1.11 (.023)	.823 (.012)	.765 (.010)	.741 (.009)	.749	.658

Weibull model: $p(x) = c\, p\, x^{p-1} \exp\{-cx^p\}$

P_0: $c = 2$, $p = 3$ $\quad$ P_1: $c = 1$, $p = 3$

.040	3.01 (.078)	2.49 (.048)	2.38 (.042)	2.42 (.041)	2.48	2.17
.018	1.67 (.039)	1.31 (.022)	1.24 (.019)	1.26 (.019)	1.35	1.11
.007	.826 (.018)	.618 (.009)	.575 (.008)	.581 (.008)	.660	.510

Laplace (double exponential) model: $p(x) = (2b)^{-1} \exp\{-|x-a|.b^{-1}\}$

P_0: $a = 2$, $b = 1$ P_1: $a = 0$, $b = 1$

.049	3.48 (.093)	2.92 (.058)	2.79 (.051)	2.77 (.050)	2.30	2.55
.020	1.98 (.047)	1.58 (.028)	1.50 (.024)	1.48 (.023)	1.26	1.35
.009	1.01 (.022)	.767 (.012)	.715 (.010)	.708 (.010)	.619	.634

Rayleigh-Rice model: $p(x) = x\sigma^{-2} \exp\{-(x^2+\mu^2)(2\sigma^2)^{-1}\}\left[1+\mu^2x^2(4\sigma^2)^{-1}\right]$

P_0: $\mu = 0$, $\sigma^2 = 1$ P_1: $\mu = 1.5$, $\sigma^2 = 1$

.049	5.84 (.122)	4.68 (.064)	4.39 (.053)	4.38 (.051)	4.10	3.71
.026	3.95 (.076)	2.97 (.036)	2.73 (.029)	2.68 (.028)	2.61	2.26
.011	2.13 (.038)	1.47 (.016)	1.32 (.012)	1.30 (.012)	1.30	1.06

Remark 2. Using approximation (4) of SD(t,n) instead of its precise value given by (3) one avoids evaluation of Q^{n-1}. The evaluation of Q^n (computed in order to establish a critical region) is not inevitable too. Instead of computing Q^n we may utilize a direct approximation of its quantiles (see Víšek (1984d)).

The Rayleigh-Rice model appears in detection of radar signals: an envelope of a noise (the hypothesis) or of a noise plus a signal (the alternative) is a random process which, sampled with proper frequency, yields approximately independent data distributed, under a strict linearity of transformations generating the envelope, according to the Rayleigh or Rice law respectively. Since in real systems the strict linearity rarely takes place, it is practically convenient to apply robust tests and to know their power dependency curves.

References

Andrews D. F. et al (1972), Robust Estimates of Location, Princeton University Press.

Boos D. D. (1981), Minimum distance estimators for location and goodness of fit, J. Amer. Statist. Assoc. 76, 663-670.

Eplett W. J. R. (1980), An influence curve for two-sample rank tests, Roy. Statist. Soc., Ser. B, 42, 67-70.

Feistauerová J. (1984), Robust estimators of parameters of autoregression models. Compstat 84 - Short Communications, Summaries of Posters.

Grim J. (1981), An algorithm for maximizing a finite sum of positive functions and its application to cluster analysis. Probl. of Control and Inform. Theory 10, 427-437.

Field Ch. A., Hampel F. R. (1982), Small sample asymptotic distribution of M-estimates of location, Biometrika 69, 29-46.

Hampel F. R. (1974), The influence curve and its role in robust estimation, J. Amer. Statist. Assoc. 69, 383-393.

Hampel F. R. (1981), The change-of-variance curve and optimal redescending M-estimators, J. Amer. Statist. Assoc. 76, 643-647.

Huber P. J. (1964), Robust estimation of a location parameter, Ann. Math. Statist. 34, 73-101.

Huber P. J. (1972), Robust statistics: a review, Ann. Math. Statist. 43, 1041-1067.

Huber P. J. (1965), A robust version of the probability ratio test, Ann. Math. Statist. 36, 1753-1758.

Huber P. J., Strassen V. (1973), Minimax tests and the Neyman-Pearson lemma for capacities, Ann. Statist. 1, 251-263.

Lambert D. (1981),Influence functions for testing, J. Amer. Statist. Assoc. 76, 649-657.

Millar P. W. (1981), Robust estimation via minimum distance methods, Zeitschr. Wahrsch. 55, 73-89.

Portnoy S. L. (1977), Robust estimation in dependent situations, Ann. Statist. 5, 22-43.

Pfanzagl J. (1969), On the measurability and consistency of minimum contrast estimators, Metrika 14, 249-272.

Rieder H. (1977),Least favourable pairs for special capacities, Ann. Statist. 5, 909-922.

Ronchetti E., Rousseeuw P. J. (1979), The influence curve for test, Research Report No 21, Fachgrupe für Statistik, ETH Zürich.

Ronchetti E. (1982), Robust testing in linear models: The infinitesimal approach, Ph.D. Dissertation, SFIT Zürich.

Stiegler S. M. (1973), Simon Newcomb, Percy Daniell and the history of robust estimation 1885 - 1920, J. Amer. Statist. Assoc. 68, 872-879.

Vajda I. (1984a), Motivation, existence and equivariance of D-estimators, Kybernetika 20 (in print).

Vajda I. (1984b), Consistency of D-estimators, Kybernetika 20 (in print).

Vajda I. (1984c), Asymptotic efficiency and robustness of D-estimators, Kybernetika (submitted).

Vajda I. (1984d), Minimum weak divergence estimators of structural parameters, Proc. of the Third Prague Symp. on Asympt. Statist.(in print).

Vajda I. (1984e), Minimum χ^2-estimates of multistate communication channel models, Probl. of Control and Inform. Theory 13, N.6 (in print)

Vajda I. (1984f), Directed D-estimators and robust decoding and coding, Trans. of IEEE, IT series (to be published).

Víšek J. Á. (1983a), On second order efficiency of a robust test and approximation of its error probabilities, Kybernetika 19, 387-407.

Víšek J. Á. (1983b), On the dependence of the test error probabilities on the contamination level, submitted to Statistics.

Víšek J. Á. (1983c), Influence of contamination level deviations on the test error probabilities, submitted to Statistics.

Víšek J. Á. (1984a), Sensitivity of the test risk with respect to contamination, Commun. Statist.,Sequential Anal. 2(3), 243-258.

Víšek J. Á. (1984b), Weighted risk of test: Sensitivity to contamination,(submitted to Decision and Statistics).

Víšek J. Á. (1984c), Sensitivity of the test error probabilities with respect to the level of contamination in general model of contaminacy, (submitted to J. Statist. Plann. Inference).

Víšek J. Á. (1984d), A note on numerical aspects of robust testing, (in preparation).

Vošvrda M. (1984), Statistical experiments with efficiency, robustness and adaptability of modified directed D-estimators, Compstat 84 - Short Communications, Summaries of Posters.

Boundary Effects in Nonparametric Curve Estimation Models

H.-G. Müller, Marburg

Summary

Boundary effects disturb nonparametric curve estimates, especially of derivatives, near the boundaries of the support of the curve. Modifications of estimates near the boundaries are necessary in order to obtain global asymptotic results as well as a satisfying finite sample behavior. We describe such modifications exhibiting different degrees of smoothness and derive the rate of a.s. convergence of the supremal deviation of nonparametric kernel regression, supremum taken over the deviation on the whole interval of support of the function to be estimated. A "reference kernel method" is proposed which allows the construction of modified kernels of different degrees of smoothness. This method requires substantially less computations than previous approaches.

Key words: Nonparametric kernel regression, maximal deviation, rate of global convergence, boundary effects, smooth boundary kernels, method of reference kernels.

1. Introduction

Nonparametric curve (density, spectral density or regression function) estimators usually exhibit a sharp increase in variance and bias when points near the boundary of the support of the function are estimated. This phenomenon was pointed out by Hominal and Deheuvels (1979) and Gasser and Müller (1979) for kernel estimates, by Hall (1981) for trigonometric series estimates and by Rice and Rosenblatt (1983) for smoothing splines and is called boundary or end effect.

We focus here on kernel estimates and discuss the fixed design regression model

$$Y_i = g(t_i) + \varepsilon_i \quad , \quad i = 1 \ldots n \quad ,$$

where Y_i are measurements of an unknown function $g:[0,1] \to \mathbb{R}$ taken at points t_i fixed in advance. The measurements are contaminated with i.i.d errors ε_i , $E\,\varepsilon_1 = 0$; $E\,\varepsilon_1^2 = \sigma^2 < \infty$.
Several types of kernel estimates for g or $g^{(\nu)}$ ($\nu \geq 0$) have been proposed (compare Priestley and Chao, 1972, Gasser and Müller, 1979, Rice, 1983); asymptotically, they are equivalent. We discuss the estimate

$$g_{n,\nu}(t) = \frac{1}{b_n^{\nu+1}} \sum_{i=1}^{n} Y_i \int_{s_{i-1}}^{s_i} K\left(\frac{t-x}{b_n}\right) dx \qquad (1.1)$$

for $g^{(\nu)}(t)$. Here, b_n denotes a scaling factor called bandwidth, s_i, $i=0\ldots n$ is a sequence satisfying $s_0=0$, $t_i \le s_i \le t_{i+1}$, $i=1\ldots n-1$, $s_n=1$, and K is the kernel function. We require support $K \subset [-1,1]$, $K \in M_{\nu,k}$ for some $k>\nu$, where

$$M_{\nu,k} := \{ f \in L^2 \;:\; \int f(x) x^j dx = \begin{cases} 0 & 0 \le j < k, \quad j \ne \nu \\ (-1)^\nu \nu! & j=\nu \end{cases} \}$$

and

$$b_n \to 0 \;, \quad nb_n \to \infty \quad \text{as} \quad n \to \infty \;. \tag{1.2}$$

Once b_n has been fixed, we call $R_n = [0,b_n] \cup [b_n,1]$ boundary intervals. Within these intervals, which can be quite large ($b=0.3$ or larger is not unusual),

$$\text{support } K(\frac{t-\cdot}{b_n}) = [t-b_n, \; t+b_n] \not\subset [0,1]$$

and therefore the moment conditions $K \in M_{\nu,k}$ are not applicable for $t \in R_n$, which implies that the usual approximation of bias

$$Eg_{n,\nu}(t) - g^{(\nu)}(t) = b_n^{k-\nu} [\frac{g^{(k)}(t)}{k!} \int w_\nu(x) x^k \, dx + o(1)]$$

is no longer valid. Gasser and Müller (1979, 1984) propose to use modified kernels in the boundary intervals and show that the rate of convergence of Integrated Mean Squared Error can be saved by an appropriate modification. Rice (1983) proposes a more flexible and in the case $k=\nu+2$ ("standard kernels") also simpler method of boundary kernel construction.

In section 2 we discuss specific requirements for boundary kernels. An important feature is that boundary kernels are classified according to their smoothness. The rôle of smoothness of ordinary kernels has been discussed in Müller (1984). The derivation of the rate of uniform convergence of the estimate on the whole interval $[0,1]$ of support of g given in section 3 depends crucially on the use of smooth boundary kernels. Practical aspects of the construction of boundary kernels are discussed in section 4.

2. Smooth boundary kernels

Define $q := t/b_n$ for $t \in [0,b_n)$ resp. $q := (1-t)/b_n$ for $t \in (1-b_n,1]$. In the following, we consider only the left boundary. Treatment of the right boundary is analogous. Given $\nu \ge 0$, $k>\nu$ and $\mu \ge 0$, we consider "boundary kernels of degree of smoothness μ" K_q satisfying

$$\text{support } K_q \subset [-1,q] \tag{2.1}$$

$$K_q \in M_{\nu,k} \cap C^\mu([-1,q]) \tag{2.2}$$

$$K_q^{(j)}(-1) = K_q^{(j)}(q) = 0 \;, \quad j=0\ldots\mu-1 \tag{2.3}$$

$$\sup_{x,q} |K_q(x)| \le C \le \infty \tag{2.4}$$

$$\sup_{q} |K_q(x_1) - K_q(x_2)| \le C\, |x_1-x_2| \text{ for any } x_1,x_2 \;. \tag{2.5}$$

An example of such kernels are the solutions of the variational problems

$$\int_{-1}^{q} K_q^{(\mu)}(x)^2\, dx = \min \quad \text{under } (2.2),\ (2.3) \;. \tag{2.6}$$

The existence of solutions of (2.6) follows from Lemma 2.1, 2.2 of Müller (1984). These solutions are polynomials of degree $(k+2\mu-1)$.The coefficients may be computed from the linear system of equations generated by (2.2), (2.3). These coefficients depend continuously on q. Therefore, if $\mu\geq 1$, these kernels satisfy (2.4), (2.5), too.

Given kernels K_q satisfying (2.1)-(2.5), we define a modified estimate

$$\tilde{g}_{n,\nu}(t) = \frac{1}{b_n^{\nu+1}} \sum_{i=1}^{n} Y_i \int_{s_{i-1}}^{s_i} K_{t,n}\left(\frac{t-x}{b_n}\right) dx \quad ,$$

where kernels $K_{t,n}$ depend on n,t in such a way that

$$K_{t,n} = \begin{cases} K & t\in[0,1]\setminus R_n \\ K_q & t\in R_n \end{cases}$$

and K is the kernel used in the "interior", which is required to satisfy (2.1)-(2.5) for q=1.

3. Maximal deviation of the modified estimate on [0,1]

Under the assumptions (2.1)-(2.5) for some $\mu\geq 1$, we derive the rate of a.s. uniform convergence of $\tilde{g}_{n,\nu}$.

Notation: Lip $\alpha([0,1])$ is the subset of functions of $C([0,1])$ which satisfy

$|f(x_1)-f(x_2)|\leq L_\alpha |x_1-x_2|^\alpha$ for some $L_\alpha>0$, for all $x_1,x_2\in[0,1]$.

$\underline{\lim}$ is the limes inferior as $n\to\infty$. $\|f\|$ denotes $\sup_{t\in[0,1]} |f(t)|$.

LEMMA Let $g^{(k)}\in\mathrm{Lip}\ \alpha([0,1])$, $0<\alpha\leq 1$, and

$$\underline{\lim}\ n b_n^{k} > 0 \;. \tag{3.1}$$

Then $\| E\tilde{g}_{n,\nu} - g^{(\nu)} \| = O(b_n^{k+\alpha-\nu})$

Proof Uniform boundedness of the support of the kernels implies

$$\left\| \sum_{j=1}^{n} |s_j - s_{j-1}|\, 1_{\{|\tilde{w}_{j,\nu}(t)| \neq 0\}} \right\| = O(b_n) \tag{3.2}$$

and by (2.4) we obtain uniformly in $t\in[0,1]$:

$$E\ \tilde{g}_{n,\nu}(t) = b_n^{-\nu} \int K_q(x)\, g(t-xb_n)\, dx + O([nb_n^{\nu}]^{-1}) \;.$$

By Taylor's Theorem with integral remainder

$$\sup_{|x|\le 1} \left|g(t-xb_n)-\sum_{j=0}^{k-1} g^{(j)}(t)(-xb_n)^j(j!)^{-1}\right| \le (k!)^{-1} L_\alpha b_n^{k+\alpha}$$

and by (2.2)

$$b_n^{-\nu} \int K_q(x)\left(\sum_{j=0}^{k-1} g^{(j)}(t)(-xb_n)^j(j!)^{-1}\right)dx = g^{(\nu)}(t)$$

which completes the proof.

For the following result we use a similar method as Cheng and Lin (1981), Thm. 3, who consider the maximal deviation on intervals [a,b], $0<a<b<1$.

<u>LEMMA</u> For some $r>2$, assume $E|\varepsilon_1|^r<\infty$ and

$$\underline{\lim}\, n^{1-2/r} b_n (\log n)^{-1} > 0 \,. \tag{3.3}$$

Then $\|\tilde{g}_{n,\nu} - E\,\tilde{g}_{n,\nu}\| = O((\log n)^{1/2}(nb_n^{2\nu+1})^{-1/2})$ a.s.

<u>Proof</u> Let $\bar{\varepsilon}_i := \varepsilon_i 1_{\{|\varepsilon_i|\le i^{1/r}\}}$, $h_{n,\nu}(t) := \sum_{i=1}^n \tilde{W}_{i,\nu}(t)\varepsilon_i$,

$$\bar{h}_{n,\nu}(t) = \sum_{i=1}^n \tilde{W}_{i,\nu}(t)\bar{\varepsilon}_i, \qquad \beta_n := (nb_n^{2\nu+1})^{1/2}(\log n)^{-1/2} \,.$$

There are sets $R_n \subset [0,1]$ with cardinality $O(n^2)$ such that for all $t \in [0,1]$ there exists $\tau(t) \in R_n$, where $\|t-\tau(t)\| = O(n^{-2})$ is satisfied. Then

$$\|\tilde{g}_{n,\nu} - E\,\tilde{g}_{n,\nu}\| \le \|h_{n,\nu}-\bar{h}_{n,\nu}\| + \|\bar{h}_{n,\nu}(t)-\bar{h}_{n,\nu}(\tau(t))\|$$

$$+ \|\bar{h}_{n,\nu}(\tau(t))-E\,\bar{h}_{n,\nu}(\tau(t))\| + \|E\,\bar{h}_{n,\nu}(\tau(t))-E\,\bar{h}_{n,\nu}(t)\| + \|E\,\bar{h}_{n,\nu}-E\,h_{n,\nu}\|$$

$$=: I + II + III + IV + V \,.$$

By (1.2), (2.4) and as

$P(\omega\in\Omega$: there exists I_ω such that $|\varepsilon_i(\omega)|<i^{1/r}$ for $i>I_\omega) = 1$:

$$\beta_n I \le \beta_n \|\max_{1\le i\le n} |\tilde{W}_{i,\nu}(t)|\,\| \sum_{i=1}^n |\bar{\varepsilon}_i-\varepsilon_i| = O(1) \quad \text{a.s.}$$

$$\beta_n V \le \beta_n \|\max_{1\le i\le n} |\tilde{W}_{i,\nu}(t)|\,\| \sum_{i=1}^n i^{-(r-1)/r} = O(1)$$

by (3.3). By (2.5)

$$\beta_n II \le \beta_n \sup_{[0,1]} |\tilde{W}_{i,\nu}(t) - \tilde{W}_{i,\nu}(\tau(t))|\, n^{1+1/r} = O(1) \quad \text{a.s.}$$

and similarly $\beta_n IV = O(1)$. For III, we use the exponential moment inequality given by Lamperti (1966), p. 43. We need (3.3) in order to satisfy the requirements. For any $M>0$:

$$P\left(\beta_n\left(\sum_{i=1}^n \tilde{W}_{i,\nu}(t)(\bar{\varepsilon}_i-E\bar{\varepsilon}_i)\right) > M\right)$$

$$\le \exp\left((\log n)\left(c_1 \beta_n^2 \sum_{i=1}^n \tilde{W}_{i,\nu}^2(t)\log n - c_2M\right)\right)$$

where $c_1, c_2 > 0$. By (2.4), (3.2)

$$\| \sum_{i=1}^{n} \tilde{W}_{i,\nu}^{2}(t) \| = O([nb_n^{2\nu+1}]^{-1})$$

and therefore (with constants c_3, c_4)

$$\sum_{n=1}^{\infty} P(\sup_{\tau \in R_n} \beta_n (\bar{h}_{n,\nu}(\tau) - E\,\bar{h}_{n,\nu}(\tau)) > M)$$

$$\leq c_3 \sum_{n=1}^{\infty} n^{2+c_1 c_4 - c_2 M} < \infty$$

if M is large enough. The Borel-Cantelli lemma implies

β_n III $= O(1)$ a.s.

<u>THEOREM</u> For some $0<\alpha\leq 1$ let $g^{(k)} \in \mathrm{Lip}\,\alpha([0,1])$ and for some $r>2+1/(k+\alpha)$ let $E|\varepsilon_1|^r < \infty$. Assume (3.1), (3.3). Then

$$\sup_{[0,1]} |\tilde{g}_{n,\nu}(t) - g^{(\nu)}(t)| = O([\log n/n]^{(k+\alpha-\nu)/(2(k+\alpha)+1)}) \quad \text{a.s.}$$

<u>Proof</u> According to the lemmas the optimal bandwidth is $b_n \sim (\log n/n)^{1/(2(k+\alpha)+1)}$. Choice of this bandwidth is consistent with (3.1), (3.3) under $r>2+1/(k+\alpha)$.

4. Construction of boundary kernels

The discussion of this section applies also to other types of curve (e.g. density) estimates based on kernel methods. We consider always the left boundary.

A conceptually simple but numerically rather demanding method is to compute the solutions of (2.6). This means that one has to solve an ill-conditioned linear$(k+2\mu-1) \times (k+2\mu-1)$ system for each point of the boundary interval R_n where one needs the estimate. The approach of Gasser and Müller (1979) is a special case ($\mu=0$); for the results of section 3 to hold we need $\mu\geq 1$. If one uses in the interior kernels which are the solutions of (2.6) (compare Müller (1984)), and which cover all practical needs by choosing different values of ν,k,μ and have the numerical advantage of a compact support, this method provides a natural continuation of these kernels onto R_n.

A more flexible approach to continue any given kernel from the interior onto R_n was proposed by Rice (1983). The modified kernel within R_n is a linear combination of kernels of different bandwidths such that the moment constraints are satisfied. This requires the computation of integrals $\int_{-1}^{q} K(x)dx$ and solution of a $(k-1) \times (k-1)$-system for each $t_i \in R_n$ where the estimate is needed. Thus if $k>2$, the disadvantage remains that for any point to be estimated a linear system has to be solved.

A"reference kernel method" is proposed here which avoids this and requires just one inversion of a $(k+\mu) \times (k+\mu)$ matrix in order to compute all boundary kernels desired. The idea is as follows: Correct moments of $K \cdot 1_{[-1,q]}$ by means of "smooth reference kernels" $\tilde{K}_j$, $j=0\ldots k-1$, with the properties

$$\text{support } \tilde{K}_j \subset [-1,0], \quad \int \tilde{K}_j(x) x^\ell dx = \begin{cases} 1 & \ell=j \\ 0 & \ell<k, \ \ell \neq j \end{cases}$$

$\tilde{K}_j^{(\ell)}(-1) = \tilde{K}_j^{(\ell)}(0) = 0$, $\ell=0\ldots\mu-1$:

Define $$K_q := K \cdot 1_{[-1,q]} - \sum_{j=0}^{k-1} \alpha_j \tilde{K}_j ,$$

where $$\alpha_j = -\int_q^1 K(x) x^j dx .$$

(Since K is usually a simple function, like a polynomial, computation of the α_j is easy). These kernels K_q satisfy requirements (2.1)-(2.5) (instead of (2.3) we have the weaker but for the results of section 3 sufficient requirement $K_q^{(\ell)}(-1) = 0$, $\ell=0\ldots\mu-1$). Reference kernels $\tilde{K}_j$ are computed as solutions of (2.6). Since the side condition in this case (q=0) implies that the first μ coefficients of the corresponding polynomial are zero, it is enough to invert the $(k+\mu) \times (k+\mu)$ matrix defined by the side conditions $\tilde{K}_j^{(\ell)}(-1) = 0$, $\ell=0\ldots\mu-1$

and $\int \tilde{K}_j(x) x^\ell dx = \begin{cases} 1 & \ell=j \\ 0 & \ell \neq j, \ 0<\ell<k \end{cases}$;

only the right side of the corresponding linear system for the coefficients of $\tilde{K}_j$ depends on j, so that just one inversion of the matrix is necessary.

References

Cheng, K. F., Lin, P. E. (1981). Nonparametric estimation of a regression function. Z. Wahrscheinlichkeitstheorie verw. Geb. 57, 223-233

Gasser, Th., Müller, H. G. (1979). Kernel estimation of regression functions. Smoothing techniques for curve estimation, Proceedings Heidelberg 1979, Lecture notes in mathematics 757, 23-68

Gasser, Th., Müller, H. G. (1984). Estimating regression functions and their derivatives by the kernel method.Scand.J.Statist., in press

Hall, P. (1981). On trigonometric series estimates of densities. Ann. Statist. 9, 683-685

Hominal, P., Deheuvels, P. (1979). Estimation non-paramétrique de la densité compte-tenu d'informations sur le support. Revue de Statistique Appliqueé 27, 47-59

Lamperti, J. (1966). Probability. New York

Müller, H. G. (1984). Smooth optimum kernel estimators of regression curves, densities and modes. Ann. Statist. 12, in press

Priestley, M. B., Chao, M. T. (1972). Nonparametric function fitting. J. Royal Statist. Soc. B34, 385-392

Rice, J. (1983). Boundary modification for kernel regression. Preprint, University of California, San Diego

Rice, J., Rosenblatt, M. (1983). Smoothing splines: regression, derivatives, and deconvolution. Ann. Statist. 11, 141-156

Monte Carlo Analysis of Variance

A.P. White, Birmingham and A.W. Still, Durham

Summary

A new computational approach to analysis of variance is introduced which makes use of the principles of the permutation test. This technique is more robust than the conventional one. The method is quite general and can be used for any type of experimental design. Another aspect of the approach is that it utilises the method of unweighted means in order to cope with the problems of interpretation that arise from non-orthogonal designs.

Keywords

Analysis of variance; Monte Carlo method; non-orthogonality; permutation test; unweighted means.

Introduction

There are two fundamental difficulties with the conventional least squares approach to analysis of variance. One problem is that arising from non-orthogonal data sets, which cause difficulties of interpretation when trying to attribute various overlapping portions of the variance to the different sources of variation. The second problem is that in many areas of application, it is difficult to show that the populations sampled meet the requirements for valid use of the F test.

This paper proposes a different computational approach to analysis of variance that incorporates techniques for overcoming both these problems. The technique known as "unweighted means" analysis of variance is put forward as a way of dealing with the problem of non-orthogonality in data sets and a Monte Carlo permutation technique is offered as a method of conducting significance tests on data which do not meet the requirements for using the parametric F test. These techniques and the justification for using them are described in the following sections.

The Problem of Non-orthogonality

The problem of non-orthogonality in unbalanced experimental designs has received widespread attention in the past. No attempt will be made to review all that work in this paper, although it is worth mentioning a few of the attempts to specify how non-orthogonal linear models should be tackled. These are illustrated by considering the simplest experimental design suitable for the purpose - namely a two-way factorial, with two levels on each of the independent variables.

It seems that there are four possible least squares approaches to such an analysis that could be described as reasonable. One of these is to compute the sums of squares for each of the effects without adjusting for any other (except the mean). Such an approach disregards entirely the non-orthogonality of the design and is generally not regarded by statisticians as appropriate. We shall not consider it further. The three remaining approaches all adjust at least some of the sums of squares in various ways to take account of the other effects. It is worth noting that each of these methods is available in the SPSS ANOVA procedure, as described by Nie et al (1975). These approaches may be briefly described as follows.

In the classical experimental approach, each of the main effects is adjusted to take account of the other, but neither is adjusted for the interaction. The hierarchical approach is similar except that only one main effect is adjusted for the other. In the regression approach, each main effect is adjusted both for the other main effect and for the interaction. All three approaches adjust the interaction for both main effects.

Now, each of these approaches has its merits and its drawbacks. Of the various methods, the hierarchical approach has the virtue that the sums of squares add up to the between-cell sum of squares. (However, this is not true when the technique is applied to higher order designs, because the method operates on interactions in the same way as the classical experimental approach). The obvious shortcoming of this method is the asymmetrical way in which different effects at the same level are dealt with. What is the justification for adjusting main effect A for B, but not vice versa?

Both the remaining methods adopt a more consistent approach to the practice of adjusting the sums of squares and the regression approach seems to be the current favourite with most statisticians. However, Appelbaum & Cramer (1974) point out that if two variables are positively correlated and each, considered alone, yields a significant effect on the dependent variable, then if each main effect is adjusted for the other, we run the risk of not getting a significant effect for either and thereby fail to account for an important portion of the variance. They argue that in such a situation, we need to retain one (but not both) of the variables in the model. They propose a model testing scheme which deals with problems of this sort using both adjusted and unadjusted sums of squares for the various effects. Unfortunately, the scheme has two drawbacks. It rapidly becomes complex for higher order designs and it does not really resolve the issue concerning which of the two main effects should be included in the model in the type of case just described.

To summarise, it seems that there is no really satisfatory solution to the problem of performing least squares analysis of variance on non-orthogonal data.

Unweighted Means Analysis of Variance

Unweighted means analysis of variance is not a new technique. Indeed it has been used by psychologists for many years and is mentioned or described in statistical textbooks intended for use by psychologists, e.g. Winer (1970), Keppel (1973). However, among statisticians, few seem to have heard of it and those that have regard it with disfavour. This attitude appears to be due to it being a "nonexact approach" as Appelbaum & Cramer (op. cit.) describe it. Presumably, they mean by this that the F distribution with appropriate degrees of freedom does not have upper percentage points which give the correct values for the probability of type I errors.

Now, the authors of the present paper are not sure just how far "out" the F distribution might be in these cases but a Monte Carlo test for a 2 x 2 ANOVA with cell frequencies of 1,9,9,1 involving 200000 trials using normally distributed data yielded actual probabilities of type I error that were within the 95% confidence limits of the nominal 0.05 level and similarly for the 0.01 and 0.001 levels, for both main effects and the interaction. This suggests that, even if the F distribution is slightly "out" when used in this way, it remains a good approximation. However, as we shall see later, by combining the unweighted means analysis with the Monte Carlo permutation principle, even this possible objection can be overcome.

The unweighted means ANOVA will not be described in detail here. An account of the technique is given by Keppel (1973). Briefly, the essence of the method is that, when computing the sums of squares for the various effects, equal weight is given to the various cells in the design, rather than to the observations, as in the least squares approach. When the cell sizes are equal, the two methods yield the same answers. When the design is unbalanced, the F ratios obtained reflect the pattern of the means of the various cells and none of the problems of non-orthogonality mentioned previously are evident.

What has been done is to analyse the data, as far as possible, as if it had originated from a balanced design. The justification for doing this is as follows. When a designed experiment is analysed, we are usually not concerned with the covariance structure of the set of independent variables in the population sampled. Normally, we have made no attempt to represent this covariance structure by proportionate sampling and we would prefer equal cell sizes so that we are not embarrassed by problems arising from non-orthogonality. If by mischance we fail to achieve equal cell sizes (due to circumstances beyond our control), then we should proceed to use an unweighted means analysis because this enables us to estimate the importance of an effect as if it is uncorrelated with the other effects in the design.

It is worth noting that there is a subtle distinction between these circumstances and those where we carry out an observational study in which we are concerned with the covariance structure of the set of independent variables. In these circumstances we might well carry out a conventional least squares regression analysis. Here, the issue of non-orthogonality is intrinsic to the problem and we wish to know the importance of each independent variable when we take the presence of the others into account.

The F Test and Claims of Robustness

The parametric ANOVA relies on the F test in order to carry out tests of significance on the various effects. However, this test relies on a number of assumptions which may not be met or known to be met. One of these assumptions is that the error variance for each cell in the design should be normally distributed. It has become customary to place faith in the so-called robustness of the F test against violation of this assumption. However, critics such as Bradley (1978) claim that this faith is misplaced and that a non-parametric equivalent can be preferable in terms of both accuracy and power for certain types of error distribution. This is too big a question to examine in detail here. Let it suffice to say that the present authors believe Bradley's criticisms to be well founded and have expounded their own views on the matter at greater length elsewhere (Still & White, 1981).

What is required is a set of non-parametric counterparts to the large family of conventional ANOVA models available. However, the problems of making rank-based permutation tests sufficiently general to be useful appears insurmountable and a glance at Siegel (1956) reveals nothing more than the Kruskal-Wallis test and the Friedman two-way analysis of variance. The following section outlines a somewhat different approach that promises the same variety of models as the parametric approach, coupled with the same reduced dependence on assumptions as non-parametric tests.

The Monte Carlo Approach to Analysis of Variance

The basis of the Monte Carlo approach to analysis of variance is to construct what Edgington (1980) calls an "approximate randomization test". In essence, this involves comparing for each effect the value of the F ratio calculated from the data obtained with other values of the same test statistic computed from members of a reference set. The reference set consists of versions of the data derived from the observed data by two processes, as follows. Firstly, all effects other than the one under consideration are removed from the data (as described below). Secondly, the data are repeatedly permuted according to constraints specified by the model. The null hypothesis is rejected if the value of F obtained from the original data is greater than some proportion, α , of the whole set of F values, where α is the usual probability of type I error.

To clarify the explanation, consider a two-way factorial design. The first step is to conduct a conventional ANOVA yielding three F values - one for each main effect and one for the interaction. Next, all effects are removed from the data by subtracting from each score the mean of its cell. The marginal means for one of the main effects are then added back on. The scores are then permuted randomly without regard for their cells of origin. The only constraint is that each permutation must have same cell frequencies as the original data. After each permutation, an F ratio is computed for the main effect under consideration. After some pre-determined number of permutations, the original F value is compared with those derived from the reference set, yielding an estimate of the probability under the null hypothesis of obtaining an F value as large that obtained from the actual data. In other words, this is a non-parametric estimate of the significance level of the effect concerned, which uses the F ratio as a non-parametric test statistic, rather than in the conventional parametric manner.

The other main effect is dealt with in a similar fashion. The appropriate procedure for dealing with the interaction is similar in principle. Here, the marginal means are subtracted from the original data before entering the permutation stage, so that the only effect remaining is that due to the interaction. It is worth noting that the underlying logic behind the removal of all effects except the one under test is that, at the permutation stage, all the permutations must be equally likely under the null hypothesis of no effect (for the particular effect under consideration). If this requirement is ignored, the permutation procedure will not yield an accurate estimate of the probability of a type I error.

Generalisation to higher order factorial designs should now be obvious. Other types of design can be accommodated by appropriate modifications to the permutation procedure to render it isomorphic with the experimental design. For example, when dealing with a repeated measures factor, scores are permuted with the additional constraint that they remain within their original level of the replicate.

This approach to analysis of variance is more robust against violations of the assumptions upon which the conventional F test is based than is the parametric F itself. The present authors reported results comparing the conventional approach with the Monte Carlo approach for certain simple ANOVA models (Still & White, 1981). Power functions were computed and they showed that, in general, there was little loss of power when the error distribution was normal (and for the two-way repeated measures model, the Monte Carlo technique was superior in power). When the error component was drawn from a particular bimodal distribution, the Monte Carlo approach was found to be superior for all designs tested on both power and accuracy.

There is a further point concerning the power of the Monte Carlo approach. Hope (1968) has shown that, for a given probability of type I error, the power of a Monte Carlo test procedure increases monotonically with the size of the reference set. Thus by increasing the number of permutations, the procedure can be made more powerful, subject to the constraint that it cannot exceed the power of the corresponding uniformly most powerful test.

Conclusion

To summarise, we have made two principal points in this paper. Firstly, we claim that the unweighted means ANOVA is appropriate for analysing unbalanced designs unless the non-orthogonality is intended to reflect that in the population from which the data has been drawn. Secondly, we assert that a Monte Carlo approach to analysis of variance is appropriate for a wider range of data than the conventional parametric technique. Furthermore, by combining these approaches to give a Monte Carlo unweighted means ANOVA, we can overcome objections concerning the possible inexactness of the F ratios attributed to the unweighted means technique, yielding a robust method of dealing with balanced or unbalanced designs, with any error distribution whatsoever.

References

Appelbaum, M.I. & Cramer, E.M. (1974), Some problems in the nonorthogonal analysis of variance. Psych. Bulletin, **81**, 355-343.

Bradley, J.V. (1978), Robustness? Brit. J. Math. & Stat. Psych., **31**, 144-152.

Edgington, E.S. (1980), Randomization tests. Marcel Dekker, New York.

Hope, A.C.A. (1968), A simplified Monte Carlo significance test procedure. J. Royal Stat. Soc. (B), **30**, 582-598.

Keppel, G. (1973), Design and analysis: A researcher's handbook. Prentice-Hall, Englewood Cliffs, N.J.

Nie, N.H., Hull, C.H., Jenkins, J.G., Steinbrenner, K. & Bent, D.H. (1975), Statistical Package for the Social Sciences. McGraw-Hill, U.S.A.

Siegel, S. (1956), Nonparametric statistics for the behavioral sciences. McGraw-Hill, Tokyo.

Still, A.W. & White, A.P. (1981), The approximate randomization test as an alternative to the F test in analysis of variance. Brit. J. Math. & Stat. Psych., **34**, 243-252.

Winer, B.J. (1970), Statistical principles in experimental design. McGraw-Hill, Yugoslavia.

Computational Graphics in Statistical Data Analysis

Graphical Methods for the Manipulation and Solution of Systems of Linear Equations

J. Gordesch, Berlin

SUMMARY: Graphical computer methods for the manipulation and solution of linear systems are presented. Additionally, computing techniques and statistical applications are discussed.

KEYWORDS: computer geometry, numerical analysis, computer graphics

1. INTRODUCTION

Geometric constructions instead of arithmetic procedures in numerical analysis are often rather cumbersome and of poor accuracy. If the computer, however, performs the drawing, geometrical methods may become a reasonable alternative to purely algebraic methods: the computer is about ten to hundred times faster and - provided high quality equipment - more accurate than a skilled human being. In the sequel, three methods (due to van den Berg (1887) and Mehmke(1892,1934), van den Berg(1888), and Massau (1889)) for the manipulation and solution of systems of linear equations are modernized, modified and extended so as to be run on a computer.

2. GRAPHICAL METHODS

2.1 The Method by van den Berg and Mehmke.

Let be

$$a_0 + Ax = 0 \text{ , where a is r x 1 and A is r x n , rank(A)=r,}$$

and the indices l, m denote two arbitrary rows, the index i an arbitrary column, a_{lj} an arbitrary element of A. To construct a graph for one linear equation, the coefficients a_{lj} are plotted on the abscissa of a Cartesian coordinate system, and parallels to the ordinate axis are drawn through the endpoints of the coefficients. On the y-axis, the lengths of the x_j are plotted (Fig. 1). The scale units on the axes may be different. The y_{lk} are reached by drawing parallels to the corresponding lines Px_j. Because of the definition of the y_{lk}, the recurrence relations

$$a_{lj}y_{lj} = a_{lj-1}x_{j-1} + a_{lj}x_j$$

resp.

$$a_{lj}y_{lj} = S a_{li}x_i ,$$
$$x_0=1, \ 0<i<j, \ j=1...n, \ S \text{ summation sign,}$$

hold.
In the case of y = 0 (i.e. the x_j satisfy the inhomogeneous linear equation), the endpoints of the polygon must lie on the x-axis at a_{ln}. Hence the theorem follows:
If x is the unique solution of the inhomogeneous system $a_0 + Ax = 0$, A of rank n, the parallels to the corresponding straight lines of the pencil of rays in Fig. 1 form a polygon for each equation the endpoints of which are given by (a_{ln}, y_{ln}).
For triangular systems the construction is considerably simpli-

fied: the first (last) equation is easily solved, the value of x introduced into the next equation, etc. (Fig. 2).

2.2 Van den Berg's Method for Graphical Gaussian Elimination

Let us denote by $Ax = 0$ a linear system, where A is r x (n+1) and rank(A) <= r < n+1. Let be $y = a_l x = 0$ and $y = a_m x = 0$ two linear forms with $x_0 = 1$, a_l resp. a_m row vectors of A. The a_{lj} resp. a_{mj} are plotted on parallels, and the a_{lj} and a_{mj} are connected for equal j as in Fig. 3.
Then the following proportions are valid:

$$(a_{mj} - b_{mj}) : (b_{mj} - a_{lj}) = O_m S_m : S_m O_l = s .$$

This yields

$$b_{mj} = (a_{mj} + s a_{lj}) / (1 + s) .$$

If the x-axis is drawn through the intersection of two connecting lines, then $b_{mk} = 0$, and the segments on the x-axis become

$$b_{mk} = a_{mk} - a_{mj} a_{lk} / a_{lj} .$$

Obviously, the b_{mk} are the coefficients obtained by a Gaussian elimination of x_k.

2.3 Massau's Elimination Method.

Let us consider a linear system $Ax - a_0 = 0$, rank(A) = n-1, and A a square matrix of order n. Then n solution vectors $x^{(i)}$, i=1,2,...n, are always linearly independent, and an analogue theorem holds for the dual space of linear forms. We add an nth equation

$$a_{no} + S a_{ni} x_i = 0$$

so that the resulting system is of rank n, and introduce the linear forms

$$y_{i-1} - a_{no} = S a_{nk} x_k , \quad k=1...i-1$$
$$y_i - a_{no} = S a_{nk} x_k , \quad k=1...i$$
$$i = 1...n, \quad y_0 = a_n , \quad S \text{ summation sign.}$$

Then numbers p, q, r exist so that

$$p y_{i-1} + q y_i + r = 0 \text{ for } i=1...n, \; y_0 = a_n .$$

For $y_{i-1} = 0$ (then $y_i = s$) resp. $y_i = 0$ (then $y_{i-1} = t$) we obtain the equation of the intersection point P_i (see Fig. 4) in parallel coordinates:

$$s y_{i-1} + t y_i - st = 0 .$$

If s/t is constant, the points P_i lie on a y-parallel.
Among the polygons which go through the point P_2 ... P_n there must be one for which $y_n = 0$, i.e. one for which $y_{nn} = a_{nn}$. This polygon is uniquely determined by a_{nn}, P_n,...P_2, z_{no}, 0 (Fig. 5).
The further procedure is similar to the algebraic method of successive elimination: starting with k=n, k equations are combined to obtain k-1 solutions, etc. until k=1. Naturally, a well-planned procedure saves much effort. So it is recommended first to

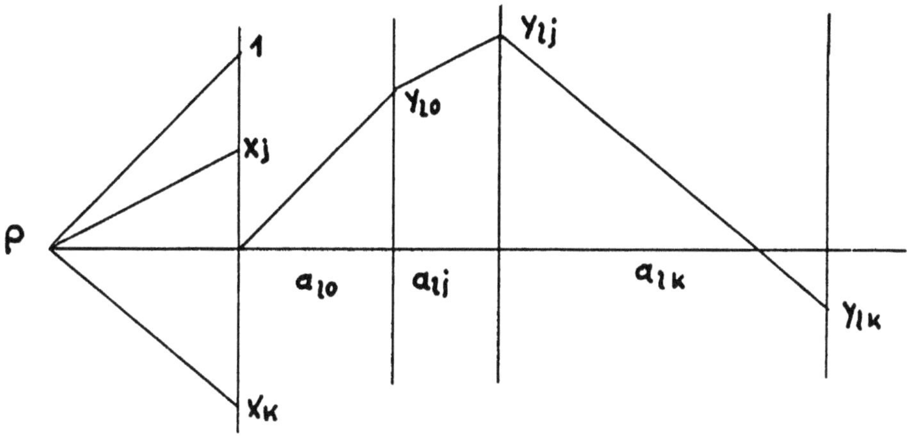

Fig. 1

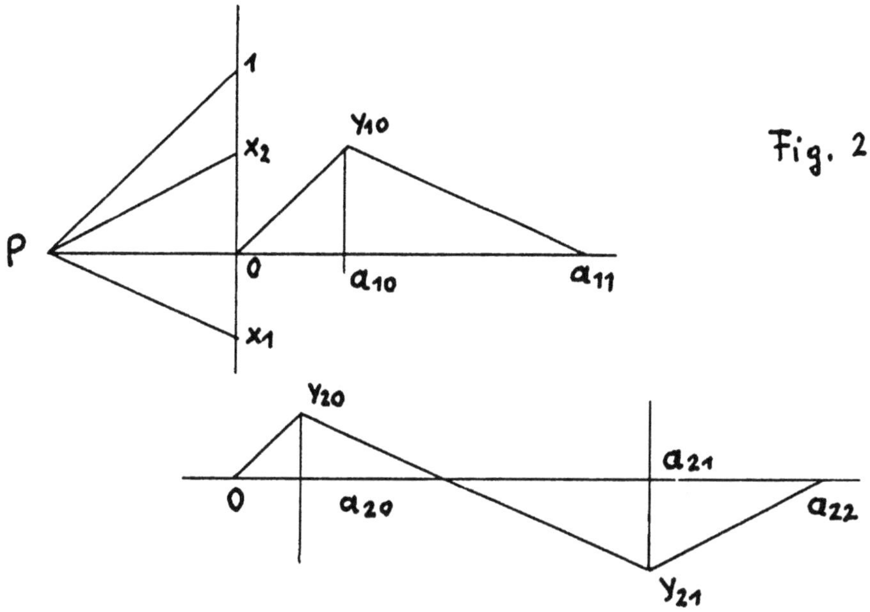

Fig. 2

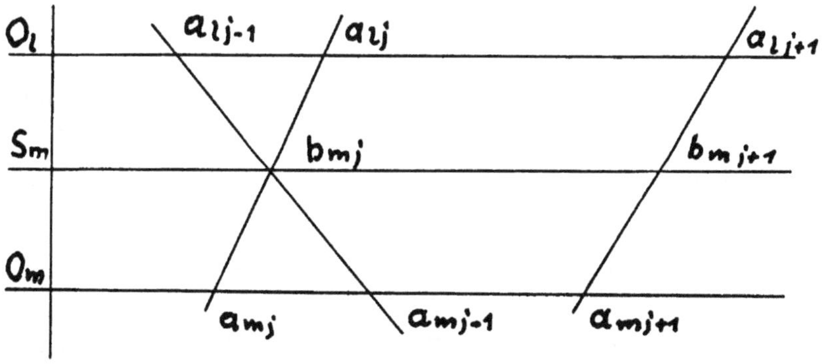

Fig. 3

rescale and to rearrange the equations conveniently, and then to combine the equations according to a strictly hierachic order. For an example refer to Fig. 6.

3. COMPUTING FACILITIES AND PROGRAMMING

Two principally different approaches are possible: Firstly, we replace pencil and paper by the computer, but still draw line after line. Secondly, the algorithm is programmed once. Then data are input as with algebraic methods, only computation (decomposition of the coefficient matrix etc.) is replaced by geometric construction.
Because of economy of time, interactive computer graphics is preferable in the first case, while in the second case also batch processing is usable. It is necessary to plot points, to draw lines connecting given points, and to determine the coordinates of points. Graphics is most likely supported in widespread programming languages as BASIC or Pascal for microcomputers, and FORTRAN or Pascal for larger systems. Programming geometric algorithms is more or less a matter of artificial intelligence, so languages as LOGO might be adequate (cf. Bundy 1978). As a rule, single commands are available that allow to draw lines or even polygons through given points. Nevertheless, user-written subroutines may give a higher resolution and save time.
Input may be done by keyboard, light pen, bit pad, or mouse. Precautions should be taken for a convenient input of frequently used instructions (by pressing function keys, moving a mouse etc.). A hard copy facility (for simple work a printer allowing for bit image mode, otherwise a plotter) may also be of great use. Sufficient resolution of the output facilities (CRT display or plotter) is essential. The simple computations needed do not require high speed computers with large memory. All the algorithms of this paper had been checked on a Fujitsu Micro 7 microcomputer using a graphics cursor (a cross-hair which can be moved by ordinary cursor keys and which permits the determination of the coordinates of the point where it had been moved to).
An excellent survey on computer graphics is Newmann (1982), where also an extensive list of references can be found. The theoretical foundations of geometric algorithms are treated in Schreiber (1975).

4. EXAMPLES OF APPLICATION

The graphical solution of large systems is a hopeless undertaking, even if an excellent CAD system is available. In special cases however, and for a restricted number of variables, graphic procedures make sense.

4.1 Massau's method is well apt for linear systems with a band structure as they arise in connection with difference equations (recurrence relations for probabilities, etc.). Take e.g. the tridiagonal system emanating from the inhomogeneous linear difference equation of second order with boundary conditions $x_0 = x_{n+1} = 0$:

$$a(j,j-1)x(j-1) + a(j,j)x(j) + a(j,j+1)x(j+1) = a(j) ,$$
$$0 < j < n$$

(recurrent events, renewal processes). The solution with four unknowns demands the drawing of about fifty lines (including axes and scales). The drawing of one line requires positioning of the graphic cursor for one point and pressing one further key for

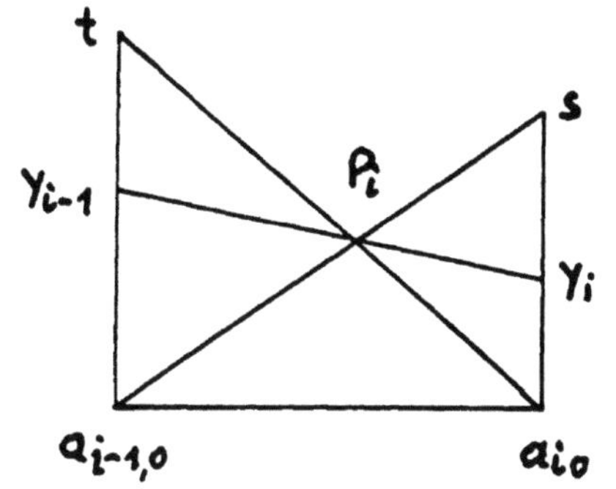

Fig. 4

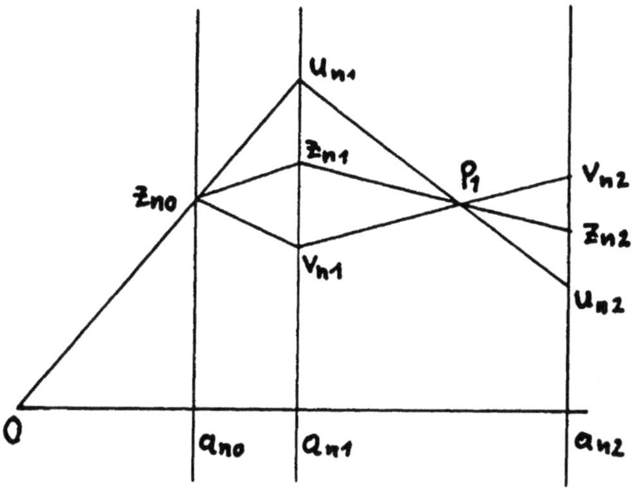

Fig. 5

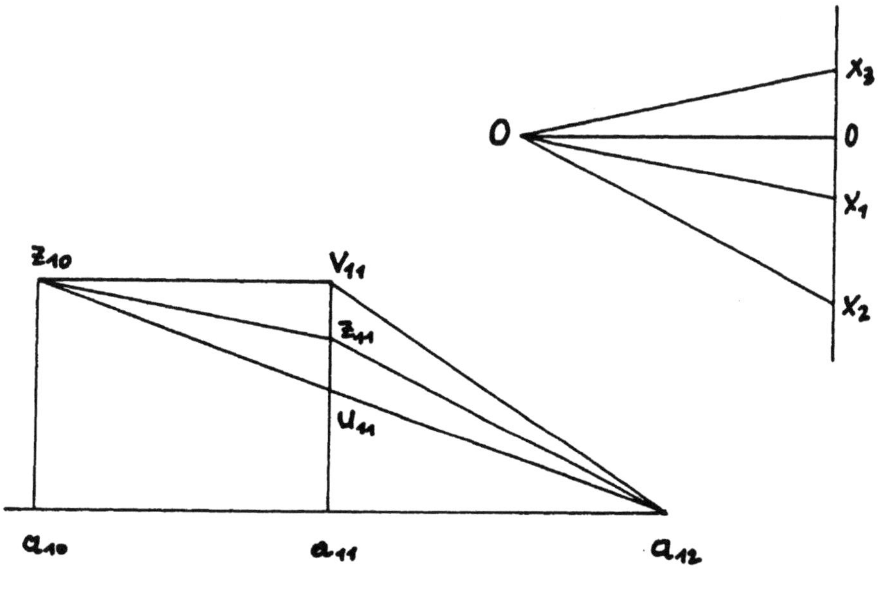

Fig. 6

drawing the line).

4.2 The methods by van den Berg and by van den Berg and Mehmke are considerably simpler in the case of systems with a triangular matrix. A system with three unknowns requires the drawing of about twenty lines (including axes and scales) if solved by van den Berg and Mehmke's method, and about thirty lines if solved by van den Berg's method. Again, drawing one line presupposes positioning of the graphic cursor for one point and pressing one function key. Van den Berg's method is preferable if only the elimination of some variables is desired (e.g. determination of boundaries of multidimensional confidence regions).

4.3 On a whole, algebraic methods are much faster than geometric, and drawing say 300 lines on a screen is highly confusing. Things however might change by the development of specialized hardware and of more efficient algorithms. So geometric methods, which had already become obselete, may find their renaissance, at least in special cases or as part of graphic techniques giving more transparency to what is done in numerical analysis and mathematical statistics.

REFERENCES

Bundy A. et al. (1978), Artificial Intelligence, Edinburgh University Press, Edinburgh

Massau (1889), Ann. de l'Assoc. des ing. sortis des écoles spéciales de Gand, Gand

Mehmke R. (1892), Moskauer math. Samml. 16, pp. 342

Mehmke R. (1934), Leitfaden zum graphischen Rechnen, 2nd ed., Berlin - Leipzig

Newman W. M. - Sproull F. (1982), Principles of Interactive Computer Graphics, 8th printing, McGraw-Hill, Auckland etc.

Schreiber P. (1975),Theorie der geometrischen Konstruktionen, VEB Deutsch. Verl. d. Wiss., Berlin

Van den Berg (1887), Amst. Akad. Versl. on Meded., 4, pp. 204.

Van den Berg (1888), Verl. med. v. d. Kon. Acad. v. Wet., 3rd Series, 6

DISTEP – Dynamic Interactive Statistical Teaching Package

J. Puranen, Helsinki

DISTEP is a simulation program package for teaching purposes. Its aim is to visualise statistical inference, and teach how to make and read graphical presentations used in statistical data analysis. This package can be used by a lecturer to produce teaching material or during lectures to produce apprioriate visualisations of theoretical concepts in statistics. The program package has been implemented on the desk-top computer Wang 2200 VP with a 64 KB memory and a graphic CRT.

Keywords: Teaching Statistics, Computer Graphics, Data analysis

The use of computers in teaching statistics

The developement of computer graphics has opened many new possibilities for the use of computers in teaching. With the help of computer graphics it is easy to create teaching material, such as transparencies, slides, and videotapes. Perhaps the most important new teaching method, offered by a computer, is the interactive computer assisted teaching.

Statistics, in particular, is a field where computer assisted teaching can be used very effectively.

On basic courses, it is possible to use computer graphics to teach the visual literacy applied in statistics.The computer is a useful media in demonstrating some abstract concepts of statistics (statistical distributions, confidence intervals ...), whenever students have learnt how to make and read the necessary graphical presentations.

Because of limited time the content of many basic courses is usually very theoretical and too little time is left for practice with real data. The students may have learnt the terms and formulae used in statistics, but their knowledge of how the statistical methods work in practice may remain very poor.

With computer assisted teaching we can avoid these problems.

As an example we shall consider concept of the confidence interval for the theoretical mean. Using the computer it is possible to generate a series of consecutive samples from a given theoretical population, display them on the graphic CRT, form the corresponding confidence intervals for the theoretical mean, and study how many of the confidence intervals, in fact cover the theoretical mean. Furthermore it is easy to investigate how the length of the interval depends on the sample size and the used confidence level.

This method makes the teaching process far more effective, whenever the simulation can be done dynamically so that the students can see the composition of each sample, observation by observation.

One major benefit with interactive computer assisted teaching is the possibility to illustrate several concepts simultaneously. In the example described above not only confidence intervals are visualised, but also the distribution of observations and sample means.

As Gentleman (1977) says: "Plots , in particular, interactively generated computer plots and especially, spontaneausly produced ones, are extremely useful as an aid to teaching statistics"

The DISTEP package

The DISTEP package, planned for interactive computer assisted teaching, includes the following programs:

DISTRIBUTIONS:

Samples from various theoretical distributions (normal, exponential, beta, binomial, geometric, and Poisson),histograms, estimates of the parameters of the distributions, goodnes-of-fit-tests ... .

SAMPLES

For normal samples: different measures of location and dispersion, confidence intervals for the theoretical mean, distribution of the sample mean in consecutive samples, various methods for estimating the probability density function, cumulative frequency plots, box & whisker plots, likelihood functions, one- and two-sample t-tests, t-test power functions, dependence between confidence intervals and t-tests ... .

LINREG

One variable linear regression models in the case of linear, heteroscedastic or nonlinear theoretical models, correlation diagrams, confidence regions for the theoretical regression line and for a future observation, residual plots, dependence of the residual-sum-of-squares on the coefficients of the regression line, distributions of residuals, various goodness-of-fit-tests ... .

The main ideas used in the development of this program package are the following:

- Recollecting that at the basic level the interactive computer assisted teaching of statistics is not dealing with the teaching of program packages, nor with that of programming and even less with that of teaching how to operate a computer. Because of this fact, all the functions, within the DISTEP package, namely, generation of observations according to the theoretical model, estimation of the parameters, hypothesis testing, building up of pictures, and special functions connected with data analysis, are performed using the same run-time unit. So jumping from one program to an other to get the job completed is almost totally avoided and the user's knowledge on computers is minimised.

- The alphanumeric and graphic CRT are used simultaneously. All numerical data, (the generated values of variables, estimated parameters and the test results) are displayed on the CRT. The generated values are presented as scatter plots or histograms on the graphic CRT. The plots are dynamically updated after each simulated observation. In this way the composition of the picture appears to be animated, which is probably the best method of visualising statistical inference.

- To give the program control flexiblity, the programs have two control systems: either control by a parameter vector at the program level or by function keys at the operating level.

- The elements of the parameter vector control the number of observations to be generated, the theoretical model to be used for the generation of the variables values, tests to be made, the form of the picture and the additional elements to be drawn on output device, such as theoretical and estimated model, confidence intervals, picture headings and so on.

  Values for the parameter vector elements can be given interactively in the conversational mode, where the program helps the user by providing him with the estimated range of variables, probable values for test-statistics and other information.

  Also it is possible to read the parameter vector from a data file prepared earlier on a disk. In this way it is possible to choose during the lecture the most suitable parameter vector in order to illustrate the actual statistical phenomena. It is not necessary to have various trials during the lesson.

  Because the values of the parameters can be chosen on the basis of true experiments, it is possible to simulate real life phenomena during lectures and tell the students about the origin of the investigated data, thus making the example more realistic.

- When the example is generated using parameter vector control, it is possible to change or complete the picture on the graphic CRT by using special features via the function keys. In this way we can visualise step by step the principles of constructing graphical presentations. Apart from tests and different elements of the picture, the function keys incorporate special functions used in data analysis. This latter feature makes the program package useful for teaching data analysis techniques.

- In addition DISTEP has also connection to the interactive statistical system SURVO 76 (ref. Mustonen 1980). Using function keys it is possible to save on disk generated data, estimated parameters, and test results in consecutive samples for further analysis with SURVO 76 modules.

Operating the package

Figure 1 shows the various steps that are needed using the package in the generation prosses.

After choosing the seed number for the random number generator (1) the parameter vector which controls generation, testing, and the construction of the picture, is initialised (2). The parameter vector (3) can be divided into four parts: general parameters, model control parameters, output control parameters, and test & picture element control parameters.

After initialisation of the parameter vector, the values of variables are generated according to model control parameters (4). Updating of the picture according to output control parameters and estimation of the parameters of model (5) are carried out every time a new observation is generated.

When all the observations in the example are generated, we have on the output device the basic picture (a scatter plot or a histogram).

The parameter vector controller (6) and the function key controller (7) form the central parts of the program.

The parameter vector controller investigates the test & picture element control part of the parameter vector. At this step all the tests are done, as defined by the user. Moreover, the picture will also be completed with those extra elements described in the parameter vector (theoretical model, estimated model, confidence intervals, picture headings and so on).

When all the functions defined by the parameter vector have been carried out, program control is taken over by the function key controller. The program pauses waiting for the user's next move which can be one of the following: starting the next example, making tests, completing or changing the picture, or entering some special function subroutine.

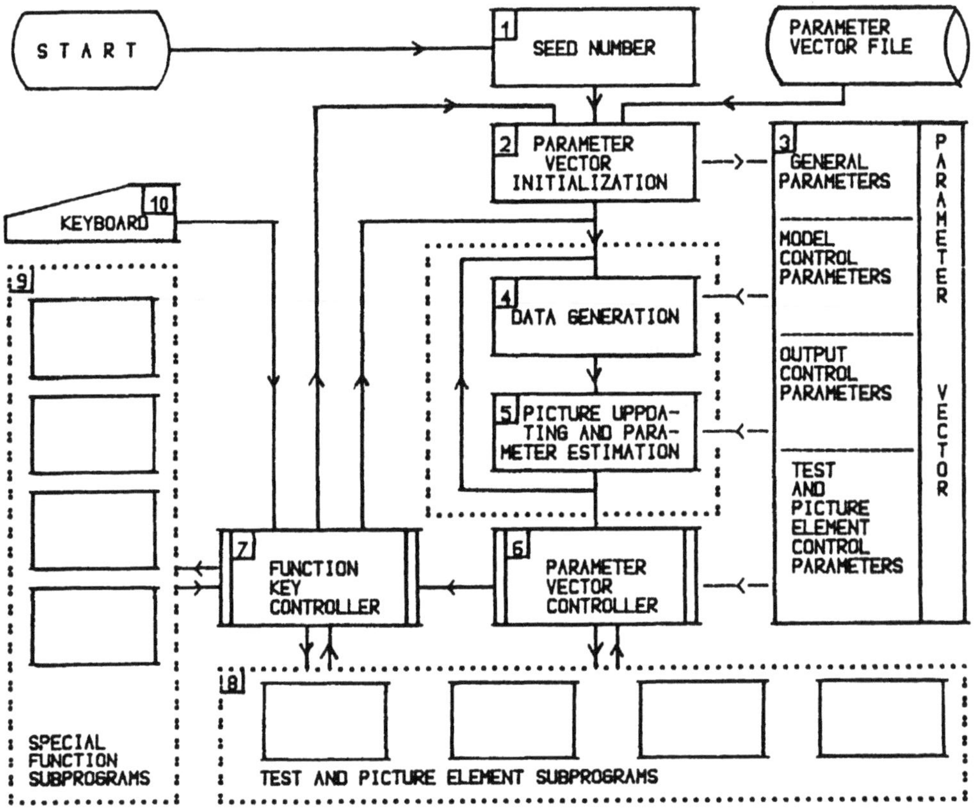

Figure 1: Functional diagram of the teaching program. (numbers are referred to in the text).

The programs have an overlay structure, but usually we have in the memory all the subroutines needed for data generation, presentation, and tests. The program calls a new subroutine into the memory merely when one needs some special function subroutine. In this way generation can be carried out very rapidly. E.g. if we visualise how the confidence interval formed by a normal sample of $n = 30$ covers the theoretical mean, the generating time for one example is approximately 20 sec, of which simulation takes 10 sec and drawing the picture the remaining ones.

Application

The use of the DISTEP package is examplified through the construction of the box & whisker - plot (ref. McNeil 1977).

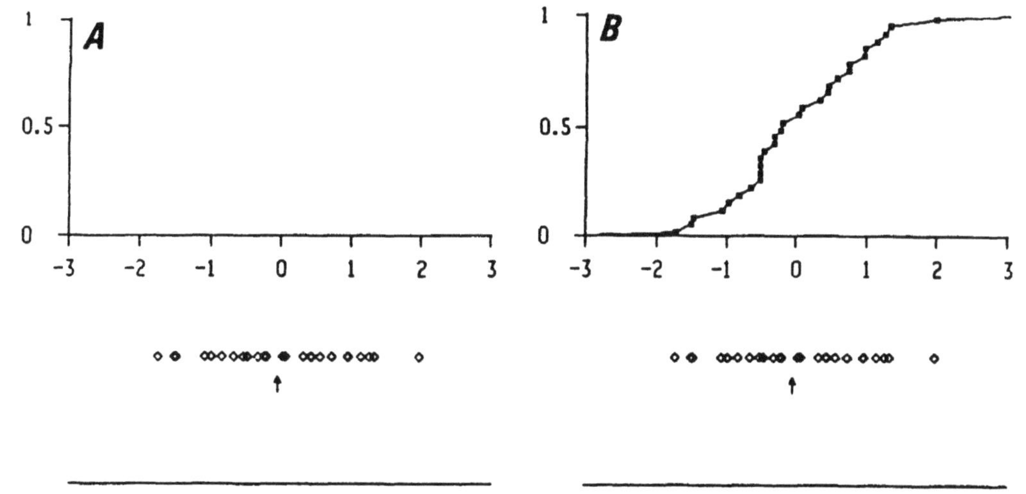

Suppose, we have a sample of size n = 30 from a normal N(0,1) distributed population. The sample is shown in Fig. A as a scatter plot. This way of presenting data is probably, in this case, not the best because many points overlap causing interpretational difficulties. To obtain the box & whisker plot one must first construct the empirical cumulative probability distribution function (Fig. B).

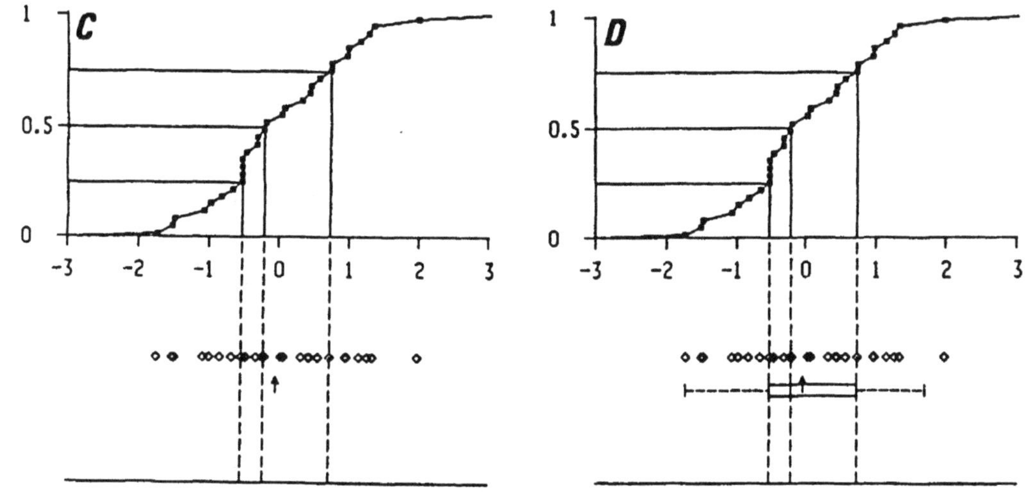

With the empirical cumulative probability distribution function we define the median Md and the quartiles Q1 and Q2 (Fig. C). Based on the quartiles, we can draw a box under the scatter plot, divide this one into two parts using the median and add the whiskers to the end points of the box (Fig. D). The length of the whiskers is formed as follows: the end points of the whiskers are defined by the marginal observations that are no further away from the median than 1.5 * (Q3 - Q1). If they are, the end points of whiskers will be given by Md + 1.5 * (Q3 - Q1). Using this rule and in case of normal samples, about 5 % of all observations will lie outside the whiskers (these are known as extreme observations).

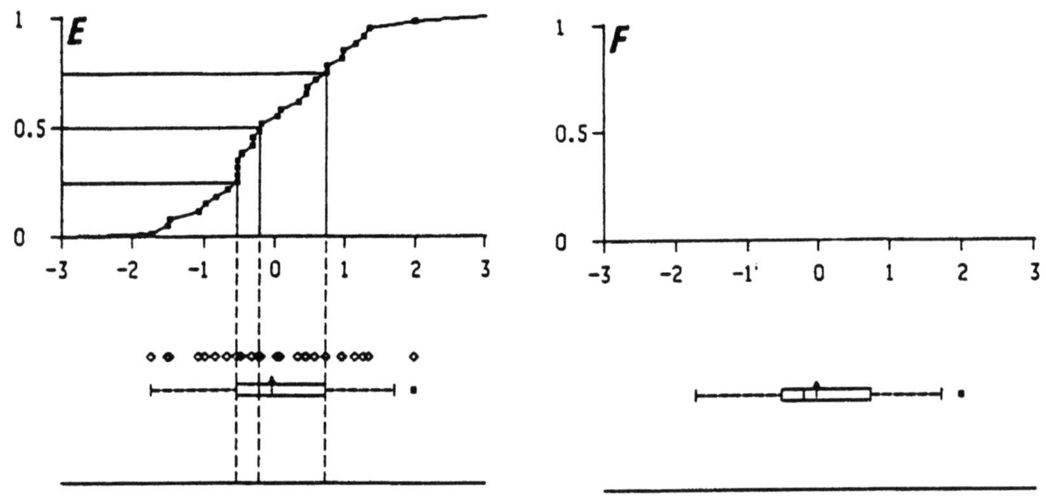

If needed, the extreme observations outside the whiskers can be added to the picture (Fig. E). Finally we erase all surplus drawings - and the box & whisker diagram is ready (Fig. F).

Concluding remarks

At the University of Helsinki we have used the DISTEP package not only to provide teaching material but also as an instrument for interactive computer assisted teaching.

At the basic courses, the number of students is usually too large (over 200) to permit the use of interactive computer assisted teaching. For these courses we have used the DISTEP package to produce transparencies and slide series, improving ordinary lectures. However, the DISTEP package has been used for interactive computer assisted teaching on data-analysis courses, where the number of students is usually less than 20. Also, and in connection to these courses we used two tv-monitors in the classroom, one for graphic CRT and one for alphanumeric CRT, which the teacher can control via the computer.

References

Gentleman, J. F. (1977) It is all Plot (Using interactive computer graphics in teaching statistics), The American Statistican, 31, 166-174.

McNeil, D. R. (1977) Interactive data analysis, Wiley, New York .

Mustonen, S. (1980) Interactive analysis in SURVO 76, Proceedings in Computational Statistics, Ed.by M.M.Barrit and D.Wishart, 253-259, Physica-Verlag, Wien.

Various Statistical and Data Analytic Methods

Local and Partial Principal Component Analysis (PCA) and Correspondence Analysis (CA)

T.A. Banet, Barcelona, and L. Lebart, Paris

This paper is devoted to the analysis of a set of multivariate observations having an a priori graph structure. Such situation occurs frequently in geography, ecology, geology. In these cases the graphical display issued from a PCA or a CA currently shows a relationship between the observations and the graph structure (for example a North-South effect).

One may be interested in keeping constant the graph effect, in order to visualize the discrepancies between local level and global level and show which level is responsible of the observed patterns. We will use the local covariance matrix, a generalization of the "within covariance matrix" and emphasize the link with partial covariance matrix. A specific software compatible with the system SPAD (1984) is available.

1) NOTATIONS

Let I be a set of n individuals characterized by p variables. I designates also the set of vertices of a graph. Let X be the (n,p) data matrix. Let G(I,E) be the graph of contiguity, E denoting the set of edges. Let M be a symmetric (n,n) matrix associated with the graph G(I,E), (m_{ij} = 1 if vertices i and j are joined by an edge, m_{ij} = 0 otherwise), and N the diagonal (n,n) matrix of degrees of vertices, ($n_{ii}=n_i=\sum_j m_{ij}$). Let $m=\sum_i n_i=\sum_{ij} m_{ij}$ (twice the number of edges).
Let T be a (m/2,n) matrix crossing the $\frac{m}{2}$ edges and the n vertices. If an edge k joins the two vertices i and j, and if i<j, $t_{ki}=1$ and $t_{kj}=-1$. $t_{ki}=0$ otherwise.

The following relation between matrices N, M and T is straightforward :

$$T'T = N-M$$

Thus N-M is positive semidefinite.
Let L be a (m/2, m/2) diagonal matrix with weights of edges as diagonal elements for an edge k joining i and j vertices, the weight is $p_i\ p_j$, where p_i (resp.p_j) is the weight of element i (resp.j).

U designates the (n,n) matrix with $u_{ij}=1$ for all i and j.

2) LOCAL COVARIANCE MATRIX

The classical empirical covariance matrix of X, is written V, with

$$v_{jj'}=\frac{1}{n}\sum_i(x_{ij}-\bar{x}_j)(x_{ij'}-\bar{x}_{j'})=\frac{1}{2n^2}\sum_{i,i'}(x_{ij}-x_{i'j})(x_{ij'}-x_{i'j'}) \qquad (1)$$

using matrix notation,

$$V=\frac{1}{n^2}X'(nI_n-U)X$$

decomposing the double sum (1) inside the graph G(I,E) and outside the graph

$\sum_{i,i'}(x_{ij}-x_{i'j})(x_{ij'}-x_{i'j'})=\sum_{(i,i')\in G}(x_{ij}-x_{i'j})(x_{ij'}-x_{i'j'})$ where the term

$\sum_{i,i'\in G}=\sum^*$ is the sum of products of local differences.

Thus we define the local covariance matrix as (see Lebart - 1969).

$V_1 = \frac{1}{m} X'(N-M)X$ with $v^1_{jj'} = \frac{1}{2m} \Sigma^*(x_{ij}-x_{i'j})(x_{ij'}-x_{i'j'})$

Note that V_1 is identical to V for a complete graph.

3) CONTIGUITY MATRIX

From local covariance matrix V_1 we define the contiguity matrix as :

$$C_c = S^{-1}V_1S^{-1} = \frac{1}{m} S^{-1}X'(N-M)XS^{-1}$$

where S is a diagonal (p,p) matrix of global standard deviations of variables

$$c^c_{jj'} = \frac{1}{2m} \Sigma^*(\frac{x_{ij}-x_{i'j}}{S_j})(\frac{x_{ij'}-x_{i'j'}}{S_{j'}})$$

Note the analogy with the correlation matrix $C=S^{-1}VS^{-1} = \frac{1}{n^2} S^{-1}X'(nI-U)XS^{-1}$.

The comparison of each term of C_c with the corresponding term of C shows the different behaviour of individuals between global level (C) and local level (C_c).

In addition, the diagonal terms of C_c coïncides with the Geary coefficient of contiguity of variables : $c = v_1/v$ (Geary - 1954), consequently we can use them to test the randomness of observations vis-à-vis the graph structure. For a large set of data, Geary coefficient has an approximate normal distribution (Cliff,Ord - 1981) with moments :

$$E(c) = 1$$

$$E(c^2) = \frac{n-1}{n+1}\left[\frac{2(\sum_i n_i^2+m)}{m^2} + 1\right]$$

From Lebart - 1973,we know that the minimum value of c is reached for the first factor issued from a CA of matrix M. Likewise the maximum of ratio c corresponds to the first inverse factor issued from a CA of matrix M.

$$\text{Max}\{c\} = 1 - \lambda_{min} \qquad \text{Min}\{c\} = 1 - \lambda_{max}$$

Thus c has an upper bound of 2.

4) LOCAL PRINCIPAL COMPONENTS ANALYSIS

The local PCA consists of the diagonalization of such matrices V_1 or C_c, (in the following we operate with V_1 without loss of generality).

$$\frac{1}{m} X'(N-M)X\,\mu_\alpha = X'T'TX\mu_\alpha = \lambda_\alpha\mu_\alpha$$

5) LOCAL CORRESPONDENCE ANALYSIS - LOCAL INERTIA MATRIX

Frequently in spatial data one must analyze matrices whose cells are frequencies. Hence the matrix to be diagonalized has as general term :

$$\sum_i \frac{f_{ij}.f_{ij'}}{f_{i.}\sqrt{f_{.j}f_{.j'}}} - \sqrt{f_{.j}f_{.j'}} = \sum_i f_{i.}(\frac{f_{ij}}{f_{i.}\sqrt{f_{.j}}} - \sqrt{f_{.j}})(\frac{f_{ij'}}{f_{i.}\sqrt{f_{.j}}} - f_{.j'})$$

and it could be written :

$$= \frac{1}{2} \sum_{i,i'} f_{i.}f_{i'.} (\frac{f_{ij}}{f_{i.}\sqrt{f_{.j}}} - \frac{f_{i'j}}{f_{i'}\sqrt{f_{.j}}})(\frac{f_{ij'}}{f_{i.}\sqrt{f_{.j}}} - \frac{f_{i'j'}}{f_{i'.}\sqrt{f_{.j'}}})$$

As before we define the local inertia matrix in taking the sum of products only over the vertices joined by an edge, that is :

$$V_1 = \frac{1}{2m_1\sqrt{f_{.j}f_{.j'}}} \Sigma^* f_{i.}f_{i'.} (\frac{f_{ij}}{f_{i.}} - \frac{f_{i'j}}{f_{i!}})(\frac{f_{ij'}}{f_{i.}} - \frac{f_{i'j'}}{f_{i'.}})$$

The local CA leads to solve, similarly :

$$V_1\mu_\alpha = \lambda_\alpha\mu_\alpha$$

These analyses may be interpreted in the following way : if the frequencies or measures are randomly distributed on the graph, the graphical displays issued from local analyses would be similar to those of the global analyses. On the contrary, if there exists a positive dependence, there will be a contraction towards origin (trace of C_c lower than trace of C, the local variance under-estimate the global variance), (although these deformations could not be preserved in projection).

Furthermore the comparison between the patterns of variables in both analyses in often of interest.

6) LOCAL COVARIANCE AND PARTIAL COVARIANCE

The goal of local analysis being to eliminate the effect of location in an euclidean space, it is tempting to relate the notion of local covariance with the classical partial covariance.

Let R be a (n,q) matrix, whose general entry r_{ik} is the value of the k-th coordinate of the i-th vertex of the graph G. For example, if G(I,E) is a planar graph representing the n counties of a country, R will be a (n,2) matrix : the i-th row of R could contain the two coordinates (in an arbitrary basis) of the centroïd of the country i.

If one wishes to eliminate the effect of location on the structure of X, one may use either :

- the local covariance matrix

$$V_1 = \frac{1}{m} X'(N-M)X \qquad (2)$$

- the classical partial covariance matrix which involves continuous coordinates instead of contiguity matrix

$$V_c = \frac{1}{n} X'(I-R(R'R)^{-1}R')X \qquad (3)$$

Let us suppose that the graph G is regular ; consequently, each vertex is contiguous to r vertices (Hence : N=rI), equation (2) is now written :

$$V_1 = \frac{r}{m} X'(I- \frac{1}{r} M)X \qquad (4)$$

Let Φ be the (n,n) orthogonal matrix containing columnwise the normalized eigenvectors of $\frac{1}{r}M$.

We can write : $\frac{1}{r}M = \phi\Lambda\ \phi'$

It has been shown that CA of matrix M yields eigenvectors likely to be used as approximate geographical coordinates of vertices, (see Lebart - 1973-1984).

In order to emphasize the link between the two approaches, we can choose as coordinates $R = \phi_q$ (ϕ_q designating the (n,q) matrix containing the q first columns of ϕ ; q = 2 if G(I,E) is a planar graph).

Note that $\phi'_q\ \phi_q = I_q$.

The matrix $R(R'R)^{-1}R'$ becomes : $\Phi_q\Phi'_q$

whereas $\frac{1}{r}M$ is written : $\Phi\Lambda\ \Phi'$

Note that if the C.A. of matrix $\frac{1}{r}M$ produces only q non-zero eigenvalues (all of them taking the value 1) the two operators involved in formulae (3) and (4) would be identical. This occurs in C.A. when the matrix to be analyzed can be divided into q diagonal blocks made of "1", the remaining blocks being made of "0".

In the general case, the two operators :

$$I - \Phi_q \Phi_q \quad \text{(partial covariance)} \qquad (5)$$

$$I - \phi\Lambda\phi' \quad \text{(local covariance)} \qquad (6)$$

are distincts.

The p-q remaining columns of ϕ are generally polynomial functions of the columns of ϕ_q. This can be observed for particular graphs : cycles, chessboard. It has been pointed out (Benzecri - 1973) that eigenfunction equations of matrices associated with such particular graphs, are the analogue in the finite case of Laplace equations. The eigenfunctions of Laplace Operators are well known ; the polynomial relationships between them have been described by Benzecri as "generalized Guttman effect".

If the eigenvalues (elements of Λ) are close to one, the operator (6) is approximately associated with a polynomial regression, whereas (5) corresponds to a linear regression upon the coordinates.

6) APPLICATION

Finally we have applied these methods to the description of the 935 counties of Catalonia from a socio-economic viewpoint by using programs "LOCAL" and "VAREF" compatible with SPAD. The (935x935) contiguity matrix M is stored in a reduced form R*, which is a (935x10) matrix giving for each county the indexes of its neighbours. In so doing, it has been shown that the patterns of socio-economic variables were almost identical, using the local covariance matrix as well as global covariance matrix. This stability means that the observed structure is not the effect of an opposition between homogeneous regions.

Remark 1 : The notion of local covariance is more general than it seems, since the graph can be artificially constructed from any set of distances or similarities between observations.
It provides an easy device for eliminating the influence of a set of variables without specifying a relationship between the so-called instrumental variables and the other variables. In the previously quoted computer programs, the graph is either given as an input data, or generated from a set of p variables, using a suitable threshold for the distances computed in the p dimensional space.

Remark 2 : The computation of the local covariance matrix is not limited to contiguous vertices. Using the boolean powers of the associated matrix M, it is possible to select a larger set of differences, e.g. those involving the vertices with distances on the graph (i.e. shortest paths) less than 3 or 4.

7) APPENDIX : BRIEF DESCRIPTION OF THE SOFTWARE

Each analysis using the software SPAD consists of a series of steps. Communication from one step to another is provided by means of external normalized sequential files. The techniques described in this paper corresponds to 5 steps. These steps widely use the library of subroutines contained in SPAD, and are consistent with the other 40 steps of the system.

a) Step "LOCAL" (geographical data)

The main functions of "LOCAL" are : to read the contiguïty matrix (reduced matrix R*), to perform a series of checks on this matrix (symmetry). A visualization of the graph as described through a C.A. of the contiguïty matrix is also possible. This matrix being very large in most applications, this C.A. is performed directly from the reduced matrix R*, using an out-of-core algorithm of diagonalization. Eventually, LOCAL creates a normalized file NF for the steps "VARCO", "VAREF", "VARNO".

b) Step "SIMIL" (general case)

This step constructs a contiguïty matrix from a set of variables (or a set of factorial coordinates). This matrix is obtained from the distances between all couples of observations, using a suitable threshold. Any couple whose distance is less than the threshold leads to an edge of the graph. The same file NF is created to serve as input for the steps below.

c) Step "VARCO" (local PCA)

This step performs a local principal component analysis using a data matrix of continuous measurements X and a graph G described by its associated matrix read on the file NF.

d) Step "VAREF" (local CA)

This step is analogous to "VARCO", the data matrix X being now a contingency table (e.g. : cross-tabulation counties - occupational groups issued from a census). A local correspondence analysis is now performed.

e) Step "VARNO" (local multiple C.A.)

This step is analogous to "VARCO", the data matrix X being made of nominal variables.

Remark 3 : If NF is created by the step SIMIL, we are dealing with a partial (instead of local) analysis (cf. remark 1).

REFERENCES

BENZECRI J.P. (1973) - L'analyse des données - Tome II - Dunod - Paris

CLIFF A.D., ORD J.K. (1981) - Spatial processes, Models and applications - Pion Limited - London

GEARY R.C. (1954) - The contiguity ratio and statistical mapping - The Inc. Statistician - p. 115 - 145

LEBART L. (1969) - Analyse statistique de la contiguïté - Publ. de l'ISUP - XVIII - p. 81 - 112

LEBART L., TABARD N. (1973) - Recherches sur la description automatique des données socio-économiques - Report CREDOC-CORDES - N°13/1971 - Chapitre III

LEBART L. (1984) - Correspondence analysis of graph structures - Bulletin technique du CESIA - Vol. 2 - N°1-2, p. 5-19

"S P A D" (1984) - Système portable pour l'analyse des données - CESIA - 82, rue de Sèvres - 75007 - PARIS.

Hierarchical Inference

E. Diday, Le Chesnay

SUMMARY

It often happens in practice, that a user wishing to make a hierarchical classification, does not know which of the panoply of dissimilarity indice will be the best one for his data. It can also happen that none of these indices satisfies the data that he must deal with.

If the user has at the outset some ideas on the classification that he wishes to obtain, our approach permits to induce an aggregation indice, from knowledge acquiring on a learning set, given by the user.

Constraint have been defined in the search of the indice, to ensure that no inversion takes place. A nearest neighbours algorithm, with constraints of validity, is used to induce the final hierarchy.

An application to pattern recognition is proposed, which has permitted to induce some aggregation indices adapted to particular two-dimensionnal repartition of points.

I - INTRODUCTION

Let Ω be a finite set, on which a hierarchy H has to be constructed.

Our approach is to use the knowledge of the user, to choose the agregation indice δ, which will permitt to obtain H from Ω.

From the data table, the user extracts a subtable, on which he can give a hierarchy or a partition which resumes the important connections between objects and the wishes of user. So we will obtain a well-adapted dissimilarity index, and induce H from the extracted learning set.

In order to get the dissimilarity index, the following formula will be used.

II - THE GENERAL RECURRENCE FORMULA

When during the unfolding of H, two new clusters h_1 and h_2 are brought together, we can express the dissimilarity between the new class formed, and all the other classes h existing at this stage of the algorithm. For this purpose, the following general recurrence formula (Jambu 1978) is used :

$$\delta(h, h_1 \cup h_2) = a_1\ \delta(h,h_1) + a_2\ \delta(h,h_2) + a_3\ \delta(h_1,h_2) + a_4\ f(h) + a_5\ f(h_1) + a_6\ f(h_2) + a_7\ |\delta(h,h_2) - \delta(h,h_1)|$$

where f(h) express the level of the class h, h_1 and h_2

This formula generalizes the classical formula of Lance an Williams (1967) which does not consider the terms in a_4, a_5 and a_6.

III - A NECESSARY AND SUFFICIENT CONDITION REGARDING THE COEFFICIENTS OF THE RECURRENCE FORMULA FOR THE NON EXISTENCE OF INVERSIONS

If the hierarchy H is indexed in such a way that the height of each level corresponds to the value of the dissimilarity index for the two levels that have formed it ; in other words, if the following condition is satisfied :

$\{$for any h_1, h_2 and h_3 in H with $h_3 = h_1 \cup h_2$, we have $f(h_3) = \delta(h_1,h_2)\}$

It can happen that f gives rise to the existence of levels h an h' such that $h \subset h'$ and $f(h) > f(h')$. In such case, we say that there is an inversion. This property is in fact equivalent to the ultrametric property for the dissimilarity index associated to H.

The following question may be raised :

Is there a necessary and sufficient condition regarding the coefficients $(a_1,\dots,a_7)$ for the non-existence of inversions in the hierarchy H indexed by f ?

The following proposition, which generalizes Ducimetière's remark (1972), Milligan's proposition (1979), and Batagelj's (1981), answer the question, in the case of the classical hierarchical ascending classification algorithm (HAC)(which consists, starting from Ω, to merge the two nearest objects, and repeat this procedure until all the clusters are merged into one)

PROPOSITION 1

A necessary and sufficient condition for any δ satisfying a recurrence formula defined by the coefficients $a_1,\dots,a_7$, to induce, by the HAC algorithm, a hierarchy indexed by f without inversions is that the following four conditions be satisfied :

(a) $a_7 \geq -\mathrm{Min}\,(a_1, a_2)$; (b) $a_1 + a_2 \geq 0$; (c) $a_1 + a_2 + a_3 \geq 1$; (d) a_4, a_5 and a_6 are positive.

IV - HIERARCHICAL INFERENCE WITH NEAREST NEIGHBOURS ALGORITH

This algorithm, which permit to obtain a CPU time clearly shorter than classical HAC, is used to induce H, with the dissimilarity index δ we have found.

The principle of this algorithm is the following one :

Two levels h_i and h_j of a hierarchy H are called "nearest neighbours" if h_j is the nearest neighbour of h_i and if h_i is also the nearest neighbour of h_j On the basic of that concept, the nearest neighbours algoritm works according to the fact that each iteration, that is to say each computation of the matrice of distance, makes it possible to aggregate not only the two less distant nearest neighbours, but all the nearest neighbours existing at this iteration. Then the question is to know if the hierarchy that we would obtain by this algorithm, is always the same than the hierarchy we would obtain with the HAC algorithm. One can show that there is no deformation of the hierarchy, if the dissimilarity index δ satisfies a property that will be called "nearest neighbours validity condition".

6.1 - Nearest neighbours validity condition

For having the same hierarchy by HAC and nearest neighbours algorithms, it is necessary and sufficient to have the following condition :

considering a partition P induced by the hierarchy : when two nearest neighbours node h_1 and h_2 merge together, we get a new partition P' such that all the nearest neighbours of P (except h_1 and h_2) stay nearest neighbours in P'.

PROPOSITION 2

A sufficient condition to get the same hierarchy by CAH and nearest neighbours algorithms is that the following condition beeing satisfied :

$$\text{(C)} \quad \begin{array}{l} \forall h \in P, \quad \delta(h_1,h_2) \le \text{Inf } (\delta(h_1,h),\ \delta(h_2,h)) \\ \implies \quad \delta(h_1 \cup h_2,h) \ge \text{Inf } (\delta(h_1,h),\ \delta(h_2,h)) \end{array}$$

6.2 - Connection between inversion and validity condition

PROPOSITION 3

If the validity condition is satisfied, a hierarchy constructed by the HAC algorithm or by the nearest neighours algorithm has no inversion.

6.3 - Connection with the recurrence formula

PROPOSITION 4

A necessary and sufficient condition for a dissimilarity index f satisfying a recurrence formula defined by the coefficients $a_1,\ldots,a_7$, to verify validity condition of nearest neighbours algorithm, is that the following four conditions be satisfied :

(a) $a_7 \geq -\text{Min}(a_1, a_2)$; (b^2) $a_1 + a_2 \geq 1$; (c) $a_1 + a_2 + a_3 \geq 1$;

(d) a_4, a_5, a_6 are positive

V - ADAPTIVE CONSTRUCTION OF DISSIMILARY INDICES

Given a recurrence formula, one can say, to simplify matters, that if conditions (a), (b), (c) and (d) of Proposition 1 are satisfied and if (b^2) is false, then there are no inversions, and CAH is avalaiblé. If conditions (a) (b^2), (c) of Proposition 4 are satisfied, then there are inversions and nearest neighbours algorithm is avalaible.

Beginning with this hierarchy given in the data, it is possible to calculate $\delta(h_i, h_j)$, where h_i and h_j are two arbitrary levels of H, by setting :

$\delta(h_i, h_j)$ = the height of the lowest level containing h_i and h_j.

With the help of these quantities and the general recurrence formula, one obtains a system of equation with 7 unknows (the a_i, for $i = 1, \ldots, 7$) and containing $(n-1)\ (n-2)/2$ equalities (n being the size of the population). As a general rule, there are more equations than unknowns, then a least square algorithm under constraints is used.

We finally find a dissimilarity index, as well adapted as possible to the learning set.

VI - THE ALGORITHM

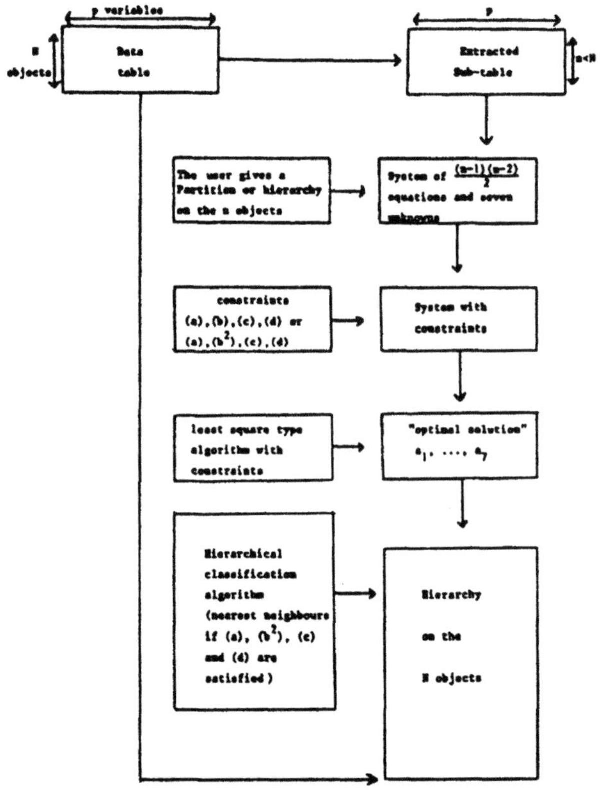

VII - EXAMPLE

The method has been tested on twelve two-dimensionnal pictures represented in Figure 1 including 185 points of the plane. In these pictures, clusters are visible to the naked eye. We have tested classical aggregation indices and inference indices to try to recognize these pictures.

Classical indices and infered indices has been tested on the twelve examples. Distance betwen objects is euclidian.

A summary of the results is given in figure 1.

BIBLIOGRAPHY

1 BATAGELJ V. "Note on ultrametric hierarchical clustering algorithms" Psychometrica, Vol. 46 n° 3 (1981)

2 DIDAY E., LEMAIRE J., POUJET J., TESTU F. "Eléments d'analyse des données" Dunod (1982)

3 DIDAY E. "Inversions en classification hiérarchique, application à la construction adaptative d'indices d'agrégations" RSA. - 1983, Vol. XXXI, n° 1

4 DUCIMETIERE P. "Les méthodes de classification numérique" Revue de Statistiques Appliquées, Volume 18, n° 4 p.5 - 25 (1970)

5 JAMBU M. "Classification automatique pour l'analyse des données", Dunod (1978)

6 JUAN J. "Le programme Hivor de classification ascendante hiérarchique selon les voisins réciproques et le critère de la variance", Cahiers d'Analyse de Données, Vol. 7 n° 2 (1982).

7 LANGE G.C., WILLIAMS W.T. "A general theory of classification sorting" Computer Journal 9.10 and Computer Journal 10.3 (1967).

8 MILLIGAN G. "Ultrametric hierarchical clustering algorithms", Psychometrica 44, 3 (1979).

Picture	CLASSICAL INDICES: Classical indices recognizing the picture	number of indices	INFERENCE INDICES: Inférence indices recognizing the picture	number of indices	TOTAL: Total number
	Minimum-link, Diametral, Average, Ward, Least Inertia, Least Variance, Centroïd	(sur 7) 7	a, c, d, e, f, g, h, i, j, k	(sur 12) 10	17
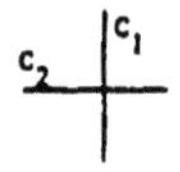		0		0	0
	Minimum-link, Diamétral, Ward, Least Inertia	4	a, c, d, e, f, h, i, j, k	9	13
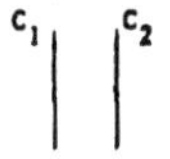	Minimum-link	1	a, c, d, e, f, h, i, j; k	9	10
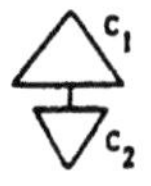	Diametral, Average, Ward, Least Inertia, Least Variance, Centroïd	6	e, f, h, i, j, k	6	12
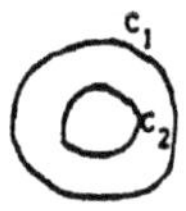	Minimum-link	1	a, d, e, f, h, i, k	7	8
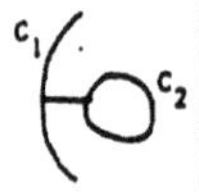	Ward, Least Inertia, Least Variance	3	g	1	4
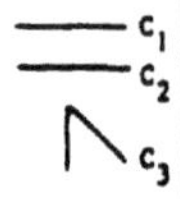		0	c, h, k	3	3
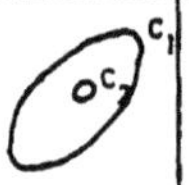		0	f, e, i	3	3
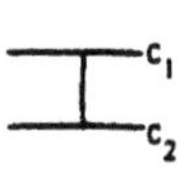		0	e, f, i, j	4	4
	Minimum-link	1	a, c, d, e, f, h, i, j, k	9	10
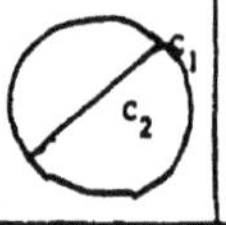		0		0	0

figure 1

An Interactive Multiple Criteria Approch to finding subjective Principal Components

P.J. Korhonen, Helsinki

ABSTRACT

In this study we have developed an interactive multiple criteria approach for finding subjective principal components for a given set of variables in a data matrix. The principal components are not determined by maximizing their variances; they are specified by a user, who try to find the most preferred correlation coefficients between principal components and the variables important to him. We have implemented an experimental version of our procedure on an APPLE III microcomputer. The graphical colour display has an essential role in the method.

Keywords:**Principal components, Multicriteria optimization, Interactive methods, Visual**

1. Introduction

In practical applications, the two main objectives of principal component analysis appear to be the reduction of dimensionality of the data or the identification of new meaningful, underlying variables. When the first few components account for most of the variation in the original data, it is often convenient to use these first few components in subsequent analyses while losing as little information as possible in the reduction.

Principal component analysis is also used in the hope of finding a few linear combinations of the original variables, which are intuitively meaningful and which lead a user to a better understanding of the correlation structure of the original variables. In this case, the user also tries to attach meaningful "labels" to principal components. The usual procedure seems to consist of looking at the correlation coefficients between principal components and the original variables and picking out the variables for which the correlations are large, either positive or negative. Having established the subset of variables which are important for a particular component, the user then tries to see what these variables have in common.

In practice, the interpretation of the principal components may be very problematic. All correlations may be quite low or it may be difficult to see which features the highly correlated variables have in common. Sometimes the first principal components are rotated in order to find a new set of components which can more easily be interpreted. In some cases, the rotation may increase the meaningfulness of the components. However, if the user has an apriori "feeling" as to which kind of principal components he would like to find or even if he has the ready labels for the principal components in advance, it may be difficult to find the rotation method, which will give him the result compatible with his intuition.

An application, described in Korhonen (1984), is a typical example in which principal component analysis is not particularly well-suited. Our problem is to derive the "total performance" for 80 business units of a Finnish organisation. It is presented as a linear combination of a set of variables describing the performance but the question of which weights are used is highly subjective. The linear combination with the maximal variance (= the first principal component) is not a very realistic solution because the user may be interested e.g. in finding the linear combination, which is highly correlated with growth and fairly correlated with profitability.

In this paper, we develop a visual interactive multiple criteria approach which helps a user to find the most preferred linear combinations of the original variables. We call these linear combinations the subjective principal components. They can be correlated or uncorrelated. As criteria we use the correlation coefficients between the linear combinations and the original variables.

The paper is in five sections. In the introduction we outlined the problem and in section 2 we describe our basic approach. In section 3 we describe some modifications and the concluding remarks are presented in section 5.

2. Development of the Basic Approach

Let us assume the sample of n observations and p variables. The sample is written as the (n x p) data matrix $X=(x_{ij})$. The rows of X standing for observations will be written $X_i=(x_{i1},x_{i2},\dots,x_{ip})$ and the columns standing for variables will be written $\mathbf{x}_j=(x_{1j},x_{2j},\dots,x_{nj})'$. That is, we may write $X = (X_1,X_2,\dots,X_n)' = (\mathbf{x}_1,\mathbf{x}_2,\dots,\mathbf{x}_p)$.

We shall use a similar notation to refer to the columns and rows of the other matrices used in this study.

For simplicity and without loss of generality, we assume that $\mathbf{x}_j$, $j=1,2,\dots,p$, has mean zero.

Let $T=(t_{ij})$, $T = X'X$, denote the matrix of the cross-products of X. By using the spectral decomposition theorem, matrix T may be written in the form $T = GLG'$, where the columns of the orthogonal matrix G, $G = (\mathbf{g}_1,\mathbf{g}_2,\dots\mathbf{g}_p)$, represent the eigenvectors of T and L is a diagonal matrix of eigenvalues of T, $l_1 \geq l_2 \geq \dots \geq l_p \geq 0$.

The principal component transformation is defined by the rotation $W = XG$. Because $T_W = W'W = G'X'XG = G'TG = G'GLG'G = L$ is diagonal, the columns of W, called the principal components, represent uncorrelated linear combinations of the original variables.

It is well-known that l_k can be given as a weighted sum

$$l_k = \sum_i t_{ii} r^2(\mathbf{x}_i,\mathbf{w}_k), \tag{2.1}$$

where $r(\mathbf{x}_i,\mathbf{w}_k)$ refers to the correlation coefficient between variable $\mathbf{x}_i$ and the principal component $\mathbf{w}_k$ (see, e.g. Mardia et. al. (1979, p. 223)). Thus, from formula (2.1), we can see that principal component analysis is one way to find orthogonal linear combinations which are highly (positively or negatively) correlated with the original variables. However, principal component analysis does not take into account subjective aspects.

In the following, we develop a multiple criteria model which helps the user to consider all reasonable solutions. Let the (n x q) matrix Y, $q \leq p$, represent a q-dimensional linear transformation Y = X B, where B is a (q x p) matrix. In addition, we also assume that the columns of Y are uncorrelated, i.e. $Y'Y$ is diagonal. Hence forward, we assume, without loss of generality, that variables x_i have been standardized such that $X'X = R$. We also assume that R is non-singular. If we scale the weight matrix B such that $B'R B = I$, so $Y'Y = B'X'X B = B'R B = I$. Now the (p x q) correlation matrix U representing the correlations between the original variables and the linear combinations can be written as

$$U = X'Y = X'X B = R B. \tag{2.2}$$

Definition 1. The variable y_k, $k=1,2,...,q$, is called the kth subjective principal component if the correlations between y_k and the original variables x_i, $i=1,2,...,p$, are the most preferred (either maximal or minimal) and y_k is uncorrelated with the first k-1 subjective principal components.

Next we proceed to construct the general model for finding the (k+1)th subjective principal component, $k> 0$, assuming that the first k components are already determined. Let us now take Y to be a (n x k) matrix representing the k first subjective principal components, $Y=XB$, where B is now a (k x p) matrix. By definition 1, y_{k+1} must be orthogonal to the columns of Y. Therefore we are only interested in the linear combinations Xb_{k+1}, for which $Y'Xb_{k+1} = 0$.

Let $P = I - Y(Y'Y)^{-1}Y'$. Then P (n x n) is a symmetric idempotent matrix of rank n-k which projects onto the subspace of R^n orthogonal to the columns of Y. In particular, note that $PY = 0$. (see, e.g. Mardia et. al. (1979, p. 158)). In our case $P=I-YY'$, because $Y'Y=I$. Now the linear combinations orthogonal to the columns of Y can be presented as

$$y_{k+1} = PXa = Xb_{k+1}, \tag{2.3}$$

where

$$b_{k+1} = R^{-1}X'P\,Xa. \tag{2.4}$$

Because $y'_{k+1}y_{k+1} = b'_{k+1}Rb_{k+1} = 1$, so also $a'X'P\,Xa=1$. The correlations u_{k+1} between the original variables and y_{k+1} can be given by formula (2.2) as

$$u_{k+1} = X'P\,Xa = X'(I-YY')Xa = (X'X - X'YY'X)a = Qa, \tag{2.5}$$

where

$$Q = R-RBB'R = R-UU' = R - \sum_i^k u_i u_i'. \tag{2.6}$$

Now we are ready to formulate the model for finding the subjective principal component $y_{k+1} = PXa = Xb_{k+1}$, where b_{k+1} is determined by formula (2.4). For simplicity, we denote $u = u_{k+1}$. Because minimizing the correlation $r(x_i,y_{k+1})$ is equivalent to maximizing $-r(x_i,y_{k+1}) = r(-x_i,y_{k+1})$, we can present the model for finding the most preferred correlations as follows:

$$\begin{aligned} &\text{"max" } Mu = M(R-RBB'R)a = MQa = MQM'Ma \\ &\text{subject to} \qquad\qquad\qquad\qquad\qquad\qquad (2.7) \\ &\qquad a'Qa = aM'MQM'Ma = 1, \end{aligned}$$

where M is a diagonal matrix defined as $m_{ii}=1$, if $r(x_i,y_{k+1})$ is to be maximized and $m_{ii}=-1$, if it is to be minimized. The vector u, representing the correlations, is an objective vector to be maximized or minimized. For the first subjective principal component $Q=R$ in the model (2.7).

The vector maximum problem (2.7) has no unique solution. Any nondominated (efficient, Pareto-optimal) solution $u=Qa$ is an acceptable and reasonable solution for the problem.

Definition 2. Solution $u^*=Qa^*$ is nondominated (efficient) if and only if there is no other feasible solution $u=Qa$, $u \neq u^*$, such that

$$Mu=MQa \geq Mu^*=MQa^*.$$

If u^* is dominated by some u, we denote $Mu > Mu^*$.

Theorem 1. (Korhonen (1984)) The solution $u=Qa$ of the problem (3.8) is nondominated iff $u \in C=\{u \mid u=Qa, Ma \geq 0\}$.

Remark. All nondominated solutions of the problem (2.7) can be found and no one solution is dominated if we add to the problem (2.7) the constraint $Ma \geq 0$:

$$\begin{aligned} &\text{"max"} \quad Mu = M(R-UU')a = MQa \\ &\text{subject to} \\ &\qquad a'Qa = 1 \\ &\qquad Ma \geq 0 \end{aligned} \tag{2.8}$$

We now present our interactive procedure for solving the model (2.8) for each subjective principal component y_k, $k=1,2,..,q$, where q is specified by a user. We assume that the user is able to define the matrix for each subjective principal component at the beginning of the search process. By multiplying any variable by -1, we can change the direction of the optimization of the corresponding correlation coefficient. Therefore, without loss of generality, we can assume that $M=I$.

We describe the method in a step-by-step manner interspersing comments and explanations between steps.

Step 1. Compute the correlation matrix R of the original variables. Set $Q:=R$ and $k:=1$.

Step 2. Solve $a:=e'/(e'Qe)^{1/2}$, where $e:=(1,1,...,1)'$ and compute $u:=Qa$.

Thus we will find a feasible (nondominated) solution.

Step 3. Present u to the user and ask which correlations he would, primarily, like to increase. Let us I, $I \subset \{1,2,...,p\}$, be an index set referring to those correlations. Construct vector Δu such that $\Delta u_i := 1$, if $i \in I$ and $\Delta u_i := 0$, if $i \notin I$.

We use the vector Δu as a reference direction vector. If desired, the strength of preferences can be taken into account in Δu, too.

Step 4. Compute $\Delta a := R^{-1}\Delta u$, $s^2 := (a + \Delta a)'Q(a + \Delta a)$, $\Delta a^* := (a+\Delta a)/s-a = (a(1-s)+\Delta a)/s$ and $\Delta u^* := Q\Delta a^*$.

Now we have $(a+\Delta a^*)'Q(a+\Delta a^*)=1$ and thus $u+\Delta u^*$ is a solution of problem (2.7) but not necessarily of the problem (2.8).

We use vector $u + \Delta u^*$ as a reference vector and shall represent to the user the projection of vector $u+t\Delta u^*$, $t \geq 0$, on the efficient frontier.

Step 5. Solve the problem:

$$\max_t \quad a + t\Delta a^* \geq 0.$$

Denote the maximum value by t_{max}.

The value t_{max} tells us to what extent it is possible to go in the direction Δa^* from the current position a until the solution is no longer feasible solution of the model (2.8).

Step 6. Present to the user the nondominated curve

$$u(t)=Q(a+t\Delta a^*)/s(t)= (u+t\Delta a^*)/s(t), \quad t:0 \to t_{max},$$

where

$$s^2(t):=(a+t\Delta a^*)'Q(a+t\Delta a^*) \quad =1+2t(t-1)a'Q\Delta a^*$$
$$=1+2t(t-1)((1+a'Q\Delta a)/s-1),$$

in the graphical mode using the graphical display as described in Korhonen and Laakso (1983). Ask him to choose the most preferred solution and denote the corresponding value of t by t_o. Set $a:=(a+t_o\Delta a^*)/s(t_o)$, $u:=(u+t_o\Delta u^*)/s(t_o)$ and return to step 3, if t_o > t. If t_o = t, ask the user if he is willing to consider other directions. If the answer is "no", go to step 7; otherwise return to step 3.

If the decision maker prefers the solution with value t_{max}, we can continue from this position by setting the corresponding $\Delta a_i = 0$ and computing a new Δu corresponding to the transformed Δa.

Step 7. Set $u_k:=u(t_o)$ and denote $U:=(u_1,u_2,...,u_k)$. Ask, whether the user is willing to consider the next component; if the answer is "no", print the correlations U and the "eigenvalues" $l_j = \sum u_{ij}^2$, $j=1,2,...,k$, and stop, otherwise, set $k:=k+1$, $Q:=Q-u(t_o)u(t_o)'$ (cf. formula (2.6)). Multiply by -1 the variables corresponding to the correlation coefficients to be minimized in the next iteration and return to step 2.

3. Some Modifications

In this section we consider problems requiring some modifications in our approach.

Let us first assume that the user is willing to decide during the search process which correlation coefficients he is willing to maximize or minimize. We can deal with that problem by making the following modifications in our procedure: Step 5 is omitted and we set t_{max} = T, where T is a given large number. In step 3 the user can also specify the correlation coefficients he would like to decrease. The latest modification means that we let the user freely change his mind from the maximization mode to the minimization mode and vice-versa. Unfortunately, in this case we cannot guarantee that all solutions are nondominated. When the user changes the direction of the optimization ($m_{ii} := -1*m_{ii}$), the corresponding a_i usually has a wrong sign and according to Theorem 1 we know that the current solution may be dominated. Therefore we should first try to find the nondominated solution dominating the current solution. However, we do not concern ourselves with that problem in this paper.

Next we consider the problem of finding the (k+1) th subjective principal component (k>0), which is not necessary orthogonal to the columns of Y. As before, the columns represent the previously determined subjective principal components, y_i, i=1,2,...,k. The correlation coefficients between y_{k+1} and y_i, i=1,2,...,k, can be given as $v = Y'y_{k+1} = Y'Xb_{k+1} = U'b_{k+1} = B'Rb_{k+1}$. Because now $v \neq 0$, usually, we formulate our model as follows:

$$
\begin{aligned}
&\text{"max"} \quad Mu = MRb_{k+1} \\
&\qquad\quad v = B'Rb_{k+1} \\
&\text{subject to} \qquad\qquad\qquad\qquad\qquad\qquad\qquad (3.1) \\
&\qquad\quad b_{k+1}'Qb_{k+1} = 1 \\
&\qquad\quad Mb_{k+1} \geq 0
\end{aligned}
$$

The user tries to maximize the correlation coefficients like in the model (2.8), but simultaneously he can consider the correlation coefficients between y_{k+1} and y_i, $i=1,2,...,k$. We can use our procedure for solving the model (3.1) by modifying step 6 such that it also gives the values of v as the function of t. Thus the user can also evaluate the correlation coefficients between the (k+1) th subjective principal components and the previously determined ones.

4. Concluding Remarks

In this paper, we have developed a visual interactive multiple criteria approach to finding the most preferred linear combinations of the original variables in a data matrix. We have assumed that the user is willing either to maximize or minimize the correlation coefficients between the linear combinations and the original variables. The user can consider all reasonable linear combinations and choose those which are correlated with the original variables in the way he likes best.

We presented a basic method which helps to find the uncorrelated linear combinations and extended that method to deal with the problems in which the linear combinations can be correlated with each other.

This approach can be applied to the problems in which the principle used in principal component analysis is too restrictive to find the linear combinations compatible with the intuition of the user. In the introduction, we mentioned as an example a problem in which this approach was well-suited.

References

Korhonen, P. and J. Laakso (1983), "A Visual Interactive Method for Solving the Multiple Criteria Problem," in Wierzbicki, A. and M. Grauer (eds.): The Proceedings of the Workshop on Interactive Decision Analysis and Interpretative Computer Intelligence, pp. 146-153, Springer-Verlag.

Korhonen, P (1983), "A Hierarchical Method for Ranking Alternatives with Multiple Qualitative Criteria," Working Paper, F-76 Helsinki School of Economics.

Korhonen, P (1984), "Subjective Principal Component Analysis," Working Paper, F-77, Helsinki School of Economics.

Mardia, K., J. Kent and J. Bibby (1979), Multivariate Analysis, Academic Press, New York.

Rao, C. (1973), Linear Statistical Inference and its Applications, 2nd. ed., Wiley.

Zeleny, M. (1982), Multiple Criteria Decision Making, McGraw-Hill.

Algorithms of Discriminant Analysis Using Parameter Restrictions for Diminishing the Error Rate

J. Läuter, Berlin

Summary:

In the first part of the paper, a modification of the discriminant analysis is described that is based on reduction of the estimated partial correlations. The second part refers to a 'factorial discriminant analysis' in which the mean values and covariances are supposed to have a joint factor structure and correspondingly, maximum likelihood estimators are determined.

Keywords:

Discriminant analysis, parameter restriction, factor structure

1. Introduction

The basic idea of the present paper shall be to show that in most applications of discriminant analysis special relations exist between the mean values and covariances of the used variables and therefore, discrimination rules with a diminished error rate can be found. In former theoretical investigations the author has proved a diminution of the mean error rate for certain models of the discriminant analysis with 2 populations (J. LÄUTER (1982, 1984)). The employed method consists in adapting the signs of the coefficient vector $S^{-1}(y_{.}^{(1)} - y_{.}^{(2)})$ of discriminant function to the signs of the difference vector $y_{.}^{(1)} - y_{.}^{(2)}$ if the given samples require this.

2. A Practical Method of Reducing the Partial Correlations

In certain technical, biological and social systems the relations between the observed variables $Y_1,\ldots,Y_p$ can have the simple form that the several variables result from an only non-observable primary variable Y_o by adding disturbances:

$$Y_i = (Y_o + V_i)\,\beta_i + \alpha_i \qquad (i = 1,\ldots,p).$$

Y_o is supposed to be normally distributed with a mean value that corresponds to one of two states of the system (e.g. healthy, ill):

$$Y_o \sim N(\mu_o^{(1)}, \sigma_o^2) \quad \text{or} \quad Y_o \sim N(\mu_o^{(2)}, \sigma_o^2).$$

The disturbances V_i are to be distributed according to $N(0, \tau_i^2)$ and to be independent of Y_0 as well as mutually. α_i and β_i are scale constants fixing the zero point and the measuring unit.

By these suppositions, random vector $(Y_1,\ldots,Y_p)'$ has the distributions $N(\mu^{(1)}, \Sigma)$ and $N(\mu^{(2)}, \Sigma)$ in the two considered states, where Σ has the structure

$$\Sigma = K + \omega(\mu^{(1)} - \mu^{(2)})(\mu^{(1)} - \mu^{(2)})' \tag{1}$$

(K positive definite diagonal matrix, $\omega \geq 0$). Then

$$\varphi_{ih} \frac{\mu_i}{\mu_h} < \varphi_{ii} \quad \text{for } i,h > p_1, \quad i \neq h, \tag{2}$$

with μ_i being the elements of $\mu = \mu^{(1)} - \mu^{(2)}$ and φ_{ih} the elements of the $p_2 x p_2$ conditional covariance matrix $\Phi = \Sigma_{(22)} - \Sigma_{(21)} \Sigma_{(11)}^{-1} \Sigma_{(12)}$ corresponding to a given partitioning of Σ into blocks $\Sigma_{(11)}, \Sigma_{(12)}, \Sigma_{(21)}, \Sigma_{(22)}$ $(p = p_1+p_2)$. This inequality is valid in arbitrary hierarchical networks of disturbances, too (J. LÄUTER (1984)).

Based on (2), a modification of the discriminant analysis is possible in which the covariance estimator S is corrected by statement

$$\hat{\varphi}_{ih} := \hat{\varphi}_{ih}/\max(1, \frac{\hat{\varphi}_{ih}\, y_{i.}}{\hat{\varphi}_{ii}\, y_{h.}}, \frac{\hat{\varphi}_{ih}\, y_{h.}}{\hat{\varphi}_{hh}\, y_{i.}}) \tag{3}$$

where $y_{i.} = y_{i.}^{(1)} - y_{i.}^{(2)}$ and $\hat{\varphi}_{ih}$ are the estimators of μ_i and φ_{ih}. This reduction of the covariances in regard to its tendency coincides with the known ridge technique of the regression analysis. P.J. DiPILLO (1976, 1979) has emphasized the importance of the ridge technique in the discriminant analysis.

The diminution of covariances is performed together with the stepwise selection of variables. We describe the algorithm in the version for J populations ($J \geq 2$). In our program, each found set of variables is assigned a matrix of the form

$$A = (a_{ih}) = \begin{pmatrix} S_{(11)}^{-1} & S_{(11)}^{-1} S_{(12)} \\ S_{(21)} S_{(11)}^{-1} & -S_{(22)} + S_{(21)} S_{(11)}^{-1} S_{(12)} \end{pmatrix}. \tag{4}$$

$S_{(11)}$ is the part of the covariance matrix S, that corresponds to the set of selected variables. We designate the sample sizes by $n^{(j)}$ and define

$$n = \sum_{j=1}^{J} n^{(j)}, \quad y_{\bullet}^{(\cdot)} = \frac{1}{n} \sum_{j=1}^{J} n^{(j)} y_{\bullet}^{(j)}, \tag{5}$$

$$Y = (y_{ij}) = (y_{\bullet}^{(1)} - y_{\bullet}^{(\cdot)}, y_{\bullet}^{(2)} - y_{\bullet}^{(\cdot)}, \ldots, y_{\bullet}^{(J)} - y_{\bullet}^{(\cdot)}) \ . \tag{6}$$

If we suppose that the variables 1,2,...,l-1 already belong to the set and in the next step variable l is to be entered, the following statements have to be used:

$$a_{gh} := a_{gh} - \frac{a_{lg}\, a_{lh}}{a_{ll}} \quad \text{for } g = l+1,\ldots,p;\ h = g+1,\ldots,p,$$

$$b_g := a_{gl} \text{ for } g = 1,\ldots,l;\ b_h := a_{lh} \text{ for } h = l+1,\ldots,p,$$

$$b_h := b_h / \min_{j=1,\ldots,J} (\max(1, \frac{a_{lh}\, y_{lj}}{a_{ll}\, y_{hj}}, \frac{a_{lh}\, y_{hj}}{a_{hh}\, y_{lj}})) \text{ for } h = l+1,\ldots,p, \tag{7}$$

$$a_{gh} := a_{gh} - \frac{b_g\, b_h}{b_l} \quad \begin{array}{l} \text{for } g = 1,\ldots,l-1;\ h = g,\ldots,l-1 \\ \text{for } g = 1,\ldots,l-1;\ h = l+1,\ldots,p, \\ \text{for } g = l+1,\ldots,p;\ h = g, \end{array}$$

$$a_{gl} := -\frac{b_g}{|b_l|} \quad \text{for } g = 1,\ldots,l-1,$$

$$a_{lh} := -\frac{b_h}{|b_l|} \quad \text{for } h = l+1,\ldots,p,$$

$$a_{ll} := -\frac{1}{b_l} \ .$$

In statement (7) the correction in question is realized. The diagonal elements of S, that is the variances, are not subjected to correction. In removing steps, no correction takes place.

The employment of statement (7) depends on two further conditions:

a) During a discriminant analysis, the corrections are started only when a probably optimal set of variables has been found and steps of swinging up and down, round the optimal set, have begun. The amplitude of swinging round the optimum is (1 + entier(0.15 p_{opt})) steps.

b) The correction of $\hat{\varphi}_{lh}$ is suppressed if a test carried out at the beginning of the entering step shows that variable l as well as variable h enable a significant improvement of the separation (at level of significance 0.05) .

In the following graph, the error rates of 8 cases of application of the discriminant analysis with 2 populations are represented. The results correspond to

(A) the discrimination with all given variables,
(B) the discrimination with automatic selection of variables by minimizing the level of significance (H. AHRENS and J. LÄUTER (1981)),
(C) the discrimination with selection of variables and with covariance reduction.

The error rates are determined by cross-validation according to the Π method (G.T. TOUSSAINT (1969)). The cross-validation concerns the whole analysis including selection of variables. The errors are plotted as percentages in relation to rate (A). For each task, the number of variables and the sample sizes (s.s.) are mentioned.

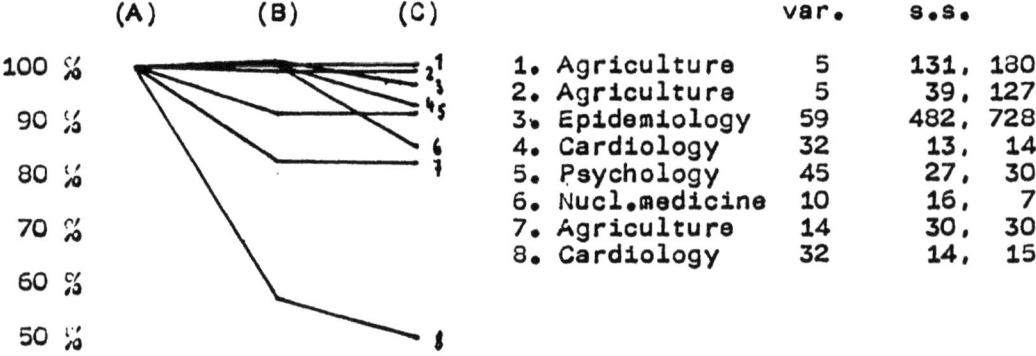

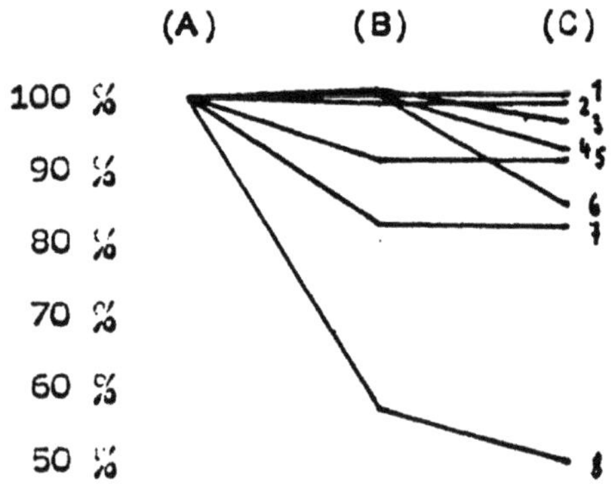

		var.	s.s.
1.	Agriculture	5	131, 180
2.	Agriculture	5	39, 127
3.	Epidemiology	59	482, 728
4.	Cardiology	32	13, 14
5.	Psychology	45	27, 30
6.	Nucl.medicine	10	16, 7
7.	Agriculture	14	30, 30
8.	Cardiology	32	14, 15

One can see that the error rate is diminished by selecting variables in 3 from 8 cases and by covariance reduction in 4 from 8 cases. No one of these 8 examples and in the same way, no one of 6 further studies with 3 or 4 populations showed a deteriorated discrimination due to the covariance reduction. Therefore, the described method can be recommended for use in universal programs of the discriminant analysis.

3. Factorial Discriminant Analysis

Equation (1) states that Σ possesses a one-factor structure in the sense of the factor analysis. In this section a more general multiple-factor structure

$$\Sigma = K + \Theta\Theta' \qquad (8)$$

is assumed, where Θ has the order $p \times f$ and K as well as $\Theta' K^{-1} \Theta$ are positive definite diagonal matrices ($0 < f < p$). For the means $\mu^{(j)}$ of the populations $1, \ldots, J$ we demand that all difference vectors $\mu^{(l)} - \mu^{(m)}$ lie in the space generated by the columns of Θ:

$$\mu^{(l)} - \mu^{(m)} \in \mathcal{M}(\Theta) \quad \text{for} \quad 1 \leq l < m \leq J. \tag{9}$$

We suppose that for $j = 1, \ldots, J$ a sample of size $n^{(j)}$ from population $N(\mu^{(j)}, \Sigma)$ is given. Then we shall set up the task to estimate the parameters K, Θ,

$$\mu^{(\cdot)} = \frac{1}{n} \sum_{j=1}^{J} n^{(j)} \mu^{(j)} \tag{10}$$

and

$$M = (\mu^{(1)} - \mu^{(\cdot)}, \mu^{(2)} - \mu^{(\cdot)}, \ldots, \mu^{(J)} - \mu^{(\cdot)}) \tag{11}$$

by the maximum likelihood method. The number f of factors is considered to be known.

This task is a generalization of the estimation problem in factor analysis that has been solved by D.N. LAWLEY and A.E. MAXWELL (1963). In this generalization a joint estimation of $\mu^{(j)}$ and Σ takes place. Therefore we can expect stabilized estimators and consequently an improved discrimination compared with the customary discriminant analysis.

Let N be the diagonal matrix of the $n^{(j)}$ and G the matrix of the sums of products of deviations

$$G = \sum_{j=1}^{J} \sum_{k=1}^{n^{(j)}} (y_k^{(j)} - y_\bullet^{(j)})(y_k^{(j)} - y_\bullet^{(j)}) \quad . \tag{12}$$

Then the maximum likelihood estimators satisfy the equations

$$\hat{\mu}^{(\cdot)} = y_\bullet^{(\cdot)}, \; \hat{M} = \hat{\Theta}(\hat{\Theta}' \hat{K}^{-1} \hat{\Theta})^{-1} \hat{\Theta}' \hat{K}^{-1} Y, \tag{13}$$

$$(G - n\hat{\Sigma})\hat{K}^{-1}\Theta + (YNY' - \hat{M}N\hat{M}')\hat{K}^{-1}\hat{\Theta}(\hat{\Theta}'\hat{K}^{-1}\hat{\Theta})^{-1}(I + \hat{\Theta}'\hat{K}^{-1}\hat{\Theta}) = 0, \tag{14}$$

$$\operatorname{diag}(G - n\hat{\Sigma} + YNY' - \hat{M}N\hat{M}') = 0 \tag{15}$$

(cf. (5) and (6)), where diag(X) designates only the diagonal part of X. For $Y = 0$, $\hat{M} = 0$, equations (14) and (15) change over into the known equations by **LAWLEY** and **MAXWELL**. The equations must be solved iteratively.

References:

Ahrens, H. and J. Läuter (1981), Mehrdimensionale Varianzanalyse, Akademie-Verlag, Berlin

Läuter, J. (1982), Improvement of the Discriminant Analysis under Parameter Restrictions, Proc. Conf. DIANA, Liblice, CSSR, 183-193

Läuter, J. (1984), Discriminant Analysis under Parameter Restrictions - Statistical and Computational Aspects, Math. Operat. Forsch. Statist., Ser. Statistics (to be published)

Lawley, D.N. and A.E. Maxwell (1963), Factor Analysis as a Statistical Method, Butterworths, London

DiPillo, P.J. (1976), The Application of Bias to Discriminant Analysis, Commun. Statist.-Theor. Meth., A5 (9), 843-854

DiPillo, P.J. (1979), Biased Discriminant Analysis: Evaluation of the Optimum Probability of Misclassification, Commun. Statist.-Theor. Meth., A8(14), 1447-1457

Toussaint, G.T. (1969), Machine Recognition of Independent and Contextually Constrained Coulor-Traced Handprint Characters. M.A. Sc. Thesis, Department of Electr. Eng., Univ. of Brit. Columbia

Constrained Multidimensional Scaling: more directions than Dimensions

J. Meulman, and W.J. Heiser, Leiden

Constrained Multidimensional Scaling is presented as a technique to analyze two sets of data simultaneously, each having a specific role. A generalization of previous methods is discussed, combining ideas from nonlinear Multivariate analysis with a convergent Multidimensional Scaling algorithm.

Keywords: Constrained Multidimensional Scaling, Nonlinear Multivariate Analysis.

1. The modelling process in MDS.

Multidimensional Scaling (MDS) is the name of a class of techniques that construct a p-dimensional spatial model from empirical dissimilarity information on a set of n objects. Each object is represented as a point and each pair of points is assigned a distance. The analysis is called nonmetric if merely the order information on the pairwise dissimilarities is preserved. In most cases external information or theory about the objects is available. The analysis is called unconstrained when this information is used only afterwards in the interpretation of the results. This defines constrained MDS as an analysis that incorporates the external information in the modelling process. Various types of restrictions are possible (see Heiser & Meulman (1983b) for a classification of constraints according to the place in the model). In this paper we concentrate on constraints on the location of the object points and, even more specifically, the external information is contained in a number of external variables that are functions on the given set of objects. The modelling process in constrained nonmetric MDS can be displayed as

$$\delta_{i\ell} \xrightarrow[\text{transformation}]{} \hat{d}_{i\ell} \xrightarrow[\text{fit}]{} d_{i\ell} \xrightarrow[\text{distance function}]{} \hat{Z} \xrightarrow[\text{constraints}]{} Z\varepsilon\Omega$$

The dissimilarities $\delta_{i\ell}$ among the objects will be transformed optimally by a monotone function and mapped in a spatial structure as distance $d_{i\ell}$ between points, whose coordinates are given in the configuration matrix $\hat{Z}$. The Euclidean distances between the rows of $\hat{Z}$ should be approximately equal to the transformed dissimilarities. Ω is the space of all configurations that satisfy the constraints. Departure from perfect fit is measured by a loss function. Before considering the minimization of this loss function over all $Z\varepsilon\Omega$ we regard

$$\sigma(X) = \sum_i \sum_\ell (\hat{d}_{i\ell} - d_{i\ell}(X))^2 = \min_{\delta_{i\ell}\varepsilon\Delta} \sum_i \sum_\ell (\delta_{i\ell} - d_{i\ell}(X))^2 \tag{1}$$

over all $X\varepsilon\mathbb{R}^{np}$. De Leeuw & Heiser (1980) describe an algorithm to minimize this loss function that guarantees a convergent series of configurations. The algorithm can be characterized as repeatedly computing the so-called Guttman transform

$$\bar{X} = \frac{1}{n} B(X)X \tag{2}$$

where the matrix B(X) is defined as

$$b_{i\ell}(X) = \frac{-\hat{d}_{i\ell}}{d_{i\ell}(X)} \text{ if } i \neq \ell;\ b_{ii}(X) = \sum_{\ell \neq i} \frac{\hat{d}_{i\ell}}{d_{i\ell}(X)};\ b_{i\ell}(X) = 0 \text{ if } d_{i\ell}(X) = 0 \tag{3}$$

Incorporating the constraints into (1) displays the basic problem as

$$\min_{Z \varepsilon \Omega} \sigma(Z) = \sum_i \sum_\ell (\hat{d}_{i\ell} - d_{i\ell}(Z))^2 \tag{4}$$

Now the key result of De Leeuw & Heiser (1980) is that the algorithm they propose remains convergent if the unconstrained update $\bar{X}$ is projected onto the configuration space Ω. The metric projection $P_\Omega(X)$ is the configuration that minimizes $tr(X-Z)'(X-Z)$ over all $Z \varepsilon \Omega$. Combining this with (2) gives the constrained update

$$X^+ \varepsilon P_\Omega(\bar{X}) \tag{5}$$

The next sections will consider the metric projection problem for the type of constraints we specifically wish to impose.

2. More directions than dimensions.

Considering the relation between external variables and dissimilarity measures, it is straight-forward to impose the constraints on the coordinate axes of the configuration. This way of handling the problem has been proposed a.o. by Bentler & Weeks (1978), Noma & Johnson (1979) and Heiser & Meulman (1983a). Since the dissimilarities are approximated in a p-dimensional space, with p preferably small, it is implied that the number of external variables m is always less or equal to p. Heiser & Meulman (1983b) try to get round this serious limitation by performing the analysis in m dimensions and by rotating the configuration to its principal axes orientation afterwards, in this way finding possibly non-orthogonal directions in the orthogonal configuration. Such an approach has the drawback of computational inefficiency and the theoretical disadvantage that the dissimilarities are approximated in a higher dimensional space than is actually intended. This is especially important because of the fact that MDS solutions are not nested, i.e. MDS methods find solutions for every p separately.

This paper generalizes an approach in which the dissimilarities are fitted in p-dimensional space, while m may possibly be much larger than p. It amounts to constraining the configuration to be a linear combination of the external variables, i.e. $Z = EB$. This type of restriction has been proposed a.o. by Carroll et al. (1976), Bentler & Weeks (1978), Bloxom (1978), De Leeuw & Heiser (1980). It can be written as

$$\min_B tr(\bar{X} - EB)'(\bar{X} - EB) \tag{6}$$

We call the optimal weights $\overline{B}$ and the result $\overline{Z}$; E are the external variables. Since the aim still remains to display the external variables as directions in the configuration, it is appropriate to realize that problem (6) results in a configuration $\overline{Z}$ *for which there are projections on some directions A that are a better approximation of the external variables than the projections of the configuration $\overline{X}$ on some directions A*. This can be seen from the following argument. Problem (6) results in $\overline{B} = (E'E)^{-1}E'\overline{X}$. When we define the projector $P_E = E(E'E)^{-1}E'$, which is symmetric and idempotent, then $\overline{Z} = E\overline{B}$ can be written as $\overline{Z} = P_E\overline{X}$ and

$$\min_A \operatorname{tr}(\overline{Z}A - E)'(\overline{Z}A - E) = \min_A \operatorname{tr}(P_E\overline{X}A - E)'(P_E\overline{X}A - E) =$$

$$= \min_A \operatorname{tr}(\overline{X}A - E)'P_E(\overline{X}A - E) \leq \min_A \operatorname{tr}(\overline{X}A - E)'(\overline{X}A - E) \quad (7)$$

The generalization mentioned above implies that we wish to find optimal transformations of the external variables. Moreover, it would be desirable to permit different transformations, e.g. monotone transformations for some variables and linear transformations for others. Minimizing (6) over functions of E means that we have to resort to some alternating least squares (ALS) algorithm, which brings us very close to the Gifi system of nonlinear Multivariate analysis (Gifi, 1981; De Leeuw, 1983). The Gifi system comprises an approach to MVA that generalizes well-known MVA techniques like Principal Components analysis and Canonical Correlation analysis, and combines MVA with insights from MDS.

3. Formulating the constraints according to the Gifi system.

The most characteristic element of the Gifi system is that it takes Homogeneity analysis, also known as Multiple Correspondence analysis, as its starting point. Homogeneity analysis deals with n observations on m variables, indexed j, each assuming at most k_j different values. These categories have no a priori order and the aim is to arrive at category quantifications that are invariant under one-to-one nonlinear transformations of the data. This invariance is accomplished by the use of indicator matrices. The indicator matrix G_j defined by variable j is an $n \times k_j$ matrix that indicates in which of the categories an object belongs, so that $g^j_{ik} = 1$ if $e_{ij} = k$ and $g^j_{ik} = 0$ otherwise. From each G_j a diagonal matrix $D_j = G_j'G_j$ can be constructed that contains the marginals. The aim of the technique can be reformulated by stating that we want to find object scores X, a $n \times p$ matrix, and optimal category quantifications Y_j, a $k_j \times p$ matrix, such that $X = G_1Y_1 = \ldots = G_mY_m$. In the perfect case the object scores are perfectly discriminating and the category quantifications are perfectly homogeneous. Departure from perfect fit is measured by the loss function

$$\sigma(X;Y_1\ldots Y_m) = \frac{1}{m}\sum_{j=1}^{m} \operatorname{tr}(X - G_jY_j)'(X - G_jY_j) \quad (8)$$

This loss function can be minimized over X *or* over Y. The optimal Y_j for given X are $Y_j = D_j^{-1}G_j'X$, while the optimal X for given Y_j is $X = \frac{1}{m}\sum G_jY_j$. These two centroid principles embody the *geometrical* emphasis. The category quantifications Y_j are called *multiple* quantifications, one for each dimension. Since Homogeneity analysis only takes into account the fact that some objects are in a certain category and others are in a different one, we call the treatment of variables *nominal* as well.

Starting from Homogeneity analysis, other techniques are developed in the Gifi system by imposing restrictions on the category quantifications. If rank-one restrictions are imposed, i.e. the category quantifications should be on a line, we call them *single*; if in addition the order of the categories should be preserved, we call the variable *ordinal*; and if the intervals between the categories should be maintained, we call it *numerical*. At this point we reintroduce the constraints in problem (6) in the form:

$$\min_{Y_1\ldots Y_m} \; tr(\bar{X} - \sum_{j=1}^{m} G_jY_j)'(\bar{X} - \sum_{j=1}^{m} G_jY_j). \tag{9}$$

We define the auxiliary matrix

$$U_j = \sum_{v\neq j} G_vY_v + \sum_{w\neq j} G_wy_wb_w' , \tag{10}$$

where multiple variables are indexed by v and single variables indexed by w. Then loss for variable j single can be partitioned into

$$tr(\bar{X} - U_j - G_j\bar{Y}_j)'(\bar{X} - U_j - G_j\bar{Y}_j) \;+\; tr(\bar{Y}_j - y_jb_j')'D_j(\bar{Y}_j - y_jb_j') \tag{11}$$

The ALS algorithm to minimize (11) is due to Gifi (1981) and is shown as part of the SMACOF-II algorithm in Table 1. Upon setting $\bar{Z} = \sum G_vY_v + \sum G_wy_wb_w'$, the Gifi system would proceed by orthonormalizing $\bar{Z}$. SMACOF-II proceeds by computing a new Guttman transform.

Table 1. SMACOF-II algorithm including constraints.

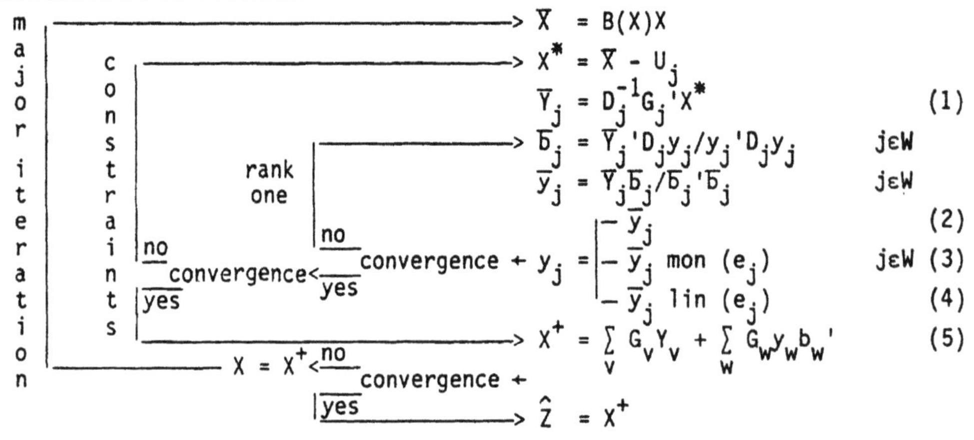

(1) multiple nominal quantification
(2) single nominal quantification
(3) single ordinal quantification by performing monotone regression
(4) single numerical quantification by performing linear regression
(5) subscript v applies to multiple, subscript w to single variables

4. Application: Parliament 1972.

The FORTRAN program SMACOF-II has been used to analyze selected data from an extensive questionnaire study among the members of the Second Chamber of the Dutch Parliament (in 1972). It concerns the *preferences* for the political parties residing in Parliament, and positions with respect to 7 political *issues*. Median ratings have been aggregated within the 11 parties, preserving a left, middle and right faction for the larger parties. From the median preference ratings the chi-squared distances known from Correspondence analysis were computed, which serve as derived dissimilarities in the MDS framework. The median ratings on the political issues were used to constrain the configuration. Both the derived dissimilarities and the issue variables were treated ordinally. The following parties are involved (ordered a priori from left to right): PSP (pacifist-socialist), PPR (radical), PVA (labor), D66 (progressive-liberal), ARP (protestant), KVP (catholic), DS70 (conservative-socialist), CHU (protestant), VVD (conservative-liberal), GPV (calvinist), SGP (calvinist). The political issues deal with: 1. Development aid, 2. Abortion, 3. Law & order, 4. Income differences, 5. Participation, 6. Taxes, 7. Western armies.

In Figure 1. the two-dimensional configuration is shown in principal axes orientation. The fit of the solution is very good, with $\sigma = .032$, from which we infer that preferences and opinions are closely related and can be fitted in some low dimensional space. Notice that the configuration of party points (labelled as indicated above) is rather flat, thus displaying a dominant first dimension, accounting for 92% of the variance. It can easily be identified as the political left-right contrast. The issues have been plotted as directions (dotted lines), with the extreme points of view indicated; the numbers refer to the descriptions given above. Three issues are evidently not consistent with the left-right contrast. These are "Income dif-

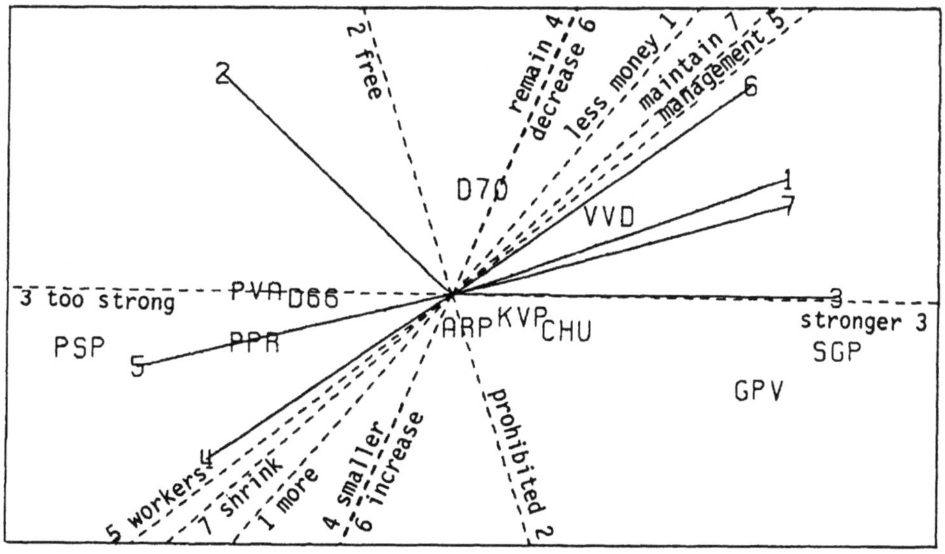

Figure 1. Joint party-issue space (explanation in the text).

ferences" and "Taxes" on the one hand, the VVD and DS70 being more extreme on these issues than the right-wing calvinist parties GPV and SGP, and "Abortion" on the other hand, which issue discriminates perfectly between the denominational parties and the non-denominational ones. "Law & order" coincides almost completely with the left-right dimension. This also applies, to a somewhat lesser extent, to "Participation", "Armies", and "Development aid".

It is important to notice that the *directions* in the configuration space cannot be interpreted as *correlations*. Unlike most MVA techniques, MDS does not have to ortho*normalize* the object configuration, as mentioned in section 3. Therefore the issues are also plotted as correlations with the axes of the configuration (solid lines), again displaying the shape of the solution.

5. Concluding remarks.

By incorporating ideas from the Gifi system we have arrived at a very general class of constraints for use in MDS models. External variables can be represented as directions in the configuration, preserving their numerical, ordinal or nominal information. In addition, some variables can be represented as a collection of p-dimensional category points. The convergent algorithm described works smoothly; future research will need to address the issue of acceleration as well.

6. References.

Bentler, P.M. & Weeks, D.G. (1978), Restricted Multidimensional Scaling models, Journal of Mathematical Psychology, *17*, 138-151.

Bloxom, B. (1978), Constrained Multidimensional Scaling in N spaces, Psychometrika, *43*, 397-408.

Carroll, J.D., Green, P.E. & Carmone, F.J. (1976), CANDELINC (CANonical DEcomposition with LINear Constraints), Paper presented at the American Psychological Association meeting, San Francisco.

De Leeuw, J. (1983), The Gifi system of nonlinear Multivariate Analysis, In: E. Diday et al. (Eds.), Data Analysis and Informatics, North-Holland Publishing Company, Amsterdam.

De Leeuw, J. & Heiser, W.J. (1980), Multidimensional Scaling with restrictions on the configuration, In: P.R. Krisnaiah (Ed.), Multivariate Analysis, Vol. 5, North-Holland Publishing Company, Amsterdam.

Gifi, A. (1981), Nonlinear Multivariate Analysis, Department of Data Theory, Leiden.

Heiser, W.J. & Meulman, J. (1983a), Analyzing rectangular tables by joint and constrained Multidimensional Scaling, Journal of Econometrics, *22*, 139-167.

Heiser, W.J. & Meulman, J. (1983b), Constrained Multidimensional Scaling, including confirmation, Applied Psychological Measurement, *7*, 381-404.

Noma, E. & Johnson, J. (1979), Constrained nonmetric Multidimensional Scaling, Technical Report MMPP 1979-4, Univ. of Michigan Psychology Program, Ann Arbor.

Acknowledgements.

Comments by Jan de Leeuw on a earlier version of the manuscript have been very useful. The data of the Parliament Study were made available to us by the Department of Political Science at Leiden University.

A Review of Fast Techniques for Nearest Neighbour Searching

F. Murtagh, Ispra

ABSTRACT

The nearest neighbour or closest point problem is of importance in many areas. Recent research in hierarchical clustering has shown how efficient algorithms for this problem may be incorporated into the clustering algorithms, leading to a more efficient implementation of the latter. In this review, a wide range of nearest neighbour searching algorithms are briefly described, and their chief properties for use in clustering algorithms appraised.

INDEX TERMS: algorithms, computational complexity, hierarchical clustering, applications.

1. INTRODUCTION

Given n points in $\mathbb{R}^m$, the nearest neighbour (NN) problem is the finding of the closest point to a given point. In the best match problem we may alternatively wish to determine which of the n points is closest to a new point. A brute-force approach to these problems may be carried out by determining the dissimilarity between the given vector and each of the n-1 (or n) vectors to be searched through. Most commonly-used dissimilarities require O(m) operations, so that the NN of a point is obtained in O(nm) time.

The algorithms reviewed in this article either carry out a preprocessing of the n vectors to be searched, in order to reduce the O(n) search time requirement; or they make use of a bounding approach which leaves unaltered the O(n) requirement but reduces the constant of proportionality. The former approach is particularly satisfying if a reasonable preprocessing computational requirement (e.g. O(n log n)) brings about fast search times (e.g. O(log n)). However this is often done at the cost of an exponential term in m. The 'curse of dimensionality' arises: the algorithms work very well in low-dimensional spaces, but become impractical as the dimensionality increases.

The NN problem is important in many areas. In information retrieval and in database design, the best match of a query is required (see Perry and Willett, 1983; Bentley, 1979). Recent research in cluster analysis has shown how many of the widely-used hierarchical methods may be implemented by repeatedly finding NNs (see Murtagh, 1983a; 1984). Incorporation of efficient NN-finding algorithms can lead to significant computational gains over currently-implemented algorithms, while producing the same exact output as before.

In the clustering area, the dimensionality of the data can be of crucial importance as to whether or not a particular NN-finding algorithm can be fruitfully employed. In the following short review of practical approaches, NN algorithms will be approximately ordered according to the dimensions of data on which they have been used, ranging from a dimensionality of 2 to many thousands. This will indicate, in broad terms, the range of dimensions for which the NN-finding algorithms can be profitably applied.

It will be seen that the most important open problem in this field is to have available generally-applicable strategies for NN-searching in high dimensional spaces.

2. HASHING

The multidimensional space is partitioned into hypercubes of side-length p. Point x_i is mapped onto cell $(\lfloor x_{i1}/p \rfloor, \lfloor x_{i2}/p \rfloor, \ldots, \lfloor x_{im}/p \rfloor)$. This mapping of all n points takes O(nm) time. Each cell is identified by an n-tuple of integers, and the index numbers of all points mapped onto a cell may be stored in a linked list. The choice of p, determining cell size, should be such that a small number of points is assigned to each cell.

The search for the nearest neighbour of point x_i proceeds by searching for the closest point which is mapped onto the same cell. If no such point exists, or if a closer point could possibly lie outside the cell (checked for by determining the distance between point x_i and the closest cell wall), the distances to points in all 3^m-1 adjacent cells are determined. If the current NN could be bettered by a point in an adjacent cell to these, then the next layer of cells must be examined.

The number of adjacent cells evidently increases exponentially with dimensionality, and so this approach is attractive for low dimensions only. Rohlf (1978) reports work in dimensions 2, 3 and 4, and Murtagh (1983b) in $\mathbb{R}^2$. Rohlf also mentions the use of the first 3 principal components to approximate a set of points in $\mathbb{R}^{15}$.

A powerful theoretical result regarding this approach is as follows: for uniformly distributed points, the NN of a point is found in O(1) expected time (Delannoy, 1980; Bentley et al., 1980). This approach is therefore best used when approximate uniformity can be assumed.

3. MULTIDIMENSIONAL BINARY SEARCH TREE

One variant of a MDBST is as follows: Halve the set of points, using the median of one of the points' coordinates. Continue halving the resulting sets of points, using the medians of all coordinates in succession in order to define a hierarchical decomposition of the set of points. When all coordinates have been exhausted, recycle through the set of coordinates. Halt when the number of points associated with the nodes at some level are smaller than or equal to a prespecified constant (Friedman et al. (1977) suggest a value between 4 and 32 based on empirical study). Clearly the tree is kept balanced: therefore there are O(log n) levels, at each of which O(n) processing is required, so that the construction of the tree takes O(n log n) time.

The search for a NN proceeds by a top-down traversal of the tree. On arrival at a terminal node, all associated points are examined and a current NN selected. The tree is then traversed in the opposite direction to the downward traversal: if the points associated with any node could furnish a closer point then such points must be examined.

All variables need not be considered for splitting the data if it is known that some are of greater interest than others. Choosing the variable with the greatest variance as the discriminator variable at each node may therefore allow repeated use of certain variables. This has the added effect that the hyperrectangular cells which the terminal nodes divide the space into will be approximately cubical in shape. In this case Friedman et al. (1977) show that search time is O(log n) on average for the finding of a NN. Results obtained for dimensionalities of between 2 and 8 are reported on in Friedman et al. (1977) and Bentley and Friedman (1978). Zolnowsky (1978) gives an expression of $O((\log n)^m)$ for m-dimensional space.

The batching together of coordinates could allow the MDBST approach to be employed in high-dimensional spaces. This is attractive in an information retrieval environment where the data is binary: a point with zero coordinates on a certain number of variables is directed towards the right subtree, and otherwise is directed towards the left subtree. The variables which define the subdivision at each level ought to be related, and to be such that the decomposition tree is kept balanced. These general guidelines are given by Weiss (1981) and Eastman and Weiss (1982). Results reported on for 1400 points (documents) and 5000 variables (terms) led to only 10% reduction in distance calculations, and it is suggested that larger n might be more fruitful.

4. BOUNDS USING PROJECTIONS

All remaining algorithms to be discussed make use of bounds: some lower bound on dissimilarity is efficiently calculated in order to dispense with the full calculation in many instances. Such an approach requires O(n) calculations for the finding of each NN, but the lowering of the constant of proportionality can substantially improve on brute-force searching.

Using projections on a coordinate-axis allows the exclusion of points in the search for the NN of point x_i. Points x_j, only, are considered such that $(x_{ik} - x_{jk})^2 \leq c$, where $x_{ik}(x_{jk})$ is the kth coordinate of $x_i(x_j)$ and c is a prespecified constant. Alternatively more than one coordinate may be used. The prior sorting of coordinate values on the chosen axis or axes will expedite the finding of points whose full distance calculation is necessitated.

The preprocessing required with this approach involves the sorting of up to m sets of coordinates, i.e. $O(mn \log n)$ time. Friedman et al. (1975) show that expected NN computation time, where uniformly distributed points are assumed, is $O(mn^{1-1/m})$. This approaches the brute-force $O(nm)$ as n gets large. Reported empirical results are for dimensions 2 to 8.

Marimont and Shapiro (1979) extend this approach by the use of projections in subspaces of dimension greater than 1 (usually about m/2 is suggested). This car be further improved if the subspace of the first principal components is used. Dimensions up to 40 are examined.

5. BOUNDS ON THE EUCLIDEAN DISTANCE

The Euclidean distance is most widely used in practice. For any two points, x and x', we have

$$d_1(x,x') \geq d_2(x,x') \geq d_\infty(x,x')$$

where the 'city block' distance, d_1, is given by $\sum_j |x_j - x_j'|$ (for coordinates j); d_2 is given by the square root of $\sum_j (x_j - x_j')^2$; and the Chebyshev distance is given by $\max_j |x_j - x_j'|$.

Kittler (1978) makes use of the more efficiently calculated d_1 distance in order to reject all points y such that $d_1(x,y) \geq \sqrt{m}\delta$ where x is the point whose NN is required, and δ is the current NN d_2-distance. In 10-dimensional space, 90% of points were rejected.

Yunck (1976) presents a theoretical analysis for the similar use of the Chebyshev metric: reject y such that $d_\infty(x,y) \geq \delta$. Richetin et al. (1980) propose the use of both bounds. For dimensions 2 to 5, they report up to 80% reduction in CPU time when the rejection rule based on the d_∞ metric precedes the rule based on the d_1 distance.

6. BOUNDS FOR DISTANCES

The triangular inequality is satisfied by distances and gives rise to the following rejection rule. Using a reference point, z, the full distance calculation between x (whose NN is sought) and y is obviated if

$$|d(y,z) - d(x,z)| \geq \delta$$

where δ is the current NN distance. Both d(y,z), for all $y \neq z$, and d(x,z) are calculated in a preprocessing stage.

The set of distances to a reference point may be sorted in the preprocessing stage so that the above rejection rule will not need to be applied in many cases.

The single reference point approach, due to Burkhard and Keller (1973) was generalized to multiple reference points by Shapiro (1977). The first reference point is used as a preliminary bound, and distances to further reference points (calculated in the preprocessing stage) further reduce the number of points whose full distance requires calculation. In 10-dimensional simulations, Shapiro found that at best 20% of full distance calculations were required although this was very dependent on the choice of reference points.

Fukunaga and Narendra (1975) make use of both a hierarchical decomposition of the dataset (they employ repeatedly the k-means iterative/relocatory algorithm), and bounds based on the triangular inequality. For each node in the decomposition tree, the centre and maximum distance to the centre (the 'radius') of associated points are determined. All points associated with a non-terminal node can be rejected in the search for the NN of point x if

$$d(x,g) - r_g \geq \delta$$

where δ is the current NN distance, g is the centre of the cluster of points associated with the node, and r_g is the radius of this cluster.

For a terminal node, which cannot be rejected on the basis of this rule, each associated point, y, can be tested for rejection using the following rule:

$$d(x,g) - d(y,g) \geq \delta$$

The previous reference details an algorithm for the traversal of the decomposition tree which allows an estimate of the NN to be obtained; in backtracking up through the tree each node may be tested using the above two rules to see if offspring nodes must also be checked out.

7. NN-SEARCHING IN INFORMATION RETRIEVAL

In information retrieval, m may typically be of the same order of magnitude as n or greater. Values in excess of 10,000 are not uncommon for test collections. The particular nature of the data (very sparse binary document-term associations) has allowed efficient NN-searching. Such data would be stored as compact linked lists (the sequence number of the document is followed by the sequence numbers of the terms which are associated with it). The inverted file (mapping terms onto documents) is often available also.

Using data in this form stored on direct access devices, a sequence of algorithms proposed by Croft (1977), Smeaton and van Rijsbergen (1981), and Murtagh (1982) have successively improved upon two basic ideas: using the inverted file to indicate only those documents which share at least one term with the document whose NN is wanted, thus cutting out consideration of many documents; and since this strategy is implemented by processing successive terms associated with the document whose NN is wanted, the number of associated terms remaining can be readily made available and this information can be used to provide a lower bound on the dissimilarity or distance. Murtagh (1982) reports that less than 7% of full distances required calculation when finding NNs; and when queries were used (with fewer terms associated), this was improved to less than 1%.

Since reading information from direct access peripheral storage is often the largest cost in handling large amounts of data, Perry and Willett (1983) describe a powerful approach which economises as much as possible in this area. Most commonly used dissimilarities are functions of the total number of terms associated with the documents, and the number of terms common to both. It is the latter whose calculation presents the greatest problem. The approach proposed is, as before, to access all terms associated with the document whose NN is required; and to use the inverted file to access documents associated with each term so obtained. A counter is incremented, for each document, to give the number of terms shared by the document whose NN is wanted, and every document in the collection. Since the data is sparse, we obtain in this manner the number of shared terms with computation time O(nm), but with a very low constant of proportionality.

REFERENCES

1. J.L. Bentley (1979), Multidimensional binary search trees in database applications, IEEE Transactions on Software Engineering, SE-5, 333-340.
2. J.L. Bentley and J.H. Friedman (1978), Fast algorithms for constructing minimal spanning trees in coordinate spaces, IEEE Transactions on Computers, C-27, 97-105.
3. J.L. Bentley, B.W. Weide and A.C. Yao (1980), Optimal expected time algorithms for closest point problems, ACM Transactions on Mathematical Software, 6, 563-580.
4. W.A. Burkhard and R.M. Keller (1973), Some approaches to best-match file searching, Communications of the ACM, 16, 230-236.
5. W.B. Croft (1977), Clustering large files of documents using the single-link method, J. of the American Society for Information Science, 28, 341-344.
6. C. Delannoy (1980), Un algorithme rapide de recherche de plus proches

voisins, RAIRO Informatique/Computer Science, 14, 275-286.

7. C.M. Eastman and S.F. Weiss (1982), Tree structures for high dimensionality nearest neighbour searching, Information Systems, 7, 115-122.

8. J.H. Friedman, F. Baskett, L.J. Shustek (1975), An algorithm for finding nearest neighbours, IEEE Transactions on Computers, C-24, 1000-1006.

9. J.H. Friedman, J.L. Bentley, R.A. Finkel (1977), An algorithm for finding best matches in logarithmic expected time, ACM Transactions of Mathematical Software, 3, 209-226.

10. K. Fukunaga and P.M. Narendra (1975), A branch and bound algorithm for computing k-nearest neighbours, IEEE Transactions on Computers, C-24, 750-753.

11. J. Kittler (1978), A method for determining k-nearest neighbours, Kybernetes, 7, 313-315.

12. R.B. Marimont and M.B. Shapiro (1979), Nearest neighbour searches and the curse of dimensionality, J. Inst. Maths. Applics., 24, 59-70.

13. F. Murtagh (1982), A very fast, exact nearest neighbour algorithm for use in information retrieval, Information Technology, 1, 275-283.

14. F. Murtagh (1983a), A survey of recent advances in hierarchic clustering algorithms, The Computer Journal, 26, 354-359.

15. F. Murtagh (1983b), Expected time complexity results for hierarchic clustering algorithms which use cluster centres, Information Processing Letters, 16, 237-241.

16. F. Murtagh (1984), Complexities of hierarchic clustering algorithms: State of the art, Computational Statistics Quarterly (submitted).

17. S.A. Perry and P. Willett (1983), A review of the use of inverted files for best match searching in information retrieval systems, J. of Information Science, 6, 59-66.

18. M. Richetin, G. Rives and M. Naranjo (1980), Algorithme rapide pour la détermination des k plus proches voisins, RAIRO Informatique/Computer Science, 14, 369-378.

19. F.J. Rohlf (1978), A probabilistic minimum spanning tree algorithm, Information Processing Letters, 7, 44-48.

20. M. Shapiro (1977), The choice of reference points in best-match file searching, Communications of the ACM, 20, 330-343.

21. A.F. Smeaton and C.J. van Rijsbergen (1981), The nearest neighbour problem in information retrieval: An algorithm using upperbounds, ACM SIGIR Forum, 16, 83-87.

22. S.F. Weiss (1981), A probabilistic algorithm for nearest neighbour searching, in: R.N. Oddy et al. (eds.), Information Retrieval Research, Butterworths, London, 325-333.

23. T.P. Yunck (1976), A technique to identify nearest neighbours, IEEE Transactions on Systems, Man and Cybernetics, SMC-6, 678-683.

24. J. Zolnowsky (1978), Topics in computational geometry, Report STAN-CS-78-659, Department of Computer Science, Stanford University.

A Study of Selection Criteria for Constructing Identification Keys

R.W. Payne, and T.J. Dixon, Harpenden

Summary
Selection criteria used in heuristic algorithms for key construction were compared using simulations based on real and artificial data. Similar results were obtained from each type of data. These showed some criteria to be better than others but, nevertheless, each of the criteria produced the best key at least some of the time. Programs for key construction should thus offer users a choice of criteria.

Keywords: Diagnostic Keys, Identification Keys, Selection Criteria, Test Selection

1. Introduction

An identification (or diagnostic) key is a device for identifying a specimen from a known set of taxa (for example species of plant, strains of bacteria, types of machine fault etc.) by applying tests sequentially in a hierarchical manner.

It can thus be represented as a tree, with one test at each branch point, and with the identified taxa at the endpoints of the branches.

2. Notation

Suppose that there are N taxa in the key and that the prior probability of a specimen being from taxon j is α_j. Let m_i be the number of different results (denoted by $1,2,\ldots,m_i$) obtained from test i, c_i be the cost of test i and β_{ijk} be the probability of obtaining result k to test i with a specimen of taxon j. Taxon j is said to have a fixed response f to test i if $\beta_{ijk}= 1$ for $k = f$, $= 0$ for $k \neq f$ (i.e. members of taxon j always give result f to test i); otherwise (if more than one result can occur) the response is said to be variable. The efficiency of a key is usually assessed by its expected cost per identification:

$$\kappa = \sum_{e=1}^{d} \{ (\sum_{(i,k)\in R_e} c_i) \times (\sum_{j=1}^{N} \alpha_j \prod_{(i,k)\in R_e} \beta_{ijk}) \}$$

where d is the number of endpoints in the key, and R_e is the set of pairs of indices of the test results on the branch to endpoint e.

3. Key construction

If all conceivable tests are available (i.e. if tests are available with all possible patterns of results over the taxa) algorithms from Coding Theory, like that of Huffman (1952), enable the optimal key to be constructed. However, in most practical applications, a limited set of tests is available and the only algorithms that guarantee to find an optimal key essentially enumerate all possible keys for the tests and taxa under consideration. This is impracticable for even moderate numbers of tests and taxa (Garey 1972a).

Consequently heuristic methods have been devised. These construct the key sequentially, selecting first the test that "best" divides the taxa into sets (set k for test i contains the taxa that can give the k'th result of the test - i.e. those taxa j with $\beta_{ijk} > 0$) then selecting the best test to use with each set, and so on, until the sets each contain only one taxon. The "best" test is usually chosen by some selection criterion function (Payne & Preece, 1980).

Many early functions were based on purely ad-hoc principles, for example that the best test should be one that divides the taxa into sets of nearly equal size or that it should have few variable responses. Of course it may not be possible to satisfy both these principles simultaneously and so different criteria will select different tests, depending on the way in which they weight these (and other) requirements. Few ad-hoc criteria extend readily to tests with more than two different results nor do they handle tests with unequal costs in a satisfactory way (Gower & Payne, 1975). An example is the function OT^2 of Gower & Barnett (1971), which is defined for binary tests as

$$(OT^2)_i = (p_{i1} - 1/2)^2 + (p_{i2} - 1/2)^2 + r_i^2$$

where p_{ik} is the proportion of the taxa in the current set with fixed response k to test i, and

r_i is the proportion with variable responses.

A better strategy was suggested by Dallwitz (1974) whose function, K, represents the expected cost of completing the identification of a specimen, starting at the current point of the key; i.e. (assuming that test i is used next) it has the form:

$$c_i + \sum_{k=1}^{m_i} \{\text{probability of result } k\} \times \{\text{estimate of the cost of completing the identification if result } k \text{ is obtained.}\}$$

For K, Dallwitz assumed that the key would be completed by a subkey with branches of equal length, containing tests of minimum cost and no variable responses.

$$K_i = c_i + c' \sum_{k=1}^{m_i} q_{ik} \log_2 n_{ik} \quad [\text{ + an optional term to bias against tests with variable responses }]$$

where c' is the minimum cost of the tests,

n_{ik} is the number of taxa, in the current set, that can give result k to test i, and

q_{ik} is the probability of obtaining result k to test i.

Assuming that taxa with variable responses are equally likely to give any of the different results, then $q_{ik} = p_{ik} + r_i / m_i$.

Payne (1981) derived several similar functions, including

$$(CM_e)_i = c_i + (\bar{c}/\log \bar{m}) \{ \sum_{k=1}^{m_i} (p_{ik} + r_i/ m_i)\log(p_{ik} + r_i/ m_i) + r_i \log m_i \}$$

$$(CM_v)_i = c_i + (\bar{c}n/2) \{ - \sum_{k=1}^{m_i} (p_{ik} + r_i/ m_i)(1 - r_i - p_{ik}) \}$$

where $\bar{c}$ is the average cost of the available tests,

$\bar{m}$ is their average number of different results, and

n is the number of taxa in the current set.

For CM_e, it is assumed that identification will be completed in an optimal way; whereas for CM_v, identification is assumed to be completed by simple binary tests (Garey 1972b) - tests with two possible results, one of which is given by only one of the taxa and for which no taxon is variable. CM_v contains an approximation to avoid calculating both p_{ik} and n_{ik}; the exact form is

$$(CM_v') = c_i + (\bar{c}n/2)\left\{\sum_{k=1}^{m_i}(p_{ik}+r_i/m_i)(n_{ik}^2+n_{ik}-2)/2n_{ik}\right\}$$

Based on their derivations, Payne (1981) recommended that, if a reasonable number of "good" tests are available, CM_e should be used, otherwise (if the tests are mainly "poor") use CM_v. In intermediate situations, K might be appropriate.

4. Study of the selection criteria in practice

It is fairly easy to simulate sets of data for assessing the criteria but it is unclear how relevant the conclusions would be to real life. Consequently simulations were done using both real and artifical data.

The first study used data from Barnett, Payne & Yarrow (1983) concerning 469 species of yeast, the largest suitable set of real data. For the yeasts there were 97 binary tests, all of equal cost, concerning the ability of the yeasts to grow in particular conditions, as well as aspects of their appearance and reproduction; 4 tests involving ascospores were omitted, making 93 tests in all. Keys were constructed, using each of the 5 criteria described above, for 25 randomly selected sets of 50 and of 100 yeasts.

In the second study, artificial data were used. For comparability with the first study, the sets of data for each set of 5 keys again contained 93 binary tests of equal cost and were for 50 and 100 taxa. The responses to each test were generated from uniform random numbers, determining first whether or not the response should be variable or fixed then, if fixed, whether it should be the first or second of the two available results; 25 sets of tests were generated for the 9 factorial combinations of 3 different probabilities of variable responses (0.1, 0.3, 0.5) by 3 different values of the ratio $p_{i1}:p_{i2}$ (1:1, 1:3, 1:9).

Finally, the third study used mixtures of tests, 10 from each of the 9 combinations in study 2, as well as 3 tests with no variable responses, one for each value of $p_{i1}:p_{i2}$. Again, 25 sets of tests were generated for each number of taxa (50 and 100). In all these studies, the prior probabilities α_i were assumed to be equal, as were the probabilities β_{ijk} for variable responses.

The random data and the selections of yeasts were produced by the random number generator and data handling facilities of the statistical package Genstat. For each set of data, keys were constructed by the program Genkey (Payne 1975, 1978) and the expected costs per identification, κ, were recorded. These were then analysed by Genstat. For each study, the expected costs were transformed to logarithms (base 10) and an analysis of variance was done, as a split plot with keys from the 5 criteria nested within data sets (Tables 1, 3 and 5). Also, for the 25 replicate data sets formed for each factor combination, the smallest cost achieved by any of the criteria was determined and the number of sets (out of 25) for which each criterion achieved that smallest cost outright or with other criteria was tabulated (Tables 2, 4 and 6).

In study 2 there were problems in that, for 100 taxa, the tests generated with the probability of variable responses = 0.5 and the ratio $p_{i1}:p_{i2}$ = 1:9 were unable to distinguish all the taxa. Some endpoints of the keys thus contained several taxa and the expected costs per identification of these incomplete keys were <u>less</u> than those from some complete keys generated with more efficient tests, causing difficulties of interpretation. Thus this combination was omitted.

5. Results

In all the analyses, differences between criteria were detected. In studies 1 and 3, these arose because the ad-hoc criterion OT^2 produced keys with higher expected cost than the other theoretically-based criteria; however this difference decreased (on a log scale) with increasing numbers of taxa (Tables 1 and 5).

In study 2, there were differences also between the theoretically-based criteria, as well as interactions with the numbers of taxa, the probability of variable responses and the ratio $p_{i1}:p_{i2}$ (Table 3); this pattern is also reflected in Table 4. This gives some support to the idea that the derivation of a criterion can be used to predict when it will be most useful; for example, with a probability of variable responses of 0.1 and $p_{i1}:p_{i2}$ = 1:9, one would expect identification to be completed with tests similar to simple binary tests - and CM_v and CM'_v do indeed seem best. However the results are far from clear cut.

Table 1 Random selections of yeasts: mean log10 (expected cost of identification)

	Criteria				
No. yeasts	$OT^2{}_e$	CM_v	CM_v	CM'	K
50	.77470	.77023	.77033	.77026	.77048
100	.85275	.84646	.84684	.84541	.84612

Standard error for differences of means in the same row .000615;
for means in different rows .001598.

Table 2 Random selections of yeasts: Number of times each criterion gives the best/equal best keys

	No. times best key produced (Criteria)					/	equal best key produced (Criteria)				
No. yeasts	OT^2	CM_e	CM_v	CM'_v	K		OT^2	CM_e	CM_v	CM'_v	K
50	3	1	1	5	4		3	10	9	13	12
100	0	5	2	10	7		0	6	3	10	7

Table 3 Tests with randomly generated results: mean log10 (expected cost of identification)

Prob. variable	Ratio $p_{i1}:p_{i2}$	No. taxa	Criteria OT^2	CM_e	CM_v	CM'_v	K
0.1	1:1	50	.75982	.75837	.75837	.75837	.75837
		100	.83240	.83089	.83102	.83097	.83077
	1:3	50	.77174	.76854	.76878	.76860	.76857
		100	.84896	.84657	.84669	.84642	.84658
	1:9	50	.83302	.82989	.82884	.82983	.83171
		100	.92303	.92136	.92038	.92174	.92363
0.3	1:1	50	.79095	.78814	.78859	.78777	.78776
		100	.87529	.87317	.87304	.87216	.87153
	1:3	50	.81440	.81074	.81128	.81100	.81091
		100	.89991	.89728	.89804	.89733	.89748
	1:9	50	.88329	.87835	.87790	.87839	.88267
		100	.97684	.97426	.97335	.97644	.98944
0.5	1:1	50	.87326	.87035	.87029	.86319	.86424
		100	.97478	.97131	.97189	.96264	.96321
	1:3	50	.90338	.89857	.89832	.89331	.89477
		100	1.00541	1.00163	1.00129	.99559	.99606
	1:9	50	.98366	.97960	.97807	.96618	.96761

Standard error for differences of means in the same row .000660;
for means in different rows .001901.

Table 4 Tests with randomly generated results: Number of times each criterion gives the best/equal best keys

Prob. variable	Ratio $p_{i1}:p_{i2}$	No. taxa	No. times best key produced: Criteria OT^2	CM_e	CM_v	CM'_v	K	/ equal best key produced: Criteria OT^2	CM_e	CM_v	CM'_v	K
0.1	1:1	50	0	0	0	0	0	10	25	25	25	25
		100	0	0	0	0	6	4	19	15	16	24
	1:3	50	0	1	1	2	5	1	14	14	15	13
		100	0	1	4	7	8	0	5	7	8	10
	1:9	50	2	3	10	6	2	2	4	12	7	3
		100	2	6	13	2	1	2	7	14	2	1
0.3	1:1	50	1	3	1	5	6	1	7	5	10	11
		100	0	2	3	3	15	0	2	3	5	17
	1:3	50	1	8	2	7	6	1	9	3	7	6
		100	1	6	4	8	5	1	6	4	9	6
	1:9	50	0	8	9	8	0	0	8	9	8	0
		100	1	11	11	2	0	1	11	11	2	0
0.5	1:1	50	0	1	1	15	8	0	1	1	15	8
		100	0	0	0	15	10	0	0	0	15	10
	1:3	50	0	2	1	16	6	0	2	1	16	6
		100	0	0	1	17	7	0	0	1	17	7
	1:9	50	0	0	0	14	11	0	0	0	14	11

Table 5 Mixtures of tests with randomly generated responses: mean log10 (expected cost of identification)

No. taxa	Criteria OT^2	CM_e	CM_v	CM'_v	K
50	.76579	.76200	.76231	.76210	.76213
100	.83831	.83524	.83538	.83526	.83540

Standard error for differences of means in the same row .000319;
for means in different rows .000781.

Table 6 Mixtures of tests with randomly generated responses: Number of times each criterion gives the best/equal best keys

	No. times best key produced					/	equal best key produced				
	Criteria										
No. taxa	OT^2	CM_e	CM_v	CM'_v	K		OT^2	CM_e	CM_v	CM'_v	K
50	0	1	0	0	1		0	20	17	19	20
100	0	3	2	2	6		0	13	8	10	15

6. Conclusion

Overall, the general impression is that the best key, out of the 5 constructed for each set of tests, was produced most often by CM'_v, followed by K and then CM_e and CM_v (Tables 2, 4 and 6). However, all of the criteria constructed keys better than those of the other criteria at least some of the time. Consequently, there are two main conclusions. Firstly, the similarities of Tables 1 and 2 with Tables 5 and 6 show that artificial data can be useful in the study of selection criteria. Secondly, as no one criterion produced the best key all of the time, programs for constructing identification keys should offer a choice of criteria so that users may try several with their data. - The differences in log (expected cost of identification) in Tables 1, 3 and 5 may seem fairly small. Nevertheless, the maximum difference in expected cost of the keys constructed for any set of tests was .69 of a test; if the keys are designed for repeated used, such savings can be important.

References

Barnett, J.A., Payne, R.W. & Yarrow, D. (1983). Yeasts: Characteristics and identification. Cambridge: Cambridge University Press.

Dallwitz,M.J. (1974). A flexible computer program for generating identification keys. Systematic Zoology 27, 50-57.

Garey, M.R. Optimal binary identification procedures, SIAM J. Appl. Math. 23 (1972) 173-186.

Garey, M.R. Simple binary identification problems, I.E.E.E. Trans. Computers C-21 (1972) 588-590.

Gower, J.C. & Barnett, J.A. (1971). Selecting tests in daignostic keys with unknown responses. Nature 232, 491-493.

Gower, J.C. & Payne, R.W. (1975). A comparison of different criteria for selecting binary tests in diagnostic keys. Biometrika, 62, 665-671.

Huffman, D.A. A method of construction of minimum-redundancy codes, Proc. I.R.E. 40 (1952) 1098-1101.

Payne,R.W. (1975). Genkey: a program for constructing diagnostic keys. In Biological Identification with Computers, pp. 65-72. Edited by R.J. Pankhurst. London: Academic Press.

Payne,R.W. (1978). Genkey: a program for constructing and printing identification keys and diagnostic tables. Rothamsted Experimental Station.

Payne, R.W. & Preece, D.A. (1980). Identification Keys and Diagnostic Tables: A Review (with discussion). Journal of the Royal Statistical Society, Series A, 143, 253-292.

Payne, R.W. (1981). Selection criteria for the construction of efficient diagnostic keys. Journal of Statistical Planning and Inference, 5, 27-36.

"Barycenter" of a Set of Probability Measures and its Application in Statistical Decision

A. Perez, Praha

After introducing the concept of "barycenter" of a set S of probability measures as a probability measure from a given projection family minimizing its maximal discrepancy with respect to the probability measures of the set S, we give an algorithm for constructing the barycenter in some important cases. The barycenter concept plays an important role in statistical decision, namely in minimal-discrepancy estimating and its improvement if more estimates are available, in determining discrimination rates and LFP of distributions in hypothesis testing, in approximating a multidimensional distribution by a set of its marginals.

Keywords: barycenter of a set of probability measures, discrepancy, f-divergence, minimal-discrepancy estimates, testing statistical hypotheses, discrimination rate, least favorable pair of distributions, unfitted decision procedures.

1. Introduction. The intuitive idea which lead us to introduce the concept of "barycenter" of a set, S, of probability measures (p.m.), P, defined on the same measurable space $(X,\mathfrak{X})$ arises from the need to represent as well as possible this set by only one p.m. we call the *barycenter* of S. For this purpose one may proceed as follows: choose a suitable discrepancy measure, D, of a p.m. with respect to a second one and then try to construct on $(X,\mathfrak{X})$ such a p.m. that its maximal discrepancy with respect to the p.m. of the set S is minimal, whereas there may be present some additional constraints so that the "projection family", T, from which the barycenter must be choosen will be a subset of M, the set of all the p.m. on $(X,\mathfrak{X})$. In such a case we shall speak of a *constrained barycenter* of S on T.

The constrained barycenter concept for a one-element $S = \{P\}$ and a projection family T reduces to some $R \in T$ minimizing $D(R,P)$. This approach in the case D is of the general *relative f-entropy* (*f-divergence* of Csiszár) type was first introduced in Perez (1967) and applied for risk estimating in statistical decision. The same approach, if applied in the case P is an "empirical" version of the actual p.m., leads to the concept of minimal-discrepancy (-distance, -divergence) statistical estimate.

For a two-element S, the barycenter concept is implied in determining the discrimination rate of simple statistical hypotheses by a maximum likelihood (ML) decision procedure adapted to the actual hypotheses or to some deviated ones; cf. Perez (1983a). Similarly, it is implied in the discrimination rate analysis of composite statistical hypotheses as in Perez (1983b) and serves as a powerfull tool for obtaining LFP (:least favorable pairs) of distributions.

For a two-or more-element S, we shall use the barycenter concept for improving our approximation (estimate) of a p.m. in the case we have at our disposal more than one original approximations (estimates) of it. Here belongs the case of approximating (estimating) an unknown multidimensional probability distribution on the base of some (estimates of) less dimensional marginals permitting to construct a set S of (estimates of) dependence structure simplifications of the original distribution as in Perez (1977).

2. Barycenter of a set of probability measures. Let M be the set of all the p.m. on $(X,\tilde{X})$ and let S and T be two subsets of M. By (minimax) D-*barycenter* of S constrained on (the projection family) T we understand (provided that it exists) a p.m. $_T R_S^D \in T$ defined by

(1) $$\sup_{P \in S} D(_T R_S^D, P) := \min_{R \in T} \sup_{P \in S} D(R, P),$$

where $D(R,P)$ is some real-valued discrepancy measure of R with respect to P.

In particular, $D(R,P)$ may be any f-divergence in the sense of Csiszár as in Perez (1967) including its different modifications or surch divergence augmented by a functional of P independent of R. If $T=M$ we shall write R_S^D instead of $_M R_S^D$ for the (unconstrained) D-barycenter. If $D(R,P):=H(R,P)$, i.e. the Shannon's *relative entropy* defined, for $R \ll P$ with $(dR/dP)=u$, by $\int u \log u \, dP$ and, otherwise, by $+\infty$, (i.e. f-divergence with as convex function $f(u):=u \log u$), we shall write $_T R_S$ instead of $_T R_S^H$ in the constrained case and simply R_S in the unconstrained case.

In general, up to an additive component independent of R the discrepancy $D(R,P)$, as defined in $M \times M$, should have some properties characteristic for the class of f-divergences (f strictly convex), namely, (P1) $D(R,P) \geq 0$ with "=" iff $R=P$; (P2) If R', P' are restrictions of R and P on $\tilde{X}' \subset \tilde{X}$, then $D(R',P') \leq D(R,P)$ with "=" iff the σ-algebra $\tilde{X}'$ is *sufficient* with respect to $\{R,P\}$; (P3) $D(R,P)=\sup_{\tilde{X}' \subset \tilde{X}} D(R',P')$ whereas the $\tilde{X}'$'s are generated by finite partitions of $(X,\tilde{X})$. In particular, there exists a growing sequence $\{\tilde{X}^i\}_{i \geq 1}$ of such algebras for which $\lim_i D(R^i, P^i) = D(R,P)$; (P4) If $\underline{\alpha}=(\alpha_1,\ldots,\alpha_m)$ is a probability vector, $P_{\underline{\alpha}}:=\Sigma_{i=1}^m \alpha_i P_i$ and $R_{\underline{\alpha}}:=\Sigma_{i=1}^m \alpha_i R_i$ where $P_1,\ldots,P_m$, $R_1,\ldots,R_m \in M$, then $D(R_{\underline{\alpha}}, P)$ and $D(R, P_{\underline{\alpha}})$ are convex functions of $\underline{\alpha}$ for any R and $P \in M$ and $D(R_{\underline{\alpha}}, P_{\underline{\alpha}}) \leq \Sigma_{i=1}^m \alpha_i D(R_i, P_i)$.

The mentioned convexity of $D(R,P_{\underline{\alpha}})$ may be extended in order to hold even if $P_{\underline{\alpha}}$ is replaced by $P^{\underline{\alpha}}$ defined by $dP^{\underline{\alpha}}:=dP_1^{\alpha_1}\dots dP_m^{\alpha_m}/N_{\underline{\alpha}}$, where the normalizing denominator $N_{\underline{\alpha}}=\int dP_1^{\alpha_1}\dots dP_m^{\alpha_m}=:H_{\alpha_1\dots\alpha_m}(P_1,\dots,P_m)$ is what we call (generalized) *alpha-entropy* of $P_1,\dots,P_m$ which in the literature is known for $m=2$. Namely, the above holds for the important case $D:=H$ mentioned before. The necessary condition $N_{\underline{\alpha}}>0$ is here ensured by the assumption that $P_1,\dots,P_m$ dominate R and, thus, have a set of equivalence with positive probability.

These convexity considerations are important in connection with the following general theorem we shall apply in the sequel.

Theorem 1. (Theorem 4.4.1. in Gallager (1968)). Let $F(\underline{\alpha})$ be a convex ∪ function of $\underline{\alpha}=(\alpha_1,\dots,\alpha_m)$ over the region V when $\underline{\alpha}$ is a probability vector. Assume that the partial derivatives $\partial F(\underline{\alpha})/\partial\alpha_k$ are defined and continuous over the region V with the possible exception that $\lim_{\alpha_k\to 0}(\partial F(\underline{\alpha})/\partial\alpha_k)$ may be $+\infty$. Then

(i) $\partial F(\underline{\alpha})/\partial\alpha_k=\lambda$ for all k such that $\alpha_k>0$

(ii) $\partial F(\underline{\alpha})/\partial\alpha_k\geq\lambda$ for all k such that $\alpha_k=0$

are necessary and sufficient conditions on a probability vector $\underline{\alpha}$ to minimize F over the region V.

H-barycenter of $S:=\{P_1,\dots,P_m\}$. By applying the Lagrange multiplicator's method one finds that the barycenter R_S is given by $P^{\underline{\alpha}}$ defined as above (provided that the set of equivalence of $P_1,\dots,P_m$ has a positive probability) with $\underline{\alpha}\in V$ minimizing the corresponding alpha-entropy. Since the latter as a function of $\underline{\alpha}$ is convex ∪ one may apply Theorem 1. Thus, one obtains that necessary and sufficient conditions for the minimizing $\underline{\alpha}$ are the following: (Hi) $H(R_S,P_k)=-\log H_{\alpha_1\dots\alpha_m}(P_1,\dots,P_m)$ for all k such that $\alpha_k>0$; (Hii) $H(R_S,P_k)\leq-\log H_{\alpha_1\dots\alpha_m}(P_1,\dots,P_m)$ for all k such that $\alpha_k=0$.

The following algorithm for solving the above minimization problem was proposed by the author and implemented by O. Kříž. We denote in the sequel by $R_{ij\dots k}$ the H-barycenter of $\{P_i,P_j,\dots,P_k\}\subset S$. $(A1)$ Let $\{P_{i1},P_{i2}\}\subset S$ such that $H(R_{i1i2},P_{i1})=H(R_{i1i2},P_{i2}):=\max_{1\leq i,j\leq m}H(R_{ij},P_i)$. If $H(R_{i1,i2},P_{i3}):=\max_{1\leq i\leq m}H(R_{i1i2},P_i)=H(R_{i1i2},P_{i1})$ stop since $R_S=R_{i1i2}$. If no, then pass to step $(A2)$ If $H(R_{i1i2i3},P_{i4}):=\max_{1\leq i\leq m}H(R_{i1i2i3},P_i)=$ $=\max_{1\leq j\leq 3}H(R_{i1i2i3},P_{ij})$ stop since $R_S=R_{i1i2i3}$. If no, then pass to step $(A3)$ and so on up to $(Ak-1)$ where $\max_{1\leq i\leq m}H(R_{i1\dots ik},P_i)=\max_{1\leq j\leq k}H(R_{i1\dots ik},P_{ij})$

(and, thus, $R_S=R_{i1\ldots ik}$) whereas $H(R_{i1\ldots ik-1},P_{ik})>\max_{1\le j\le k-1} H(R_{i1\ldots ik-1},P_{ij})$.

A similar algorithm may be applied for other discrepancy measures. At the end of this section we mention some general results without proofs.

Theorem 2. If ${}_TR^{\mathcal{D}}_{S_1}={}_TR^{\mathcal{D}}_{S_2}$ then both are equal to ${}_TR^{\mathcal{D}}_{S_1\cup S_2}$.

Theorem 3. There exists a subset $\bar{S}$ of S such that ${}_TR^{\mathcal{D}}_{\bar{S}}={}_TR^{\mathcal{D}}_S$ with $\mathcal{D}_{\bar{S}}:=\sup_{P\in\bar{S}}\mathcal{D}({}_TR^{\mathcal{D}}_S,P)=\mathcal{D}_S$ whereas for any $P\in\bar{S}$ it holds $\mathcal{D}({}_TR^{\mathcal{D}}_S,P)=\mathcal{D}_S$.

Theorem 4. If $S:=\{P\in M:\mathcal{D}(R,P)\le\varepsilon\}$ and if $\{P_1,P_2\}\subset S$ such that $R^{\mathcal{D}}_{12}=R$ with $\mathcal{D}(R,P_1)=\mathcal{D}(R,P_2)=\varepsilon$, then $R^{\mathcal{D}}_S=R$, $\mathcal{D}_S=\varepsilon$.

Theorem 5. Let $S_0\subset S$ be the set of P's on which actually $R^{\mathcal{D}}_S$ depends (S_0 is of the type $\bar{S}$ above) and let $\mathfrak{X}'\subset\mathfrak{X}$ be *sufficient* with respect to $S_0\cup\{R^{\mathcal{D}}_S\}$. Then $R^{\mathcal{D}}_{S'}=(R^{\mathcal{D}}_S)'$, i. e. the $\mathcal{D}$-barycenter of the set S' of restrictions on $\mathfrak{X}'$ coincides with the restriction of $R^{\mathcal{D}}_S$ on $\mathfrak{X}'$, provided that $\mathcal{D}$ has the property (P2).

3. The barycenter concept in discriminating simple statistical hypotheses. Let $P_1\ne P_2$ of M be two statistical hypotheses to be discriminated by a growing number, n, of i.i.d. observations on $(X,\mathfrak{X})$ by a ML decision procedure adapted either (i) to the actual hypotheses P_1 and P_2 of (ii) to a deviated pair Q_1, Q_2 of M; cf. Perez (1983a).

Case (i). Then the two ML error probabilities converge to zero as e^{-nE} with an asymptotic rate $E=H(R_S,P_1)=H(R_S,P_2)$, where $S=\{P_1,P_2\}$. If, however, one fixes the rate $E_1=r$ of the first-type error, then the second--type-error asymptotic rate E_2 is given by $H({}_TR_S,P_2)$ with $S:=\{P_2\}$ and $T:=\{R\in M:H(R,P_1)=H(R,P_2)+\gamma\}$, where γ (if exists) is a constant depending on r; cf. Perez (1984).

Case (ii). Then $E_j=H({}_TR_{S_j},P_j)$ with $S_j=\{P_j\}$ and $T:=\{R\in M:H(R,Q_1)=H(R,Q_2)\}$, $j=1,2$, under some general conditions. If, however, one fixes the rate $E_1=r$, then $E_2=H({}_TR_S,P_2)$ with $S:=\{P_2\}$ and $T:=\{R\in M:H(R,Q_1)=H(R,Q_2)+\gamma\}$, where γ (if exists) is again a constant depending on r.

4. The barycenter concept in discriminating composite statistical hypotheses. Let $H_i\subset M$, $i=1,2$, be the two disjoint hypotheses to be discriminated by a ML decision procedure adapted to some pair $(Q_1,Q_2)\in H_1\times H_2$.

Parting from (ii) above, the discrimination rate analysis made in Perez (1983b) lead in particular to the minimax concept of *Maximum Discrimination Rate Least Favorable Pair* (MDR-LFP) of distributions $(\hat{P}_1,\hat{P}_2)\in H_1\times H_2$ defined (if it exists) by

$$H(R_{\{\hat{P}_1,\hat{P}_2\}},\hat{P}_1) := \sup_{(Q_1,Q_2)\in H_1\times H_2} \ \min_{i} \ \inf_{P_i\in H_i} H({}_T R_{\{P_i\}},P_i) := \hat{D}R ,$$

where $T:=\{R\in M: H(R,Q_1)=H(R,Q_2)\}$.

Theorem 6. (Theorem 1 of Perez (1983b)). A necessary and sufficient condition for a pair $(P_1,P_2)\in H_1\times H_2$ to be MDR-LFP is $H(R_{\{P_1,P_2\}},P_1)=\min_{Q_i\in H_i} H(R_{\{P_1,P_2\}},Q_i)$, $i=1,2$.

On the base of Theorem 6, $(\hat{P}_1,\hat{P}_2)$ may be approximated from above as follows: Take any pair $(P_1,P_2)\in H_1\times H_2$ and compute its barycenter $R_{\{P_1,P_2\}}$. Let $(P_1^o,P_2^o)\in H_1\times H_2$ be defined by $H(R_{\{P_1,P_2\}},P_i^o):=\min_{Q_i\in H_i} H(R_{\{P_1,P_2\}},Q_i)$, $(i=1,2)$, construct their barycenter $R_{\{P_1^o,P_2^o\}}$ and repeat the preceding step by taking (P_1^o,P_2^o) instead of (P_1,P_2) obtaining, thus, a pair (P_1^1,P_2^1) and so on. We, thus, obtain for $i=1,2$

$$H(R_{\{P_1,P_2\}},P_i)\ge H(R_{\{P_1,P_2\}},P_i^o)\ge H(R_{\{P_1^o,P_2^o\}},P_i^o)\ge H(R_{\{P_1^o,P_2^o\}},P_i^1)\ge\ldots\ge$$
$$\ge H(R_{\{P_1^{k-1},P_2^{k-1}\}},P_i^k) \xrightarrow[k]{} H(R_{\{\hat{P}_1,\hat{P}_2\}},\hat{P}_i) = \hat{D}R.$$

Similarly, it holds for $T_k:=\{R\in M: H(R,P_1^k)=H(R,P_2^k)\}$,

$$\min_i H({}_{T_0}R_{\{P_i^1\}},P_i^1)\le\min_i H({}_{T_1}R_{\{P_i^2\}};P_i^2)\le\ldots\le\min_i H({}_{T_{k-1}}R_{\{P_i^k\}},P_i^k) \xrightarrow[k]{} \hat{D}R.$$

5. The barycenter concept in improving minimal-discrepancy estimating (approximating) if more estimates (approximations) are available.

Let $P\in M$ be the p.m. to be approximated (estimated) by some $R\in T\subset M$ minimizing the discrepancy $D_0(R,P)$ (resp. $D_0(R,\hat{P})$ where $\hat{P}$ is a suitable empirical version of P), whereas for a suitable choice of T there are available more approximations $P_1,\ldots,P_m$ (estimates $\hat{P}_1,\ldots,\hat{P}_m$) of P. For instance, P_i ($\hat{P}_i$) may be such that $D_i(P_i,P):=\min_{R\in T_i\subset M} D_i(R,P)$ ($D_i(P_i,\hat{P}):=\min_{R\in T_i\subset M} D_i(R,\hat{P})$). It is natural to try to choose T above of the form $T:=\{R\in M: R$ construct of $P_1,\ldots,P_m$, resp. of $\hat{P}_1,\ldots,\hat{P}_m$, with some free parameter $\pi\in\Pi\}$ in the aim to achieve some $R_{\pi_0}\in T$ such that

$$D_0(R_{\pi_0},P):=\min_{\pi\in\Pi} D_0(R_\pi,P) < D_0(P_{i_0},P) := \min_{1\le i\le m} D_0(P_i,P)$$

and similarly for $\hat{P}$ and $\hat{P}_1,\ldots,\hat{P}_m$.

Restricting us here to $D_0(R,P):=H(P,R)$ and R of the type $P^{\underline{\alpha}}$ we, thus, have to minimize $H(P,P^{\underline{\alpha}})$ (which is a convex $\cup$ function of $\underline{\alpha}\in V$). Applying Theorem 1, one finds for the minimizing $\underline{\alpha}$ the necessary and sufficient condition

(H') $\quad H(P^{\underline{\alpha}},P_k)-H(P,P_k) \le \max_{1\le i\le m} (H(P^{\underline{\alpha}},P_i)-H(P,P_i)),$

where for all k such that $\alpha_k > 0$ the sign "=" relation holds.

Moreover, for such $\underline{\alpha}$ the *improvement* relation I holds

(I) $$H(P,P^{\underline{\alpha}}) \le H(P,P_{i_0}) - H(P^{\underline{\alpha}},P_{i_0}),$$

whereas "=" holds if $\alpha_{i_0} > 0$ and $H(P,P_{i_0}) := \min_{1 \le i \le m} H(P,P_i)$.

From the barycenter point of view we see that for such $\underline{\alpha}$ $P^{\underline{\alpha}} = {}_T R^{H'}_{\{P\}}$ where $H'(R,Q) := H(Q,R)$. By introducing the discrepancy measure $D(R,Q) := = H(R,Q) - H(P,Q)$ one easily finds that the (unconstrained) D-barycenter, R^D_S, of $S := \{P_1,\ldots,P_m\}$ is of the form $P^{\underline{\alpha}}$ with $\underline{\alpha} \in V$ satisfying (H') as well as the improvement relation (I). Similarly for the $\hat{P}$'s.

If P is especially given by a multidimensional distribution and if there are available some less dimensional marginals of P one has the possibility to construct a set $\bar{S} = \{\bar{P}_1,\ldots,\bar{P}_m\}$ of *dependence structure simplifications* of P as in Perez (1977) for which it holds $H(P,\bar{P}_i) = I(P) - I(\bar{P}_i)$, where $I(Q)$ is what we call the *dependence tightness* of the random variables distributed according to Q. In particular, the improvement relation (I) takes the form

$(\bar{I})$ $$H(P,\bar{P}^{\underline{\alpha}}) \le H(P,\bar{P}_{i_0}) - H(\bar{P}^{\underline{\alpha}},\bar{P}_{i_0}),$$

whereas the condition (H') for the minimizing $\underline{\alpha}$ reduces to a similar one taking $I(\bar{P}_k)$ instead of $-H(P,\bar{P}_k)$, not implying, thus, the unknown P. By introducing $\bar{D}(R,Q) := H(R,Q) + I(Q)$ we again find that the $\bar{D}$-barycenter, $R^{\bar{D}}_{\bar{S}}$, of $\bar{S}$ is of the form $\bar{P}^{\underline{\alpha}}$ with $\underline{\alpha} \in V$ satisfying the latter condition as well as $(\bar{I})$. Similar results are obtained if instead of the marginals considered one has available some estimates of them. The connesponding $\hat{\bar{P}}^{\underline{\alpha}}$ as minimizing $H(\hat{P},\hat{\bar{P}}^{\underline{\alpha}})$ is a ML estimate of P constrained to be in $T := \{R \in M : R = \hat{\bar{P}}^{\underline{\alpha}}, \underline{\alpha} \in V\}$.

References

Gallager R. (1968), Information Theory and Reliable Communication, John Wiley, New York.

Perez A. (1967), Information-Theoretic Risk Estimates in Statistical Decision, Kybernetika 3, 1-21.

Perez A. (1983a), Discrimination Rate Loss of Simple Statistical Hypotheses by Unfitted Decision Procedures, Studies in Probability and Related Topics - Papers in Honour of Octav Onicescu, Nagard Publisher, 381-390.

Perez A. (1983b), Discrimination Rate Least Favorable Pairs of Distributions for ε-Contaminated Statistical Hypotheses or with f-Divergence Like Neighborhoods, Transactions Ninth Prague Conf. on Information Theory, etc., Academia, 53-66.

Perez A. (1984), Second-type error exponent given the first-type error exponent in testing statistical hypotheses by unfitted procedures, (in preparation).

POSCON – a Decision-Support System in Diagnosis and Prognosis Based on a Statistical Approach

D.M. van der Sluis, and W. Schaafsma, Groningen

Summary: It is natural to consider the probability that an individual, given its characteristics, belongs to a certain population. Such individual-dependent probabilities can often be defined as posterior probabilities. Basic assumptions are that k mutually exclusive populations are involved and that the individual belongs to one of these k populations or, more precisely, that it is a random drawing from one of them. The posterior probabilities can then be defined by means of Bayes' theorem. To obtain numerical values for the thus defined individual-dependent probabilities we not only need the scores for the characteristics of the individual but also numerical values for the k prior probabilities and for the values of the k densities in the score vector x. These values are usually based on some estimation procedure. Accordingly there are uncertainties involved. A nice way to express uncertainties is by means of confidence intervals. The interactive program POSCON (compute POSterior CONfidence intervals) has been written for it.

Keywords: posterior probability, computer-assisted diagnosis, classification, pattern recognition.

1. Introduction

Crucial decisions are often made with respect to isolated situations, e.g. patients in clinical decision making or future years in policy making. Apart from contemplating about the data to be incorporated, one can often distinguish the following phases:

(1) statistical inference in the sense that statements are formulated with respect to "populations" of similar situations,
(2) predictive analysis of the situation under consideration, e.g. in the form of a differential diagnosis or of statements about what is going to happen if certain policies are adopted,
(3) making the actual decision in the face of all involved uncertainties.

The theory of statistical decision functions was originated by Wald(1902-1950) as an attempt to cover all aspects of the decision-making process. In most applications statistical uncertainties, e.g. standard errors of estimates, are carefully expressed in phase 1 but more or less ignored in the phases 2 and 3. The present program is an attempt to improve upon this situation. It will be introduced by focussing on clinical decision making.

The philosophy behind POSCON can be introduced by considering the discrete case. Given the vector of characteristics of the patient under observation, one can discuss the corresponding reference population and reference sample. The probability that this patient with its vector of scores belongs to a certain diagnostic or prognostic group is defined as the corresponding proportion of the reference population. This unknown parameter can be estimated by means of the corresponding relative frequency in the reference sample. The corresponding uncertainties can be expressed by providing a standard error, or by using a confidence interval instead of the point estimate. The philosopher Reichenbach(1949) already discussed the problem of the reference class. He remarked that the probability of a single case depends on the characteristics taken into account. One should look for "the narrowest class for which reliable statistics can be compiled". This idea is interesting because, contrary to most selection of variables techniques, it admits the possibility that the selection depends on the vector of scores of the individual under consideration.

To obtain quantitative theory and to extend the domain of application beyond the discrete case, several approaches can be derived from the areas of discriminant analysis, logistic regression, survival analysis, etc.

POSCON belongs to the areas of discriminant analysis and supervised pattern recognition. Previously acquired knowledge is presented in the form of training samples from various diagnostic or prognostic categories. The tasks of POSCON are visualized below.

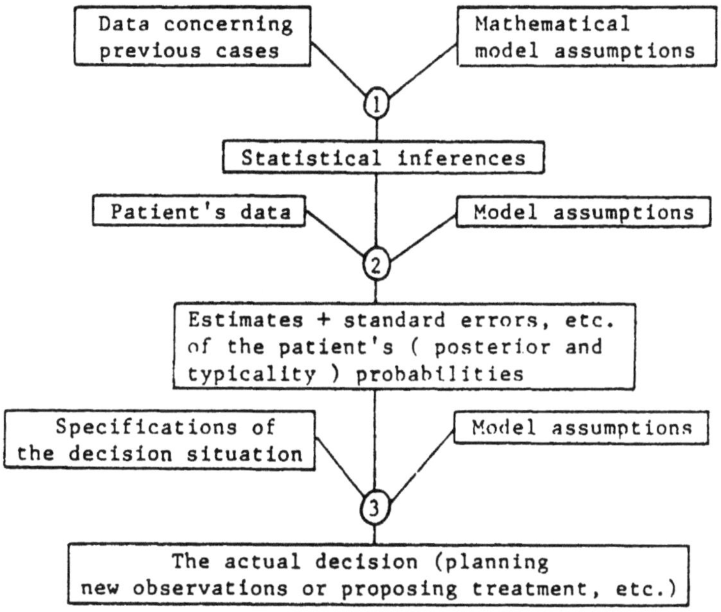

The idea is that the computer is used to perform the operations 1 and 2. Note that POSCON expresses the uncertainties due to sampling fluctuations if we regard the context, i.e. the mathematical model together with the choice of variables, as given. The advantages of this approach are as follows:

(1) informal cognitive processes are replaced by reproducible computations,
(2) knowing the uncertainties is pertinent because sample sizes are often so small that standard errors of posterior probability estimates are considerable,
(3) computation by hand is only feasible in the discrete case,
(4) it is easy to study the context-dependency of the results by repeating POSCON with other choices of model or variables.

The disadvantages are also obvious: one has to do a lot of work in order to collect reliable training samples. POSCON has no options for missing data. If for certain patients some variables could not be measured, then either these patients or these variables are deleted. (Note that the vector of scores for the individual under investigation is allowed to be incomplete.)

Applying POSCON as indicated above, leads to the following methodological problems:

(1) which of the POSCON outputs should be preferred,

(2) how should operation 3 be carried out.

It is obvious that the solution of both problems should depend on the uncertainties expressed by the POSCON output. Proposals have been made in Schaafsma(1983) but, in essence, this part of the work is still in progress.

2. <u>The theory behind POSCON</u>

Given the vector of scores of a patient, the corresponding reference sample will often be almost empty. One needs the information provided by the case histories of neighbouring score vectors. A mathematical model is needed to convey this information. One of the possible approaches is that based on the formula of Bayes where the posterior probability of belongingness to group t, given the vector of scores x, is expressed by

$$\rho(t|x) = \rho_t f_t(x) / \sum_{h=1}^{k} \rho_h f_h(x).$$

In this formula ρ_h denotes the prior probability of population h while f_h is the probability density of the vector X of observations of a random drawing from population h (h=1,...,k).

The prior probabilities to be introduced will either be postulated or be estimated from other data. In the case that they are estimated, the corresponding uncertainties expressed by standard deviations can be incorporated in the analysis. The probability density functions will usually be unknown. We need training samples from each of the k populations. In practice this will mean that the cases belonging to population h are regarded as an independent random sample. This assumption can often be criticized. However, without this assumption nothing seems possible. Using a mathematical model for the densities $f_1,...,f_k$, one obtains estimates of the posterior probabilities. Complicated theory in e.g. Ambergen-Schaafsma(1983) provides asymptotic distributions of the underlying estimators. These results are used to obtain approximations for the standard deviations of and correlation coefficients between the employed estimators. This task of POSCON is expressed below.

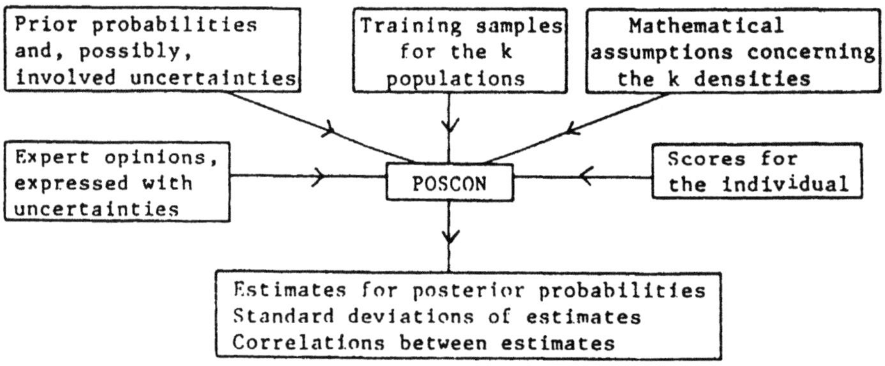

The novelty of this approach in comparison with earlier ones is that the uncertainties in the estimated posterior probabilities are expressed by means of standard errors. These can naturally be used to construct approximate confidence intervals for the posterior probabilities separately.

The program offers some extra possibilities under normality . In that case an F statistic is computed which can be used to test the null-hypothesis H_0 that the score vector x of the individual under investigation comes from the same p-variate normal population as the score vectors of the individuals in training sample h. This is a simple application of the Hotelling two-sample test, one sample size being equal to one. This approach is not limited by the assumption that the individual under investigation belongs to exactly one of the k categories.

In addition to the F statistic, the program provides also approximate confidence intervals for the so-called typicality probabilities $\alpha_1, \dots, \alpha_h$. Their definition is obvious: if the vector of scores has the p-variate normal distribution $N_p(\mu_h, \Sigma_h)$ for individuals taken at random from population h, then the typicality probability α_h of the individual under investigation with given vector of scores $x \in \mathbb{R}^p$ is defined as the probability that a random drawing X from $N_p(\mu_h, \Sigma_h)$ provides a larger value for the probability density than the observed vector x. In formula

$$\alpha_h = P(\chi^2_p > \Delta^2_{h,x})$$

where χ^2_p is a random variable having the chi-square distribution with p degrees of freedom and

$$\Delta^2_{h,x} = (x - \mu_h)^T \Sigma_h^{-1} (x - \mu_h)$$

is the Mahalanobis distance between x and the vector μ_h of expectations for population h. Note that $x = \mu_h$ implies that $\Delta^2_{h,x} = 0$ and hence $\alpha_h = 1$: an individual with this vector of scores is indeed very typical for population h. The procedure is based on an unbiased estimate for $\Delta^2_{h,x}$ and results in an approximate confidence interval for α_h $(h=1,\dots,k)$.

Next the ingredients to be used by POSCON are discussed in some detail.

(1) The prior probabilities, possibly with involved uncertainties. POSCON uses either posulated prior probabilities, without uncertainties , or estimated prior probabilities in which case standard deviations of estimates can be introduced. These uncertainties are incorporated in the analysis by replacing them, together with the estimated prior probabilities, by the outcome of a simple fictitious experiment (in which case the user has to agree with some modification of his input standard deviations) or by means of a more complicated fictitious experiment (in which case a degeneracy may appear).

(2) Expert opinions, expressed with uncertainties. In clinical medicine it may happen that an independent expert is consulted. POSCON offers two ways to incorporate expert opinions. Way 1 is similar to the way prior probabilities are introduced together with uncertainties. Way 2 is such that the expert opinion is expressed in the form of estimated likelihoods together with corresponding standard errors.

(3) Training samples for the k populations. Handling the training samples is the task of two special parts of POSCON called CREATE and CHANGE. An important requirement is that the training samples should be complete. Note that POSCON's aim to express uncertainties in estimated probabilities is only of interest in situations where populations are so closely related that it makes sense to require that all variables are observed for all individuals in the training

samples (though, of course, often something goes wrong).
(4) Mathematical assumptions concerning the k densities. An important flexibility of POSCON is that the set of variables can be partitioned into a number of subsets. These subsets will then be regarded as stochastically independent for each of the k populations. For the separate subsets one of the following probabilistic models has to be chosen.

DIS - discrete data;
NEC - normal data with equal variance/covariance matrices;
NOR - normal data with unequal variance/covariance matrices;
NON - nonparametric data;
MIX - mixed data.

At the moment the first three options have been implemented and the fifth is underway. Theory for nonparametric density estimation is available, but not implemented because it is expected to produce too large standard errors.
(5) Scores of the individual. Contrary to the training samples, it is allowed that some scores are missing. The available scores are read in one run. Partitioning into independent subclasses is part of the interactive process. Users often worry about measurement errors in the scores. If measurement errors are caused by the same random mechanism for the training samples as for the individual under study, then relief is obtained by noting that one is working with the posterior probabilities with respect to the variables as they are measured, instead of with respect to their true but latent values.
(6) F-tests, Mahalanobis distances, lower and upper bounds of typicality probabilities. These results were defined under normality . Accordingly they are only presented in the sub-loops NOR and NEC. The results refer to the set of variables selected before entering the sub-loop. The F-tests refer to testing the null-hypothesis that this individual with its vector of scores is from the same population as training sample h. The Mahalanobis distance is estimated by using the square root of the best unbiased estimator of its square. The confidence bounds for the typicality probabilities are rather inaccurate if sample sizes are small. More accurate iterative procedures have not yet been implemented.

3. Action and use of the program

- POSCON is an interactive program (FORTRAN77) with fixed or dynamic memory allocation.
- POSCON has a short(FAST) and an extensive(SLOW) input mode. In the fast mode more than one answer can be given on the same input line and the program does not generate the requests, which belong to the answers.In the slow mode each set of data is asked for by an explicit request. For example, if POSCON has to read a sequence of sample names, in the fast mode this can be done in one run, but in the slow mode POSCON asks for each sample its name separately. If the user makes some mistake the request is repeated. If more than one piece of data is needed for the request then the fast mode is replaced by the slow mode when the user makes some mistake. All input is checked. Before computation the program asks the user to verify the input. If this check fails the user is invited to produce the right data .

In the short mode the user can give, if he knows the next questions, more answers on one line. In that case POSCON deletes the question and goes on. For the experient user a very useful extension.

- For almost all input menu selection is applied. This means that the request will be followed by a list of all possible answers.
- Before computations are carried out, a POSCON system file has to be created. This consists of samples of individuals, each individual characterized by the scores for a given number of variables. For each variable upper and lower bounds are required as the basis of certain checks. Apart from these training samples there is also a facility to introduce an array consisting of a vector of means and a matrix of sample covariances. This facility is sometimes of interest as a mean to represent the population of normal individuals.It is possible to change the system file by replacing, deleting or adding cases and/or even complete samples.
- Names and numbers of variables and samples can be used interchangeably. A sequence of names or numbers can be mentioned in a short way by giving the first and the last name, or number, linked by a dash(-). A dash can only link names or numbers. For example, var1,var2,var3,var4 or var1 - var4 or 1,2,3,4 or 1 - 4 or 1,var2 - var4 have the same meaning.
- POSCON uses a number of scratch files. Output not essential for the computation but containing extra information will not be written on the terminal but on a scratch file. The contents of this scratch file can be made visible and can also be written to a line printer.
- POSCON writes all input to a backup file and this file can be used as input file for a next run. POSCON can read all the instructions from input and from file. Also in the case of creating and changing the POSCON system file the data can be read from files.
- POSCON has the possibility to give an overview of all update steps, which have been done. This overview can be written to a line printer and can also be made visible on the terminal.

4. References

Aitchison,J.,Habbema,J.D.F. and Kay,J.W.(1977), A critical comparison of two methods of statistical discrimination. Applied Statistics 26, no 1, page 13.

Ambergen,A.W. and Schaafsma,W.(1983), Interval estimates for posterior probabilities in a multivariate normal classification model. Report SW96183, Mathematical Center, Kuislaan 413, 1098 SJ Amsterdam(to be published in Journal of Mult. An.), the Netherlands.

Reichenbach,H.(1949), The theory of probability. University of California Press, Berkeley and Los Angeles

Schaafsma,W.(1983), Standard errors of posterior probabilities and how to use them. Sixth Symposium on Multivariate Analysis, preprint can be obtained by writing the author, Department of Mathematics, P.O. Box 800, Groningen, the Netherlands.

Sluis,D.M. van der (1984), User's Manual for POSCON. Computer Center Groningen University, P.O. Box 800, Groningen, the Netherlands.

Wald,A.(1950), Statistical decision functions. Wiley, NewYork.

Simulated Powers Under Two Different Approximations to the Exact Distribution of MRPP Statistics

D.S. Tracy, and I. H. Tajuddin, Windsor

SUMMARY

The MRPP test statistic for randomness of classification is studied when the distance function is the Euclidean distance between ranks of observations and the population is classified into two equal sized groups. Simulated powers of the test statistic against location shift are examined under different approximations from the Pearson system of curves. When the population size is large, Pearson Type VI approximation seems to be better for the four underlying distributions concerned.

KEY WORDS: Multiresponse permutation procedure, simulated power, location shift, Pearson curves, underlying distributions.

1. INTRODUCTION

Mielke, Berry and Johnson (1976) proposed an exact permutation procedure for the analysis of multi-response data. The multi-response permutation procedure (MRPP) test, so developed, requires a minimum of assumptions, and considers the distance between a pair of observations, rather than the observations themselves.

Let $\{X_I : I = 1,\ldots,N\}$ be a population of N observations. Let K of these observations be classified into g subgroups, according to some a priori scheme, with n_i observations in the ith subgroup. Let

$$\Delta_{I,J} = |f(X_I)-f(X_J)|^{\nu} ,$$

$\nu > 0$, f some function, be a measure of distance between X_I and X_J . Then the average distance within the ith subgroup is

$$\xi_i = \binom{n_i}{2}^{-1} \sum_{I<J} \Delta_{I,J} S_i(I) S_i(J)$$

where $S_i(M) = 1$ if X_M is in the ith subgroup and 0 otherwise, and the summation $\sum$ is over the set $\{(I,J): 1 \le I < J \le N\}$. The MRPP test statistic δ is then defined as

$$\delta = \sum_{1}^{g} c_i \xi_i$$

where c_i are weights with $\sum_{1}^{g} c_i = 1$.

The test statistic δ tests the null hypothesis H_0 : Classification of observations into the g subgroups is random. Under H_0 , each of the

$\frac{N!}{(N-K)!n_1!\ldots n_g!}$ classifications is equally likely. This gives an exact probability distribution of δ, but with the increase in N, the number of classifications is enormous. To circumvent this difficulty, Mielke, Berry, Brockwell and Williams (1981) adopt a Pearson Type III approximation to the exact distribution of δ. They obtain the simulated power of δ for various underlying distributions and show its superiority over Wilcoxon test in case of some underlying distributions.

We study the behaviour of δ for small samples using the exact probability distribution. We study the dichotomous case with $n_1 + n_2 = K = N$, letting $f(X)$ be the rank of X in the combined sample, and weight $c_i = \frac{n_i}{N}$. Calculating the four moments, we fit one of the Pearson curves in Section 2. The results motivate us to find an expression for the fourth moment of δ, and use it to study two different approximations from the Pearson system for its null distribution. Simulated powers under these are studied in Section 3. The conclusions are presented in Section 4.

2. SOME RESULTS UNDER EXACT DISTRIBUTIONS

We study the behaviour of the test statistic δ for the case of two groups, when n_1, n_2 are small and $n_1 + n_2 = K = N$. We take $c_i = \frac{n_i}{N}$ and $f(X)$ to be the rank of X in the combined sample. The underlying distributions are taken to be normal, uniform and U-shaped. The four moments are computed by calculating all possible values of δ. Three replications are done for each of the cases $(n_1,n_2) = (5,5),(10,5)$ and Pearson's criterion

$$\kappa = \frac{\beta_1(\beta_2+3)^2}{4(4\beta_2-3\beta_1)(2\beta_2-3\beta_1-6)}$$

(Elderton and Johnson, 1969, p.41) is calculated. These, as well as values of $2\beta_2 - 3\beta_1 - 6$, required to be zero for Pearson Type III, are presented in Table 1.

Table 1. Computed Values of κ and $2\beta_2 - 3\beta_1 - 6$ for δ for Samples from Various Underlying Distributions

Underlying Distribution	Replicate No.	$(n_1,n_2) = (5,5)$		$(n_1,n_2) = (10,5)$	
		κ	$2\beta_2-3\beta_1-6$	κ	$2\beta_2-3\beta_1-6$
Normal	1	-1.09	-3.48	-1.11	-0.73
	2	-1.61	-3.18	-0.32	-1.46
	3	-0.80	-6.60	-4.74	-1.28
Uniform	1	-1.43	-2.41	-3.94	-1.90
	2	-1.06	-3.08	-5.50	-0.97
	3	-1.83	-2.81	-1.85	-0.29
U-shaped	1	-5.82	-1.83	-3.58	-2.64
	2	-2.80	-2.71	-3.95	-2.22
	3	-2.60	-3.22	-3.99	-2.13

Table 1 indicates that different Pearson curves may be suitable under different situations. We examine the goodness of fit of Pearson Type I and Type III curves to the null distribution of δ for replicate no. 3 from the normal distribution for sample size (5,5). To apply the χ^2 goodness of fit test, the number of classes is chosen in accordance with the recommendations of Williams (1950) and Cochran (1952). Since the distribution of δ is highly skewed, the equal probability classification recommended in recent literature is not appropriate. We use (a) the pooling criterion given by Yarnold (1970), and (b) the usual pooling when expected frequencies are ≤ 5. The results are presented in Table 2 below.

Table 2. Results of Goodness of Fit Tests

	Yarnold Criterion		Usual Pooling	
	Type I Fit	Type III Fit	Type I Fit	Type III Fit
χ^2	25.2379	26.6463	23.8087	22.7867
d.f.	14	14	12	11
p-value	0.0323	0.0214	0.0216	0.0189

We notice that with either criterion the p-value for Pearson Type I fit is larger than that for Type III fit.

The conclusions above motivate us to obtain the expression for the fourth moment of δ. We do this for the case $\Delta_{I,J} = |R(X_I)-R(X_J)|$ where R denotes the rank in the combined sample. The test statistic is designated δ_1 for this case. When $c_1 = c_2 = \frac{N}{2}$, we obtain

$$E(\delta_1^4) = \frac{N+1}{14175\, N^2(N-2)^3}\,(175N^8-525N^7+315N^6-1035N^5+4818N^4-6180N^3+1640N^2+8640N-4608).$$

This expression, with the results of Mielke et al. (1981, p.722), yields $\mu_4(\delta_1) = \frac{16(N+1)}{4725N^2(N-2)^3}(31N^4-175N^3+180N^2+260N-96)$. Using this with the expressions for the first three moments (Mielke et al., 1981), we obtain values of Pearson's criterion κ and of $2\beta_2-3\beta_1-6$ for various values of N in Table 3 below.

Table 3. Values of κ and $2\beta_2-3\beta_1-6$ for the Null Distribution of δ_1

N	κ	$2\beta_2-3\beta_1-6$	N	κ	$2\beta_2-3\beta_1-6$
4	-0.125	-4.500	40	84.865	0.126
10	-2.640	-2.077	50	38.002	0.292
20	-12.973	-0.670	60	28.176	0.404
30	-68.287	-0.146	100	18.975	0.631
34	-541.067	-0.019	1000	13.566	0.944
36	303.732	0.034	∞	13.175	0.980

We note from Table 3 that $2\beta_2-3\beta_1-6$ is an increasing function of N, and differs from 0, the Type III value, by more than 0.6 for $N \leq 20$ and $N \geq 100$. The table also indicates that Pearson Type I is a competitive fit to Type III for $N \leq 34$

Table 4. Simulated Powers against Location Shifts for Different Approximations to the Null Distribution

Large Samples

$n_1 = n_2 = 40$

Location Shift		.03σ			.05σ			.03σ			.05σ		
Statistic		δ_1		δ_2	δ_1		δ_2	δ_1		δ_2	δ_1		δ_2
Type Used		VI	III	III	VI	III	III	VI	III	III	VI	III	III
Rep	α												
		NORMAL						UNIFORM					
	.01	.095	.093	.093	.316	.314	.341	.066	.065	.091	.233	.232	.282
1	.05	.239	.238	.255	.551	.549	.584	.194	.190	.231	.461	.457	.526
	.10	.360	.357	.376	.622	.662	.693	.291	.291	.356	.589	.587	.659
	.01	.088	.087	.091	.320	.318	.335	.070	.069	.098	.242	.242	.312
2	.05	.241	.237	.260	.531	.528	.562	.199	.198	.251	.486	.482	.565
	.10	.349	.348	.365	.656	.656	.679	.295	.295	.358	.616	.616	.683
	.01	.090	.089	.089	.293	.292	.323	.071	.071	.094	.258	.256	.325
3	.05	.215	.213	.244	.528	.525	.562	.205	.199	.266	.504	.500	.579
	.10	.325	.324	.346	.648	.647	.674	.324	.324	.384	.645	.643	.691
		LOGISTIC						LAPLACE					
	.01	.131	.128	.126	.383	.378	.394	.183	.182	.161	.536	.535	.501
1	.05	.283	.279	.293	.634	.629	.643	.360	.357	.348	.770	.767	.741
	.10	.397	.396	.417	.728	.727	.754	.478	.478	.468	.850	.850	.832
	.01	.120	.120	.131	.382	.380	.386	.177	.176	.154	.558	.557	.508
2	.05	.273	.273	.289	.633	.631	.642	.362	.360	.351	.767	.767	.745
	.10	.384	.383	.394	.731	.731	.752	.505	.505	.485	.854	.851	.840
	.01	.115	.113	.119	.373	.372	.381	.201	.197	.181	.581	.578	.543
3	.05	.270	.269	.283	.599	.598	.625	.419	.415	.392	.777	.777	.748
	.10	.375	.374	.389	.719	.719	.741	.530	.530	.524	.858	.858	.849

Table 5. Simulated Powers against Location Shifts for Different Approximations to the Null Distribution

Small Samples

$n_1 = n_2 = 10$

Location Shift		0.3σ			0.5σ			0.3σ			0.5σ		
Statistic		δ_1		δ_2	δ_1		δ_2	δ_1		δ_2	δ_1		δ_2
Type Used		I	III	III	I	III	III	I	III	III	I	III	III
Rep	α												
		NORMAL						UNIFORM					
	.01	.022	.022	.024	.057	.057	.070	.017	.017	.022	.035	.035	.048
1	.05	.099	.099	.106	.176	.176	.181	.080	.080	.090	.127	.127	.164
	.10	.171	.171	.170	.268	.268	.299	.136	.136	.172	.225	.225	.283
	.01	.030	.030	.031	.065	.065	.070	.018	.018	.021	.037	.037	.048
2	.05	.108	.108	.106	.176	.176	.188	.075	.075	.093	.126	.126	.167
	.10	.171	.171	.184	.270	.270	.289	.140	.140	.157	.222	.222	.263
	.01	.036	.036	.037	.067	.067	.079	.025	.025	.031	.049	.049	.055
3	.05	.113	.113	.116	.193	.193	.196	.084	.084	.101	.162	.162	.200
	.10	.178	.178	.182	.274	.274	.295	.168	.168	.187	.277	.277	.321
		LOGISTIC						LAPLACE					
	.01	.039	.039	.033	.074	.074	.076	.033	.033	.038	.111	.111	.112
1	.05	.102	.102	.108	.192	.192	.209	.148	.148	.148	.274	.274	.269
	.10	.173	.173	.181	.291	.291	.304	.238	.238	.232	.383	.383	.387
	.01	.024	.024	.022	.056	.056	.067	.036	.036	.044	.093	.093	.104
2	.05	.094	.094	.102	.210	.210	.223	.129	.129	.131	.249	.249	.255
	.10	.174	.174	.196	.309	.309	.322	.198	.198	.211	.354	.354	.355
	.01	.021	.021	.028	.068	.068	.076	.039	.039	.044	.106	.106	.111
3	.05	.106	.106	.109	.196	.196	.202	.121	.121	.127	.235	.235	.237
	.10	.178	.178	.182	.275	.275	.300	.202	.202	.200	.345	.345	.344

while Type VI is competitive for $N > 34$. We study the power of δ_1 under two approximations next.

3. SIMULATED POWERS OF δ_1 UNDER DIFFERENT APPROXIMATIONS

We conduct a simulation study to examine the power of δ_1 against a location shift of 0.3σ and 0.5σ, where σ is the standard deviation of the underlying distribution - taken to be normal, uniform, logistic and Laplace distributions. For large samples of size 40 each, the null distribution of δ_1 is approximated by Pearson Type VI and Type III curves and the simulated power is presented in Table 4 when $\alpha = .01,.05,.10$. For small samples of size 10 each, we approximate the null distribution of δ_1 by Pearson Type I and Type III curves, and present the simulated power in Table 5. The simulation study is based on 3000 independent samples from the underlying distributions. To indicate the sampling variations, the results are presented as three independent replicates of 1000 each. We also present the corresponding power of δ_2, when $\Delta_{I,J} = [R(X_I)-R(X_J)]^2$, a test equivalent to Wilcoxon test (Mielke et al., 1981), under Type III approximation.

4. CONCLUSION

From Table 4, we note that the power of δ_1 is higher under Type VI approximation than under Type III for large samples. For small samples, we find in Table 5 that the powers are essentially the same under Type I and Type III approximations. While this is true of δ_1, the behaviour of the MRPP statistic δ in general may be different.

We also note that for large samples from Laplace distribution (i.e. double exponential), δ_1 has higher power than δ_2, whereas under normal and uniform distributions, δ_2 performs better.

REFERENCES

Cochran, W.G. (1952). The χ^2 test of goodness of fit, Ann. Math. Statist., *23*, 315-345.

Elderton, W.P. and Johnson, N.L. (1969). Systems of Frequency Curves, Cambridge University Press, Cambridge.

Mielke, P.W., Berry, K.J. Brockwell, P.J. and Williams, J.S. (1981). A class of nonparametric tests based on multi-response permutation procedures, Biometrika, *68*, 720-724.

Mielke, P.W., Berry, K.J. and Johnson, E.S. (1976). Multiresponse permutation procedures for a priori classifications, Comm. Stat.-Theor. Math. A, 5, 1409-1424.

Williams, C.A. (1950). On the choice of the number and width of classes for the χ^2 test of goodness of fit, J. Amer. Statist. Asso., *45*, 77-86.

Yarnold, J.K. (1970). The minimum expectation in χ^2 goodness of fit test and the accuracy of approximations for the null distribution, J. Amer. Statist. Asso., *65*, 864-886.

Classification and Recognition

N.G. Zagoruiko, Novosibirsk

Summary

Proposed is a unique approach to solving problems of automatic classification and pattern recognition. Described is a new class of decision functions, based on this approach, - taxonomic decision functions (TDF) possessing high stability against the breaking the representation law of training sampling.

Key words: recognition, classification, pattern, cluster, decision functions

1. Among all kinds of problems of statistical analysis of data (Zagoruiko, 1981) the two problems - pattern recognition and automatic classification (cluster analysis) - are most frequent and have been studied well enough.

In the set-ups of the problems there is much in common, but their characteristic peculiarities result in the fact that principally different methods are used for their solution.

The goal of the present paper is to describe the problems of cluster analysis and of pattern recognition in the same terminology and to formulate methods of solution, common for these problems.

2. Describing both these problems we imply or indicate explicitly the following elements:

a) general totality G . This is a set of objects of the world under study, whose power is unknown and is usually assumed to be infinite;

b) training sampling $\mathcal{A}_o$ - a finite set of objects taken occasionally from G . This sampling is supposed to be "representable", i.e. it is supposed that regularities that can be found on the sampling $\mathcal{A}_o$ will exist for unbounded random supplement of $\mathcal{A}_o$ by new objects from G ;

c) test sampling $\mathcal{A}_K$ - a finite set of objects occasionally taken from the set $G / \mathcal{A}_o$;

d) description space X which is formed by a finite selection of signs $(x_1, \dots, x_j, \dots, x_n)$ characterizing measurable properties of objects of the set G . The signs of X may be of various types (qualitative, ordinal nominal, etc.). Dimensionality (n) of the space X may be large, which makes it difficult for a man to perceive regular connections between various objects of the set A_0 and A_K or between various signs from X ;

e) space of specific signs $Y=(y_1,\ldots, y_l)$ characterizing properties the interest of which induces the set-up of all the problems of data analysis. In recognition and classification problems the space Y contains a sign y_0 measured in the denomination scale. It acquires a finite number (K) of values: $y_{01},\ldots, y_{0i},\ldots, y_{0K}$.

f) cluster (or taxon) - a subset of objects from A_0 , the relations between them only obey some conditions F . Most often, the requirements of "compactness", maximal similarity of cluster objects are included into the conditions F ;

g) pattern - a subset of objects from A_0 having the same values of the specific sign y_0 . A pattern may consist of one or several clusters;

h) decision function D - a function assigned analytically or algorithmically and connecting the values of the signs of X and a specific sign of y_0 :

$$y_0 = D(X).$$

By means of the function D the description space X is divided into K non-intersecting subspaces S_i $(i=1,2,\ldots, K)$ which are not necessarily simply connected. The objects of the space G which are mapped into one of such subspaces S_i are considered to belong to the i-th pattern;

i) subspaces S_i may be described by their boundaries or by a list of objects (of "standards"), by the distance to them one can decide to which subspace S_i any point of the space X belongs.

3. In the traditional set-up the problem of pattern recognition consists of two stages: training and recognition. At the training stage, by the values of X and y_0 presented in the training sampling A_0 , one must construct a function D of some class of decision functions D^* , such that, by the values of signs of X , allows one to determine the values $(\hat{y}_{i0})$ of a specific sign y_0 , the number of errors at the training sampling (i.e. the number of non-coincidences between y_{0i} and $\hat{y}_{0i}$)

should be minimal. Unlike the training sampling, the test sampling A_K does not contain information about the values of the specific sign. These values should be predicted, therefore the second stage - recognition - implies the application of the function D to take decision about the belonging of any object of the test sampling to this or that pattern, i.e. to predict $\hat{y}_{oi}$ by the values of its properties X .

Pay attention to the fact that the information about the recognizable objects from the test sampling A_K is in no way used at the training stage. The decision rule D is constructed by A_0 and does not depend on what properties the objects of the set A_K have.

4. Describe now the usual set-up of classification problems. The value of the specific sign is unknown here for the objects of the set A_0 as well. The problem is to predict y_0 for all the objects of A_0 . Objects, having the same values of the sign y_0 , refer to one and the same cluster. Various clusterization variants are compared by some quality criterion, and the variant is chosen for which a criterion F reaches the extremal (for instance, maximal) value.

There exist different kinds of the criterion F , but their principal general idea of construction is that the "similarity" maximum between the objects from one and the same taxon should be provided, as well as the "difference" maximum between objects of different taxons. Information of similarity and difference is contained in the signs X of the objects of A_0 .

The classification stated as the result of cluster analysis is not only destined to obtain more economical description of the set A_0 . If the sampling A_0 reflects the structure of the general totality G , then the classification obtained on A_0 coincides with the classification on G . Consequently, clusters S_i may be used as patterns to recognize any objects of the set G / A_0 , in particular, to recognize the objects of the test sampling A_K.

However, at such recognition, the artificiality of division of a sampling into the training one A_0 and the test one A_K becomes noticeable: in the initial data the specific sign y_0 is unknown both in A_K and in A_0 . It will be natural if we combine the samplings A_0 and A_K , and make the classification on this combination. Such a classification will automatically better reflect the cluster structure of the general totality G.

5. Let us turn back to the recognition problem and see if it is reasonable to perform the training and recognition procedures at the simultaneous usage of training and test samples.

Remind that in training and recognition we are interested in natural connections between X and y_0 which are formally expressed by the function D and are based on the fact "similar" by X objects have the same values of the sign y_0.

"Similarity" of objects of one pattern and "difference" of objects of different patterns should manifest themselves on both A_0 and A_K, therefore it is reasonable to disclose and use these relations between objects, analysing the samplings A_0 and A_K together. In this case the training and recognition procedures merge into one procedure whose contents are as follows.

Mixture of the samplings A_0 and A_K undergoes cluster analysis to obtain K_1 clusters, $K_1 \gg K$, where K is a number of patterns. The objects of A_K, that are included into the cluster containing only the objects of A_0 of the i-th pattern, refer to the i-th pattern. If in some cluster there are representatives of A_0 belonging to different patterns, such a cluster undergoes additional division into smaller clusters.

Clusters containing only the objects from A_K either are said to belong to the $(K-1)$-th pattern which is not represented in the training sampling or are joined to the "nearest" pattern.

6. The above rules of taking decisions in recognition are called "taxonomic decision functions" (TDF) (Zagoruiko, 1972).

Fig. 1 shows a training sampling for two patterns (circles stand for the 1st pattern, crosses - for the 2nd one). Dark dots denote test sampling objects. The rule D, obtained as a result of traditional training, is shown by the solid line, the solutions, obtained by means of TDF, are shown by the dotted line.

It is seen from Fig. 1 that the rule of TDF using information not only from A_0, but from A_K as well allows one to show the regularity of the general totality better, even if the sampling A_0 is non-representable.

The rule of TDF demands great consumption of computer time and storage for constant keeping of the training sampling A_0. But is does not possess (as compared to the rule D') larger resistance to the breaking of the sampling A_0 representability condition which is practically uncheckable. Such an effect is achieved by means of more complete use of information on the structure of general totality which is contained in both training and test samplings.

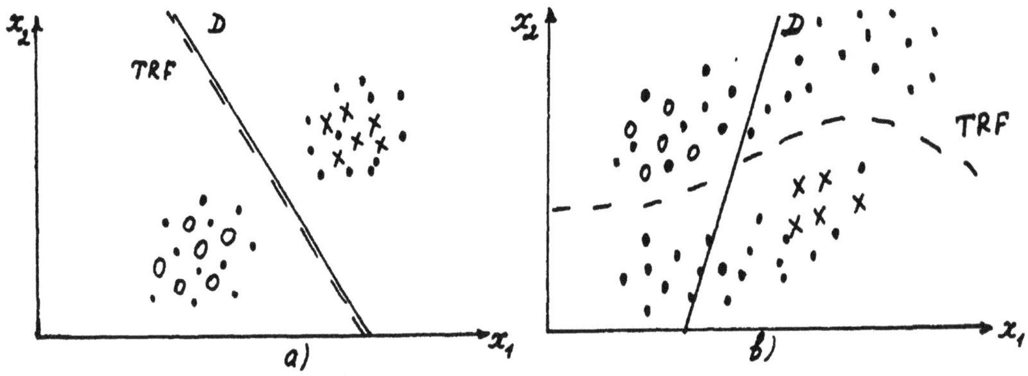

Fig. 1. The rule D, constructed by A_0, and the TDF rule constructed by $A_0 \cup A_K$, under the condition that the sampling A_0 is representable (a) and non-representable (b).

7. What cluster precedures may be used to work by the TDF method? Probably, any procedures suitable in every particular case.

The wider the range of the problems of cluster analysis being studied, the oftener the necessity to deal with still more complicated initial data: the amount of data increases; the same data array may contain data of various types (cardinal, ordinal, nominal); to adequately describe the structure of sampling, simple geometric elements (hypershehres or hypersurfaces) are not sufficient; cluster boundaries are not denoted uniquely (groups of points against the uniform "grey" background); data tables have gaps (there are gaps in data tables) - some properties X of a number of objects turn out to be unknown. These peculiarities of real problems of data analysis show the way of development of the cluster analysis technique. Let us describe some algorithms applied when solving the recognition problems by TDF methods.

8. For the preliminary stopping the gaps in data tables let us use the ZET algorithm (Zagoruiko, 1983). Its main idea is that the gap (b_{ij}) belonging to a sign x_j on an object a_i is stop-

ped according to linear regression equations between the values of the sign x_j and all the rest of the signs of the system X. The same gap is stopped by using linear dependences between the properties of the object a_i and of all other objects of the sampling under analysis. The values of b_{ij} computed by all such regression equations are weighted when obtaining the final decision. The weight of every particular solution is taken to be proportional to correlation coefficient between two series of numbers participating in the construction of the given regression. As a result of such weighting, the main contribution to the estimation of the element b_{ij} is paid by the signs which are most closely connected with the sign X_j, and the objects which are most like the object a_i. The ZET algorithm can be used to discover errors in the initial data arrays. An error is usually indicated by a great difference between the value of b_{ij} in the data table and the value of b_{ij} predicted by the ZET algorithm. We apply this algorithm to solve the problem of dynamic series continuation (subsequent stopping of gaps showing the next in turn value of the dynamic series).

9. The samplings A_0 and A_K (which do not contain gaps) are classified in our programs by means of algorithms of the types FOREL and KRAB. The algorithms of the FOREL type (Zagoruiko, 1972) handle the spherical clusters. The algorithm COLAPS allows one to obtain clusters as spheres of various diameters which is equivalent to the construction of quadratic decision functions. Besides, it solves the problem of pointing clusters out in the "grey" background (see Fig. 2).

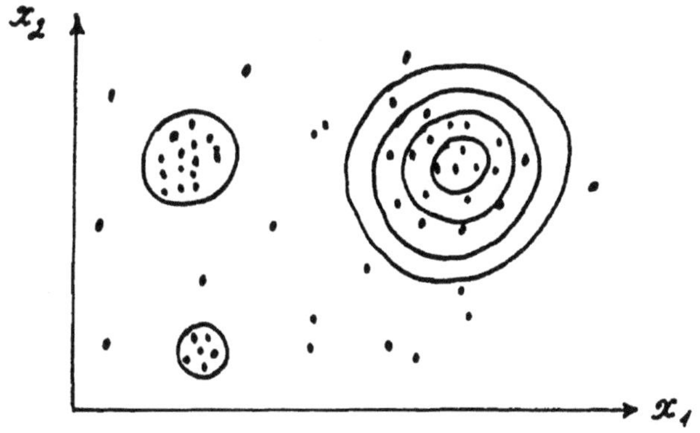

Fig. 2. The result of the algorithm COLAPS operation.

To get an answer whether the given local group of points is a cluster or not, the radius of the sphere is successively enlarged, and the density of objects in the obtained sphere is considered. The threshold value of the rate of density decrease of objects allows one to choose the cluster's boundaries.

10. Clusters of a more complicated (arbitrary) form can be obtained by means of the algorithm KRAB (Zagoruiko, 1972; Yolkina, Zagoruiko, 1978; Zagoruiko, Yolkina, 1982).

The criterion F of the quality of its operation is based on the use of the measure (ρ) near the objects inside the cluster, on the distance (d) between clusters, on the measure (h) of equivalence of clusters, and on the magnitude (λ) estimating local changes in the distribution of objects.

The solutions obtained by means of the criterion $F = \frac{\rho \cdot h}{d \cdot \lambda}$, usually coincide with those taken by a man in the manual (visual) way. Boundaries between clusters may have any form as is shown in Fig. 3.

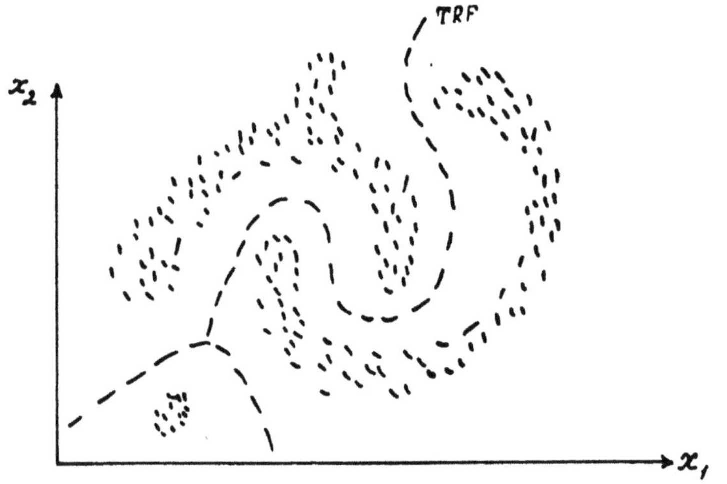

Fig. 3. The result of operation of the algorithm KRAB.

11. Two ways are possible to analyse data of different types. The former consists in reducing the signs measured in "weak" scales (the nominal and the ordinal ones) by means of "figuring" to the signs with a stronger scale (to "quantitative" ones). After such a strengthening of scales one may use algorithms destined to work in metric spaces.

The latter way consists in making the information more rough, in the rendering of the initial information into the language of the first-level predicates. By the values of these predicates ("true", "false") one may make conclusions about the "nearness", "similarity" of objects, and, consequently, point out clusters and construct decision functions. The transition of non-binary signs to predicates, if not done occasionally, is connected with the analysis of an unacceptable great number of variants. The directed search for a good combination of predicate values is investigated in the papers by G.S.Lbov (Lbov, 1982). The "logical decision functions" developed by him possess high statistical stability, they are obvious and simple in manipulation.

12. It has been mentioned above that taxonomic decision rules are complicated for computer realization. If one is sure that the sampling A_0 is representable enough, then it is reasonable to construct the usual rule D, for instance, as a logical decision rule. On the basis of the abovesaid the rational scheme of solving classification and recognition problems can be thought of as follows.

For a small size of the training sampling A_0, when one is not sure that the sampling A_0 is representable, it is reasonable to apply taxonomic decision rules to the union of samplings A_0 and A_k. Representability tests are usually done by repeated tests on the test sampling objects. If the number of errors is smaller than some $\delta\%$, then the sampling A_0 is considered to be representable. In this case one may pass over from taxonomic rules to more economic logical decision rules. They can describe both the boundaries of clusters in classification and the boundaries of patterns in recognition.

REFERENCES

Zagoruiko N.G. (1981), Classification of forecast problems by the tables "object-property", In: "Computer methods of discovering regularities" (Vychislitelnye sistemy, 88), Novosibirsk, p.3-8 (Russian).

Zagoruiko N.G. Recognition methods and their application. "Sovetskoje radio", Moscow, 1972. 207 p. (Russian).

Yolkina V.N., Zagoruiko N.G. (1978), Some classification algorithms developed at Novosibirsk. R.A.I.R.O. Informatique/Computer Science. Vol. 12, No 1, p.37-46.

Zagoruiko N.G. and Yolkina V.N. (1982), Inference and Data tables with missing values. Handbook of Statistics, vol. 2, North-Holland Publishing Company, p.493-500.
Lbov G.S. (1982), Logical Function in the Problems of Empirical Prediction. Handbook of Statistics, vol. 2, North-Holland Publishing Company, p.479-492.

Numerical Aspects and Complexity of Statistical Algorithms

Computational Aspects of Monte Carlo Tests

K.-H. Jöckel, Bremen

Summary: Especially in the field of spatial analysis Monte Carlo tests have attracted considerable attention. This paper is intended to discuss the relative merits of Monte Carlo tests and give some recommendations for the needed number of simulations based on a unified approach developed in Jöckel(1982) which includes the case of an arbitrary null distribution of the test statistic. Furthermore a short cut procedure is proposed which offers a considerable saving of simulation effort.
Keywords: Monte Carlo tests, testing hypothesis, simulation sample size

1. Introduction

In hypothesis testing problems especially in the field of spatial analysis statisticians often want to use unconventional test statistics the null distribution of which are either unknown or too complicate to evaluate. Thus since their proposal by Barnard(1963) the so-called Monte Carlo tests have become a well-established tool in this area, cf. Besag,Diggle(1977), Ripley(1977) and for the case of conditional testing Lotwick,Silverman(1982). Independently, but not recognized in this field, there was an earlier proposal of Monte Carlo tests due to Dwass(1957), who considered the special case of the Pitman two sample permutation test(this having a discrete null distribution). Most of the theoretical papers, however, as e.g. Hope (1968), Birnbaum(1974) consider the case of a continuous null distribution. The considerations in this paper are based on a unified approach developed in Jöckel(1982) allowing for an arbitrary distribution of the test statistic.
This approach is briefly described in section 2. In section 3 we discuss the choice of simulation sample size which is a critical question for the design of Monte Carlo tests. The recommendations are based on the study of power and asymptotic efficiency. Another point which is discussed in some detail is the interpretability in terms of different results of Monte Carlo and exact test. In section 4 we describe a short cut procedure (exact, not sequential as in Wall,Meystre(1974)) which may lead to an essentially smaller number of simulations.

2. Description of the Monte Carlo test procedure

Consider a test statistic T which we want to use for testing the hypothesis Θ_o against the alternative $\Theta_1 = \Theta - \Theta_o$. For simplicity of notation we assume w.l.o.g. small values of T to be significant. Let $F_\theta(t) = P_\theta(T \le t)$ denote the distribution function of T if the parameter $\theta\varepsilon\Theta$ applies. Furthermore we assume that there exists a parameter θ_o in the hypothesis, s.th. $F_\theta(t) \le F_{\theta_o}(t)$ for all $\theta\varepsilon\Theta_o$ (in many situations $\{F_\theta : \theta\varepsilon\Theta_o\}$ consists of one single element).^{o}
The exact test will then be based on the critical value of the distribution function F_{θ_o} which is often not easy (sometimes even impossible) to obtain. We assume,

however, that it is possible to generate a random sample of simulation size m $t_1,\ldots,t_m$, where the t_i's are realizations of i.i.d. random variables distributed according to F_θ. If $F_{\hat\theta}$ is continuous and the desired testing level α is an integer multiplier of $1/(m+1)$ ($\alpha=k/(m+1)$) the Monte Carlo test (MC test of simulation size m) ϕ_m is: reject Θ_0 if the observed value of the test statistic t is less of equal to $t_{k:m}$ (the k-th order statistic of the t_i's), accept otherwise, equivalently:

$$(2.1)\quad \phi_m = \begin{cases} 1 \\ 0 \end{cases} \text{if } \sum_{i=1}^{m} 1_{(-\infty,t]}(t_i) \begin{matrix} \leq \\ > \end{matrix} k-1 .$$

This procedure is of exact size α, cf. Birnbaum(1974), Hope(1968). For the case of an arbitrary null distribution (e.g.conditional tests as permutation tests, cf. Dwass(1957)) this is no longer the case. As has been shown in Jöckel(1982) the appropriate Monte Carlo test then is

$$(2.2)\quad \phi_m \begin{cases} 1 & \sum_{i=1}^{m} 1_{(-\infty,t]}(t_i) \leq k-1 \\ 0 \quad \text{if} & \sum_{i=1}^{m} 1_{(-\infty,t)}(t_i) \geq k \\ \gamma & \text{otherwise} \end{cases}$$

where the randomizing constant is $\gamma = \dfrac{k - \sum_{i=1}^{m} 1_{(-\infty,t)}(t_i)}{\left(\sum_{i=1}^{m} 1_{\{t\}}(t_i)\right) + 1}$.

People minding the use of randomized tests may instead of (2.2) take (2.1),which may result in a loss of power, cf. Jöckel(1982). It must, however, be emphazised that the MC test is a randomized test (being randomized everywhere, so that an additional randomization should not make too much troubles.

If $\beta(\alpha)$ denotes the power of the exact test of size α based on T under a fixed alternative parameter $\theta_1 \varepsilon\, \Theta_1$ and $\beta_m(\alpha)$ denotes the respective quantity for the MC test ϕ_m then

$$(2.3)\qquad \beta_m(\alpha) = \int_0^1 \beta(\xi)\; b\,(\alpha,m,\xi)\; d\,\xi \quad ,$$

where $b(\alpha,m,\cdot)$ is the density of a Beta distribution on the unit interval with parameters $p=\alpha(m+1)$ and $q=(1-\alpha)\;(m+1)$. Although $\beta_m(\alpha)$ is defined only on the grid $\{1/(m+1), 2/(m+1),\ldots.m/(m+1)\}$ it has turned out to be extremely useful to extend the range and regard (2.3) as a definition of the power of the MC test as a function of the level α for all $\alpha\varepsilon(o;1)$. For a more detailed discussion we refer to Jöckel(1982).

3. Considerations for the number of simulations

It has been mentioned earlier that the MC test is a test that is randomized everywhere. To make this more precise let us for the moment assume that $F_0(t):= F_{\theta_0}(t)$

is a continuous function. Then the exact test ϕ will reject the null hypothesis if $F_0(t)$ (t being the observed value of the test statistic) is less or equal to α, whereas the MC test ϕ_m decides only with a probability of $\sum_{j=0}^{k-1} \binom{m}{j} (F_0(t))^j (1-F_0(t))^{m-j}$ against the null hypothesis, viz.

$$\phi(t)=1_{[0,\alpha]}(F_0(t)),\quad \phi_m(t)=\sum_{j=0}^{k-1} \binom{m}{j} (F_0(t))^j (1-F_0(t))^{m-j} = \int_{F_0(t)}^{1} b(\alpha,m,\xi)\, d\xi \ .$$

Clearly as the simulated sample size increases, the blurring of the MC test(as Marriott(1978)puts it) diminished, viz.

$$\phi_m(t) \to \phi(t) \text{ as } m \to \infty, \text{ if } F_0(t) \neq \alpha \ .$$

As to the question how many simulations should be executed the discussion normally is based only on the power of the MC test as compared to the exact test, making distributional assumptions on the test statistic, see for example Hope(1968).Here we want to draw attention to a different point. The probability of achieving different decisions by the exact and the MC test (having observed t) is just $|\phi_m(t)-\phi(t)|$. Table 1 illustrates that this probability may be considerably high even for relatively large simulation samples.

Table 1: Probability[1] that the MC test decides different from the exact test depending on the level attained $F_0(t)$

α = 10 %

		$F_0(t)$							
m	$\alpha(m+1)$	0.02	0.05	0.08	0.09	0.10	0.11	.15	0.20
39	4	.008	.129	.381	.475	.557	.365	.143	.033
99	10	.001	.027	.280	.414	.536	.340	.059	.002
999	100	.000	.000	.013	.145	.506	.146	.000	.000

α = 5 %

		$F_0(t)$							
m	$\alpha(m+1)$	0.02	0.03	0.04	0.05	0.06	0.07	0.08	0.10
39	2	.183	.328	.466	.588	.311	.231	.169	.087
99	5	.049	.177	.363	.545	.284	.169	.094	.025
999	50	.000	.000	.065	.517	.079	.004	.000	.000

1) The calculations of $\int_{\mu}^{1} b(\alpha,m,\xi)\, d\xi$ have been done using an approximation of Peizer,Pratt(1968), the approximation error is $\leq$0.001 for the values given here.

If the true level attained (which of course we won't know) of the data is 0.08 for a test at the 5% level and the investigator decided to do only 99 simulations ($k=\alpha(m+1)=5$) then the probability of rejecting the null hypothesis opposite to what the exact test would have done is still 9.4%, giving two different statisticians a 17% chance of reporting different results for the same data. In our opinion this illustrates the risk of using only few simulations, even if the MC test is applied in a more explorative setting: Most frequently an unconventional test statistic will be used because the investigator hopes that it will reveal specific patterns of the data with a high sensitivity. This, however, makes sense only if this sensitivity is not lost because the MC test gives different answers than the exact one would have done. From a purely decision theoretic point of view of course only the 'on average' outcome of the MC test is of relevance. In the sequel this approach will be used to obtain some more concrete recommendations for the sample size of the simulation experiment.

Clearly, if $\beta(\cdot)$ is continuous at α then asymptotically ($m\to\infty$) the power of the MC test approaches that of the exact one, cf. Jöckel(1982), and under more restrictive conditions Birnbaum(1974),Hope(1968). Not so clear, however, is the question whether an increased simulation size always results in an increased power of the MC test. The answer to this question is positive, if $\beta(\cdot)$ is a concave function in α, cf. Jöckel(1982). In view of Jensen's inequality this concavity condition (which seems to be a natural one, since it guarantees that there does not exist a better test based solely on T) implies that the MC test is never better than the exact one, see Foutz(1981) and Jöckel(1981) for a discussion on this point. Since an increase of m gives a higher power of the MC test it would be helpful to determine inf $(\beta_m(\alpha)/\beta(\alpha))$ where the inf should be taken over a broad class of testing problems (here the only information on a testing problem relevant to our question is the power $\beta(\alpha)$).

This problem has already been tackled by Dwass(1957) for the special case of the Pitman two sample permutation test. We cite here without proof

<u>Theorem</u> (Jöckel(1982)): Assume that the power of a testing problem $\beta(\cdot)$fulfills the following conditions:

(i) $\lim_{\alpha\to 0} \beta(\alpha) = 0, \quad \lim_{\alpha\to 1} \beta(\alpha) = 1$

(ii) There exists $\alpha \in (0;1)$, s.th.

$$\frac{\beta(u)}{u} \geq \frac{\beta(\alpha)}{\alpha} \qquad \text{for all } u \in [0;\alpha].$$

Then the following inequality holds

$$\frac{\beta_m(\alpha)}{\beta(\alpha)} \geq \frac{1}{\alpha}\int_0^{\alpha}\left(\int_u^1 b(\alpha,m,\xi)\, d\xi\right) du =: e^D_{m,\alpha} \ .$$

$e^D_{m,\alpha}$ is called the Dwass efficiency (for simulation sample size m).

<u>Remark:</u> Conditions (i) and (ii) are rather natural, (ii) guarantees that for samll levels $\alpha' < \alpha$ the original test ϕ_α is at least as good as the randomzied test $\frac{\alpha'}{\alpha}\phi_\alpha$(if $\beta(\cdot)$ is concave this is trivially fulfilled). It may be shown, Jöckel(1982), that the Dwass efficiency may be approximated by

$e^D_{m,\alpha} \doteq 1-((1-\alpha)/(2\pi m\alpha))^{1/2}$.

As a consequence the simulation sample size $m(\alpha,e^D)$ to achieve Dwass efficiency of at least e^D with a MC test of size α may be approximated by

(3.1) $m(\alpha,e^D) \doteq \frac{1}{2\pi}\,\frac{1-\alpha}{\alpha}\,\frac{1}{(1-e^D)^2}$

For some selected values of α, e^D the corresponding values of $m(\alpha,e^D)$ are given in Table 2.

Table 2: Simulation sample size $m(\alpha,e^D)$ to achieve Dwass efficiency e^D for different levels α

α \ e^D	o.7	o.8	o.9	o.95	o.99
1 %	199	399	1599	6399	157599
5 %	39	79	319	1219	3o239
10 %	19	39	149	579	14329

From Table 2 we see that for a MC test of size 5 % in order to have at least 95 % of the power of the exact test it is necessary to do 1219 replicates. Of course in many situations the power loss will be less than 5%, since the Dwass efficiency considers the worst case possible.

Another way to get some notion of the behaviour of MC tests is the calculation of asymptotic resp. local asymptotic relative Pitman efficiency (ARPE resp. LARPE). For a detailed discussion of the (L)ARPE of MC tests the reader is referred to Jöckel(1982), where also explicit formulas are derived. Some values of LARPE in the case of an asymptotically normal test criterion are given in Table 3.

Table 3: Local asymptotic relative Pitan efficiency (LARPE) (in %) of MC tests (for simulation size m if the test criterion is asymptotically normal)w.r.t. the exact test

m	size α = 1 %	2.5 %	5 %	10 %
39	/	/	89.8	93.0
99	/	/	95.6	97.1
999	98.7	99.4	99.5	99.7

The results obtained so far can be summarized that it is not advisable to use too small simulation samples. The material presented here, especially Tables 1 and 2 should be helpful for users of MC tests in choosing their sample size. Of course, the actual choice is always a compromise between accuracy and computer costs. In our opinion, however, if m gets too small ($\alpha(m+1)< 10$) it must be asked whether it is wise to use MC tests or whether there exist alternative procedures.

Another point is that in many cases it will not be necessary to execute all m simulations as will be shown in the next section.

4. Efficient application of MC tests: a short cut procedure

Consider for the case of a continuous null distribution ($\alpha= k/(m+1)$) the MC Test as written in equation (2.1). Then the "test statistic" $F_m(t) = \frac{1}{m}\sum_{i=1}^{m} 1_{(-\infty,t)}(t_i)$ yields a strongly consistent, unbiased and asymptotically normal estimator of $F_o(t)$ (the true level attained). Thus we ware enabled to estimate the level attained consistently and unbiasedly and furthermore to construct asymptotic confidence intervals. For m sufficiently large we thus get - at least from a practical point of view - some more insight into the original testing problem. In our opinion this evidence should always be compiled. It remains, however, unsatisfactory that we have to draw a lot of random samples which may be very expensive and computer time consuming. But this needs not necessarily be the case:

Assume that during the simulation process for some index $l \leq m$ the number of simulated t_i's less or equal t already exceeds k, that is $\sum_{i=1}^{l} 1_{(-\infty,t]}(t_i) \geq k$. In this case the null hypothesis cannot be rejected whatever the results of the remaining simulations are. On the other side if $\sum_{i=1}^{l} 1_{(t,\infty)}(t_i) \geq m-k+1$ the null hypothesis has to be rejected.

If we define

$$z_0 = \inf\left\{l : \sum_{i=1}^{l} 1_{(-\infty,t)}(t_i) \geq k\right\}, \quad z_1 = \inf\left\{l : \sum_{i=1}^{l} 1_{(t,\infty)}(t_i) \geq m-k+1\right\}$$

then $z = \min(z_0, z_1)$ gives us the number of simulations that are necessary to obtain a final decision. Let us denote by z_0, z_1, z the corresponding random variables and observe that z_0-k, $z_1-(m-k+1)$ have a negative binomial distribution with parameters $k, (1-F_0(t))/F_0(t)$ and $m-k+1$, $F_0(t)/(1-F_0(t))$ respectively, then the distribution of z is easily derived, cf. Jöckel(1982). The expected simulation effort may be estimated by $E\,Z \leq (m+1)\min(\alpha/F_0(t), (1-\alpha)/(1-F_0(t)))$.

Observing the distribution results cited above, it is still possible to give estimates (based on the ML principle) for the level attained, viz.

k/l if the MC test could not reject the null hypothesis

$(l-m+k-1)/l$ if the MC test rejected the null hypothesis

after l simulations.

References:

Barnard,G.A.(1963), Discussion of Professor Bartlett's paper,J.R.Statist.,Soc., B 25, 294.
Besag,J.,Diggle,D.J.(1977), Simple Monte Carlo tests for spatial pattern, Appl. Statist., 26, 327-333.
Birnbaum,Z.W.(1974), Computers and unconventional test-statistics, in:Reliability and Biometry, SIAM, Philadelphia.
Dwass,M.(1957), Modified randomization tests for non-parametric hypothesis,Ann.Math. Stat., 28, 181-187.
Foutz,R.V.(1981), On the superiority of Monte Carlo tests, J.Statist.Comput.Simul., 12, 135-137.
Hope,A.C.A.(1968), A simplified Monte Carlo test procedure, J.R.Statist.Soc.,B 30, 582-589.
Jöckel.,K.-H.(1981), A comment on the construction of exact tests from test statistics that have unknown null distributions, J.Statist.Comput.Simul., 12, 133-134.
Jöckel,K.-H.(1982), Eigenschaften und effektive Anwendung von Monte-Carlo-Tests, Dissertation, Abteilung Statistik der Univ., Dortmund.
Lotwick,H.W., Silverman,B.W.(1982), Methods for analysing spatial processes of several types of points, J.R.Statist.Soc.,B 44, 406-413.
Marriott,F.H.C.(1979), Barnard's Monte Carlo tests: How many simulations?, Appl. Statist., 28, 75-77.
Peizer,D.B., Pratt, J.W.(1968), A normal approximation for binomial, F, Beta and other common related tail probabilities, JASA, 63, 1416-83.
Ripley,B.D.(1977): Modelling spatial patterns, J.R.Statist.Soc.,B 39, 172-212.
Wall,M.,Meystre(1974), Testing simple hypothesis by a Monte Carlo method with sequential decision procedure. Compstat 1974, 36-46.

On NP-Hardness in Hierarchical Clustering

M. Křivánek, and J. Morávek, Praha

ABSTRACT

We consider a class of optimization problems of the hierarchical-tree clustering, and prove that these problems are NP-hard. The sequence of polynomial reductions and/or transformations used in our proof is based on graph-theoretical techniques and constructions, and starts in the NP-complete problem of 3-dimensional matching.

Keywords

Hierarchical-tree clustering, NP-theory, Graph-theoretical techniques.

I. INTRODUCTION

The problems of single clustering, e.g. Fischer and vanNess(1971), Diday et al.(1979), were studied partially from the view-point of the computational complexity, see e.g. Brucker(1977), Gonzales (1982). On the other hand, to our knowledge, no similar results concerning the hierarchical clustering have been published. Our aim is to present a result in this respect. For the terminology and notation concerning the computational complexity see Garey and Johnson (1979).

Let us review and formalize the concept of hierarchical clustering, cf. e.g. Johnson(1967). Throughout this paper, n will denote an integer, $n \geq 3$, $\Omega = \{\omega_1, \omega_2, \dots, \omega_n\}$ an n-element set, and $D = (d_{i,j})$ a symmetric, real $n \times n$ -matrix with the property

$$d_{i,j} \geq 0 \text{ if } i \neq j, \text{ and } d_{i,j} = 0 \text{ if } i = j .$$

Elements of Ω are called <u>objects</u> (these are to be clustered), and D is called a <u>dissimilarity matrix</u>.

A <u>hierarchical tree</u> T <u>over</u> Ω is defined as a finite sequence of pairs

$$T = ((P_1, l_1), (P_2, l_2), \dots, (P_q, l_q)) ,$$

where (i) $P_1, P_2, \dots, P_q$ are partitions of Ω ;

(ii) $l_1, l_2, \dots, l_q$ are integers, $0 = l_1 < l_2 < \dots < l_q$;

(iii) P_k is a proper refinement of P_{k+1} $(1 \leq k \leq q-1)$;

(iv) $P_1 = \{\{\omega_1\}, \{\omega_2\}, \dots, \{\omega_n\}\}$ and $P_q = \{\Omega\}$.

Integer q is called the <u>height</u> of T and integer l_k the kth <u>level</u> of T. It follows from this definition that $2 \leq q \leq n$.

As an objective function for measuring the "quality" of the

hierarchical-tree clustering we use the function of Jardine and Sibson(1971)

$$T \longmapsto F(T) \stackrel{df}{=} \sum_{1 \le i < j \le n} | d_{i,j} - u_T(\omega_i, \omega_j) | ,$$

where

$$u_T(\omega_i, \omega_j) \stackrel{df}{=} \min \{ l_k \mid \text{there exists } M \in P_k \ (1 \le k \le q) \text{ such that } \{\omega_i, \omega_j\} \subseteq M \} ;$$

u_T is an ultrametric on Ω , cf. Jardine and Sibson(1971) .

Using F we introduce the following optimization problems :

(I) Problem HIC :

INSTANCE : Set of objects Ω and dissimilarity matrix D ;

PROBLEM : Determine a hierarchical tree T_* over Ω such that $F(T_*) = \min F(T)$, where the minimum is extended over the set of all hierarchical trees T over Ω .

(II) Problem HIC_q ($q \in \{2,3,\dots\}$) :

INSTANCE : Set of objects Ω and dissimilarity matrix D with $n \ge q$;

PROBLEM : Determine a hierarchical tree T_+ over Ω , with height q , and such that $F(T_+) = \min F(T)$, where the minimum is extended over the set of all hierarchical trees over Ω and with height q .

It is evident that problem HIC_2 has a polynomial time complexity. Our result is :

Theorem . Problems HIC and HIC_q for $q \ge 3$ are NP-hard. ∎

II. PROOF OF THE THEOREM

Our theorem will be obtained from the following series of lemmas:

Lemma 1. Problem HIC_q is reducible to problem HIC_{q+1} for $q = 2,3,\dots$ ∎

By virtue of Lemma 1 it is sufficient to prove the NP-hardness of HIC_3. In fact, we obtain a little stronger result. A dissimilarity matrix $D = (d_{i,j})$ will be called binary if each off-diagonal element of D equals either 1 or 2 . Now, let ^bHIC resp. $^bHIC_q$ denote the computational problem introduced in the precisely same way as HIC resp. HIC_q except that D is binary. It follows immediately from this definition that

(1) $\qquad {}^bHIC \propto HIC \quad$ and $\quad {}^bHIC_q \propto HIC_q \quad (q \ge 2).$

Lemma 2 . It holds that ${}^bHIC_3 \propto {}^bHIC$. ∎

In the next lemma we investigate the levels of a hierarchical tree, solving ${}^bHIC_3$.

Lemma 3 . If $T = ((P_1,0), (P_2,l_2), (P_3,l_3))$ is a solution of ${}^bHIC_3$ then $l_2 = 1$ and $l_3 = 2$. ∎

In the next lemma, problem ${}^bHIC_3$ is restated.

Lemma 4. For an arbitrary partition $\{I_1, I_2, \ldots, I_r\}$ of Ω let us set $i_\rho = \mathrm{card}(I_\rho)$ and $j_\rho = \mathrm{card}\ \{\{i,j\} \subseteq I_\rho \mid d_{i,j} = 1\}$.

Problem ${}^{b}HIC_3$ can be stated equivalently as follows : Find a partition $\{I_1, I_2, \ldots, I_r\}$ of Ω such that

$$\sum_{\rho=1}^{r} \left(\binom{i_\rho}{2} - 2j_\rho \right) \quad \text{is minimum .} \quad \blacksquare$$

For proving the NP-hardness of ${}^{b}HIC_3$ we shall use the following decision problem :

Problem EC5 (exact cover by ordered 5-tuples) :

INSTANCE : 1) a finite set X with $\mathrm{card}(X) = 5m$ for some positive integer m ;

2) a finite indexed family

$$\mathcal{C} = ((x_{\alpha,1}, x_{\alpha,2}, \ldots, x_{\alpha,5}))_{\alpha=1}^{\mathrm{card}(\mathcal{C})}$$

of ordered 5-tuples of elements of X with the property that each element of X occurs at least in one 5-tuple of $\mathcal{C}$;

QUERY : Decide whether $\mathcal{C}$ contains a subfamily $\mathcal{C}'$ such that each element of X occurs in exactly one 5-tuple of $\mathcal{C}'$.

Using the fact that the problem of 5-dimensional matching is NP--complete, see Kirkpatrick and Hell(1978), we obtain immediately :

Lemma 5. Problem EC5 is NP-complete . $\blacksquare$

Our next aim is to transform polynomially EC5 to ${}^{b}HIC_3$. For this aim we assign to each instance $(X, \mathcal{C})$ of EC5 an instance (Ω, D) of ${}^{b}HIC_3$. First, let us set

(2) $$n \overset{\mathrm{df}}{=} 5m + 25\ \mathrm{card}(\mathcal{C})$$

and

(3) $$\Omega \overset{\mathrm{df}}{=} X \cup \{y_{\alpha,\beta,\gamma} \mid 1 \le \alpha \le \mathrm{card}(\mathcal{C})\ ;\ 1 \le \beta, \gamma \le 5\},$$

where $y_{\alpha,\beta,\gamma}$ are 25 card($\mathcal{C}$) "new" objects joined to X. (It is appropriate to index these objects by triple subscripts.)

The dissimilarity matrix D will be introduced using certain graphs (we consider in the sequel finite undirected graphs without loops and parallel edges) : For each $\alpha \in \{1, 2, \ldots, \mathrm{card}(\mathcal{C})\}$ let us consider the graph $G_\alpha = (V_\alpha, E_\alpha)$, where

(i) The vertex-set is

$$V_\alpha \overset{\mathrm{df}}{=} \{x_{\alpha,1}, x_{\alpha,2}, \ldots, x_{\alpha,5}\} \cup \{y_{\alpha,\beta,\gamma} \mid 1 \le \beta, \gamma \le 5\} ;$$

(ii) The edge-set is (edges are represented as two-element subsets of the vertex-set) :

$$E_\alpha \overset{\mathrm{df}}{=} \{\{x_{\alpha,\beta}, y_{\alpha,\beta,\gamma}\} \mid 1 \le \beta \le 5,\ 1 \le \gamma \le 3\} \cup$$

$$\cup\{\{y_{\alpha,\beta,\gamma},\ y_{\alpha,\beta,\gamma'}\} \mid 1 \leq \beta \leq 5,\ 1 \leq \gamma < \gamma' \leq 5,\ \gamma' \neq 2\gamma + 1\} \cup$$

$$\cup\{\{y_{\alpha,\beta,5},\ y_{\alpha,\beta',5} \mid (1 \leq \beta < \beta' \leq 5) \text{and} (\beta' + \beta + 2 \text{ or } \beta \geq 3).$$

The graph G_α is shown in Figure 1.

Using graphs G_α we introduce the graph

$$G \overset{df}{=} (\bigcup V_\alpha,\ \bigcup E_\alpha) = (\Omega,\ \bigcup E_\alpha) \quad ,$$

where the union is extended over $\alpha \in \{1,2,\dots,\mathrm{card}(\mathcal{C})\}$.

Matrix $D = (d_{i,j})$ will be now defined as follows. We shall consider an arbitrary fixed numbering $\omega_1, \omega_2, \dots, \omega_n$ of elements of Ω and put

(4) $\qquad d_{i,j} = 0 \qquad$ if $\qquad i = j$,

(5) $\qquad d_{i,j} = 1 \qquad$ if $\qquad i \neq j$ and $\{\omega_i, \omega_j\} \in \bigcup E_\alpha$,

(6) $\qquad d_{i,j} = 2 \qquad$ otherwise.

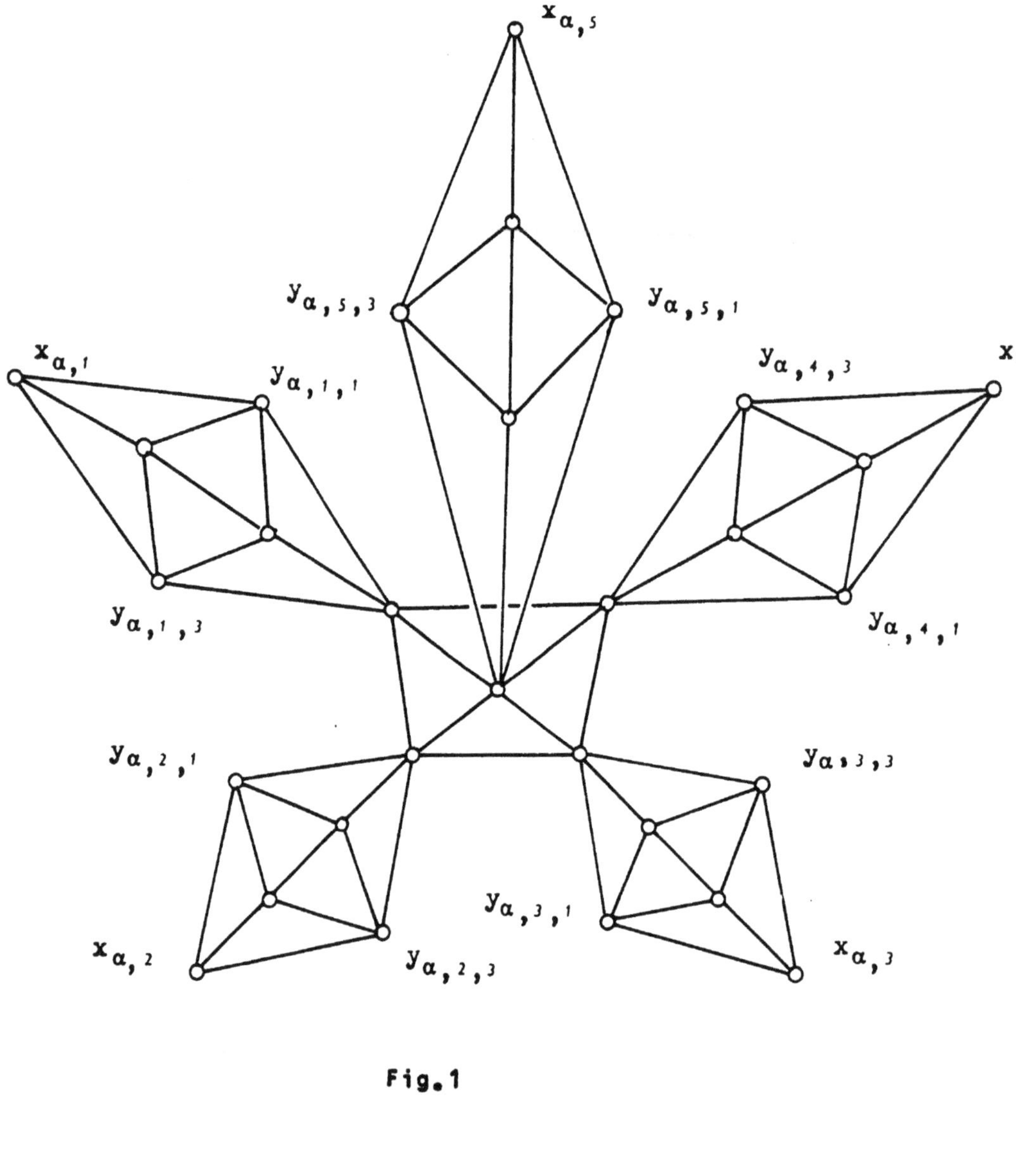

Fig.1

For continuing our path of presentation, we shall use some additional graph-theoretical definitions and notations : The term *subgraph* will allways denote an induced subgraph ; the subgraph of G induced by the subset $I \subseteq \Omega$ will be denoted by $G(I)$. A subgraph of G with 5 vertices and 8 edges, isomorphic to the graph shown in Figure 2, will be called *envelope* .

Fig.2 :

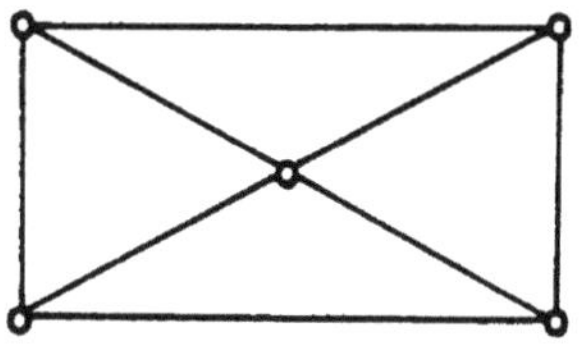

Let $\mathcal{E} = (W_\iota , H_\iota)$ be a finite set of envelopes ; W_ι is the vertex-set and H_ι the edge-set of the envelope (W_ι, H_ι) . $\mathcal{E}$ will be called a *vertex-partition* of the graph G *into envelopes* if $\{ W_\iota \}$ is a partition of Ω and $\bigcup H_\iota \subseteq \bigcup E_\alpha$.

Lemma 6. A solution of EC5 exists if and only if there exists a vertex-partition of G into envelopes. ∎

For $i \in \{1,2,\dots, \mathrm{card}(\Omega)\}$ let $j_{max}(i)$ denote the maximum number of edges in a subgraph $G(I)$ with $\mathrm{card}(I) = i$.

Lemma 7. It holds

$$j_{max}(1) = 0 , \quad j_{max}(2) = 1 , \quad j_{max}(3) = 3 ,$$
$$j_{max}(4) = 5 , \quad j_{max}(5) = 8 , \quad j_{max}(6) = 11,$$

and

$$j_{max}(i) \le \frac{11}{4} \left(i - \frac{18}{11} \right) \qquad \text{if} \quad i \ge 7 . \quad ∎$$

Using Lemma 7 we prove

Lemma 8. Let $I \subseteq \Omega$, $i = \mathrm{card}(I)$, and let j be the number of all edges in $G(I)$. Then

$$\binom{i}{2} - 2j + \frac{6}{5} i \ge 0 .$$

Moreover, $\binom{i}{2} - 2j + \frac{6}{5} i = 0$ if and only if $i = 5$ and $G(I)$ is an envelope . ∎

Now, let us put for each partition $\{ I_1, I_2, \dots, I_r \}$ of Ω

$$\psi (\{ I_1, I_2, \dots, I_r \}) \overset{\mathrm{df}}{=} \sum_{\rho=1}^{r} \left(\binom{i_\rho}{2} - 2 j_\rho \right) ,$$

where i_ρ , j_ρ are defined as in Lemma 4. It is easily observed that j_ρ equals the number of all edges in the subgraph $G(I_\rho)$.

Using Lemma 8 we prove :

Lemma 9. Let $\{ I_1, I_2, \dots, I_r \}$ be a partition of Ω . Then

$$\psi (\{ I_1, I_2, \dots, I_r \}) \ge -6(m + 5 \,\mathrm{card}(\mathcal{C})) .$$

Moreover, $\psi (\{ I_1, I_2, \dots, I_r \}) = -6(m + 5 \,\mathrm{card}(\mathcal{C}))$ if and only if

$$\{ G(I_\rho) \mid \rho = 1,2,\dots,r \}$$

is a vertex-partition of G into envelopes. ∎

Using Lemmas 6 and 9 we obtain :

Lemma 10. EC5 is polynomially transformable to $^{b}HIC_3$. ∎

Now, we can prove the announced result as follows : The NP-hardness of HIC_q for $q \geq 3$ follows from (1), Lemma 1, Lemma 5 and Lemma 10. The NP-hardness of HIC follows from (1), Lemma 2, Lemma 5 and Lemma 10 .

REFERENCES :

Brucker P.(1977), On the Complexity of Clustering Problems, in R.Henn, B.Korte, W. Oletti(eds.): Optimization and Operations Research, Springer-Verlag,Berlin, 45 - 54.

Diday E. et al.(1979), Optimisation en classification automatique, INRIA,Rocquencourt.

Fischer L., vanNess J.W.(1971), Admissible Clustering Procedures, Biometrika, 58 , 91-104.

Garey M.R.,Johnson D.S.(1979), Computers and Intractability : a Guide to the Theory of NP-completeness, W.H.Freeman, San Francisco.

Gonzales T.(1982), On the Computational Complexity of Clustering and Related Problems, in R.Drenick, F.Kozin(eds.): System Modelling and Optimization, Springer-Verlag, Berlin, 174 - 182.

Jardine N., Sibson R.(1971), Mathematical Taxonomy, Wiley, London.

Johnson S.C.(1967), Hierarchical Clustering Schemes, Psychometrika, 32 , 241 - 254.

Kirkpatrick D.G., Hell P.(1978), On the Complexity of Generalized Matching Problem, Proc. 10th Ann. ACM Symp. on Theory of Computing, 240 - 245.

Linear Models

Linweight: A Programm for Weighting Sample Survey Data

J.G. Bethlehem, Voorburg

Summary

In order to improve the quality of estimates in sample surveys often some kind of weighting is carried out. Post-stratification is a popular way to do this. Two major problems can make application of post-stratification difficult: empty strata and lack of enough population information. This paper presents a program which implements a general method for weighting which solves both problems. Weights are obtained from a linear model which relates the target variables of the survey to auxiliary variables.

Keywords
Post-stratification, weighting

1. Introduction

A sample survey is an instrument for making inferences about a finite population using observations on only a part of the elements in the population. If sufficient population information is available, estimates of population parameters can be improved. Weights are assigned to the observed elements in such a way that proper estimates are obtained by simple summation of the weighted observations.

This paper presents a general framework for weighting based on a estimator constructed from a linear model. It will be shown that classical weighting is a special case of the theory. Due to the generality of the theory it presents a number of other possibilities for weighting which are especially useful in situations where practical problems prevent ordinary weighting.

In this paper inference is based on probability sampling where randomization is introduced by the sampling design. Information about the population can be used to improve the efficiency of the estimates. If the sample size is large enough, this will never increase the variance. The variance reduction is determined by the extent to which the auxiliary information is able to explain the values of the target variable. More details on the theory can be found in Bethlehem and Keller (1983a, 1983b).

2. The general regression estimator

Let the finite population consist of N elements. Let Y be the N×q-matrix of values of q target variables and let X be the N×p-matrix of values of p auxiliary variables. Objective of the sample survey is assumed to be estimation of the population mean vector

$$\bar{y} = Y'\iota/N \tag{2.1}$$

where ι is a vector of N one's. Likewise the p-vector of means of the auxiliary variables is denoted by

$$\bar{x} = X'\iota/N \ . \tag{2.2}$$

In case of sampling without replacement the sample can be represented by an N×N-diagonal matrix T, the k-th diagonal element of which assumes the value 1 if element k is in the sample, and 0 otherwise. The expected value of T is equal to the N×N-diagonal matrix Π of first order inclusion probabilities.

If auxiliary variables are correlated with the target variables, they can be used to construct precise estimators. Such a relationship implies that for a suitably chosen p×q-matrix B of regression coefficients the residuals in the N×q-matrix E=Y-XB vary less than the target variables themselves. Application of ordinary least squares produces

$$B = (X'X)^{-1}X'Y \ . \tag{2.3}$$

A sample based estimator for B is equal to

$$\hat{B} = (X'\Pi^{-1}TX)^{-1}X'\Pi^{-1}TY \ . \tag{2.4}$$

Using (2.4) the population matrix Y can be predicted by $X\hat{B}$. The generalized regression estimator is defined as the mean of the predicted values:

$$\hat{\bar{y}}_R = \hat{B}'X'\iota/N = \hat{B}'\bar{x} \ . \tag{2.5}$$

Bethlehem and Keller (1983a) show that the bias of this estimator is of order n^{-1} if there exists a p-vector c of fixed numbers such that $Xc=\iota$. In that case the estimator can be rewritten as

$$\hat{\bar{y}}_R = \hat{\bar{y}}_{HT} + \hat{B}'(\bar{x} - \hat{\bar{x}}_{HT}) , \tag{2.6}$$

where $\hat{\bar{x}}_{HT}$ and $\hat{\bar{y}}_{HT}$ are the vectors of Horvitz-Thompson estimates for $\bar{x}$ and $\bar{y}$. The estimator is also proposed by a number of other authors, see e.g. Robinson and Särndal (1980) and Isaki and Fuller (1982), but, in contrast to our approach, its distributional properties are mostley studied under assumed superpopulation models.

The variance-covariance matrix of regression estimator (2.6) can be approximized by

$$V(\hat{\bar{y}}_R) \doteq E'\Delta E/N^2, \tag{2.7}$$

where element $\Delta_{k\ell}$ of Δ is equal to $(\pi_{k\ell}/\pi_k\pi_\ell)-1$, $\pi_{k\ell}$ is the second order inclusion probability, and π_{kk} is defined to be equal to the first order inclusion probability π_k. The variance-covariance matrix can be estimated by

$$\hat{V}(\hat{\bar{y}}_R) = \hat{E}'T\Lambda T\hat{E}/N^2, \tag{2.9}$$

where $\hat{E}=Y-X\hat{B}$, and element Λ_{kl} of Λ is equal to $(1/\pi_k\pi_\ell)-(1/\pi_{k\ell})$.

3. Complete multiway weighting

Regression estimators are usually applied in situations in which the auxiliary variables are quantitive variables. However, they can equally well be applied in case of qualitative auxiliary variables. Then regression estimation turns into stratification after selection of the sample. In case of one qualitative auxiliary variable this result is obtained by replacing the qualitative variable by as much dummy variable as it has categories. The h-th dummy variable assumes the value 1 if the element is in the h-th category, and 0 otherwise. Working out (2.5) results in the well-known post-stratification estimator, see e.g. Holt and Smith (1979). In case of simple random sampling the variance approximation turns out to be

$$V(\hat{\bar{y}}_R) \doteq \frac{1-f}{n} \sum_{h=1}^{L} \frac{N_h-1}{N-1} S_h^2 , \tag{3.1}$$

in which f is the sampling fraction, L is the number of strata/categories, N_h is the size of the h-th stratum and S_h^2 is the variance in this stratum. The approximation differs slightly from the approximation given in the sampling literature, see e.g. Cochran (1977). This type of stratification will be referred to as oneway weighting.

Often more than one auxiliary variable is used for stratification. The strata are constructed by crossing these variables. Weighting based on this type of stratification will be called complete multiway weighting. When carrying out complete multiway weighting two problems may present itself:

(1) If too many strata are constructed relative to the sample size, it is very well possible that there are strata without observations (thus preventing computation of weights) or strata with a single observation (thus preventing variance estimation).
(2) Application of stratification requires knowledge of the population totals in all strata. This information is not always available.

The first problem is usually dealt with by manually merging strata until all remaining strata have a sufficient number of observations. This can be a time consuming process. The second problem can be avoided by leaving out as much auxiliary variables as is necessary in order to arrive at a stratification about which population information is available. The next section proposes a more satisfactory solution to both problems. .

4. Incomplete multiway weighting

Incomplete multiway weighting can be carried out by not considering the complete crossing of all auxiliary variables. Instead a number of subsets of auxiliary variables is selected and for every subset complete multiway weighting is carried out (based on less variables than original complete multiway weighting). By proper specification of the designmatrix X these complete multiway sub-weightings can be carried out simultaneously. Regression estimation then turns out to be equal to assignment of weights which are equal to

$$w = \Pi^{-1}TX(X'\Pi^{-1}TX)^{-1}\bar{x} \ . \qquad (4.1)$$

Application of these weights to the auxiliary variables indeed produces the population means of these variables.

5. The program LINWEIGHT

The theory of weighting, as developed in the previous sections, is implemented in the program LINWEIGHT. The program automatically selects a suitable weighting scheme. To perform this task the program requires two data files: a file with sample data (target variables, auxiliary variables, inclusion weights) and a file with available population information. Furthermore the user must specify the sampling design. The program can handle stratified two-stage samples. Given the population data and the specified minimal number of observations per post-stratum the program selects a proper weighting scheme and computes estimates of population means and their estimated variance-covariance matrix.

The deck-setup of LINWEIGHT agrees to a large extent to the SPSS conventions. This facilitates the use of the program in an environment where SPSS is a frequently used data analysis package. An example of a deck-setup is given below.

```
RUN NAME          ESTIMATION OF THE MEAN INCOME
SAMPLE FILE       VARIABLES=PROV, MUN, MUNWGT, SEX, AGE, INCOME, EDUC, WGT /
                  CASES=10240
POPULATION FILE   VARIABLES=PROV, SEX, TOT / COUNT=TOT / CASES=22 /
                  VARIABLES=PROV, AGE, TOT / COUNT=TOT / CASES= 110
SAMPLING DESIGN   STRATUM=PROV /
                  CLUSTER=MUN / CLUSTERSAMPLING=PPS / CLUSTERWEIGHT=MUNWGT
                  SAMPLING=SRS / WEIGHT=WGT
STRATIFICATION    FILLING=15 /
                  AUXILIARIES=PROV(11), SEX(2), AGE(10) /
                  TARGETS=INCOME, EDUC
FINISH
```

The RUN NAME card specifies an identifying text. The SAMPLE FILE card gives the names and order of appearances of the variables in the sample file, and the number of cases. The POPULATION FILE card describes the structure of the population information. In the above example population totals are apparently available for each combination of province and sex, and for each combination of province and age (and not for each combination of province, sex and age). The variable TOT contains the population total for each combination of values of the auxiliary variables in the population file. The SAMPLING DESIGN card specifies the sampling design. In this case a stratified two-stage sample is selected. Strata are identified by the variable PROV, clusters by the variable MUN. Clusters are selected with unequal probabilities and the inclusion weights (i.e. the inverse inclusion probabilities) are contained in the variable MUNWGT. Within clusters elements are selected simple random without replacement and the inclusion weights are contained in the variable WGT. The STRATIFICATION card indicates how the weighting scheme must be selected:

PROV, SEX and AGE are auxiliary variables (number of categories within brackets), and each post-stratum must contain at least 15 observations.

Construction of a weighting scheme proceeds in two steps. First, those complete multiway sub-weighting schemes are selected, for which population totals are available. Secondly, from each such weighting scheme those sub-weighting schemes are selected for which each stratum contains at least the specified number of observations.

For the resulting weighting scheme the design matrix X is constructed. Application of the theory enables computation of estimates of population means and their estimated standard errors. Furthermore the program produces a file of weights. If these are added to the sample data file, a package like SPSS can be used to produce tables of weighted estimates of population totals.

The program is written in PASCAL. Interested readers may contact the author.

References

Bethlehem, J.G. and W.J. Keller, 1983a, Weighting sample survey data using linear models. Internal CBS-report (Netherlands Central Bureau of Statistics, Voorburg).

Bethlehem, J.G. and W.J. Keller, 1983b, A generalized weighting procedure based on linear models. Proceedings of the Section on Survey Research Methods of the American Statistical Association, pp. 70-75.

Cochran, W.G., 1977, Sampling techniques, 3rd edition (Wiley, New York).

Holt, D. and T.M.F. Smith, 1979, Poststratification. Journal of the Royal Statistical Society A 142, pp. 33-36.

Isaki, C.T. and W.A. Fuller, 1982, Survey design under the regression superpopulation model. Journal of the American Statistical Association 77, pp. 89-96.

Robinson, P.M. and C.E. Särndal, to appear, Asymptotic properties of the generalized regression estimator in probability sampling. Sankhyā B.

Parametric Link Functions in Generalized Linear Models

R. Gilchrist, London, and A. Scallan, Loughborough

Summary.

In their original formulation of a generalised linear model, Nelder and Wedderburn assumed a known relation ('link') between the mean of each observation and its corresponding linear predictor. It is possible to allow unknown parameters in this link function and Scallan, Gilchrist and Green(1984) discuss a stable 2-stage algorithm for estimating these extra parameters. This paper notes how an extension of the concept of a link, namely the 'composite' links of Thompson and Baker (1981), can be further extended by treating them as having unknown parameters. This, for example, allows the fitting of models with a correlated error structure.

KEYWORDS : GLIM; PARAMETRIC LINK FUNCTIONS; COMPOSITE LINKS; TIME SERIES; CHOLESKY DECOMPOSITION

1. Introduction.

Generalised linear models provide a powerful tool for data analysis. The Nelder and Wedderburn(1972) formulation can be generalised to allow unknown parameters in the link function (i.e. the mean - linear predictor relation), and a 2-stage algorithm provides a stable way of estimating the parameters of this function (Scallan et al, 1984). This technique is related to the idea of direct likelihood inference (Aitkin, 1982).

Composite links (Thompson and Baker, 1981) widen the class of models within the generalised linear model framework; parametric composite links offer further advantages and we here illustrate an application of these to models with correlated errors. Use of quasi-likelihood with correlated errors is discussed.

2. Parametric Link Functions.

A convenient specification of a generalised linear model may be expressed in terms of :-

1) A vector of independent observations $y_1,...,y_n$, with the distribution of each y_i a member of the exponential family.

2) A set of explanatory variables $x_{k1},\ldots,x_{kp}$ available on each observation describing the linear predictior $\eta_k=\Sigma x_{kj}\beta_j$. where the β_j's are unknown parameters to be estimated. Here η_k is used to predict μ_k, the mean of y_k.

3) A known differentiable function which relates the mean of each observation to its linear predictor; this is the 'link function'.

In Scallan et al (1984), we discuss the idea of a parametric link function. Thus we set $\mu_k=h(\eta_k,A_1,\ldots,A_r)$, with h(.) a known function and $A_1,\ldots,A_r$ unknown parameters not contained in the linear predictor. This generalised form of link may, for example,be used to estimate the Generalised Logistic Curve or Box-Cox link with unknown exponent, i.e. where $\eta_k=(\mu_k^A - 1)/A$, $A \neq 0$ or $\eta_k=\ln(\mu_k)$ for $A = 0$.

2.2 Estimation of Parametric Link Functions.

The method of weighted least squares is known to provide a convenient way of finding maximum likelihood estimates for quite general models (see Jorgensen, 1983 and McCullagh, 1983). Thus, for example, Pregibon(1980) indicates how the parameters of a parametric link function might be estimated by fitting an extra explanatory variables to the model for each extra parameter A_i, $i=1,\ldots,r$. A related technique is given in Baker et al(1980). These techniques can be unstable and we have noted that a 2-stage algorithm can be more effective (further confirmation is given in Green, 1984).

This 2-stage technique distinguishes between the parameters β_j in the linear predictor and the extra parameters A_i. The β_j are estimated with the the A_i kept fixed and the likelihood is treated as a function of $\hat{\beta}(\underline{A})$ and $\underline{A}$.; this can then be maximised with respect to the A_i. This is effectively maximising the likelihood by moving up the profile likelihood as defined by Aitkin(1982). Of course this technique requires being able to evaluate $\partial\hat{\beta}(\underline{A})/\partial\underline{A}$. These quantities can also be found using weighted least squares. For example, for one extra parameter we obtain $X'F(A)X(\partial\underline{\beta}/\partial A) = X'F(A)\underline{Z}$, where $F(A)$ is an iteratively calculated diagonal weight matrix and $\underline{Z}$ is a modified dependent variable. Full details are given in Scallan et al(1984). The joint maximum likelihood estimates of $\underline{\beta}$ and $\underline{A}$ are thus found iteratively in turn.

2.3 Transformation of the Independent Variables.

Parametric links can be extended to the case where the explanatory variables are assumed to be of a general parametric form. This is the case originally considered by Box and Tidwell(1962). Thus, for example, we might consider whether time or some function of time is the more appropriate explanatory variable.

In this case the parameterc link is written as $\underline{\mu} = h(\underline{\eta}(\underline{\beta},\underline{A}))$ and, using the technique discussed in the previous section, we find that for one extra parameter,

$$X'F(A)X(\partial\hat{\underline{\beta}}/\partial A) = X'F(A)\underline{Z} \ ,$$

where $\underline{Z} = F(A)^{-1}(dF^*(A)/dA).(\underline{y}-\underline{\mu}) - \partial\underline{\eta}/\partial A)$, with $\partial\underline{\eta}/\partial A$ being interpreted as the strict partial derivative of $\underline{\eta}$ w.r.t. A and $F^*(A) = diag[\partial\eta_k/\partial A].F(A)$. Hence we find $\partial\hat{\underline{\beta}}/\partial A$ using weighted least squares and the estimation of $\underline{\beta}$ and $\underline{A}$ proceeds as in section 2.2.

3.Parametric Composite Links.

3.1.Definition.

Composite link functions were introduced by Thompson and Baker(1981) and may be explained briefly as follows. For some models we may wish to associate more than one set of systematic effects with each observation; so we write $E[\underline{y}] = \underline{\mu} = U\underline{\gamma}$, $\underline{\gamma} = h(\underline{\eta})$ and $\underline{\eta} = X\underline{\beta}$, where U is a known nxm matrix. This is the linear form of the composite link as the mean of each observation, μ_i, is a linear combination of the $\underline{\gamma}$'s.

Thompson and Baker showed that, if the observations come from a member of the exponential family, the maximum likelihood estimates of $\underline{\beta}$ can be found using weighted least squares.The iterative explanatory variable is $X^* = UFX$ and dependent variable is $\underline{Z}^* = UF\underline{\eta}+\underline{y}-\underline{\mu}$, where F is the usual diagonal weight matrix defined by the link function h(.) and the mean-variance relationship.

We define a parametric composite link in an analogous manner to the parametric link function of section 2.1, i.e. we assume that the matrix U contains certain unknown parameters which can be regarded as fixed for the purpose of estimating $\underline{\beta}$, but are themselves to be estimated from the data.

3.2.An Application.

Suppose we have observations $y_1,\ldots,y_n$ s.t. $\underline{y} \sim N(\underline{\gamma},\varphi\Sigma)$, where Σ is an nxn symmetric matrix which can be regarded as fixed except for certain unknown parameters and, as before, $\underline{\gamma} = h(\underline{\eta})$ with $\underline{\eta} = X\underline{\beta}$.

A convenient way of representing this model in the traditional g.l.m. framework is to find the Cholesky decomposition of Σ^{-1}, i.e. a lower triangular matrix, U, such that $\Sigma^{-1} = U'U$. It is then easy to show that $\underline{w} = U\underline{y} \sim N(\underline{\mu},\varphi I_n)$, where $\underline{\mu} = U\underline{\gamma}$. Thus, after this transformation, we have a parametric composite link as defined in the previous section and the parameters $\underline{\beta}$ can be estimated using weighted least squares.

In this case the explanatory variable is $X^* = UFX$ and dependent variable is $\underline{z}^* = UF\underline{\eta}+\underline{y}-\underline{\mu}$, where F is the usual weight matrix defined by the link function and the mean-variance relation. Thus, the composite link function has arisen following a transformation of the original data rather than being assumed inherent in the model. The estimates of the unknown parameters in U can be updated given the current estimates of $\underline{\beta}$ and the joint maximum likelihood estimates thereby found iteratively.

The above general technique was employed by Scallan(1984) as a means of estimating the correlation parameters of an autoregressive process. As noted in that paper, the required Cholesky transformation matrix can be found explicitly for AR(p), MA(q) and ARMA(p,q) processes of low order using the method of Booth and Smith(1982). However, in many cases, it may be easier to employ a direct computational algorithm rather than find these explicit expressions.

The advantage of having an explicit expression for U is that this can be used directly in the GLIM package (Baker and Nelder,1978), although in the next issue of GLIM a facility may be provided to allow users to employ their own FORTRAN subroutines. Users of high-level languages such as APL can, of course, tackle the Cholesky decomposition directly, however they then suffer from not having available the other facilities of a package such as GLIM (e.g. the model syntax of Wilkinson and Rogers, 1973)

3.3.Quasi-Likelihood.

It would, of course, be desirable to extend this technique to other distributions in the exponential family. However, there are few tractable likelihood techniques for dependent non-normal observations One possibility is to consider the idea of quasi-likelihood (Wedderburn, 1974) and its extension to dependent observations (McCullagh, 1983). We assume that $cov(\underline{y}) = \varphi V(\underline{\mu})$, where the elements of $V(\underline{\mu})$ are known functions of the mean vector $\underline{\mu}$. The quasi-likelihood for dependent observations is then defined as a function Q satisfying $\partial Q/\partial \underline{\mu} = V(\underline{\mu})^{-}(\underline{y}-\underline{\mu})$, where $V(\underline{\mu})^{-}$ is a generalised inverse of $V(\underline{\mu})$.

For this model, maximum quasi-likelihood estimates and their approximate variance-covariance matrix can be found using weighted least squares as it can easily be shown that the function Q satisfies many of the properties of a likelihood function. For details of this procedure see McCullagh(1983). The main difficulty appears to be in defining an appropriate form for the covariance matrix, $V(\underline{\mu})$.

In many situations it may be possible to consider the correlation structure and the mean-variance relationship of the observations separately. Thus, for example, we might consider models in which the covariance matrix can be written as $V(\mu)^{1/2}\Sigma.V(\mu)^{1/2}$, where $V(\mu)^{1/2}$ $(=\text{diag}[V(\mu_i)^{1/2}])$ defines the mean-variance relationship and Σ, which does not depend on μ, defines the correlation structure

A technique which is commonly used for independent observations is to look for a transformation which removes the dependence between the mean and variance, i.e. we look for a function f(.) s.t. var(f(y)) = const. . This technique can also be used for dependent observations if we assume the form of the covariance matrix given above, since we have var(f($\underline{y}$)) = Σ. The methods of section 3.2 can then be used to fit the model to the transformed data. It may be noted that an appropraite transformation may not exist for more general forms of covariance structure (see Seng, 1982).

4.References.

AITKIN, M.(1982). Direct likelihood inference. In GLIM82, ed. R.Gilchrist, New York: Springer Verlag.

BAKER, R.J., PIERCE, C.B. and PIERCE, J.M.(1980). Wadley's problem with controls. GLIM Newsletter, Dec, pp. 32-35.

BOX, G.P. and TIDWELL, P.W.(1962). Transformations of the independent variables. Technometrics, No. 4, pp.531-550.

GREEN, P.J.(1984). Iteratively reweighted least squares for maximum likelihood estimation and some robust and resistant alternatives. J.R.Statist.Soc., B, 46.

JORGENSEN, B.(1983). Maximum likelihood estimation and large sample inference for generalised linear and non-linear regression models. Biometrika, 70.

McCULLAGH, P.(1983). Quasi-likelihood functions. Ann.Statist., 11, pp. 59-67.

NELDER, J.A. and WEDDERBURN, R.W.M.(1972). Generalised linear models. J.R.Statist.Soc., A, 135, pp.370-384.

PREGIBON, D.(1980). Goodness of link tests for generalised linear models. Appl. Statist., 29, pp.15-24.

RICHARDS, F.S.G.(1961). A method of maximum likelihood estimation. J.R.Statist.Soc., A, 135,pp. 370-384.

SCALLAN, A.J.(1984). Fitting autoregressive processes in GLIM. GLIM Newsletter, No. 8.

SCALLAN, A.J., GILCHRIST, R. and GREEN, M.(1984). Fitting parametric link functions in generalised linear models. To appear in Computational Statistics and Data Analysis.

SENG, C.C.(1982). Covariance stabilizing transformations and a conjecture of Holland. Ann.Statist., 10, pp.313-315.

THOMPSON, R. and BAKER, R.J.(1981). Composite link functions in generalised linear models. Appl.Statist., 30, pp.125-131.

WEDDERBURN, R.W.M.(1974). Quasi-likelihood functions, generalised linear models and the Gauss-Newton method. Biometrika, 61, pp.439-47.

WILKINSON, G.N. and ROGERS, C.A.(1973). Symbolic description of factorial models for analysis of variance. Appl.Statist., 22, pp.392-399.

Statistical Analysis of Mixed Linear Models with Inhomogeneous Variances

J. Kleffe, Berlin

Summary: We present an easy way of extending estimation of unknown parameters and prediction of random effects to mixed linear models where random effects have inhomogeneous variances. This method is part of an interactive computer programm AMP 84 which handles arbitrarily unbalanced nested classification ANOVA models.

Key words: mixed linear models, ANOVA, MINQUE, estimation

1. Introduction: Homogeneity of variances in statistical models is often assumed for convenience not by conviction. This matters little for relatively small carefully designed data sets. But more often statisticians try to incoporate all data available by using powerful computer facility. Also computational ease is no longer the all important question. The new problem, however, which comes with this development, is that large data sets are rarerely homogeneous in their precision and that usual statistical analysis fails if it does not take such differences into consideration. This experience invited investigation of slightly generalized linear models

$$Y = X\beta + U_1e_1 + \dots + U_pe_p \, , \tag{1.1}$$

where the covariance matrices of the random effects vectors e_i are not longer a multiple of an identity matrix, but they are $n_i x n_i$-diagonal matrices of heteroscedastic variances σ_{ij} for at least some $i = 1,\dots,p$. Not all diagonal elements are different so that we assume

$$Ee_ie_i' = \sum_{j=1}^{c_i} \sigma_{ij} D_{ij} = \Sigma_i \, , \tag{1.2}$$

where D_{ij} is a diagonal matrix which satisfies

$$D_{ij}^2 = D_{ij} \, , \; D_{ij} D_{ij_*} = 0 \;\; j \neq j_* \tag{1.3}$$

for any given $i = 1,\dots,p$. The matrices $X, U_1 \dots U_p$ are known and β and σ_{ij} are the unknown parameters. Model (1.1) and (1.2) may be considered a special case of the general mixed linear model which is discussed in Rao and Kleffe (1980). But expressing general results in term of our particular model yields an easy and interesting way to handle the inhomogeneous variances. This way admits application of recently development result by Kleffe and Seifert (1984) on computation of

$$Y_0 = X'V^{-1}Y,\ t_i = U_i'\ (MVM)^+Y,\ i = 1,\dots,p\ , \tag{1.4}$$

where V is the covariance matrix of Y and $M = I-XX^+$. Weighted least squares estimators for fixed effects and best linear predictors for random effects vectors are simple functions of (1.4) and given by

$$\hat{\beta} = [Y_0(X)]^-Y_0;\ \hat{e}_i = \sum_i t_i\ , \tag{1.5}$$

where the l-th column of the matrix $Y_0(X)$ follows by computation of Y_0 for the l-th column of X. It makes no essential difference for computation of (1.4) whether $\sum_i$ is a diagonal matrix or a multiple of an identity matrix. But estimation of variance components needs some extra consideration.

<u>2. Main result:</u> Let $Z_{ii_*} = t_i(U_{i_*}) = U_i'(MVM)^+U_{i_*}$ be the matrix with columns obtained by computing t_i for each column of the matrix U_{i_*} and introduce $\tau_{ij} = D_{ij}t_i$ and

$$Z^{ii_*}_{jj_*} = D_{ij}Z_{ii_*}D_{i_*j_*} = \tau_{ij}(U_{i_*}D_{i_*j_*}). \tag{2.1}$$

$Z^{ii_*}_{jj_*}$ contains a certain block of the matrix Z_{ii_*} and zeros in all other positions. The following theorem shows how compute C. R. Rao's Minimum Norm Quadratic Unbiased Estimator (MINQUE) for the variances σ_{ij} based on τ_{ij}.

<u>Theorem:</u> Let σ_{ij} be estimable by invariant quadratic forms. Then, the MINQUEs for the individual variances σ_{ij} are given by the unique solution to

$$\sum_{ij} \| Z^{i,i_*}_{j,j}\|^2 \hat{\sigma}_{ij} = \|\tau_{i_*j_*}\|^2 \text{ for all } i_*j_*. \tag{2.2}$$

where $\|\cdot\|$ denotes Euclidian norm.

<u>Proof:</u> Let rank $(D_{ij}) = r_{ij}$ and construct matrices S_{ij} of order $r_{ij} \times n_i$ such that

$$D_{ij} = S_{ij}S_{ij}'\ ;\ S_{ij}'S_{ij} = I_{r_{ij}}\quad . \tag{2.3}$$

Then necessarily $S_{ij}'S_{ij_*} = 0$ for $j \neq j_*$ and every i. We may now express model (1.1) as

$$Y = X\beta + \sum_{ij} U_iS_{ij}S_{ij}'e_i = X\beta + \sum_{ij} U_{ij}e_{ij} \tag{2.4}$$

with $U_{ij} = U_i S_{ij}$ and $e_{ij} = S'_{ij} e_{ij}$. It follows from (2.3) and (2.1) that $Ee_{ij}e'_{ij} = \sigma_{ij} I_{r_{ij}}$ and $Ee_{ij}e'_{i_*j_*} = 0$ if $i \neq i_*$ or $j \neq j_*$.

Therefore, application of note 5 following theorem 5.1.1 in Rao and Kleffe (1980) to model (2.4) yields the MINQUE equations

$$\sum_{ij} \| U'_{ij}(MVM)^+ U_{i_*j_*} \|^2 \hat{\sigma}_{ij} = \| U'_{i_*j_*}(MVM)^+ Y \|^2 .$$

(2.2) follows by noting that $\| U_{ij}(MVM)^+ Y \| = \| D_{ij} t_i \|$ and $\| U'_{ij} (MVM)^+ U_{i_*j_*} \| = \| D_{ij} Z_{ii_*} D_{i_*j_*} \|$.

3. How to establish the MINQUE equations? For ordinary variance components models holds $Ee_ie'_i = \sigma_i I_{n_i}$ and the well-known MINQUE equations are

$$\sum_i \|Z_{ii}\|^2 \hat{\sigma}_i = \|t_i\|^2 \text{ for all } i_* \;. \tag{3.1}$$

But if the i-th random effect has inhomogeneous variances σ_{ij}, it is convenient to number the components of e_i as $e^{(i)}_{jk}$ so that

$$E(e^{(i)}_{jk})^2 = \sigma_{ij} \quad k=1,\dots r_{ij}. \tag{3.2}$$

The same numbering should be introduced for the components of t_i and the columns of U_i being $t^{(i)}_{jk}$ and $U^{(i)}_{jk}$, respectively. This convention makes it easy to express

$$\|\tau_{ij}\|^2 = \sum_{k=1}^{r_{ij}} (t^{(i)}_{jk})^2 \tag{3.3}$$

as a subsum of squares over all components of t_i. The quadratic form $\| t_i \|^2$, as it appears in (3.1), is obtained by summation of (3.3) over $j=1,\dots,c_i$. If the components of e_i and e_{i_*} both have inhomogeneous variances, then the element in the j,k-th row and j_*, k_*-th column of Z_{ii_*} is $Z^{ii_*}_{jk,j_*k_*} = t^{(i)}_{jk} (U^{(i_*)}_{j_*k_*})$. If follows from (2.1)

$$\|Z^{ii_*}_{jj_*}\|^2 = \sum_{k_*=1}^{r_{i_*j_*}} \sum_{k=1}^{r_{ij}} \left[t^{(i)}_{jk}(U^{(i_*)}_{j_*k_*})\right]^2 = \sum_{k=1}^{r_{i_*j_*}} \|\tau_{ij}(U^{(i_*)}_{j_*k_*})\|^2 . \tag{3.4}$$

This expression is obtained by computing the subsums of squares $\| \tau_{ij}\|^2$ for a subset of columns of U_{i_*}, only. $\| Z_{ii_*}\|^2$ follows by summation of (3.4) over j and j_*.

Let us denote the components of e_{i*} by $e_{k_*}^{(i*)}$, $k_*=1,\dots,n_{i*}$ if the components of e_{i*} have homogeneous variances σ_{i*}. Then (3.4) reduces to

$$\|z_t^{ii}\|^2 = \sum_{k=1}^{n_i} \sum_{k=1}^{r_{ij}} \left[t_{jk}^{(i)} U_{k_*}^{(i*)}\right]^2 = \sum_{k=1}^{n_i} \|\tau_{ij}(U_{k_*}^{(i*)})\|^2 \,, \tag{3.5}$$

where $U_{k_*}^{(i*)}$ is the k_*-'th column of U_{i*}. This way of a looking at deriving the MINQUE equations (2.2) is nicely applicable in ANOVA where the components of random effects vectors e_i are usually numbered by sets of indices describing the levels of those factors, the random effects depend on. It is then always possible to arrange that the variances of the single random effects depend on the levels of one or more given factors so that we have a natural indexing as required by (3.2). The subcripts j and k may stand for sets of subcripts. It turns out that computations for inhomogeneous variances differ from those for homogeneous variances by just omitting summation over the subcript gathered up in j.

<u>AMP 84:</u> The method of deriving the more complexe MINQUE equations (2.2), as outlined in section 3, forms the heart part of an interactive computer program AMP 84. It handles arbitrarely unbalanced nested classification models with inhomogeneous variances and provides weighted least squares estimators, best linear predictors, and several estimators for variance components inducing MINQUE and the maximum likelihood method. Commonly these methods involve heavy computations, requiring the inversion of large matrices. But AMP 84 is based on a new numerical algorithm with requires a minimum of computer memory by avoiding storage and inversion of matrices. The extension to inhomogeneous variances anables AMP 84 to analyse combined data from different sources by using an optimal way of pooling. A short description of the program is given by Hermann at al (1984). More insight into the mathematical background of AMP 84 and generalizations to multivariate models are given by Kleffe (1984).

References:

Hermann, K., Kleffe, J. and Seifert, B. (1984): Estimation, Prediction and Testing Hypotheses in Mixed Linear Models by using AMP84.

Kleffe, J. (1984): On heteroscedasticity in variance components estimation.

Kleffe, J. and Seifert, B. (1984): Computation of variance components by the MINQUE method.

all above: Report, Institute of Mathematics, Academy of Sciences of the GDR, 1086 Berlin, Mohrenstr. 39, PF 1304.

Rao, C.R. and Kleffe, J. (1980): Estimation of variance components. P.R. Krishnaiah ed. Handbook of Statistics Vol 1, 1-40 North Holland Publishing Company.

Reliability of Calculations in the Linear Estimation and some Problems of the Analysis of a Multistage Regression Experiment

L. Kubáček, Bratislava

Summary: The generalized matrix inversion is the base for:
- methods for checking the numerical correctness of estimates in a mixed regression model;
- proposal for a special type of an efficient stage estimator in a multistage regression model, numerically more stable than the classical one;
- estimating the variance components in a mixed regression model.

Key words: mixed regression model, multistage regression model, variance components, generalized matrix inversion, linear and quadratic estimation.

1. INTRODUCTION

The task to secure the numerical correctness of estimates is in general very important and complicated. In a regression experiment trobles grow with the growing dimension of the design matrix.

The aim of this paper is to point out some possibilities in checking the numerical correctness of estimates in regression experiments, further to show how to improve the numerical reliability in a multistage regression experiment and mainly to emphasize the importance of the program for generalized matrix inversion.

It may be said that this paper is an ode on the program for generalized matrix inversion, mainly on possibilities of its many-sided utilization in the linear and quadratic estimation.

2. SYMBOLS AND FORMULATION OF PROBLEMS

The following symbols will be used:

R^k k-dimensional real linear space,

A' transposition of the matrix A,

$r(A)$ rank of the matrix A,

$Tr(A)$ trace of the matrix A,

A^- generalized inversion of the matrix A (A^- is any solution of the equation $AXA = A$; see Rao and Mitra (1971)),

$M(A)$ column space of the matrix A; if A is n x k matrix then $M(A) = \{Au: u \in R^k\}$,

$A \otimes B$ tensor product of the matrices A and B,

$vec(A) = (a_1', a_2', \dots, a_k')'$ if $A = (a_1, \dots, a_k)$; a_i is the i-th column of the matrix A.

If the m x m matrix A is symmetric with (i,j)-th element $\{A\}_{i,j} = a_{i,j}$ then

$\text{vech}(A) = (a_{1,1}, a_{1,2}, \dots, a_{1,m}; a_{2,2}, a_{2,3}, \dots, a_{2,m}; \dots; a_{m-1,m-1}, a_{m-1,m}; a_{m,m})'$.

If the p x m^2 matrix A is $A = (a_{1,1}, a_{1,2}, \dots, a_{1,m}; a_{2,1}, a_{2,2}, \dots, a_{2,m}; \dots; a_{m,1}, a_{m,2}, \dots, a_{m,m})$, where $a_{i,j}$ is a column of the matrix A, then

$(cC)(A) = (a_{1,1}, a_{1,2} + a_{2,1}, \dots, a_{1,m} + a_{m,1}; a_{2,2}, a_{2,3} + a_{3,2}, \dots, a_{2,m} + a_{m,2}; \dots; a_{m-1,m-1}, a_{m-1,m} + a_{m,m-1}; a_{m,m})$;

(cR)(A), A is a m^2x p matrix, is defined as $(cR)(A) = ((cC)(A'))'$,

$E_\theta(Y)$ mean value of the random vector Y where θ is the considered parameter of its probability distribution,

$D_\theta(L'Y)$ dispersion of the random variable $L'Y$,

$\text{Var}_\theta(Y)$ covariance matrix of the random vector Y.

A regression model is denoted by the triplet $(Y, X\beta, \text{Var}(Y))$, where Y is a random vector, X is a given n x k design matrix, $E_\beta(Y) = X\beta$, β is an unknown parameter (1-st order parameter), $\beta \in R^k$ or $\beta \in \underline{\beta} \subset R^k$ ($\underline{\beta}$ is a linear variety in R^k). The covariance matrix $\text{Var}_\delta(Y)$ is considered in the form $\text{Var}_\delta(Y) = \sum_{i=1}^{p} \delta_i V_i$, where V_i, $i = 1,\dots,p$, are given symmetric matrices and δ_i, $i = 1,\dots,p$, are unknown variance components (2-nd order parameters). The parametric space in the estimation problem is $\underline{\theta} = R^k \times \underline{\delta}$ or $\underline{\beta} \times \underline{\delta}$, where $\underline{\delta} \subset R^p$ and the topological interior of the set $\underline{\delta}$ is not empty.

The structure of the matrix X, except its general form, will be considered in so called multistage form (see part 4 of this paper).

Following problems will be dealt with:

- verification of the numerical correctness of the estimation $\hat{\beta}$ in the model $(Y, X\beta, \text{Var}(Y))$;
- reduction of the amount of the computing in estimating the stage parameters in the multistage regression model;
- utilization of the program for generalized matrix inversion for estimating the parameter δ and for simultaneous estimating the parameters β and δ.

3. VERIFICATION OF THE NUMERICAL CORRECTNESS OF THE ESTIMATION IN THE UNIVERSAL REGRESSION MODEL

The regression model $(Y, X\beta, Var(Y))$ is universal when no restrictions on ranks of the matrices X and Var(Y) are given (Rao (1971b)); the matrix Var(Y) is at least positive semidefinite. It is easy to show that all situations occuring in the practice are special cases of the universal model (in detail see Kubáček (1978b)).

A linear function $g(.) : \underline{\beta} \to R^1$ of the form $g(\beta) = c'\beta$, where $c \in R^k$ is given, is unbiasedly estimable iff $c \in M(X')$ and that is why the vector function $g(\beta) = X\beta$, $\beta \in \underline{\beta}$, which represents all unbiasedly estimable functions is considered in the following.

Statement 1. The best linear unbiased estimator of the function $g(.)$ is $\tau_g(Y) = X((X')^-_{m(V)})'Y$, where $Var(Y) = \sigma^2 V$, $(X')^-_{m(V)}$ is minimum V-seminorm generalized inversion of the matrix X'. The matrix $(X')^-_{m(V)}$ is a solution of the system $X'(X')^-_{m(V)}X' = X'$, $((X')^-_{m(V)}X')'V = V(X')^-_{m(V)}X'$.

Proof see in Rao (1971b), Kubáček (1983) p.16.

A M-LS solution of the estimation problem for the function $g(\beta) = X\beta, \beta \in \underline{\beta}$, is $X\hat{\beta}$, where $X'MX\hat{\beta} = X'MY$, provided such a vector $\hat{\beta}$ exists.

For the verification of the numerical correctness three following statemats will be utilized.

Statement 2. The system $X'MX\hat{\beta} = X'MY$ is solvable and the estimator $X\hat{\beta}$ is the best linear unbiased estimator of the function $g(.)$ iff

(a) $M = (V + XUX')^- + K$,

(b) $r(X'MX) = r(X')$,

where the matrices U, K fulfil the following conditions

(c) $M(V,K) = M(V + XU'X') = M(V + XUX')$,

(d) $VK'X = 0$, $X'KX = 0$.

Proof see in Rao (1971b) .

Statement 3. In the universal model $(Y, X\beta, Var(Y) = \sigma^2 V)$ the following statements (A) and (B) are equivalent

(A): The system $X'MX\hat{\beta} = X'MY$ is solvable for each realization y of the vector Y and $X\hat{\beta}$ is the best linear unbiased estimator of the function $g(\beta) = X\beta, \beta \in \underline{\beta}$. Simultaneously $\hat{\sigma}_1^2 = (y - X\hat{\beta})'M(y - X\hat{\beta})/(r(V,X) - r(X))$ is such an estimate of the factor σ^2 which is equal to the realization of the estimator

$$\hat{\sigma}_2^2 = \{Y - X((X')^-_{m(V)})'Y\}'V^-\{Y - X((X')^-_{m(V)})'Y\}/(r(V,X) - r(X))$$

with probability one.

(B): The matrices U, K, Z satisfy the following conditions

(a) $M = (V + XUX')^- + K$,

(b) $r(X'MX) = r(X)$,

(c) $VK'X = 0$, $X'KX = 0$,

(d) $M(V,X) = M(V + XU'X') = M(V + XUX')$,

(e) $Z'V(K + K')VZ = 0$,

where Z is a matrix with the maximal rank such that $X'Z = 0$.

Further the "Pandora-Box" denotation

$$\begin{pmatrix} V, & X \\ X', & 0 \end{pmatrix}^- = \begin{pmatrix} C_1, & C_2 \\ C_3, & -C_4 \end{pmatrix}$$

is used.

Statement 4. In the universal model $(Y, X\beta, Var(Y) = \sigma^2 V)$ the best linear unbiased estimator of the function $g(.)$ is XC_3Y ($= XC_2'Y$ with probability one), $\hat{\sigma}_3^2 = Y'C_1Y/Tr(VC_1)$ is unbiased estimator of the factor σ^2, $\hat{\sigma}_1^2 = \hat{\sigma}_3^2$ with probability one ($\hat{\sigma}_1^2$ is the estimator from Statement 3). The covariance matrix of the estimator XC_3Y (and $XC_2'Y$ respectively) is $\sigma^2 XC_4X'$.

Proofs of the last two statements (author is Rao (1971b)) are in details given in Kubáček (1983) pp.153-160.

Statements 1 - 4 give the possibility to calculate the vector $X\hat{\beta}$ and the factor $\hat{\sigma}^2$ by infinitely many different ways, which is the base for checking their numerical correctness.

In actual cases the checking can sometimes continue. For example in the case $E_\beta(Y) = X\beta$, $\beta \in \underline{\beta} = (\beta : b + B\beta = 0)$, $Var(Y) = \sigma^2 V$, $r(X) = k \leq n$, $r(V) = n$, $r(B_{q,k}) = q \leq k$ the best linear unbiased estimator $\hat{\beta}$ of the parameter β is

$$\hat{\beta} = ((X', B')^-_{m\begin{pmatrix} V,0 \\ 0,0 \end{pmatrix}})' \begin{pmatrix} Y \\ -b \end{pmatrix} =$$

$$= \{I - (X'V^{-1}X)^{-1}B'(B(X'V^{-1}X)^{-1}B')^{-1}B\}\hat{\beta}_0 - (X'V^{-1}X)^{-1}B'(B(X'V^{-1}X)^{-1}. \; . B')^{-1}b,$$

where $\hat{\beta}_0 = (X'V^{-1}X)^{-1}X'V^{-1}Y$ (in details see Kubáček (1983) p.134).

The well known procedure of Hoerl and Kennard (1970a, 1970b) gives an another possibility in this domain of consideration. In this case, however, the estimators are biased.

4. THE MULTISTAGE REGRESSION MODEL

The structure of the individual elements of the triplet $(Y, X\beta, Var(Y))$ is now given by the relationships:

$$E_{\beta}(Y) = \begin{pmatrix} E_{\beta}(Y^{(1)}) \\ E_{\beta}(Y^{(2)}) \\ \vdots \\ E_{\beta}(Y^{(m)}) \end{pmatrix} = X\beta = \begin{pmatrix} X_1 & , 0 & ,\dots, 0 \\ C_{2,1} & , X_2 & ,\dots, 0 \\ \dots & \dots \dots & \dots \dots \\ C_{m,1} & , C_{m,2} & ,\dots, X_m \end{pmatrix} \begin{pmatrix} \beta^{(1)} \\ \beta^{(2)} \\ \vdots \\ \beta^{(m)} \end{pmatrix} ;$$

$$\mathrm{Var}(Y) = \begin{pmatrix} \Sigma_{1,1} & , 0 & ,\dots, 0 \\ 0 & , \Sigma_{2,2} & ,\dots, 0 \\ \dots & \dots \dots & \dots \dots \\ 0 & , 0 & ,\dots, \Sigma_{m,m} \end{pmatrix} ,$$

where $M(C'_{i,j}) \subset M(X'_i)$, $i = 1,\dots,m$, $j = 1,\dots, i-1$. (The meaning of the another condition $M(C_{i+1,i}) \subset M(X_i)$ see in Volaufová [15].)

This especial structure is typical for several application domains (metrology, geodesy, geophysics etc.).

In the multistage model it is required for the estimator of the i-th stage parameter $\beta^{(i)}$ to be based on the vectors $Y^{(1)},\dots,Y^{(i)}$ only. (The estimate of the i-th stage parameter $\beta^{(i)}$ must not be influenced by measurement in the j-th stage, for $j > i$.)

The following denotation

$$H_i = \begin{pmatrix} \Sigma_{1,1} & , 0 & ,\dots, 0 \\ 0 & , \Sigma_{2,2} & ,\dots, 0 \\ \dots & \dots \dots & \dots \dots \\ 0 & , 0 & ,\dots, \Sigma_{i,i} \end{pmatrix} , \quad i = 1,\dots,m,$$

will be used.

With respect to Statement 1 the best linear unbiased estimator of the i-th stage parameter $\beta^{(i)}$ based on $Y^{(1)},\dots,Y^{(i)}$ is

$$(*) \qquad \hat{\beta}^{(i)} = (0,0,\dots,0,I)\left[\left(\begin{pmatrix} X_1 & ,0 & ,0 & ,\dots,0 & ,0 \\ C_{2,1}, X_2 & ,0 & & ,\dots,0 & ,0 \\ \dots & \dots & \dots & \dots \dots & \dots \\ C_{i,1}, C_{i,2}, C_{i,3}, & \dots, & C_{i,i-1}, X_i \end{pmatrix}^{-}\right)'_{m(H_i)}\right]' \begin{pmatrix} Y^{(1)} \\ Y^{(2)} \\ \vdots \\ Y^{(i)} \end{pmatrix}.$$

This procedure is numerically instable and hardly realizable if the number of the stages is large. That is why the procedure characterized by relationships

$$\tilde{\beta}^{(i)} = (X'_i\Sigma^{-1}_{i,i}X_i)^{-1}X'_i\Sigma^{-1}_{i,i}\{Y^{(i)} - \sum_{j=1}^{i-1} C_{i,j}\tilde{\beta}^{(j)}\}, \quad i = 1,\dots,m,$$

is preferred in the practice. However it can shown that the estimate $\tilde{\beta}^{(i)}$ can be substantialy worse than the estimate $\hat{\beta}^{(i)}$ from (*). The way out from this dilemma is in changing the matrices $\Sigma_{j,j}$, $j = 1,\ldots,i$, in the expression for $\tilde{\beta}^{(i)}$ and in the suitable design of the multistage regression experiment. Let us start with $m = 2$.

Statement 5. If the estimators

$$\hat{\beta}^{(1)} = ((X_1')^{-}_{m(\Sigma_{1,1})})'Y^{(1)},\quad \hat{\beta}^{(2)} = ((X_2')^{-}_{m(D_{2,2})})'(Y^{(2)} - C_{2,1}\hat{\beta}^{(1)}),$$

where $D_{2,2} = \Sigma_{2,2} + C_{2,1}((X_1')^{-}_{m(\Sigma_{1,1})})'\Sigma_{1,1}(X_1')^{-}_{m(\Sigma_{1,1})}C_{2,1}'$, are used in the two-staged regression model, then

$$(**)\quad ((X_2')^{-}_{m(D_{2,2})})'(Y^{(2)} - C_{2,1}\hat{\beta}^{(1)}) = (O,I)\left[\left(\begin{pmatrix} X_1 & ,O \\ C_{2,1}, & X_2 \end{pmatrix}'\right)^{-}_{m(H_2)}\right]'\begin{pmatrix} Y^{(1)} \\ Y^{(2)} \end{pmatrix}.$$

Proof see in Kubáček [9] .

In the case of regularity of matrices $\Sigma_{1,1}$, $\Sigma_{2,2}$ two "small" systems with k_1 and k_2 unknowns

$$X_1'\Sigma_{1,1}^{-1}X_1\hat{\beta}^{(1)} = X_1'\Sigma_{1,1}^{-1}Y^{(1)}$$

$$X_2'D_{2,2}^{-1}X_2\hat{\beta}^{(2)} = X_2'D_{2,2}^{-1}(Y^{(2)} - C_{2,1}\hat{\beta}^{(1)})$$

are to be solved for determination of the left hand side of the(**); determination of the right hand side of the(**) requires the solution of a "big" system with $k_1 + k_2$ unknowns

$$(X_1'\Sigma_{1,1}^{-1}X_1 + C_{2,1}'\Sigma_{2,2}^{-1}C_{2,1})\hat{\hat{\beta}}^{(1)} + C_{2,1}'\Sigma_{2,2}^{-1}X_2\hat{\beta}^{(2)} = X_1'\Sigma_{1,1}^{-1}Y^{(1)} + C_{2,1}'\Sigma_{2,2}^{-1}Y^{(2)}$$

$$X_2'\Sigma_{2,2}^{-1}C_{2,1}\hat{\hat{\beta}}^{(1)} + X_2'\Sigma_{2,2}^{-1}X_2\hat{\beta}^{(2)} = X_2'\Sigma_{2,2}^{-1}Y^{(2)}.$$

In the "big" system an unpleasant fact occurs in addition; the vector $\hat{\hat{\beta}}^{(1)}$ is an estimate based on the both vectors $Y^{(1)}$, $Y^{(2)}$ and thus it cannot be used.

It is not necessary to analyze the economy of the proposed procedure. The cut of calcutation naturaly implies the improvement of the numerical stability.

Statement 6. If $m > 2$ and the design of the multistage regression experiment is prepared in the following way

$$C_{i,j} = O \iff i - j \geq 2,\ i = 1,\ldots,m,\ j = 1,\ldots,i-1,$$

then $\hat{\beta}^{(i)} = (X_i'D_{i,i}^{-1}X_i)^{-1}X_i'D_{i,i}^{-1}(Y^{(i)} - C_{i,i-1}\hat{\beta}^{(i-1)})$ is equal to the estimator from (*). The matrices $D_{i,i}$, $i = 1,\ldots,m$, are given by the equalities:

$D_{1,1} = \Sigma_{1,1}$,

$D_{2,2} = \Sigma_{2,2} + C_{2,1}((X_1')^-_{m(D_{1,1})})' D_{1,1} (X_1')^-_{m(D_{1,1})} C_{2,1}'$,

$\vdots$

$D_{m,m} = \Sigma_{m,m} + C_{m,m-1}((X_{m-1}')^-_{m(D_{m-1,m-1})})' D_{m-1,m-1} (X_{m-1}')^-_{m(D_{m-1,m-1})} C_{m,m-1}'$

Proof see in Kubáček [9] .

The condition $C_{i,j} = 0 \iff i - j \geq 2$ is fulfilled in metrology almost allways. However in geodesy and geophysics to fulfil this condition is sometimes impossible. In this case the estimator

$$\tilde{\beta}^{(i)} = (X_i' D_{i,i}^{-1} X_i)^{-1} X_i' D_{i,i}^{-1} (Y^{(i)} - \sum_{j=1}^{i-1} C_{i,j} \tilde{\beta}^{(j)})$$

is not as good as (*) but from the point of the practice it is still well applicable and substantially better than the estimator

$$\tilde{\beta}^{(i)} = (X_i' \Sigma_{i,i}^{-1} X_i)^{-1} X_i' \Sigma_{i,i}^{-1} (Y^{(i)} - \sum_{j=1}^{i-1} C_{i,j} \tilde{\beta}^{(j)}).$$

It is necessary to remark that the characterization of the precision of the last estimator $\tilde{\beta}^{(i)}$ is sometimes misrepresented by considering the matrix $(X_i' \Sigma_{i,i}^{-1} X_i)^{-1}$ as the covariance matrix.

5. UTILIZATION OF GENERALIZED INVERSION FOR ESTIMATION OF THE SECOND ORDER PARAMETERS

For determination of the estimate $t'X((X')^-_{m(Var(Y))})'y$ of the function $f(\beta) = t'\beta$ and for evaluating its precision it is necessary to know the matrix Var(Y). If $Var(Y) = \sum_{i=1}^{p} \delta_i V_i$ and the second order parameters $\delta_1, \ldots, \delta_p$ are unknown then they are to be estimated as well. In the following the possibility of utilization of the program for the generalized matrix inversion is pointed out with respect to estimation of just mentioned second order parameters.

For the sake of simplicity all parameters $\delta_1, \ldots, \delta_p$ are considered unbiasedly estimable. It means the regularity of the matrix $K = V'(I - P \otimes P)V$, where $V = (vec(V_1), \ldots, vec(V_p))$, $P = X(X'X)^- X'$; the (i,j)-th element of the matrix K is $\{K\}_{i,j} = Tr(V_i V_j - PV_i PV_j)$ in details see Seely (1970a, 1970b), Rao and Kleffe (1980), Kubáček [8] .

<u>Statement 7</u>. Let matrices A and B be of the type $m_1 \times n$ and $m_2 \times n$ and let a symmetric matrix D be of the type $n \times n$ and positive semidefinite; then

$$\begin{pmatrix} A \\ B \end{pmatrix}^-_{m(D)} = (C,E),\ C = (A'A + B'B)^-_{m(D)} A',\ E = (A'A + B'B)^-_{m(D)} B'.$$

Proof. It is sufficient to verify the validity of the relations $(A',B')'(C,E)(A',B')' = (A',B')'$, $((C,E)(A',B')')'D = D(C,E)(A',B')'$.

<u>Statement 8</u>. The locally best unbiased estimator of the vector δ, which is of the form $\hat{\delta} = (Y'A_1Y,\dots,Y'A_pY)'$, where A_i, $i = 1,\dots,p$, are $n \times n$ matrices, is

$$\hat{\delta} = (O,I)\left(\begin{pmatrix} X'\otimes X' \\ \tilde{V}' \end{pmatrix}^-_{m(\mathrm{Var}(Y^{2\otimes}))}\right)' Y^{2\otimes} =$$

$$= \{((XX')\otimes(XX') + \tilde{V}\tilde{V}')^-_{m(\mathrm{Var}(Y^{2\otimes}))}\tilde{V}\}' Y^{2\otimes},$$

where

$$\mathrm{Var}(Y^{2\otimes}) = \psi + \Sigma\otimes X\beta\beta'X' + X\beta\beta'X'\otimes\Sigma + ((X\beta)\otimes I)\Sigma(I\otimes(X\beta)') +$$

$$+ (I\otimes(X\beta))\Sigma((\beta'X')\otimes I) + \phi(I\otimes(\beta'X')) + \phi((\beta'X')\otimes I) +$$

$$+ (I\otimes(X\beta))\phi' + ((X\beta)\otimes I)\phi' - \mathrm{vec}(\Sigma)(\mathrm{vec}(\Sigma))',$$

$$\Sigma = \mathrm{Var}(Y) = \sum_{i=1}^{p}\delta_i V_i\,,\quad \phi = E\{(Y - X\beta)\otimes((Y - X\beta)(Y - X\beta)')\},$$

$$\psi = E\{((Y - X\beta)(Y - X\beta)')\otimes((Y - X\beta)(Y - X\beta)')\}$$

Proof. The expression for $\mathrm{Var}(Y^{2\otimes})$ can be obtained in the routine way; the relationships $E(Y^{2\otimes}) = (X\otimes X)\beta^{2\otimes} + \tilde{V}\delta$, $E(Y'A_iY) = (\mathrm{vec}(A_i))'E(Y^{2\otimes}) = (\mathrm{vec}(A_i))'(X\otimes X)\beta^{2\otimes} + (\mathrm{vec}(A_i))'\tilde{V}\delta = \delta_i$, $i = 1,\dots,p \iff \begin{pmatrix} X'\otimes X' \\ \tilde{V}' \end{pmatrix}(\mathrm{vec}(A_1),\dots,\mathrm{vec}(A_p)) = \begin{pmatrix} O \\ I \end{pmatrix}$ is to be considered. As $D((\mathrm{vec}(A_i))'Y^{2\otimes}) = (\mathrm{vec}(A_i))'\mathrm{Var}(Y^{2\otimes})\mathrm{vec}(A_i)$ the minimal dispersion is attained by choosing $\mathrm{vec}(A_i) = \begin{pmatrix} X'\otimes X' \\ \tilde{V}' \end{pmatrix}^-_{m(\mathrm{Var}(Y^{2\otimes}))}\begin{pmatrix} O \\ e_i \end{pmatrix}$, where $e_i = (0_1,0_2,\dots,0_{i-1},1_i,0_{i+1},\dots,0_p)'$. To finish the proof is elementary when Statement 7 is used (in details see Kubáček [8]).

<u>Remark 1</u>. When $Y \sim N(X\beta, \sum_{i=1}^{p}\delta_i V_i$ then ϕ and ψ are dependent on β and δ only and thus "the locally best" is connected with the point (β,δ) used in the individual expressions in $\mathrm{Var}(Y^{2\otimes})$. If Y is not normally distributed then the matrices ϕ and ψ depend on parameters of the third and fourth order. With respect to Kleffe (1976) and for the sake of simplicity in this case the parametric space is $\underline{\theta} = \underline{\beta}\times\underline{\delta}\times\underline{\phi}\times\underline{\psi}$, where $\underline{\phi}$ and $\underline{\psi}$ are classes of matrices ϕ and ψ, respectively. The estimator from Statement 8 is the locally best with respect to the point (β,δ,ϕ,ψ).

Remark 2. A realization of the estimator $\hat{\delta}$ from Statement 8 depends on the parameter β . Sometimes this is not desirable. That is why the class of all unbiased estimators is restricted to the class of all unbiased and invariant (with respect to the vector β) estimators. In this case the condition $X'A_iX = 0$ for matrices A_i from Statement 8 is substituted by condition $A_iX = 0$.

Statement 9. The locally best unbiased and invariant estimator of the form $\hat{\delta}^{(I)} = (Y'A_1^{(I)}Y,\dots,Y'A_p^{(I)}Y)$ is

$$\hat{\delta}^{(I)} = (0,I)\left(\begin{pmatrix} X'\otimes I \\ \tilde{V}' \end{pmatrix}^{-}_{m(D_{2,2}^{(I)})}\right)' Y^{2\otimes} = \{((XX'\otimes I + \tilde{V}\tilde{V}')^{-}_{m(D_{2,2}^{(I)})}\tilde{V}\}'Y^{2\otimes},$$

where $D_{2,2}^{(I)} = \psi - \mathrm{vec}(\Sigma)(\mathrm{vec}(\Sigma))'$.

This estimator exists iff the matrix $K^{(I)} = \tilde{V}'(M\otimes M)\tilde{V}$, where $M = I - X(X'X)^{-}X'$, is regular.

Proof see Kubáček [8], simultaneously taking into account Statement 7.

Considerable cut of calculation can be obtained when symmetric matrices A_i are considered in Statements 8 and 9 and operation vech(.), (cC)(.) and (cR)(.) are applied. The following statement can serve as an example.

Statement 10. The locally best invariant and unbiased estimator of the vector δ is

$$\hat{\delta}^{(I)} = \{((cR)(X\otimes I)(cC)(X'\otimes I) + (cR)(\tilde{V})(cC)(\tilde{V}'))^{-}_{m((cC)(cR)(D_{2,2}^{(I)}))}\cdot$$
$$.(cR)(\tilde{V})\}'(cR)(Y^{2\otimes}) \ .$$

Remark 3. In Statements 8,9 and 10 allways the minimum seminorm generalized matrix inversion is used. Statements 1,2 and 4 show how to verify the numerical correctness of estimates in the case of the second order parameters. For example, let us consider Statement 8. The regression model can be written in the form $\left(Y^{2\otimes}, (X\otimes X,\tilde{V})\begin{pmatrix}\beta^{2\otimes}\\ \delta\end{pmatrix},\right.$ $\left.\mathrm{Var}(Y^{2\otimes})\right)$. Thus

$$\hat{\delta} = (0,I)\left(\begin{pmatrix} X'\otimes X' \\ \tilde{V}' \end{pmatrix}^{-}_{m(\mathrm{Var}(Y^{2\otimes}))}\right)'Y^{2\otimes},$$

with respect to Statement 2, is the solution of equations

$$\begin{pmatrix} X'\otimes X' \\ \tilde{V}' \end{pmatrix} M(X\otimes X,\tilde{V})\begin{pmatrix}\hat{\beta}^{2\otimes}\\ \hat{\delta}\end{pmatrix} = \begin{pmatrix} X'\otimes X' \\ \tilde{V}' \end{pmatrix} My^{2\otimes},$$

where $M = (\mathrm{Var}(Y^{2\otimes}) + (X\otimes X,\tilde{V})U\begin{pmatrix} X'\otimes X' \\ \tilde{V}' \end{pmatrix})^{-} + K$;

the matrices U and K fulfil conditions (b), (c), (d) from Statement 2. With respect to Statement 4 $\hat{\delta} = (0,I)C_3 Y^{2\otimes}$, where

$$\begin{pmatrix} \mathrm{Var}(Y^{2\otimes}), & (X \otimes X, \tilde{V}) \\ \begin{pmatrix} X' \otimes X' \\ \tilde{V}' \end{pmatrix}, & 0 \end{pmatrix}^{-} = \begin{pmatrix} C_1, & C_2 \\ C_3, & -C_4 \end{pmatrix}.$$

Remark 4. For especial structure of the matrix X a remarkable algorithm for δ is developed by Kleffe and Seifert [4] .

Utilization of the program for generalized matrix inversion for simultaneous estimation of parameters β and δ is sketched in the following.

Statement 11. The locally best unbiased estimator of the parameter $(\beta', \delta')'$, which is considered in the form

$$(***) \quad \begin{pmatrix} \hat{\beta} \\ \hat{\delta} \end{pmatrix} = \begin{pmatrix} A_1 \\ B_1 \end{pmatrix} Y + \begin{pmatrix} A_2 \\ B_2 \end{pmatrix} Y^{2\otimes},$$

is given by equations

$$A_1 = ((X')^{-}_{m(*)})';$$

$$A_2 = -((X')^{-}_{m(*)})' D_{1,2} D_{2,2}^{-} \{ I - (X \otimes X, \tilde{V}) \left(\begin{pmatrix} X' \otimes X' \\ \tilde{V}' \end{pmatrix}^{-}_{m(D_{2,2})} \right)' \};$$

$$B_1 = -\tilde{V}' \{ ((XX') \otimes (XX') + \tilde{V}\tilde{V}')^{-}_{m(**)} \}' D_{2,1} \Sigma^{-} (I - X((X')^{-}_{m(\Sigma)})');$$

$$B_2 = \tilde{V}' \{ ((XX') \otimes (XX') + \tilde{V}\tilde{V}')^{-}_{m(**)} \}',$$

where

$$\Sigma = \mathrm{var}(Y),$$

$$D_{1,2} = \mathrm{cov}(Y, Y^{2\otimes}),$$

$$D_{2,2} = \mathrm{var}(Y^{2\otimes}),$$

$$(*) = \Sigma - D_{1,2} D_{2,2}^{-} (D_{2,2} - D_{2,2} \begin{pmatrix} X' \otimes X' \\ \tilde{V}' \end{pmatrix}^{-}_{m(D_{2,2})} \begin{pmatrix} X' \otimes X' \\ \tilde{V}' \end{pmatrix}) D_{2,2}^{-} D_{2,1}$$

$$(**) = D_{2,2} - D_{2,1} \Sigma^{-} (\Sigma - \Sigma (X')^{-}_{m(\Sigma)} X') \Sigma^{-} D_{1,2}.$$

Proof see Kubáček [8].

Remark 5. Considerable cut of calculation can be attained again by using operations vech(.), (cC)(.) and (cR)(.).

Remark 6. From the given it is clear haw to continue in utilazing the program for generalized matrix inversion in calculation of the best unbiased and invariant estimators and how to develop procedures for estimation of the third order parameters $\phi_1, \ldots, \phi_r$, where

$\phi = \sum_{j=1}^{r} \phi_j F_j$ and matrices F_j, $j = 1,\dots,r$, are given. Similarly it can be continued in the case of the fourth order parameters $\psi_1,\dots$ $\dots,\psi_s$, where $\psi = \sum_{j=1}^{s} \psi_j G_j$ and matrices G_j, $j = 1,\dots,s$, are given, etc.

References:

[1] Hoerl,A.E.,Kennard,R.W.(1970): Ridge regression: Biased estimation for nonorthogonal problems. Technometrics 12,55-68

[2] Hoerl,A.E.,Kennard,R.W.(1970): Ridge regression: Application to nonorthogonal problems. Technometrics 12,69-82

[3] Kleffe,J.(1978): Simultaneous estimation of expectation and covariance matrix in linear models. Math. Operationsforsch.Statist., Ser.Statistics 9,443-478

[4] Kleffe,J.,Seifert,B.: Matrix free computation of C.R.Rao's MINQUE and its efficiency under unbalanced nested classification models (submitted to Computational Analysis and Data Analysis)

[5] Kubáček,L.,Bartalošová,L.,Pecár,J.(1977): "Pandora-Box" matrix in the calculus of observations. Studia geoph.et geod.21,227-235

[6] Kubáček,L.(1978): Universal model for adjusting observed values. Studia geoph.et geod.20,103-113

[7] Kubáček,L.(1983): Základy teórie odhadu Foundation of Estimation Theory . Bratislava, Veda

[8] Kubáček,L.: Locally best quadratic estimators (to appear in Mathematica Slovaca)

[9] Kubáček,L.: Multistage regression model (submitted to Aplikace matematiky)

[10] Rao,C.R.,Mitra,S.K.(1971): Generalized Inverse of Matrices and Its Applications. N.York, J.Wiley

[11] Rao,C.R.(1971): Unified theory of linear estimation. Sankhyā, A 33,370-396

[12] Rao,C.R.,Kleffe,J.(1980): Estimation of variance components. In: Krisnaiah,P.R.,ed.Handbook of Statistics, Vol.I,1-40, North Holland,N.York

[13] Seely,J.(1970): Linear spaces and unbiased estimation. Ann. Math.Stat.,41,1725-1734

[14] Seely,J.(1970): Linear spaces and unbiased estimation - Application to the mixed linear model. Ann.Math.Stat.,41,1735-1748

[15] Volaufová,J.: Estimation of parameters of mean and variance in two-steps linear models (submitted to Aplikace matematiky)

Intelligent Software in Statistical Data Analysis

Constructing an Expert System for Data Analysis by Working Examples

A. Gale, and D. Pregibon, Murray Hill

ABSTRACT

Our experience in building an expert system for regression analysis (REX) convinced us that acquiring statistical knowledge (strategy) empirically by working examples and keeping diaries is a useful endeavor. However, we realized that statisticians need tools to aid them in

- developing strategy in a systematic and rigorous fashion, and
- encoding strategy into software for general use.

This paper expands the motivating ideas listed above for building such a system and outlines our design plans for doing it. We named the new system "Student" as it is designed to learn data analysis techniques from an experienced statistician (teacher) by observing how actual analyses are done and by asking questions.

1, Introduction

Student is being designed to allow a professional statistician to construct a knowledge based consultation system in a data analysis area by selecting and working examples and by answering questions. An advantage of such a system is that the statistician will not need to know the internal representation of the knowledge base, and will not need to know how to program.

Another main advantage of the system as conceived is the ease with which it can be modified for a local environment. This will allow constructing systems with domain specific vocabulary more accessible to infrequent users. It could also be highly specialized to do a very thorough job on some restricted range of target problems.

Student is also to serve as the vehicle for the constructed knowledge based consultation systems in data analysis. REX (Gale and Pregibon, 1982, 1984) is a working demonstration of the kind of consultation that Student will provide. It is an interface to the *S* statistical system (Becker and Chambers, 1984) which allows a novice to use advanced regression techniques safely by systematically checking the assumptions of the techniques. It provides guidance to what tests need to be done and when, interpretation of the results of tests and plots, and instruction in statistical concepts. It has appeared that this system, while designed for use by novices, is of interest to expert statisticians, because it makes explicit much knowledge that has not been formalized. Most experts have also expressed interest in using such a consultation system because it automates many tasks that they know they want to do, but don't always do.

Student is still on the drawing board. This paper serves to provide the motivation for proceeding to build Student and our design plans to carry it out. The following section describes one main area of REX's stregth, while section three focuses on its limitations. The remainder of the paper presents some of the main issues encountered in designing Student.

2. CALIBRATION OF REX

The best way to describe the statistical content of REX is to discuss the kinds of problems it does and does not solve. For a more detailed description including the user interface, see Pregibon and Gale (1984, this volume).

2.1 The Kinds of Problems REX Solves

The strategy upon which REX is based was constructed by abstracting common features from a collection of about three dozen regression analyses performed manually (using S). Most of the data sets were taken from standard textbooks on regression analysis. All the data sets violate at least one assumption of the classical regression model, and in many cases multiple violations are present. The most serious limitation we see with our choice of data sets is that all were of fairly limited size: Most had fewer than 50 observations, and only a few had more than 100 observations. The comments below are therefore subject to this qualification.

REX is capable of preventing most of the disasters facing inexperienced users of regression software. In our view, the single most important component of the analysis is the linearity of the relationship between the dependent and explanatory variables. REX is particularly adept at both detecting the presence of nonlinearity and suggesting specific transformations of the data to remove it. Furthermore, REX seems to perform these tasks equally well in the presence of other problems such as outliers and nonhomogeneity of variance.

An important reason why REX does so well in this difficult area is that a preliminary processing of the data is routinely performed. We call this preliminary processing "first aid" since obvious imperfections in the data are "bandaged up" before proceeding to the next analysis stage. Included here are checks for extreme asymmetry and gross outliers. Corrections for these anomalies include taking logarithms and setting aside outlying observations. Even though the procedures used in the linearity assessment stage are moderately robust against most problems addressed here, it helps considerably to get the scale of the variables about right prior to fitting.

The remaining assumptions are dealt with only after linearity has been established (possibly for a transformed version of the original variables). REX has demonstrated good performance in detecting heterogeneity of variance and suggesting ways to either fix it or accommodate it. Similarly with the problem of unduly influential points in the regression. It has limited capabilities to deal with serially correlated data other than being able to detect and accommodate linear time trends, and first order autoregressions. Thus we don't see REX being very useful on time series data without much expanding of the basic strategy.

2.2 The Kinds of Problems REX Doesn't Solve

In discussion with colleagues and other interested parties, REX was repeatedly challenged by a variety of hypothetical situations. A favorite situation was data distributed along two intersecting lines. Another was data distributed along the perimeter of a circle. None of the diagnostic methods for assessing linearity included these possibilities. REX would most probably not even be able to detect the pattern, let alone identify it. In both of these cases we *can* derive appropriate diagnostic methods. In both of these cases we chose not to do so. We feel that these situations are so infrequent in occurrence that special methods to handle them are not warranted. The performance of the system would degrade considerably if a large number of special cases are processed. The guiding principle we used was "demand and supply": a real example must be supplied to substantiate the need. For the two cases outlined above, this principle indicated no further action.

It is important to note that expert software can identify cases which it cannot solve. REX identifies such cases by *requiring* a transformation while there is no transformation acceptable to the user. We would only consider it a "failure" of the system if REX reported a regression which still violated underlying assumptions.

3. REX Critique

The following points are our major reasons for moving beyond REX.

3.1 REX is Too Rigid

The strategy encoded in REX is composed of two parts:

- a hierarchical network of problems and fixes which are specific to regression analysis
- a collection of rules associated with each node in the network.

Both the network and the rule set are static from one analysis to another. The dynamic aspect of the system is attained because different data sets result in different rules "firing", resulting in different traversals of the network.

Currently there is no way to deviate from the programmed network. This is not overly restrictive for nonstatisticians who have come to expect being told how to do things by statistical software. Statisticians find it frustrating, especially when they see something at an early point of the analysis, and they must wait until REX whittles away at tasks which come before it.

For statisticians using REX, a point of disagreement rather than frustration can occur, because in some problems, there is no single "right" analysis. REX might choose one, and this may not agree with the choice of a knowledgeable user. In cases where the user has access to additional subject matter information, REX's solution will usually be the inferior one.

3.2 REX's Current Strategy is *Ad Hoc*

We call the knowledge base in a data analysis expert system a *strategy*. The current strategy in REX was developed empirically by

1. selecting a small number of data sets
2. analyzing each
3. maintaining an annotated diary of each analysis
4. abstracting common analysis features
5. iterating 1-4.

Loosely, the strategy describes how we (Daryl) do regression. (Actually it describes how Daryl thinks he does regression though for our purposes this is not important.)

Because we were just learning to grope with the concept of strategy, the induction process we used was not very systematic. There are many decisions to make once one gets started in this business: which problems to address, when to address them, which procedures to use, how to tune the procedures, which fixes to apply, *etc*. Unfortunately, these decisions interact with each other, making the strategy abstraction even harder.

The *ad hoc* development of strategy led to inconsistencies in repeated analyses which were difficult to avoid. As an example, consider the problem of choosing a cutoff for a particular test (say for outliers). Initial values were borrowed from operating characteristics for that procedure assuming all other assumptions were satisfied. These were gradually adjusted up or down so that the current example would be worked correctly. However this could ruin the optimized behavior of the procedure on previous examples. An effort was made to maintain consistency over the examples in the early stages, but when their number grew to about twenty this was abandoned. In retrospect, forcing ourselves to ask (and answer!) why consistency should be violated in a particular example would have led to a richer, more flexible strategy.

3.3 Strategy is Hard to Derive

The manual induction process outlined in the previous section would not be necessary if statisticians were as good at describing what they do, as they are at doing it. The corollary is that even after observing themselves on numerous instances, they still are not able to extract the "method" from the "madness".

In our case, a major simplification was attained when three major data analysis stages were identified for regression:

- model independent scrutiny of dependent and explanatory variables
- assessment of the linearity of the relationship
- assessment of second-order assumptions (including homogeneity, independence, and normality).

This was a major simplification which came after much introspection. However, strategy within these stages could now be developed relatively independently of each other. This allowed the rules for directing traffic over the network to be simpler: fewer *premise* clauses and fewer *action* clauses. However in cases where several tests were being considered simultaneously, we still found it difficult to sort through the alternatives logically. This reinforced our thesis that statisticians require support to aid development of strategy.

3.4 REX is Hard to Modify

The regression strategy was developed by Daryl, the inference engine by Bill. This is an undesirable situation since each person must learn a considerable amount of each other's specialty and it limits the facility with which the system can be modified. Modifications of the system are required for a number of reasons:

- to correct bugs in the strategy
- to extend the system to handle special cases not previously considered
- to generalize the system to handle a wider class of problems than originally planned.

Each of these requires a considerable expenditure of time on the part of both the statistician and the knowledge engineer. In our case, we have incurred the overhead so each is familiar enough with the other's area that joint effort is not required. In general this is asking too much, especially since the style of programming used in knowledge-based systems is very different from anything statisticians are familiar with.

3.5 REX is Hard to Specialize

A particularly important case of modifying a consultation system is specializing it to a local environment with specialized interests and expertise.

We have previously pointed out that statistics is a domain with knowledge well insulated from generalized informal knowledge (Pregibon and Gale, 1984). However, there is a broader issue of how well statistics is separated from other areas of knowledge. In a broader context, the answer is not as favorable. This is because statistics is a domain of knowledge which will be applied to data from another domain, which we call the "ground" domain. The ground domain may be physics, psychology, business analysis, or agriculture, for instance.

REX uses a strategy which is domain independent, and does not use knowledge from any ground domain. To change this would require reprogramming the strategy used by REX, which would require learning the language used to express the strategy. The effort would be a substantial fraction of the original effort to produce the REX strategy.

If it were fairly easy to specialize strategy to a ground domain, the utility of a consultation system would be greatly increased, because infrequent users could use it more easily. A particularly important group of infrequent users is managers, who haven't the time to learn statistics procedures and vocabulary, but can be expected to know their own specialty well.

Another mode of use for an easily specialized system would be exemplified by an agricultural extension station. Here, one statistical expert could specialize the strategy by working cases from the files. The agricultural experts could then use a system tailored to their specialty.

A specializeable system is also particularly useful for analysing restricted sets of data. An important special case is repeated analysis of updated data. Such stylized analyses would be well suited to a rule based system.

4. Design Issues for Student

This section presents some of the main issues that we have been struggling with as we thought about the design of a system to accomplish the objectives described in the previous section. We have found suggestions from the literature to be of some help, but that no system seems to do all that we think will have to be done.

4.1 Previous Systems Have Solved Some Problems

But not all problems. When we were considering building REX, we could see that many knowledge based consultation systems had been demonstrated and a few had been used regularly. We are not aware of any system that demonstrates all that Student will have to do to be useful.

The closest system, we believe, is Randy Davis's Teiresias (Davis, 1979). This was built as an addition to Mycin, with the goal of acquiring rules, just as Student is an addition to REX with the goal of acquiring strategy. Teiresias was able to acquire an additional rule consistent with an existing knowledge base. However, there are two major problems besides this that need to be solved.

First, the system must support the acquisition of the *first* example or rule. The first rules to be acquired are typically different from later rules, because a rule based system uses a core of rules to encode control information. A subject matter expert will not be able to provide control information.

Second, the system must support deliberate changes to strategies over a long period of time. Current technology, such as used for REX, results in a "compiled" strategy, which is dificult to change. There are some analogies to reason maintenance systems (Doyle, 1978) but these thechniques have not been used to support rule acquisition in expert systems.

There are some systems that support programming by example, but none of them are for programming knowledge based consultation systems. Tinker (Lieberman, 1983), PHD (Attardi and Simi, 1983), SBA (Zloof and De Jong, 1977), and a system by Biermann and Krishnaswamy (1976) are examples. Attardi and Simi review several of these systems. They cluster around programming business applications, because every office has its own procedures, so programs do not get wide distribution to lower software costs. Tinker appears to be the closest to our ideas for Student.

In using Tinker, the programmer selects a concrete typical example of data for the procedure. He then performs the procedure step by step. The system is therefore able to learn how to do the first example. As more examples are supplied, the program required for them is compared with the already constructed program. If the two differ, the user is queried for a predicate that will distinguish the two cases. Therefore, the user ultimately provides one example for each branch of the final program. Tinker seems to assume that the user knows how each example should be worked; there is no means to change the program by deleting an example already worked. Since the way a particular data analysis should be done is not so cut and dried, Student will have to have a capability to change an example already worked.

Moreover, Tinker does not (nor does it intend to) insulate the programmer from the underlying language, Lisp. The demonstration is in terms of a mixture of pointing, menu selection, question answering, and specifying Lisp operations. During program construction, bugs may occur which require the user to know Lisp to understand what the bug was. Rather, the thrust of Lieberman's (1983) description is that the programmer get a feeling of the user's viewpoint while constructing the program. A primary goal of Student is that the user be unaware of the underlying representation language.

On the other hand, Tinker is tackling a harder project in that it hopes to support programming any Lisp codable procedure. Its level of success in this is shown by Lieberman by creating a simple editor. It is an encouraging demonstration. Tinker's use of menus, pointing, and question answering are suggestive techniques, and conveying the user's viewpoint to the statistician is a useful goal.

4.2 Why We Think Strategy Acquisition Can Be Done

Since we don't see any systems that accomplish what we think Student must be able to do to be successful, \ have tried to understand why we think such a system can be built.

First, there is a big advantage to restricting the system to acquisition of *data analysis* strategies. Student is not a general purpose knowledge acquisition program for a general purpose inference engine. This restriction provides a meta-framework within which the program will be built. We know we have to deal with data sets, and we will provide representations to deal with them. We know that the analysis consists of looking for violated assumptions, and if found, curing them. We know that we look for violated assumptions by making tests and by showing the user plots. We have learned these things by building REX, and by considering extension of the methods to other data analysis areas. Having this meta-framework provides enough structure for Student to guide the user through the first analysis of a given kind, and to acquire additional consistent examples.

Despite being limited to data analysis, the invention of strategy acquisition techniques will be worth while, because there are a number of areas of data analysis which a complete consultation system should cover, and because the specialization ability is very useful.

Second, data analysis has the property that it can be done entirely on a computer. Modern statistical packages provide the expert statistician with sufficiently powerful tools that many data sets can be analyzed interactively. They provide graphics and data management as well as computations. It is therefore possible for another computer program to observe all that a statistician does to analyze a data set.

Third, we do not require any greater performance as a consultant than that provided by REX. We intend to try to improve the explanation capabilities, and to provide for capability for recognizing multiple solutions. But these could be built using REX as a vehicle, and are independent of the strategy acquisition aspects.

In particular, for the level of performance of REX, numerical tests appear to be sufficient to guide the strategy. We find plots are very useful for explanations to the user, and we can accept user decisions based on graphs, but a reasonable level of performance can be achieved by a novice who is using the system to learn how to read the graphs. Furthermore, it appears that statisticians, even if they make a decision based on examining a graph, are willing to describe a numerical test to detect what they have seen. Since learning to read graphs is difficult for a computer, while learning to make numerical tests is easy, Student can learn what it needs to. Research in learning to "read" graphs is planned, but Student's success does not depend on it.

4.3 How Will Student Know What the Statistician is Doing?

In their normal use of S, statisticians are accustomed to totally free access and complete control. Clearly this is sufficient for them to do the required data analyses. However, Student needs to know what is going on, and we don't see any way of giving statisticians free access to S and still following what they are doing. For instance, *we* cannot always figure out what other statisticians are doing just by watching their commands. This suggests that it would be hopeless for a program to try this approach. Therefore Student will have to control the statistician's access to S.

Can we then provide *sufficiently* free access to S to allow the statisticians to do the required data analyses? The approach we will try with Student will be to provide a menu system that will support a particular model of data analysis, derived from our experience with REX. The model assumes that there is a specific "simple" computation that the user wants to do, such as ordinary least squares regression or a fourier transform. However, using the "simple" computation requires making a number of assumptions. The broad outline of the style of data analysis that our model supports is the systematic checking of assumptions underlying the use of a simple calculation. Because there is a limited number of ways to check an assumption, most notably tests and plots, we believe that a menu system will provide sufficient flexibility. If this proves to be too restrictive, we will iterate the design.

4.4 How Will Student Support a Change of Strategy?

We foresee that statisticians will want to change their analysis of some data sets. There are two reasons. First, data analysis is a reasonably dynamic area and new techniques are published regularly. Reading about a new method, or research on a similar problem could lead to revising a specific analysis. Second, the strategy of data analysis is not formalized. Therefore we expect that statisticians may change their minds about the organization of their tests. We want to support changes to the way one data set is analyzed without requiring reanalysis by hand of every data set.

The basic approach to supporting strategy change is to produce a structured trace of each analysis. After a number of analyses have been done, there will be an ordered set of traces, each *consistent* with those before it. This means we will have sufficient information in the traces that the ordered set can be reprocessed to produce the current integrated strategy, just as if each data set were being analyzed by hand. This much appears feasible, if difficult.

Then we need the property that we can drop one (actually any subset of the ordered set) of the traces and automatically process the remainder. In the processing of the remainder, we will allow rejection of one of the traces as not consistent with the prior set of analyses processed. These would be data sets that depended for their analysis on tests introduced first in a data set deleted by hand. If there remain traces in which the deleted tests are not used, these remaining traces should recompile to an integrated strategy.

We do not know whether this is feasible. We plan to try a simple method first: we will mark each assumption-test the first time we learn it. In automatic analysis, if we find in a trace an assumption-test which is not marked as a first time, for which the assumption is violated, and which is not in the current strategy, then the data set represented is inconsistent with the strategy.

The data sets that are marked as inconsistent will then need to be reanalyzed by hand. We would like to use as much of the old analysis as possible, however. This reanalysis requirement is both natural and helpful in establishing a strategy, because a new method can not be said to be understood until it has been used successfully on a variety of data sets. By pointing out the inconsistent data sets, Student identifies a group to try the new method on. The use of a new method on the data sets will depend on the experience and expertise of the statistician, and it is not reasonable to expect that a program could recomplete the analysis.

If this simple method is not sufficient, then we anticipate a major redesign to use reason maintenance methods developed in Artificial Intelligence research (Doyle, 1978).

4.5 How Much Will Student Know About S?

It would be possible for Student to assist the statistician in using S. For instance, it could know what functions were provided by S, what arguments they needed, and it could provide some indexing assistance. In order to understand the statistician's use of S, and to learn how to make extensions for the data analysis method, this knowledge would be useful. We have decided that we will focus on representing statistics knowledge, however.

One reason for this is that we would like the underlying statistical system to be changeable without totally rebuilding the system. While we do not have plans for using any system besides S, we believe the methods we are using could be used with other statistical packages. But learning a lot about S functions without actually doing the same for another package would probably destroy any modularity which we might achieve.

Another reason for the decision is that what we may learn in building Student could have applications beyond statistics, while learning about S would be restricted to this system.

Therefore, we are expecting the statistician to know how to use S. Since our goals include that the statistician not know about the internal language for the consultation system, it may appear that we have traded one language for another. It is true, but we have traded a language that has been shaped by and for statisticians (S) for a language that has concepts and expressions foreign to the statistician. Also, statisticians having access to S would probably want to learn to use it for their own research. Thus, the trade seems to be a reasonable one.

4.6 Will Everything Have to be Demonstrated?

In using Tinker, it appears that every branch must be demonstrated. While we think that programming by demonstration is more convenient for a non-programmer than the usual prescriptive style, we also feel that complete reliance on it will be difficult. Since statisticians must select data sets to demonstrate with, they might wind up spending time contriving data sets to demonstrate a branch. For example, any time a logarithm or square root transform is considered, it should be easy to instruct that the action is only applicable if the data is positive or non-negative.

One way to allow this might be to provide a facility for the statistician to input specific rules. It should be possible to support a few limited kinds of rules, such as preventive checks.

Another possibility would be to allow the statistician to create a variant data set on the spot, say by changing the sign of one number. A test to distinguish the variant data set could then be demonstrated, while the main effort would be on the analysis of the original data set.

A more ambitious effort would attempt to associate the knowledge of a non-negativity requirement with the concept of a logarithmic transform. Then, whenever a logarithmic transform was demonstrated, Student would silently add a test for non-negativity. However, to do this requires understanding the commands given to the underlying statistical package, which (as discussed above) we do not want to do now.

4.7 Will Student Learn Generic Information?

The case just cited, of learning preconditions for a transformation, is an example of learning *generic* information, that is, information which will be useful in more contexts than the one it was first learned in. In that example, we do not plan to support learning the generic information, because we don't want to have to understand S commands, and because we don't see enough similar examples to know where the information should be stored.

However, there are two cases in which we think we do know enough to plan to store generic information. One is information about data types, and the other is information about plot types.

The first phase of learning about a new kind of analysis is to learn about the data required for the analysis. There will be one or more data sets required for any given kind of data analysis. Each data set may be one of a relatively few S data types, such as a vector or a matrix. For each data set required for the analysis, there will usually be some preprocessing, such as checking that a matrix is square, or allowing a vector instead of a matrix, but coercing it to a one column matrix. We plan to store the techniques for preprocessing of specific data sets indexed by the data type. We will then be able to suggest preprocessing options once we know what data type a new data set is to have.

In looking for violated assumptions, we envision plots as being the primary means used by statisticians. Yet Student is blind, and needs to use tests. Therefore, we will index possible features and corresponding tests by the type of plot used. For example, outliers or clusters are possible features of a scatter plot. We will then be able to suggest a test to use if a statistician finds a problem based on visual examination of a plot.

5. Summary

REX is a working demonstration of the feasibility of expert systems for data analysis. It has several strengths: a convenient user interface, ability to solve standard textbook regression problems, and a modest ability to explain the reasons for its suggestions. However, it also has limitations, mainly in supporting strategy acquisition, modification, and specialization.

Student has been designed to build upon REX's strengths while ovecoming its limitations. Student will allow statisticians to construct or extned knowledge based consultation systems by working examples and answering questions. If successful, this will provide easier and faster construction of better consultation systems in data analysis.

6. References

Attardi, G. and Simi, M. (1983). "Extending the Power of Programming by Examples." Appearing on pp. 3-26 in *Integrated Interactive Computing Systems*, ed. Degano and Sandewall, North-Holland Publishing Company, Amsterdam.

Becker, R.A. and Chambers, J.M. (1984). "S: An Interactive Environment for Data Analysis and Graphics." Wadsworth Advanced Book Program, Belmont, California.

Biermann, A.W. and Krishnaswamy, R. (1976). "Constructing Programs from Example Computations." *IEEE Transactions on Software Engineering SE-2*, 3, 141-153.

Davis, R. (1979). "Interactive Transfer of Expertise: Acquisition of New Inference Rules." *Artificial Intelligence* 12, 121-157.

Doyle, J. (1978). *Truth Maintenance for Problem Solving*, MIT AI Laboratory Technical Report 419.

Gale, W.A. and Pregibon, D. (1982). "An Expert System for Regression Analysis." Appearing on pp. 110-117 in *Proceedings of the 14th Symposium on the Interface*, Ed. Heiner, Sacher, and Wilkinson, Springer-Verlag, New York.

Lieberman, H. (1983). "Designing Interactive Systems from the User's Viewpoint." Apperaing on pp. 45-59 in *Integrated Interactive Computing Systems*, ed. Degano and Sandewall, North-Holland Publishing Company, Amsterdam.

Pregibon, D. and Gale, W.A. (1984). "REX: An Expert System for Regression Analysis." *Proceedings COMPSTAT 84*, Prague, Czechoslovakia.

Zloof, M.M. and DeJong, S.P. (1977) "The System for Business Administration (SBA): Programming Language." *CACM 20*, 6, 385-396.

Analysis of Suppositions

A. Kowalski, Warsaw, and A. Schütt, Osnabrück

Summary

The goal of this paper is to present "intelligent software" that avoids the well-known disadvantages of computational statistics and expert system approaches. An expert's knowledge is formulated as some kind of suppositions rather than decision rules, and, in addition to that, a choice of a relevant data set. Those clearly defined suppositions are then transformed into rules typical for expert systems.

Keywords: Expert systems, Intelligent Software, Statistical Software.

0. Introduction

The current state of statistical software can be characterized by the existence of, at least, two different lines which seem to be considered as practically independent from each other.
There is the line of computational statistics processing information contained in data sets, often using methods of hypotheses testing. On the other hand, expert systems are coming along, belonging to the fields of AI methods (Raufels (1981)).
The latter ones are often based on so-called certainty factors (CF's) being an experts measure of subjective probability with virtually no reference to data sets. However, in many cases, it is not easy or even possible for experts to declare CF's, what severely limits the possibilities of those systems.

1. CONSUL and its goal rules

CONSUL[x/] is an interactive expert system specially designed and

[x/] CONSUL is a private property of Dr. Adam Kowalski and Dipl.Ing. Ryszard Tyrcha. Therefore there are NO formal obstacles to sell it, exchange it, or use it as a basis for any joint project. CONSUL is ready for presentation and everyday usage in several computer centres in Warsaw. It is also possible to tailor the Warsaw version to your standards. In the above case the actual PASCAL documentation is necessary.

implemented for microcomputers. With exchangable knowledge bases, it is applicable practically in every domain of human activities.

To make software portable, CONSUL has been written in the PASCAL language (Jensen - Wirth standard), but with clearly defined know-how it is quite easy to re-program it in other languages. There are various implementations of CONSUL, including PDP 11/34, SM4, Cromemco etc.

CONSUL has been heavily tested in research environments, now reaching commercial standards (CONSUL 3.0, 26.MAR.1984).

Although it requires less than 32Kb memory, the main expert systems features like:

- utility;
- high performance;
- transparency

are fully provided, what enables even a completely unprepared user to handle a consultation session.

For a knowledge base designer only a skill of creating text files under a text editor and understanding of the BNF notation is required, as knowledge-base in the sense of CONSUL consists of the following three text files:

- rules;
- attributes;
- comments.

For the purpose of this paper we restrict our attention only to simple non-deterministic, goal rules. Such rules, being also rules in the sense of other expert systems, particularly MYCIN (Shortliffe (1976)) are of the form:

$$(0)\ (F_{1,i_1} \wedge F_{2,i_2} \wedge F_{3,i_3} \wedge \cdots \wedge F_{n,i_n} \longrightarrow U_j,\ CF)$$

where F_{k,i_k} stands for a fact, in constatation of which the i_k-th value of the k-th attribute is involved. U_j stands for the j-th value of the goal attribute, often interpretable as a classification unit. (In the MYCIN system U is a set of therapies). CF is an abbreviation of certainty factor, which is understood here as a measure of subjective probability.

2. Definitions and method

The problem starts in the case (experienced by the authors)

when experts are not certain on the CF specification, or simply disagree. Simple ad hoc solutions like, for instance, defining CF belonging to the set $\{0.5, 0.7, 1.0\}$ with interpretations — perhaps, possibly, sure-does not contribute significantly to the problem. Generally speaking, the current trend is to allow an expert not only to propose ready-to-use rules, but also to SUPPOSE, i.e. to be not sure on certain elements of rules (see for instance Politakis and Weiss (1984)). That lack of knowledge is supplemented through case experience (in the form of stored cases with known conclusions) leading to rules of the form (0). The process of transforming a set of such suppositions into a set of rules is called by us ANALYSIS of SUPPOSITIONS.
We allow an expert to "violate" rule (0) in the following sense:

- lack of knowledge concerning CF's
- partial knowledge concerning the output attribute
- only partial knowledge on which values of an input attribute a rule is based
- only partial knowledge on which attributes the rule 0 will be formed

More formally, we define the 0-th level of supposition as a rule of the form (0).
The supposition of the 1-st level is defined as

$$(1)\quad (F_{1,i_1} \wedge F_{2,i_2} \wedge F_{3,i_3} \longrightarrow U_j,\ \ ?)$$

where ? describes lack of knowledge (here concerning CF).
The supposition of the R2nd level is defined as

$$(R2)\ (F_{1,i_1} \wedge F_{2,i_2} \wedge F_{3,i_3} \longrightarrow \{U_j\}_{j \in J},\ \ ?)$$

where $\{U_j\}_{j\in J}$ describes a partial no knowledge on which value values of the output attribute will constitute the rule/rules of the form (0).
The supposition of the L2-nd level is defined as

$$(L2)\ (F_{1,\{i_1\}_{i_1 \in I_1}} \wedge F_{2,\{i_2\}_{i_2 \in I_2}} \wedge F_{3,\{i_3\}_{i_3 \in I_3}} \longrightarrow U_j,\ \ ?)$$

where $F_{1,\{i_1\}_{i_1 \in I_1}}, \ldots, F_{3,\{i_3\}_{i_3 \in I_3}}$ describe a partial/no knowledge on which value/values of the particular attributes will constitute the rule/rules of the form (0).
The supposition of the L3-rd level is defined as

$$(L3)\ (\{F_{k_1}\}_{k_1 \in K_1} \wedge \{F_{k_2}\}_{k_2 \in K_2} \wedge \{F_{k_3}\}_{k_3 \in K_3} \longrightarrow U_j,\ \ ?)$$

where $\{F_{k_1}\}_{k_1 \in K_1}, \ldots, \{F_{k_3}\}_{k_3 \in K_3}$ describe a partial/no knowledge on which attribute/attributes will constitute the rule/rules of the form (O).

Obviously, it is possible to define L2-R2, L3-R2 suppositions etc.,and some of them may appear in mixed forms,for instance

$$(*)\ (\{F_{k_1}\}_{k_1 \in K_1} \wedge F_{2,\{i_2\}_{i_2 \in I_2}} \wedge F_{3,i_3} \longrightarrow \{U_j\}_{j \in J'},\ \ ?)$$

To built up some intuitions, let us consider the following supposition being a sentence from the natural language:
If people are seriously engaged in their personal hobby and are not too old and live in small cities (= with less than 0.5 mln inhabitants) then their state of health is at least fair. This sentence is the supposition of the form
We have:

F_{k_1} - hobby, suppose further F_1 - tennis
$F_{1,1}$ - tennis = twice a week

F_2 - age , suppose further $F_{2,1}$ - age = young
$F_{2,2}$ - age = medium

F_3 - city , $F_{3,1}$ - city = small

U - state of health, suppose further U_2-state of health=good

Analysis of suppositions consists of three stages.

- transforming expert's supposition into a set of the 1-st level suppositions
- calculating CF's using methods of automatic testing similar to those used within the GUHA system (Hajek, Havranek(1978), Francis (1981))
- checking consistency of obtained rules with rules belonging already to a knowledge base.

Following our considerations on the natural language sentence, analysis of supposition may result under the assumption of relevant data sets on people hobbies in rules of the form:
If tennis = twice a week &age = young &city = small then state of health = good.

3. Conclusions

On the basis of that briefly sketched philosophy, the SUGGEST[xx]

system have been created (up to now with the exclusion of the L3-rd level suppositions, or supposition components).
The rules being the SUGGEST output can be used as input to expert systems, in particular to the CONSUL system.
In out opinion, the SUGGEST software is well-equipped with the following properties:
- based on routinely accepted statistical methods
- avoids disputable aspects of statistical inference
  see for instance (Cox (1978))
- avoids "black box" effects
- definitions of input and output data are clear in terms of common-sense reasoning
- supports expert systems
- user-friendly
- to be executed also on micro-computers.

---

xx SUGGEST is a private property of Dr. Adam Kowalski and Dipl.Ing. Ryszard Tyrcha. Therefore there are NO formal obstacles to sell it, exchange it, or use it as a basis for any joint project. This product will be ready for distribution starting from January 1985.

<u>Literature:</u>

1. Cox, D.R. (1978) Foundations of Statistical Inference: The Case for Ecclecticism
Austral. J. Statist. 20 1 , 1978, 43-49

2. Francis, I. (1981) Statistical Software: A Comparative Review
North Holland, New York 1981

3. Hajek, P., Havranek, T. (1978) Mechanising Hypotheses Testing
Springer Verlag, Berlin - Heidelberg - New York

4. Politakis,P., Weiss,S.M.,(1984) Using Empirical Analysis to Refine Expert Systems Knowledge Bases
Artificial Intelligence vol 22, No 1, 23-48

5. Raulefs, P. (1981) Expert Systems: State of the Art and Future Prospects
in: Brauer, W.: Informatik Fachberichte Nr.47: GWAI-81
Springer, Berlin 1981, pp. 98-111

6. Shortliffe, E.H.(1976) Computer-Based Medical Consultations MYCIN
Elsevier, New York 1976

REX: an Expert System for Regression Analysis

D. Pregibon, and W.A. Gale, Murray Hill

ABSTRACT
Existing statistical software is quite powerful, but requires considerable statistical knowledge to be used effectively. The knowledge required is not available as theorems or algorithms, but is nevertheless used regularly by statisticians. A new programming methodology, expert systems, has been developed by artificial intelligence research for this kind of problem. REX is an exploration of the feasibility of expert system techniques for making advanced statistical techniques available safely to novices. This paper explains why we undertook the development of REX and illustrates how an interactive session with REX looks to the user.

We believe that REX demonstrates the feasibility of using expert system techniques in the data analysis area.

1. WHY WE BUILT REX

Existing statistical software, while quite powerful, requires considerable statistical knowledge to use it effectively. The capability to serve more people would be valuable, because it would increase productivity and reduce training requirements. But the extension of statistical software to serve more people will require incorporating symbolic knowledge in addition to the numerical algorithms now available. Expert systems are a new technique for representing symbolic knowledge. Their demonstration in widely diverse fields has led us to study their feasibility in data analysis.

1.1 Current Statistical Software is Powerful

Over the last thirty years, statistical software has shown two major trends — more power and wider useability. More power has resulted from computation intensive algorithms, from better data management, and from better graphical capabilities. Wider useability has resulted from elaborating individual subroutines to subroutine libraries to statistical computation systems, and from the wider availability of computing power including personal computers and hand held calculators.

Current statistical software comes in two flavors:

1. a collection of high level stylized programs for doing a variety of statistical analyses, each with a number of options to control testing, reporting, and plotting; and
2. a collection of low level functions which alone have limited capability, but as a whole, provides users with powerful tools for doing both standard and nonstandard analyses.

The first of these types is that which is typically available and used by nonstatisticians. Examples include SPSS, BMDP, and MINITAB. Software of this sort is gradually evolving toward the second type which is currently used most by statisticians. Examples are SAS and S. The general feeling is that if it is good for statisticians, it must be good for nonstatisticians. This is *not* in fact true.

We believe that software of the first sort is too inflexible to be generally useful for providing high quality analyses of data, while we believe that software of the second sort is too *flexible* to be generally useful for nonstatisticians.

The reason is that, for all its power, a novice may not get much more from S than from a hand held calculator. It is possible to do some regressions, for instance, on a hand held calculator. In S, the equivalent work can be done with one command which is easy to learn. The problem is that more consideration of the data than this is required to do a regression safely.

1.2 Regression Analysis Requires Care and Experience

Regression analysis is a dynamic discipline with new methods continually being developed. Early work centered on optimality properties of the least squares estimates. Recently research has turned to assess the importance of underlying assumptions and their effect on the optimality properties. The research is ongoing but one clear result is that a careful regression analysis requires more than the estimates and summary statistics provided by commonly available packages. A model must be challenged using a battery of numerical and graphical diagnostic procedures.

The movement to flexible, general purpose software, while more convenient to the statistician for exploring new techniques, has made things worse for the nonstatisticians, who are also generally nonprogrammers. They no longer have any guidance as to how to analyze their data. Our motivation in building REX is to bridge this gap.

We want to build a system which provides *guidance, interpretation,* and *instruction* for doing regression analysis. Guidance is important since it is no longer clear at the start of an analysis which statistical techniques will be required to complete the analysis, and more importantly, in which order they should be applied. Interpretation is required both for diagnosing problems encountered while using a particular technique (*e.g.* violations of assumptions), and suggesting possible ways to ameliorate the problem (*e.g.* transformations). Instruction is required to explain terms used in interpretation and guidance.

1.3 Why We Selected Expert System Techniques

The knowledge for guidance could be provided in a program constructed along standard lines. Such a program would produce a numerical output, briefly indicating changes that were made to the problem to achieve a satisfactory analysis. The symbolic knowledge would be represented as program control structure, a black box that the user could not see into.

This black box approach is certainly not acceptable in the short run, perhaps not in the long run. Many data analysis problems do not have a single best solution. Expert statisticians may disagree over the severity of a problem, or about how it can be taken into account. The subject matter analyst may have constraints which preclude using some techniques. In short, the problem is one of sufficient complexity that statisticians may disagree about what constitutes a satisfactory solution. Given this, any software which undertakes such a problem must be prepared to explain what it has done and why. The facility to explain *why* particular suggestions are being made changes the appearance of statistical software from a "black box" to a "glass box," *i.e.* one which you can see into.

Expert system techniques, developed in artificial intelligence research, offer a rudimentary capability for a program to do symbolic reasoning and to explain itself. To us, this is the primary distinguishing characteristic of an expert system.

In addition, we have noted a number of advantages of expert systems over conventional programming styles. These include the explicit representation of knowledge the program contains (*i.e.* declarative versus procedural representations), and the incremental nature in which the knowledge can be added to the target system.

1.4 Why We Thought an Expert System for Data Analysis was Feasible

This paper does not review expert system techniques because many good reviews are available. Buchanan and Duda (1982) provide a review of the component techniques of knowledge representation, inference, and reasoning with uncertainty. Buchanan (1981) describes the application and current limitations of expert system techniques. Duda and Gashnig (1981) show a simple example along with a readable introduction. Rauch-Hindin (1983) and Kinnucan (1984) discuss the prospective commercial applications, showing the breadth of impact that the techniques have achieved. *Building Expert Systems*, edited by Hayes-Roth, Waterman, and Lenat provides background as well as current procedures for building expert systems.

The cited papers show a large range of disciplines in which at least demonstration expert systems have been built. These include medicine, mineral exploration, artillery targeting, and electronic trouble shooting. This range of disciplines was one of the main reasons it seemed feasible to consider building such a system for data analysis.

From several sources, but especially Davis (1982) and Barstow and Buchanan (1981), we formed impressions of the characteristics of a field to which expert system techniques could be applied.

Desired Characteristic	Unwanted Characteristics
demonstrated and recognized need for help	need for inexact inference
a formalized core of knowledge	need to reason about time
established means of introducing a new system	need for common sense knowledge
highly motivated collaborators	need for high quality explanations
task takes a human a few hours	

These hints from the pioneers led us to believe that data analysis was a good domain for expert systems.

1.4.1 Statistics shares many desired characteristics. Subsections 1.1 and 1.2 were largely devoted to demonstrating the need for help in current statistical systems. In the next subsection we point out that one reason for selecting regression analysis as a specific subdomain was the formalized core of knowledge. Statistical systems are in common use on computers, and new ones have been introduced frequently. REX has been joint work by two of us, with Bill primarily handling the inference engine and Daryl primarily the statistical strategy; so motivation has been no problem. Regression analysis is rarely disposed of in a few minutes. There is more danger that we have chosen too hard an area, in that humans may spend days on a problem.

1.4.2 Statistics lacks most unwanted characteristics. Inexact inference means reasoning about knowledge which has an appreciable "grayness" to it — that facts are not simply true or false. This might have been difficult for us, since the essence of a statistical test is the uncertainty of the relationship between a precisely observable value on the particular data set, and what might be an underlying population value. We found early that we were extremely uncomfortable with a dichotomous interpretation of tests as pass or fail. But a trichotomy (ok, mild, severe) has been usable, since we seek more involvement from the user when we find a "mild" condition.

We are reasoning about a static set of data, but we have to make transformations which introduce a time-like element. However the "time" intervals are not uncertain and not overlapping, and so they present no problem.

Common sense knowledge refers to the facts that humans learn from experience as they cope with the world. It has been found that these facts are numerous, highly connected, and hard to express. As a formal discipline, statistics is well insulated from this swamp. There is a broader issue of how well separated statistics is from other areas of knowledge, however. In particular, data analysis is a domain of knowledge which will be applied to data from another domain, which we call the "ground" domain. The ground domain may be chemistry, psychology, or agriculture, for instance. We cannot support knowledge in ground domains, because of their multiplicity.

Therefore, we rely on the users to learn statistics, and to relate their ground domain knowledge to questions REX asks using statistics terminology. This limits the potential range of users, excluding, we believe, managers and other occasional users who would not have the time to learn the required statistics.

While we have viewed the key advantage of expert systems as their ability to explain their reasoning, this is still an area of active research. We feel that current techniques provide minimally acceptable explanations. We believe the lack of high quality explanations may again rule out managers as users of near term expert systems, but that analysts, with more time to learn how a system works, will find the explanations understandable.

1.5 Why We Chose Regression

Studying the expert system criteria led us to believe that data analysis was a suitable area, but didn't give grounds for discrimination among subareas. There are many different kinds of data analysis — analysis of variance, logistic models, cluster analysis, spectrum analysis, regression analysis, and multidimensional scaling to name a few. REX is research that hopefully will lead to strategies for all of these areas and more, but why have we started with regression analysis?

First, regression analysis is a widely used data analysis technique. Software for regression analysis is widespread. Every major software package which claims to do statistical analysis includes a linear regression module. Rudimentary regression analysis is provided on some hand held calculators.

This widespread use and availability leads to our second reason for selecting regression analysis — it is widely *abused*. The simple techniques which are widely available have known conditions under which they are valid. But the simple systems do not have the capability to check that these conditions hold, and most users do not know what checks need to be made.

Thirdly, there is widespread formalized knowledge of the assumptions of least squares regression, a growing body of diagnostic tests for checking the assumptions, and an increasingly sophisticated set of alternatives to least squares if the assumptions are violated. Regression is, in short, one of the best studied areas of data analysis.

Clearly these three facets are related — the technique was invented early because something was needed; it is studied because it is used; it is abused through need. Together they provide some reason for taking regression analysis as a first data analysis area. We expect that the techniques reported here can be used for other areas of data analysis, though extensions will surely be needed.

2. HOW REX LOOKS TO THE USER

We tell this story by showing a series of snapshots of REX in action. The data set has the body weights and brain weights of 62 terrestrial mammalian species. The regression is made in order to be able to estimate the brain weight for a species from its body weight. This data set is one which shows gross problems.

References
Barstow, D, & BG Buchanan, (1983) "Maxims for Knowledge Engineering", §5.5 *Building Expert Systems* cited below.
Buchanan, BG, (1981), "Research on Expert Systems", HPP-81-1, Stanford Heuristic Programming Project.
Buchanan, BG, & RO Duda, (1982) "Principles of Rule-Based Expert Systems", HPP-82-14, Stanford HPP.
Hayes-Roth, F, DA Waterman, & DB Lenat, (1983) *Building Expert Systems*, Addison Wesley, Reading, MA.
Kinnucan, P, (1984) "Computers that Think like Experts," *High Technology*, Jan p30.
Rauch-Hinden, W, (1983) "Artificial Intelligence: A Solution Whose Time Has Come," *Systems & Software*, Dec p150.

Figure 2.1 REX shows four windows to the user on a bit mapped terminal. The top window is for dialogue, the bottom for graphics. The left window will hold brief interpretations of tests, the right window tracing information.

REX begins by personalizing the system to the user. This snapshot shows it asking the user how thorough the analysis should be. The user can answer any query by a unique prefix of a key word. Other personalizations implemented are controls on verbosity and on showing debugging information. We set these both to maximum levels for demonstration.

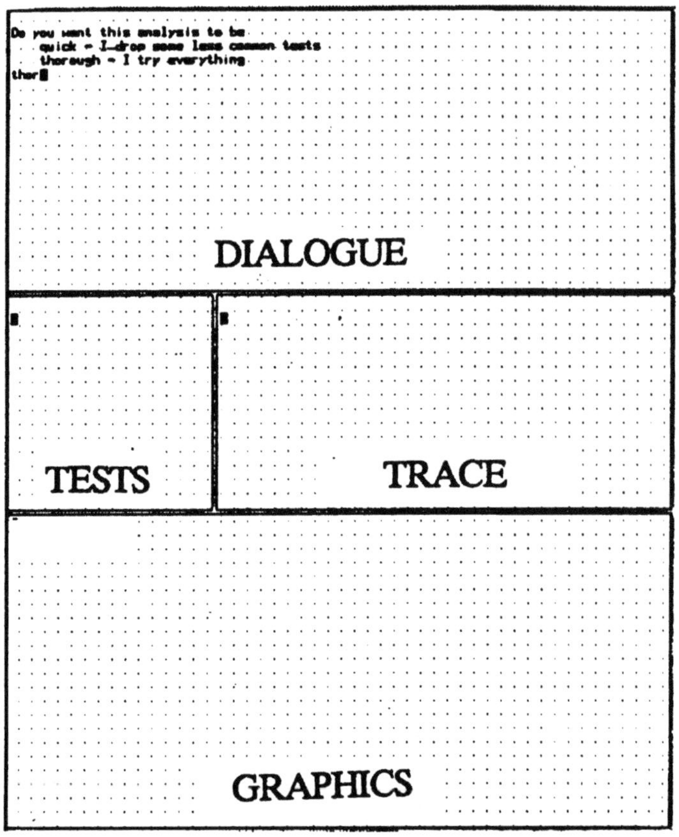

Figure 2.2 The right window now shows the first calls to *S*. These would not be shown if we had not turned on the debug flag.

The statistical analysis begins as a depth first search through the network of possible problems described in the section on REX's statistical strategy.

The initial task is to check for missing data since the algorithms require complete data. When the test for missing data is made, there is none, so no action needs to be taken. REX reports in the left window that the test result was 'ok'!

REX interprets test results as 'ok', 'mild', or 'severe'. The 'ok' interpretation means functionally that no action is needed. The 'severe' interpretation means functionally that the problem found will have to be fixed. The 'mild' interpretation means that REX does not insist on a fix but will attempt to find one if desired.

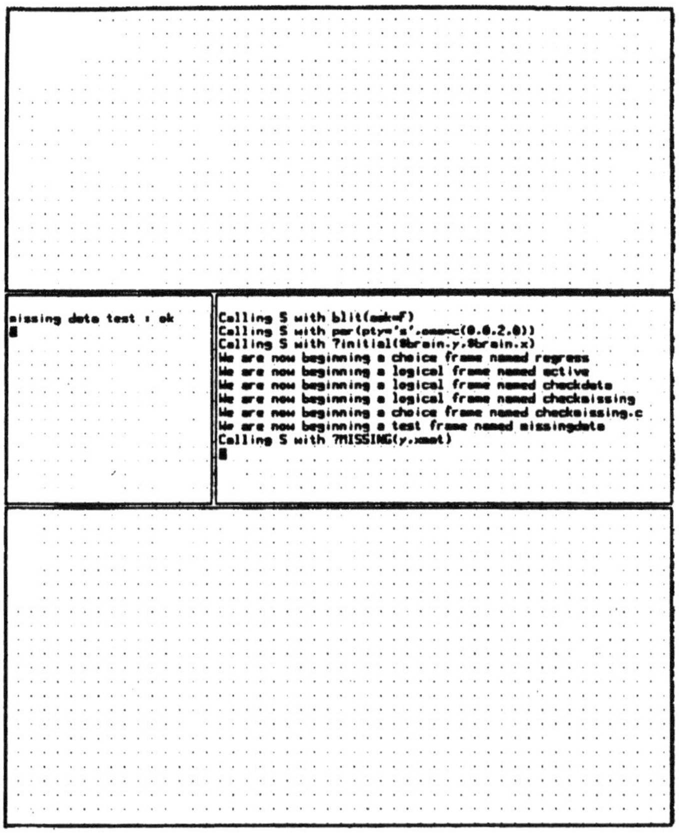

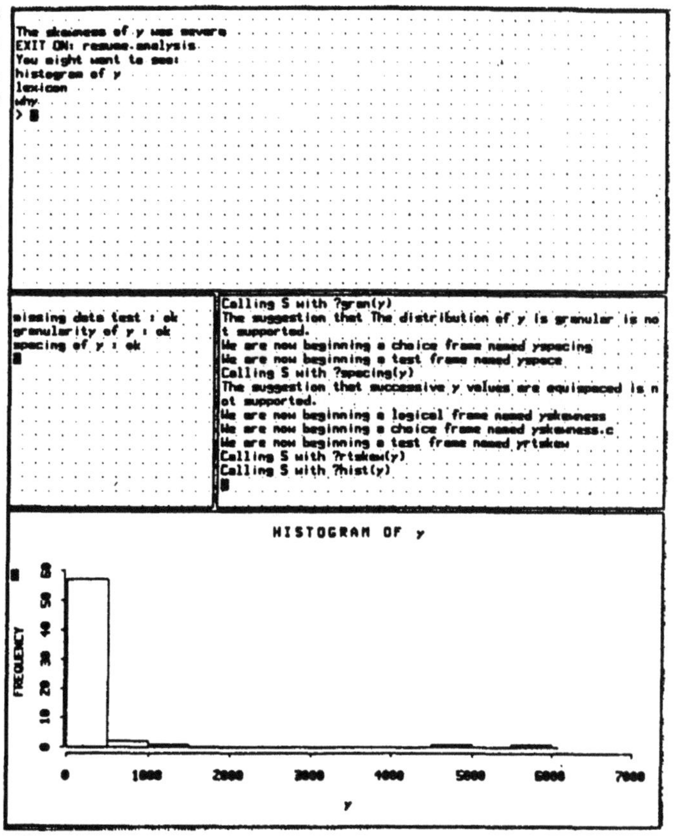

Figure 2.3 The first phase of analysis is model independent checks on each variable, the dependent variable first. The left window shows some of these tests.

Granularity is a test for the number of different values of the dependent variable. If there are exactly two, then a linear discriminant analysis is needed, not a regression analysis. Spacing tests whether the dependent variable is exactly evenly spaced.

REX finds that "skewness" of *y* is severe and offers to show the user a plot selected to show the problem. We asked for the plot which appears in the graphics window.

The plot is a histogram which shows that the brain weights of two species are much larger than others. In fact these are the two species of elephants. (If a 'mild' problem is found, the user is offered a plot and the option of declaring that the problem is really 'ok'. If that option is selected, then no attempt to fix the problem is made.)

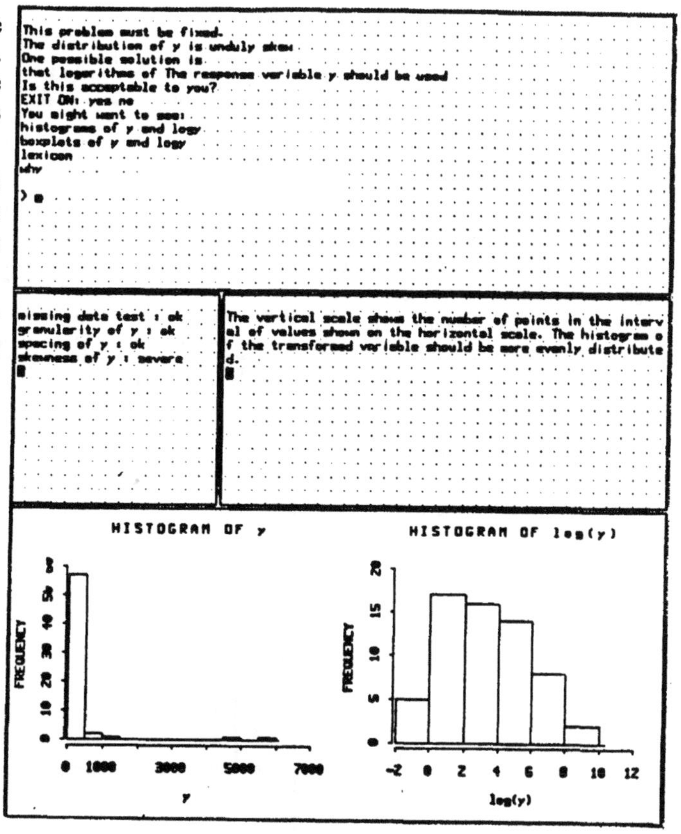

Figure 2.4 Having found a severe problem, REX seeks possible solutions. It finds that taking logarithms will cure the skewness problem. Only after this check will REX suggest the action.

At the top you can see "must be fixed." This is because the problem is severe. For a mild problem "may need to be fixed" would have been printed.

The user is now being asked whether it is acceptable to make the suggested transformation, a yes or no choice. But REX offers three aids in making the decision, which the next figures will demonstrate — figures, lexicon and why.

This figure shows the selection of a histogram as a plot. Whenever REX suggests a change it offers one or more before and after plots that show the effect of the change.

A legend in the right window describes briefly how the user should interpret the plots. The plots shown should convince the user of the value of the suggested change in this case.

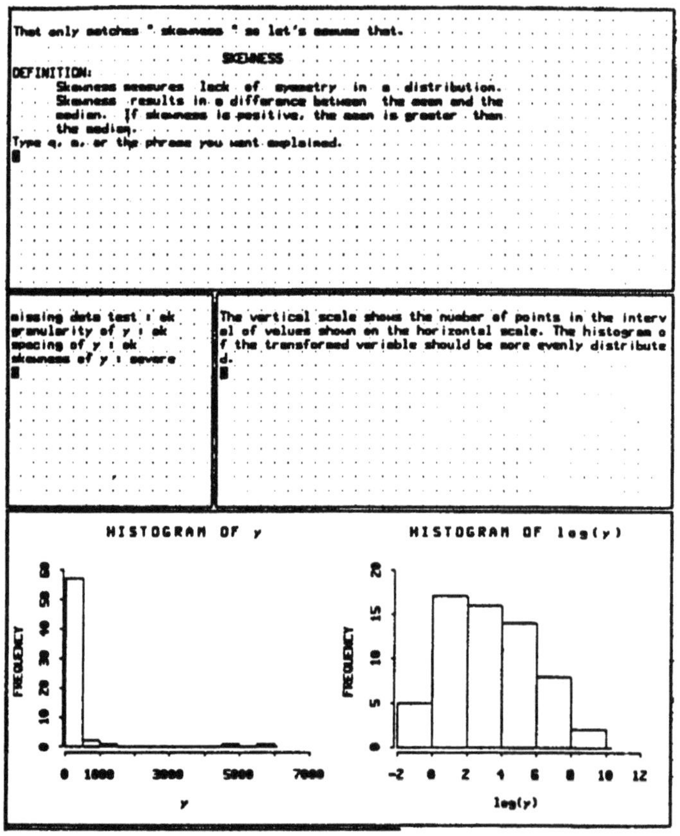

Figure 2.5 We asked for the lexicon, got brief instructions (not shown) and typed "skew" asking for a definition.

REX consults a dictionary of defined terms and finds one match to "skew," "skewness." It assumes that it is what the user wants and prints a definition found on a file.

If the user is uncertain of another word used in the definition, REX can be queried about that word. The software to support this lexicon is in place in REX, but only a demonstration set of a few dozen definitions. The construction of definitions would be a substantial undertaking. They should be constructed in an 'onion' model with a minimum of mutually referential terms.

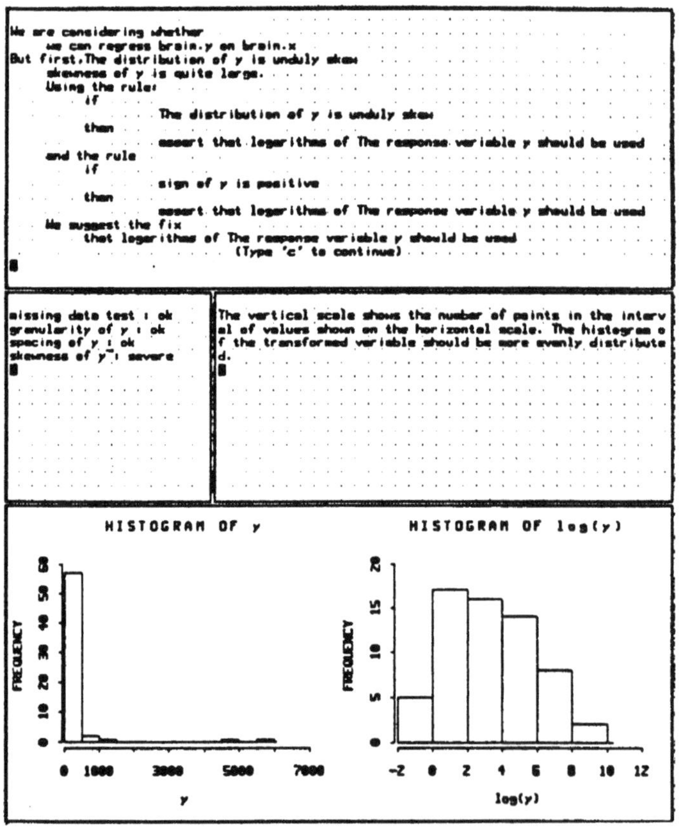

Figure 2.6 We got out of the lexicon, were presented again with the choice of Figure 3.4, and asked for why. This provides a verbal answer to why the suggestion is being made in addition to the graphical answer still showing.

To conclude that logarithms could be used at this point, REX had to check that no negative values were present in the data. If it had found zero values, it would have asked the user whether adding a small amount to each value would change the interpretation.

We terminate the demonstration at this point to save space. In this data set a total of three problems are found and corrected, leading to a log-log model.

On the Possibilities to Automate the Study of Statistical Dependence Between Two Variables

T. Pukkila, S. Puntanen, and O. Stenman, Tampere

Summary: This paper discusses the possibilities of automating the study of statistical dependence between two empirical variables when beginning to familiarize oneself with the data set. The use of STUDY, a program planned for this purpose, is described. STUDY is a part of the KONSTA system developed in the Department of Mathematical Sciences at the University of Tampere.

Keywords: Automatization of data analysis; computers; teaching of statistics.

1. INTRODUCTION

In data analysis, when investigating the statistical dependence between two empirical variables, say x and y, it occasionally happens that the procedures to be carried out are fairly similar whatever the variables x and y are. In particular this is the case at the very beginning of analysis, when one is not so familiar with the data. Hence it will not strike the data analyst as strange that the whole routine, i.e., the "*first look*" at the data, should be carried out automatically using only a few simple commands. This paper discusses the possibilities of such automatization and describes how an operation called STUDY, planned for this purpose, works in the KONSTA system. KONSTA is an interactive computer system for the illustration of statistical concepts. It was developed in the Department of Mathematical Sciences at the University of Tampere (cf. Puntanen and Pukkila 1982, and Pukkila *et al.* 1984). The system works in a time-sharing mode and the results are immediately printed by the terminal. The system was implemented in 1981 on a DEC 2060 system with more than 100 asyncronic lines. Parts of KONSTA have been in use since 1978.

One of the big questions in statistical computing is: How far can the computer take empirical statistical research without human intervention? The authors are aware that there might be sharply divided opinions on the suggested automatic procedure. The use of an operation like STUDY can be badly misleading but it can also be very helpful, particularly when the user is beginning to familiarize himself with a set of data.

The STUDY program handles three variables. The user has to define the dependent variable y, the explanatory variable x and thirdly the variable z that will be taken into account when studying the dependence between x and y. After that, the classifications for the variables can be defined, although the program can also take care of these. Then a series of statistical operations will be carried out, for example scatter diagrams, box-and-whisker plots, correlations, regression analysis, cross-classifications, frequency distributions, conditional (eliminating the effect of z) scatter diagrams etc. The paper gives examples and discusses the difficulties arising in planning this kind of automatic procedure.

2. PROBLEMS REGARDING THE LEVEL OF AUTOMATIZATION OF STATISTICAL PROGRAMS

At the moment it is very simple in most statistical packages to carry out e.g. a multiple regression analysis with several predictors. The user need give only a few simple commands and the program will produce (after not-so-simple calculations) numerous statistics related to the regression. In fact, it seems that in some cases there is a reverse ratio: Statistically "simpler" analyses may be more time-consuming and frustrating to carry out than the more "advanced" ones. By simple analyses we mean here particularly a sensible sequence of statistical calculations needed when the data analyst takes his first look at the possible dependence between two variables, i.e., when he is beginning to familiarize himself with the data set. That first look, however, is always needed and its role is very important; it may give new impulses to the user's statistical imagination.

It is our impression that the possibilities of automating the first look at data have not received as much attention as they may deserve. However, there has been some general discussion on the effect of the computer on statistics, which is of particular interest and worth a brief review in this context.

In 1979 Professor D.J. Finney stated: "The outlook I want to give to students is that good statistical practice is not a purely mechanical matter of applying a rigid set of rules. I am worried lest, at the elementary level, use of the computer for the good purpose also encourages the idea that statistical interpretation of data can be done as well by a machine as by a human." Finney's comment is particularly relevant in our case because a great proportion of the users of our programs are university students who are not majoring in mathematical sciences. We strongly emphasize to them that no program - more or less automatic - can replace their common sense.

C.R. Rao (1979, pp. 136-137) gave some interesting general comments on the role of computers and their impact on research in statistics. He said e.g. the following: "It [the computer] has also encouraged uncritical use of statistical methods through the commercially available computer package programs. ... There are always special features in individual cases which have to be taken into account in the analysis of data and interpretation of results. ... The use of an existing computer program should not be the end of statistical analysis of given data. ... I am reminded of an aphorism mentioned in a lecture by Dr. David Cox: There are no routine statistical questions, there are only questionable statistical routines." When planning the automatic statistical programs, it is surely wise to remember Professor Rao's words.

Criticizing the existent statistical programs Mustonen (1981, p. 130) writes: "Many statistical programs are too automatic or they are automatic in wrong places." This is an essential statement from our present viewpoint and it was an important factor when Mustonen developed the SURVO 76 system. One important difference between our system and SURVO 76 is that according to Mustonen (1981, p. 130): "Our [SURVO 76] main goal has been to provide suitable tools for a statistician who likes to have a quick test of his research ideas by making a computational experiment."

The typical user of our system is a non-statistician who is familiar with some basic statistical methods - but not necessarily familiar with computer science at all - who likes to have a quick look at his data by means of simple analyses and then to continue his research. Remembering the differences in users, it is natural that the stages where SURVO 76 is automatic could be rather different from the automatic stages of our program.

Joiner (1982) suggests that: "... more emphasis be put on developing software that will automatically make a variety of plots, or at least suggest a menu of plots which we might want to consider." Further, he points out: "Models should never be fit without plotting the data and the residuals. In addition, we should seek to develop software that will automatically do as much assumption checking as practical." Joiner also emphasizes the importance of data preparation and data checking; though this activity, according to Hartley (1980, p. 6), "seems, at first sight to be of a trivial nature it is in fact one of the most challenging applications of high-speed computer technology."

3. THE STUDY PROGRAM

The main question in the present context is the following: How much from the first look at the data could be automated? By the first look we mean a set of statistical operations which in most (or at least in many) cases would best be carried out when one sets out to familiarize oneself with the data. Considering all variables simultaneously, it is obvious that a listing of different univariate statistics should automatically take place. But when we study the interdependence of the variables, the situation becomes more complicated from the point of view of automatization. It does not seem reasonable or natural to consider more than three variables at the same time.

The user of the STUDY program has to define (i.e., to give the name of) the dependent variable (y), the independent variable (x) and in addition, he can define the third variable (z) whose effect will be taken into account when studying the dependence between x and y. After that the program works in the manner seen in Figure 1 and Example 1. It should be noted that STUDY is a part of KONSTA, which is an interactive computer system mainly planned for the illustration of statistical concepts in the teaching of statistics.

EXAMPLE 1. *Parts of a STUDY program: MATH = mark in long mathematics in the matriculation examination; GRADE1 = grade of a student in a statistics examination; SEX = 0 = female; ☆☆☆ = some output cancelled.*

```
run <stat.prag>study

THE DATA MATRIX
<stat.data>data79

NAMES OF THE EXPLANATORY VARIABLE AND
THE DEPENDENT VARIABLE  (X,Y)
math, grade1

OPTIONAL SPECIFICATION? (NO/YES)
no

************ REGRESSION ANALYSIS, SCATTER **************
           DIAGRAM AND FREQUENCY DISTRIBUTIONS

CORRELATION COEFFICIENT R = 0.444          R**2= 0.197

                  SS          DF          MS
REGRESSION      4898.4        1         4898.37
ERROR          19921.4       69          288.72
TOTAL          24819.8

               COEFFICIENT   STDEV        T
CONSTANT          -3.566     9.090      -0.39
MATH               8.152     1.979       4.12
```

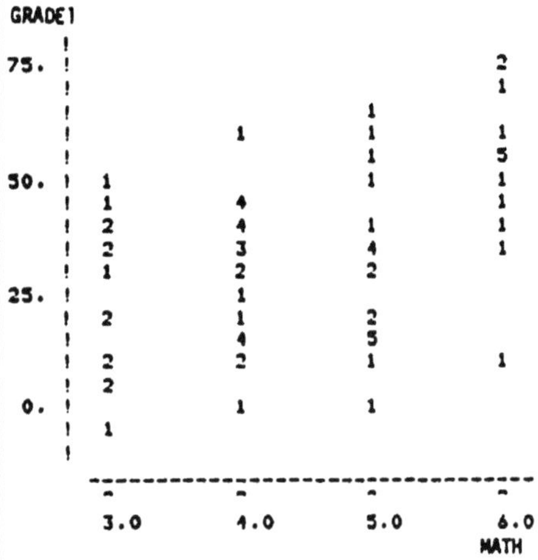

```
>GRADE1
  MEAN   =        32.944       MIN  =     -6.00
  ST.DEV.  =      18.830       MAX  =     75.00
  QUARTILES:   (  16.00,       33.00,     47.00)

  CLASS INTERVAL  FREQUENCY %

-25.00  -   -15.00<   0   0.0   :
-15.00  -    -5.00<   1   1.4   :=
 -5.00  -     5.00<   3   4.2   :===
  5.00  -    15.00<  10  14.1   :==========
 15.00  -    25.00<  12  16.9   :============
 25.00  -    35.00<  12  16.9   :============
 35.00  -    45.00<  14  19.7   :==============
 45.00  -    55.00<   8  11.3   :========
 55.00  -    65.00<   7   9.9   :=======
 65.00  -    75.00<   3   4.2   :===
 75.00  -    85.00<   1   1.4   :=
 85.00  -    95.00<   0   0.0   :
 NUMBER OF
 OBSERVATIONS   =     71        = : 1

☆☆☆
```

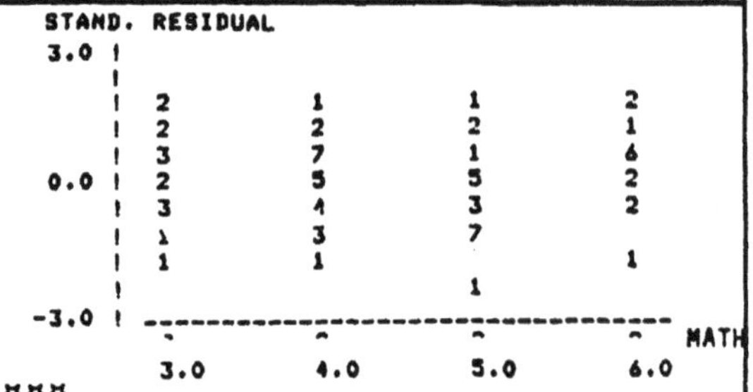

☆☆☆

```
*************** BOX-AND-WHISKER PLOTS, ****************
               ONE-WAY ANALYSIS OF VARIANCE
```

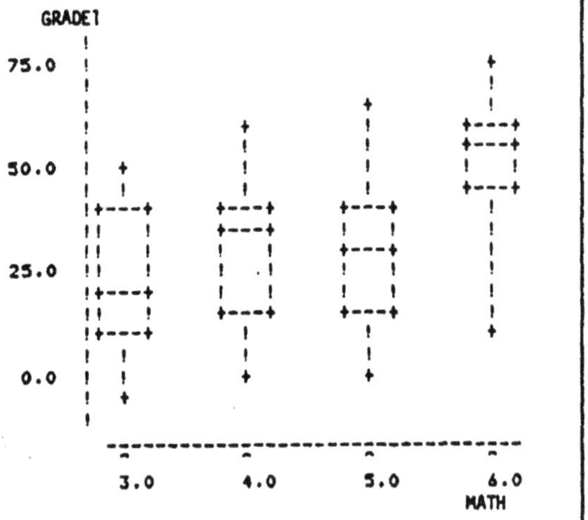

	3.0	4.0	5.0	6.0
MEAN	23.8	29.6	29.9	52.0
STDEV	17.1	15.3	17.2	16.5
N	14	23	20	14

ONE-WAY ANOVA, WHERE DEPENDENT VARIABLE IS GRADE1

SOURCE	SS	DF	MS	F
MATH	6705.97	3	2235.32	8.268
ERROR	18113.81	67	270.36	
TOTAL	24819.77	70		

KRUSKAL-WALLIS CHI SQUARE (DF= 3) : 17.359

☆☆☆

```
* VARIABLE WHICH IS TAKEN INTO ACCOUNT (OR RETURN)
* sex
```

☆☆☆

```
**** MEANS OF GRADE1   IN DIFFERENT *****************
     CLASSES OF MATH     AND SEX
```

FREQUENCIES

SEX	MATH 3.0	4.0	5.0	6.0	TOTAL
0.0	2	8	12	6	28
1.0	12	15	8	8	43
TOTAL	14	23	20	14	71

MEANS OF GRADE1

SEX	MATH 3.0	4.0	5.0	6.0	TOTAL
0.0	0.0	16.3	26.6	51.5	27.1
1.0	27.8	36.7	34.9	52.4	36.8
TOTAL	23.8	29.6	29.9	52.0	32.9

☆☆☆

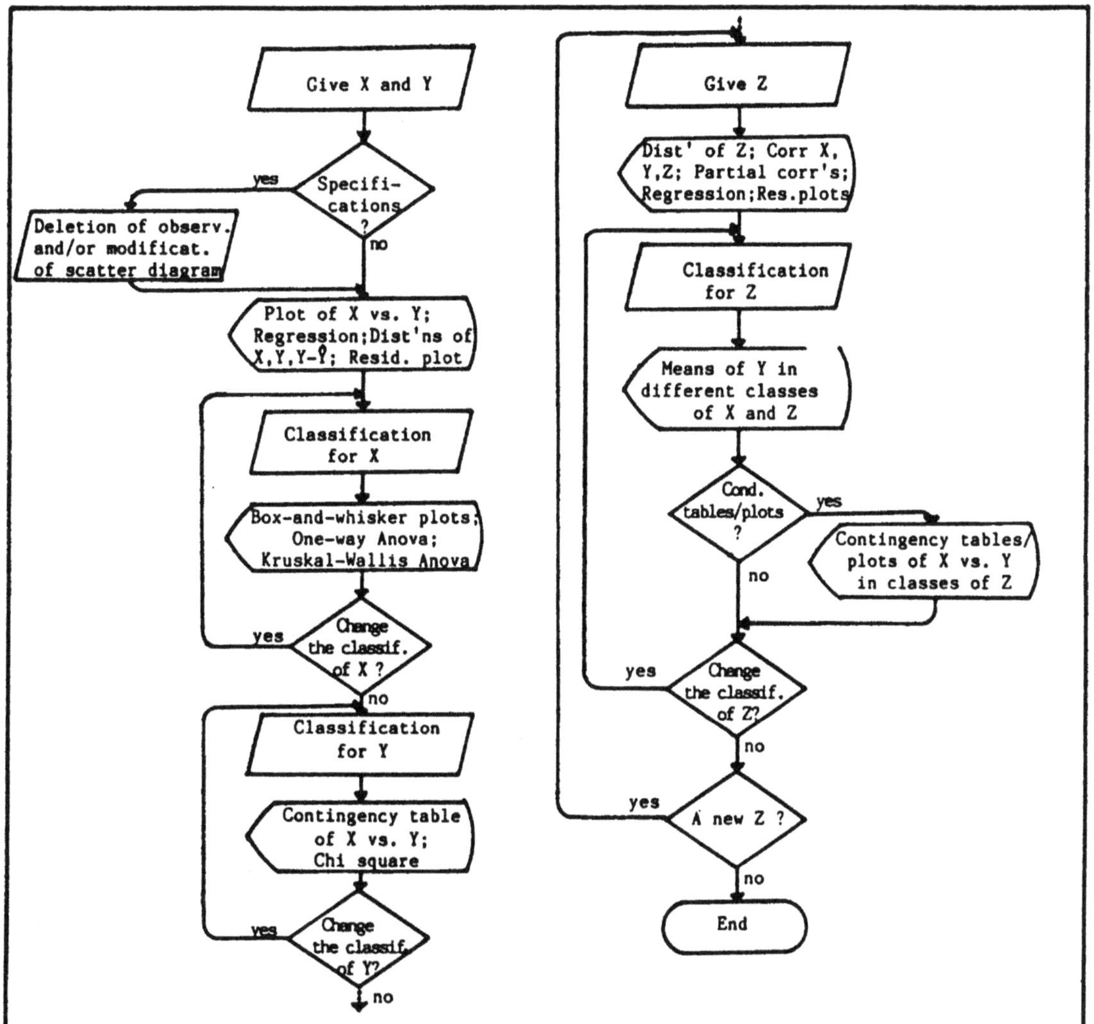

FIGURE 1. *Flow-chart describing the STUDY program:* input output question

In planning the STUDY program, the following principles have been important:

● It consists of simple statistical analyses because data analysis should, in any case, be started with simple considerations. This means that methodologically it is possible to understand the output on the basis of the first-year university course in statistics.

● The program is as intelligent and automatic as possible - in the correct places. This means that STUDY tries to help the user with the maximum effort whenever possible. For instance, as regards classification, the user of STUDY can tell that the variables are measured on a nominal scale or that the variable should have classes with an equal number of observations or the user can leave the classification to be decided by the program, which, as we hope, makes an intelligent choice.

● The program works interactively, so that the results are printed immediately on the ordinary paper terminal. Hence the quality of the pictures is not the best possible but, in any case, good enough for the purpose in hand. Conversation takes place in every day statistical English, so that no programming skill is needed.

● The output should give a good basis for further analyses. For instance, the STUDY program gives the distribution of ordinary residuals ans plots the residuals against relevant variables. If the user wants to make some special plots, then he can easily use the particular PLOT program of the KONSTA system.

REFERENCES

Finney, D.J. (1979). Personal Communication.

Hartley, H.O. (1980). Statistics as a science and as a profession. *Journal of the American Statistical Association*, 75, 1-7.

Joiner, B.L. (1982). The frontiers of statistical analysis. *COMPSTAT 1982, Proceedings in Computational Statistics, Part I.* Physica-Verlag, Wien, 278-281.

Mustonen, Seppo (1981). On interactive statistical data processing. *Scandinavian Journal of Statistics*, 8, 129-136.

Pukkila, Tarmo and Puntanen, Simo (1982). The use of KONSTA, an interactive computer system for the illustration of statistical concepts, in the teaching of statistics. Department of Mathematical Sciences, University of Tampere, Report A 73. (Abstract given at the *ICOTS Conference*, August 1982, Sheffield).

Pukkila, Tarmo, Puntanen, Simo and Stenman, Olavi (1984). The illustration of important basic concepts of statistics in the KONSTA system. Short communication to be given at *COMPSTAT 1984.*

Rao, C. Radhakrishna (1979). Perspectives in statistics. *Sankhyā*, 41, 129-137.

Integration of Statistical Data Base Management and Data Analysis

Statistical Data Analysis Using a Data Base for Experiment Design and Systems Identification

L. Borzemski, and L. Koszałka, Wroclaw

In the paper the approach to the problem-solving in the areas of experiment design and system identification is presented. The main principles of the system EXPES-NET such as functional framework, dialogue and distributed data base are discussed and an example of multistage identification at mechanical engineering laboratory using the approach proposed is given. Distributed environment gives the ability to handle the problems of increased complexity and intristic decentralised nature.

Keywords: Identification; problem-solving system; distributed data base.

1. INTRODUCTION

The approaches to the problem-solving in the area of system identification and control systems are described in the IDPAC, PIPACF, KEDDC, IMSL systems presented by Wieslander and Gustavsson /1976/, Furuta and colleagues /1979/, Schmid and Unbehauen /1979/, Craven and Wahba /1979/, respectively. The group of the Institute of Control and Systems Engineering /ICSE/ has developed problem-solving approach in experiment design and system identification. The EXPES computer systems family has been implemented /Borzemski and colleagues /1982//. The system EXPES-NET which is the latest achievement in the family is being built in a distributed environment in a local area network of mini- and microcomputers. The distributed data base supports the system. Special properties of EXPES-NET are such that /i/ it is able to handle the problems of increased complexity and intristic decentralised nature, /ii/ computations may be distributed to increase the computability and decrease processing costs, /iii/ the flexibility in problem statement can be achieved, /iv/ it employs the procedures which may be defined by users on the basis of available problem-solving programs and subroutines. The functions provided present among others the new ideas in system identification developed at ICSE and formulated by Bubnicki /1980/ as global identification, multistage identification and new concepts in control of experiments.

2. THE EXPES-NET ARCHITECTURE

Figure 1 shows the layered architecture of the EXPES-NET system. The system is built around two main layers: the DPM layer referring to distributed data base /DDB/ and program management, and the PPM layer - dealing with the distributed data and programs applications. The

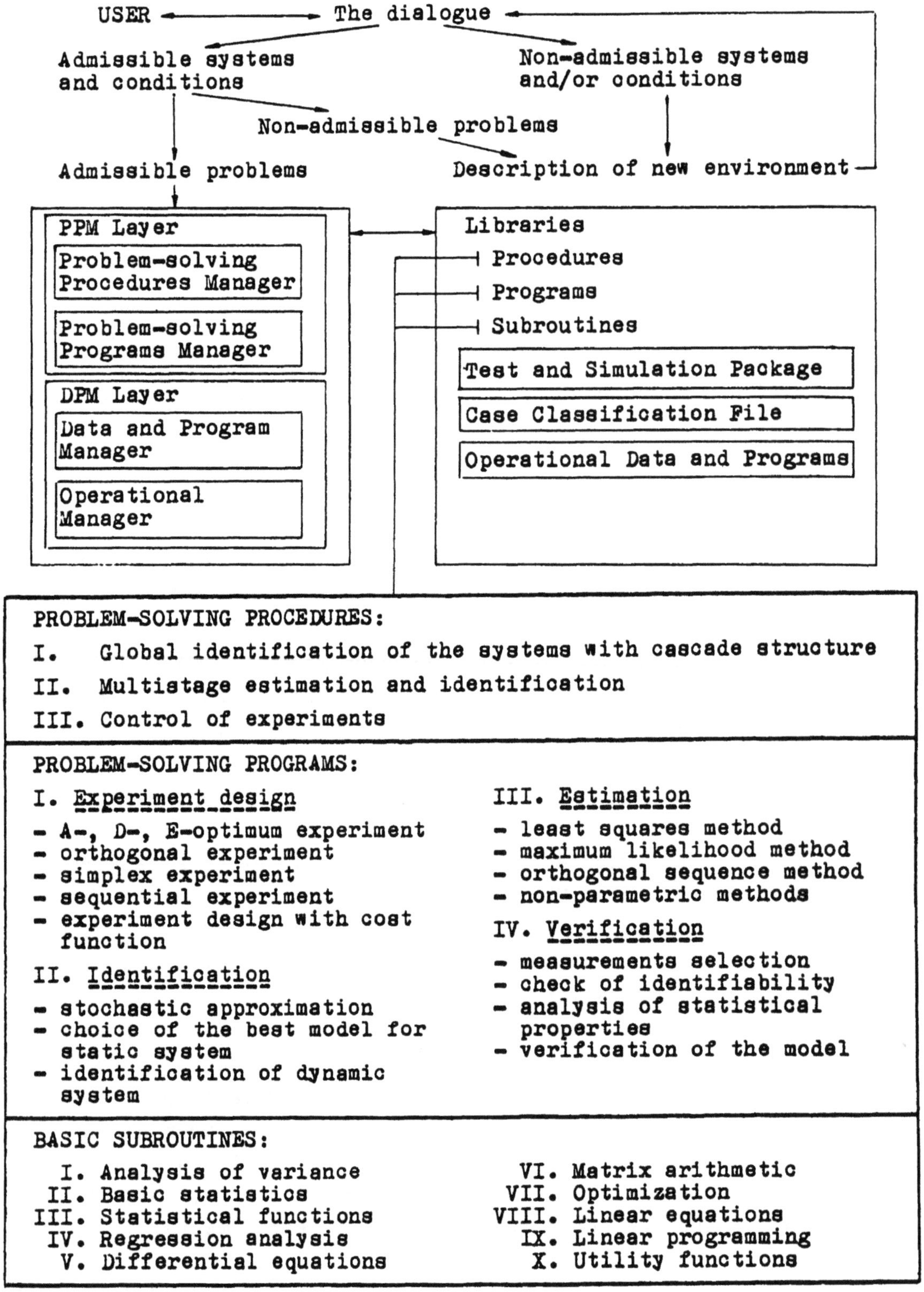

Fig. 1. The EXPES-NET architecture

procedures and programs are grouped around admissible problems. Each problem has one ore more problem-solving programs or procedures. The architecture of the EXPES-NET includes the dialogue which supervises operation of the system. The dialogue is connected with the steps of the process of identification as shown in Fig. 2 and consists of four main phases: I. Choice of the case of the problem. II. Definition of environment. III. Realization of experiments. IV. Computations. In the first phase the user informs the system about structure of the investigated system, interconnections between elements, admissible input and output sets, a priori information about disturbances, output covariance matrix, class of model, the assumed local and global criteria, possibilities and constraints in experiment design, criteria of experiment design and measurement selection, cost functions, number of experiments, etc. After then the system is able to recognize the case of the problem, presenting the methods which are to be used. In the second phase the system learns about experiment locations, local peculiarities, actions to be performed

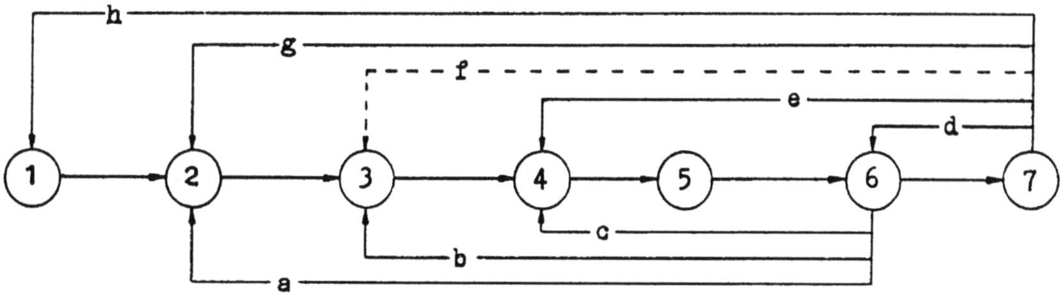

1 - Definition of the system
2 - Analysis of assumptions
3 - Analysis of experiment conditions
4 - Experiment design
5 - Measurements
6 - Identification algorithm
7 - Verification of the model

a - to estimate output variance, b - to estimate cost function, c - to control of experiment, d - to change identification algorithm, e - to change the method of experiment design, f - to complete measurements, g - to change class of models, h - to take other inputs into account.

Fig. 2. Conceptual framework of the identification

and informs the user which data files are to be used for storing measurements. In the next phases the experiment is performed and data base is filled with measurements. Appropriate problem-solving programs are executed to obtain and verify the model. At the end the system gives the suggestion to user about the results obtained. The user makes decision whether to continue or to stop the process of identification.

3. DISTRIBUTED DATA BASE

The DDB architecture in the EXPES-NET system /Borzemski /1984// introduces a global level and a local level of data abstraction. Both levels are considered to be defined in the data base aspects which are classified into three categories: /i/ conceptual, /ii/ internal and /iii/ external. Therefore, three types of data base models are considered: /i/ a conceptual model as a set of rules describing which information may enter and be stored in the data base, /ii/ an internal model as a set of rules describing how the information is physically represented on storage media, and /iii/ an external model as a set of rules describing how information is viewed by a user.
The DDB provides the users with the data structures viewed as relations defined by Codd /1970/. The DPM layer can support various external models /user's views/ which contain references to the relations, application programs and so-called actions. This results in transparent distributed use of programs described in the actions. Local operating systems handle action running accross several sites. The problem-solving procedures are implemented as actions.
The user is provided with the DDB Data Manipulation Language /DDB/DML/ as an interface to the DDB. The user is not obliged to use the DDB/DML if only standard action or program is involved. However, if he wants to define and use his own distributed environment, specific to his problem, he has to be aware of the DDB/Data Description Language /DDB/DDL/ and DDB/DML.

4. EXAMPLE

One of the typical tasks in the field of mechanics is the problem of the study of the strain of the plate clamped at one edge /Fig. 3/. The strain state of the plate can be described by tensions and moments treated as system outputs y. The inputs of the system under identification are: x_1 and x_2 - coordinates of the measurement point, x_3 - force.

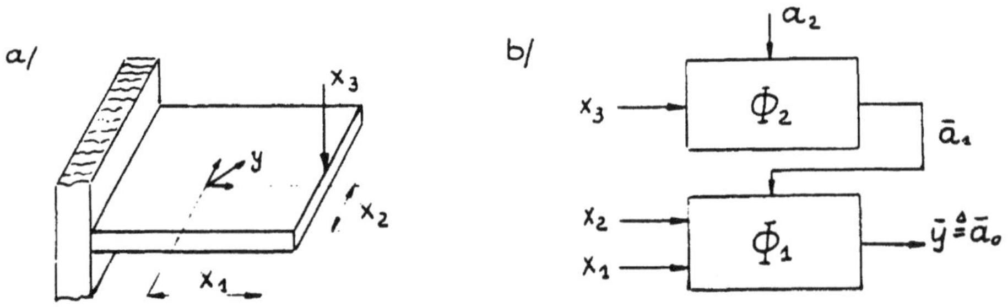

Fig. 3. a/ Restrained plate b/ two-stage model

We choose the model $\bar{y} = \Phi(x_1, x_2, x_3, a) = a^T \cdot \varphi(x)$ with 12 unknown parameters. The model can be decomposed into two-stage model: at the first stage - $\bar{y} = \bar{\Phi}_1(x_1, x_2, a_1) = a_1^{(1)} + a_1^{(2)} x_1 + a_1^{(3)} x_2 + a_1^{(4)} x_1 x_2 + a_1^{(5)} x_1^2 + a_1^{(6)} x_2^2 \triangleq \bar{a}_0$, at the second stage - $\bar{a}_1 = \bar{\Phi}_2(x_3, a_2) = a_2^{(1)} + a_2^{(2)} x_3$, where a_i - vector of unknown parameters, $\bar{a}_k$ - model output /i=1,2; k=0,1/. The measurements are carried out at two laboratories /site #1, site #2/ and the user #1 is responsible for performing two-stage experiment at both stages, whereas the user #2 performs direct identification and comparision of both approaches.
The quadratic criteria Q_1, Q_2 for multistage identification and Q_D for direct approach are assumed. The different ranges of x_3 may be chosen at both sites.
A part of DDB for this multistage experiment can be viewed as shown in Fig. 4. A partial description of the DDB in Fig. 4 in basic terms of DDB/DDL is given after the figure.

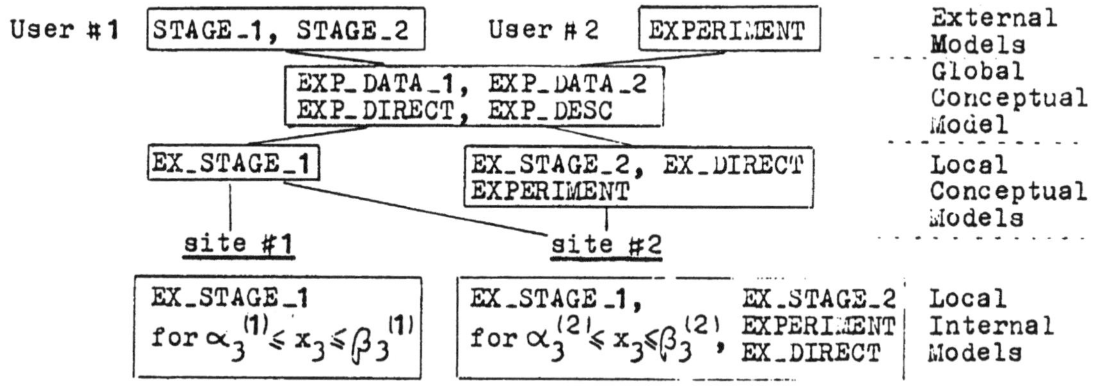

Fig. 4. Example of multistage experiment DDB

```
GLOBAL DB SECTION
DEF GR[ELATION] EXP_DATA_1 (EXP#, INPUT_1, INPUT_2, OUTPUT_1,
CRITERION_1, P_A11,..., P_A16, INPUT_3) KEY EXP# ASC
DEF GR EXP_DATA_2 (EXP#, IN_3, CRITERION_2,P_A21,P_A22) KEY EXP# ASC
DEF GR EXP_DIRECT (E#,A1,...,A12,QD_VALUE,Q_VALUE,N) KEY E# ASC
END GLOBAL
LOCAL DB SECTION
DEF LR EX_STAGE_1 (EX#, IN_1,IN_2,OUT_1,P_1,..., P_6,Q,IN_3,...)
LOCALIZATION SECTION
EX_STAGE_1 ON SITE_1 WHERE IN_3 =< "BETA" AND IN_3 >= "ALPHA"
                 ...
MAPPING SECTION
DISTRIBUTE EXP_DATA_1 ON PROJECT EX_STAGE_1 ON EX#, IN_1, IN_2,
OUT_1, P_1,...,P_6,Q,IN_3
DISTRIBUTE EXP_DIRECT ON EX_DIRECT
                 ...
USER SECTION
```

DEF VIEW USER #1
STAGE_1 (E#,X1,X2,Y,A11,...,A16,Q1,X3) AS PROJECT EXP_DATA_1 ON EXP#,
INPUT_1,INPUT_2,OUTPUT_1,CRITERION_1,P_A11,...,P_A16,INPUT_3)
...
DEF VIEW USER #2
EXPERIMENT AS EXP_DIRECT

In a query language the requests for data are formulated in the following way:

SET DATA_1 AS PROJECT (SELECT FROM STAGE_1 WHERE X3 = "10")
ON E#,A11,...,A16
SET DATA_2 AS PROJECT (SELECT FROM EXPERIMENT WHERE E# = "1"
ON QD_VALUE, Q_VALUE

Then DATA_1 includes the parameters at the first stage for $x_3 = 10$, whereas DATA_2 gives the value of criterion Q_D calculated in direct identification /QD_VALUE/ and the value of criterion Q_D where the parameters were substituted by values calculated in multistage identification /Q_VALUE/. Now both values of the criterion can be compared by the user #2.

5. CONCLUSIONS

The briefly described first version of the EXPES-NET is being implemented on the local area computer network at research laboratories at mechanical, electrical and chemical engineering departments at Technical University of Wrocław to be used by research workers and students. The distributed environment gains the flexibility in problem-solving and the system is able to help the experimenters in solving the problems in identification of the complex systems.

REFERENCES

Borzemski L., Kordecki H., Koszałka L. /1982/, EXPES and EXPES-E - computer systems for experimentation, Wyczislitielnaja Technika Socialisticzeskich Stran, no 11, Moskva

Borzemski L. /1984/, The relational distributed database architecture for minicomputers, Proc. IFIP/Network Symposium, Sofia /to appear/

Bubnicki Z. /1980/, Problems of the complex systems identification, Proc. Internat. Conference on Systems Engineering, Coventry

Codd E.F. /1970/, A relational model of data for large shared data banks, Comm. ACM 13, No. 6

Craven P., Wahba G. /1979/, Smoothing noisy data with spline functions, Numer. Math. 31, pp. 377-403

Furuta K., Hatakeyama S., Kominami H. /1979/, Structural identification and software package for linear multivariable systems, 5th IFAC Symposium on Identification and System Parameter Estimation, pp. 415-422

Schmid C., Unbehauen H. /1979/, KEDDC, a general purpose CAD software system for application in control engineering, 2nd IFAC/IFIP Symp. SOCOCO, Prague

Wieslander J., Gustavsson I. /1976/, IDPAC - An efficient interactive identification program, IV IFAC Symp. on Ident. and Syst. Par. Estimation, Tbilisi.

SICLA-PEPIN: A System Integrating Data Analysis and Relational Data Base Management System

G. Jomier, and O. Kezouit, Paris
H. Ralambondrainy, Rocquencourt

SUMMARY : In this paper we present a prototype, in development at INRIA, which integrates the multivariate analysis system SICLA and the relational data base management system PEPIN.

KEYWORDS : data analysis, system, data base, relational

I. INTRODUCTION

Two converging trends are appearing in data bases and in data analysis (And.82).

As Data Base Management Systems (DBMS) are used to store, manipulate and query large data bases, it is common to have access to some elementary statistical functions (minimum, maximum, mean etc..) called "aggregates". But the designers of those systems discover more and more the help that statistical tools (LST 83) and especially Multivariate Data Analysis (factorial analysis, clustering methods...) could provide for data base user : they allow a deeper and fuller knowledge of data, the possibility of taking into account some parameters simultaneously, and offer a more synthetical view of the data.

An other interesting consequence of the integration of statistical tools in a DBMS is the possibility of analysis of the system itself. The problem of DBMS performance is crucial, and still open, especially for Relational DBMS (RDBMS). Two aspects must be studied : the system itself and its load, data and requests. A statistical study of their behaviour (the statistical data being collected during DBMS sessions and analyzed after) would be very useful in an optimization prospect (better ressources use, response time optimization, etc.).

Moreover the statistical systems designers are more and more interested in using DBMS rather than the traditional file systems to manage the data which are to be analyzed (KFI 82). This is explained by the DBMS power : it makes easier the update and consultation of the data, allows a more powerful control of their coherence and assures their consistency maintain in a multiuser context, even when crashes occur.

Those reasons explain the present interest for systems integrating data bases and data analysis. However different possibilities exist in the choice of the DBMS model, and in the way to integrate both systems (Hex 82).

In this paper , first we point out the interest of relational DBMS for statistical analysis. Then we present a system in development at INRIA : it integrates the Interactive Data Analysis System SICLA (Ral.82), (Ral.83) and the Relational DBMS PEPIN (INRIA, ISEM Univ. Paris Sud) which are both operational (BCFJK 81), (BCFJS 83).

II. THE RELATIONAL APPROACH

The relational systems are the DBMS second generation, the zero generation designating the File Management Systems and the first one the Hierarchical and Network DBMS.

The main advantage of a DBMS, and especially of a relational DBMS, compared to a File Management System, is the abstraction level offered to the user for the manipulation of bulky data. It allows him to completely ignore the physical aspects of the implementation (files used, physical format of data, management of access paths to the data, etc..), and to realize, by calls to relational operations , powerful processing which would require, without DBMS, the development of sizeable programs.

An other characteristic of interest for the statisticians is that relational systems process large sets of data (the "relations") at once, whereas the other

DBMS access data one by one.

Before going ahead, let us recall some of the main features of relational DBMS (Dat. 82).

Relational DBMS process "relations" which may be imagined as tables. Each table has a number of columns, called attributes, fixed at the table definition, and an indefinite number of lines, the tuples. Each attribute takes its values in a "domain". The domains allowed are different from one system to an other: in PEPIN the domains may be real, integer, strings and predefined sets or intervals.

We call "schema" the description, for a data base, of its relations, attributes and domains (nota: some other informations on relations may be add to the schema).

Relational systems allow operations at the tuple level, and at the relation level. The tuple level concerns the update of relations: insertion of new tuples, suppression and modification of the tuples satisfying a condition. The results of the relation level operations are new relations. The goal of the unary relational operations is to suppress lines or columns in a relation. The selection keeps all the tuples verifying a given condition. The projection keeps only the designated attributes (and eliminates the replicated tuples; this elimination which may be undesirable in preparing data for analysis may be suppressed on some systems). In the binary relational operations we find some set operations : intersection, union and difference between two "union-compatible" relations (that is, to say it briefly, relations with the same schema), we have also the cartesian product and the division of two relations. But the most important and useful binary operation is the "join": this operation creates the tuples of the new relation in concacinating couples of tuples of the two relations to be joined if they satisfy a condition. The most common condition is the equality of two attributes values, one in each relation: it is called "equijoin".

III. DATA ANALYSIS AND RELATIONAL MODEL

Although some people try to implement DBMS specially devoted to data analysis, the relational model appears particularly well fitted to data analysis : the rectangular arrays (objects-variables) may be seen as relations (tuples-attributes), and the relational operations correspond to the data management needed by the statistician.

Moreover some qualities of relational systems are very useful in a data analysis context : automatic coherence controls of the data, existence of transactions, possibilities of multiuser, data consistency maintain even when crashes occur etc.. The extent of these properties is variable from one system to another.

However two qualities appear necessary when it is intented to use a DBMS to manage data to be analyzed. The first one is the possibility for the data base schema to evolve during the data base life. The reason is that the analyst is interested in saving in its data base not only different transformations of its original relations (=data), but also the results of its statistical analysis which, in many cases, may be seen as relations : let think for instance to partitions, or to the coordinates of objects on factorial axes. The schema dynamical evolution allows these different new structures to be integrated in the data base and manipulated by relational operations like the other data base relations.

For physical implementation reasons this facility does not exist in hierarchical and network DBMS, and in some relational ones.

The second important property is for the system to be open. This means it is written in such a way that it is easy to add new elements or functionalities which are interesting in a specific area, as data analysis for instance.

IV. PEPIN and SICLA

The relational DBMS we use, PEPIN, has all the properties we described above (BCFJK 81), (BCFJS 83). It is relational, has a dynamically evolving schema, its transactions are atomic (two phase commit, crash recovery). It has mechanisms to maintain the data consistency and moreover it is easily extensible because of its layer structuration and the fact that it is written in Pascal. It is operational on different computers, from Apple 2 to DEC 20.

Pepin as Sicla is principally used in conversational mode, but it is easy to access it via the host language Pascal.

The SICLA Multivariate Data Analysis System offers a set of elementary statistical description modules (min, max, bar charts...) and multivariate data analysis (factorial analysis, clustering methods, discriminant analysis...).An important effort is made to avoid incoherences between data and analysis methods : the users are brought to precise the data types (Burt tables, measures tables) and the statistical kind of its variables (binary, qualitative, etc..). For instance the clustering methods do not allow the choice of a khi2 distance between variables which have been declared as measures taking any values.

Sicla is written in Fortran and operational on the Multics system of INRIA (Ral 82), (Ral 83). It must be transfered on smaller computers.

V. THE INTEGRATED SYSTEM

It is possible to imagine different solutions for the cooperation of Sicla and Pepin. Our first care has been to avoid the creation of an intractable monster...

The solution we chose is based on a strict separation of the functions between PEPIN which is used to save and manage the data and the analysis results, and SICLA which is now completely devoted to statistical analysis (in earlier versions a part of Sicla's modules where used to manage data). This sharing may evolve in the future for the elementary statistical functions in extending the "aggregate" function of PEPIN.

The interface definition appears all the more important since the integrated system must be extended by a graphical module, used to visualize analysis results, and an expert system (DQR 84) used to guide the users in the choice of statistical methods.

The problems to be solved concern the data to be saved and the cooperation of both processes, written in Fortran and Pascal.

Concerning the data it appears that some of them are naturally stored in relations, whereas others concern the schema. Thus it is necessary to introduce in the DBMS new domains, like probabilities for instance, and new consistency rules to avoid their incoherent use in query and update. More generally new consistency rules must be introduced to avoid inopportune updates of data deduced, ie of data which have been inserted in the data base as results of analysis. Some informations, important for the consistency data-analysis methods must be stored in the schema , in the relation description (ex. contingency table), or in the attribute description (ex.quantitative value). Some typical values of attributes, like mean for instance, seems rather naturally appear in the schema, but other choices may be imagined.

The conclusion on the data storage is that some extensions (which are of little importance for the DBMS designer) would be nice for the statistician.

Let us examine now the communication between PEPIN and SICLA.

Pepin and Sicla exchange the data to be analyzed or to be stored in the data base via parametrizd files. When the tuples of a relation are to be analyzed, they are copied in a file, each tuple corresponding to a logical record. The description of the data files, and some other informations contained in the relation schema are accessible to Sicla in a special file called "dictionary" (an other choice would be to put that in the data file heading). The procedure is similar, in the reverse order, when data elaborated by Sicla are to be saved by Pepin.

Now Pepin and Sicla are able to exchange data, but how does the user easily go from one subsystem to the other ?

The Multics system allows the definition of a master process, Pepin for us, which may activate a slave process, Sicla. This choice master-slave is justified by considerations on transaction, initialization and languages used.

To realize the processes communication three functions have been added to the Pépin menu (and the three reciprocal ones in the Sicla menu) :

.) "write a file for Sicla" (parameter : the name of the relation)
.) "read a file from Sicla" (parameter : the name of the file)
.) "goto Sicla".

As our integrated system is not yet operational it is to early to appreciate its performances. But if you notice that multivariate statistical analysis operates on data not greater than some thousands, the choice of using an intermediate file for data exchanges does not burden a lot the system and is fundamental for its ease of implementation.

VI. EXAMPLE OF USE

Let us see on an example the interest of our integrated system .

This (real) example is a sample survey about the use of psychotrops by school boys. About 2000 questionnaires have been completed by schoolboys; they concern their alcohol, beer, medecine... consummation, their family surroundings and their previous history (at school for instance).

All those variables are qualitatives.

The first step of the processing, data acquisition and checking, is made by the DBMS. Three relations, one for each topic , are created.

The second step is the analysis of the population with regard to the consummation variables. A correpondance analysis and a clustering method "nuées dynamiques" are then realized by Sicla. The results of those analysis are a set of objects coordinates in factorial references of the population typology (clustering).

Those results are introduced in the data base as new relations and studied with the relations "family surroundings" and "previous history" to explain the partition for instance.

VII. CONCLUSION

This SICLA-PEPIN integrated system appears as a useful tool for the statisticians. It is also a first step toward new possibilities of data base queries using not only the original data of the database but also data statistically deduced, as decision functions, homogeneous classes or typical profiles for instance.

REFERENCES

(And 82) Anderson J. B.
"Software to link Database Interrogation and Statistical Analysis"
Comptstat 1982, Toulouse France

(BCFJK 81) Bouchet P., Chesnais A., Feuvre JM., Jomier G., Kurinckx A.
"Database for microcomputers: the Pepin approach"
Sigsmalls Newsletter vol.,7,2 , oct. 1981

(BCFJS 83) Bouchet P., Chesnais A.,Feuvre JM., Jomier G., Szulman S.
"Le SGBD relationnel pour microordinateurs PEPIN"
Printemps Convention 1983, Paris France

(Dat 81) Date C.J.
"An Introduction to Database Systems"
Addison Wesley 1982 , third edition

(DQR 84) Demonchaux E., Quinqueton J., Ralambondrainy H.
"Une application de l'Intelligence Artificielle à l'Analyse des données"
Rapport de Recherche INRIA Le Chesnay France 1984

(Hex 82) Hext G.R.
"A comparison on types of Databases Systems used in Statistical work"
Compstat vol. 1 , 1982 Toulouse France

(KFI 82) Kobayashi Y., Futagami K., Ikeda H.
"Implementation of a Statistical Database System: HSDB"
Compstat 1982 Toulouse France vol.1

(LST 83) Lefons E., Silvestri A., Tangorra F.
"An Analytical Approach to Statistical Databases"
Very Large Data Bases, Firenze Italy nov.83

(Ral 82) Ralambondrainy H.
"An Interactive System of Classification: SICLA"
Compstat 1982 vol.1 Toulouse France

(Ral. 83) Ralambondrainy H.
"Le système SICLA"
3ème Journée Internationale d'Analyse des Données, Versailles oct.83 Fr.

Processing of Statistical Data for Multistage Recognition System Using Data Base

M.W. Kurzyński, and S. Lebiediewa, Wroclaw

The present paper deals with multistage recognition using data base. The concept of multistage classification, logical structure of data base, data structures, and operations on data are presented.

Key words: Multistage recognition, data base, decision-tree.

I. INTRODUCTION

In many practical pattern recognition problems, the number of features and the number of pattern classes are both very large. It has been reported by Kurzyński /1983a/ that in such cases it would be advantageous to use a multistage recognition system characterized by the property that an unknown pattern is classified into a class using several decisions in a succesive manner. The action of multistage classifier can be described by means of a decision-tree which consists of a root-node associated with the entire set of classes and a number of interior nodes and terminal nodes. A nonterminal node is an intermediate decision and its immediate descendant nodes represent the outcomes of that decision. A terminal node corresponds to a terminal decision i.e. pattern class. In a multistage classifier the unknown pattern undergoes a sequence of decision rules /a strategy of multistage recognition/ on the path from the root-node to a terminal node.

In Kurzyński /1983b/ the strategies with learning for performing the classification at each nonterminal node have been derived, under assumption that so called learning sequence, i.e. a set of correctly classified samples is known. Unfortunately,these strategies are complicated and need processing of a great number of statistical data included in learning sequence /different for different nodes/. Thus, in practical applications of multistage classifier,using computer technique, the relevant role plays data base.

Generally, data base for multistage recognition system must contain the following data: decision logic /decision tree/, recognition strategy, subsets of features for each nonterminal node and learning sequence.

In this paper we present organization of data base for multistage recognition system, its content and structure at the logical level.

Moreover, operations on data are defined.

II. LOGICAL STRUCTURE OF DATA BASE

The data base for a multistage pattern recognition contains the following data: names of nodes of the decision-tree /names of groups of classes/ No_i, $i = 1,\dots,k$, where k is the number of all nodes, the vector of names of all features $C = [C_1,C_2,\dots,C_n]$ used in the multistage recognition process, the vectors of names of features for every decision-node, matrices of values of pattern features and aggregated learinig sequence $S = \langle (X_1,1),(X_2,2),\dots,(X_M,M)\rangle$. The elements of learning sequence are pairs (X_i, i), X_i is the value matrix of the features of learning objects, belonging to the i-th class. The data base reflects the structure of the decision-tree: interior nodes correspond to decision nodes and terminal nodes are connected with pattern classes. The set of data for one decision node will be refered to as <u>data segment</u> and designated with SNo_i where i is the number of the node.

From among the data contained in a segment two types of data can be distinguished: information data and measurement data. To the former belong: identifier of SNo_i segment, l, which is the number of features used in decision taking at a given node, $l \leq n$, r, which is the number of direct successors of No_i node, vector of names of features used in a given node $C_i = [C_{i1},C_{i2},\dots,C_{il}]$, vector $V_i = [i_1,\dots,i_r]$ of names of direct successors of No_i node, information in which node the C_{im} $m = 1,\dots,l$ was measured for the first time. Besides, all segments, except the one corresponding to the root-node of the decision tree, contain the vector of names of predecessors. All components of vector of features C_i are elements of C sequence. The names of direct successors of node No_i correspond either to the decision algorithms connected with nonterminal nodes or to the names of pattern classes. To the measurement data belong the value matrix of features T_i and learning sequence $S_i = \langle (X_{i_1},i_1),(X_{i_2},i_2),\dots,(X_{i_r},i_r)\rangle$. The learning sequence S_i is a subsequence of the learning sequence S. The value matrix of features of the objects has the form of array of records where the first column contains the identifiers of object Id_i, the last one lists the names of successors /the result of decision for a given node/, while the remaining columns present the values of particular features of objects. Fig.1 presents fragments of segments SNo_1, SNo_2, SNo_3. Note, that certain features of objects occur in more than one segment, e.g., features C_1 and C_2 are found in segments SNo_1 and SNo_2, the C_3 feature occurs in segments SNo_2 and SNo_3.

All data occuring in data segment has a regular format, i.e. array of records, vectors or scalars. The following operations are performed on the elements of the data base: reading one or more rows of the table, reading part of a row, concatonation of rows belonging to different segments.

Due to a regular character of data and to operations performed, a relational model of data base seems to be the most convenient. However, it must be noted, that the data contained in particular segments are not independent, and the decision concerning which features of the object will be measured in No_i node depend on the decision taken in predecessing nodes which, in its turn, depends on the values of features of the object contained in other data segments. As it has already been mentioned, the data base for a multistage recognition reflects the structure of the decision tree and, it is more convenient for the description of the decision-tree to employ hierarchical structures than relational ones. Finally, the following solution was accepted: the data base has a hierarchical model added on the relative model. The model of the user working with one or more segments is the relative model with operations SELECT, PROJECT, MASK /superposition of SELECT and PROJECT/ and CONCATENATION /concatenation of rows having the same identifiers but belonging to different segments/. An example of CONCATENATION is forming the matrix of features of objects 01 and 04 from matrices T1, T2, and T3.

Id	c_1	c_2	c_3	c_5	c_6	c_7	c_9	c_{10}
01	2	0	11	1.5	3.1	0.7	4.8	4.8
02	2	1.3	19	1.8	9.0	0.4	0.4	3.0

For the designer or the manager and also for the user of the data base who must know the structure of the decision tree, the data base has the hierarchical model added over the relative model. Every change in the structure of the decision tree leads the way to the re-organization of the data base.

1) Matrix T1

Id	C_1	C_2	C_6	C_{10}	Name of a successor
01	2	0	3.1	7.5	No2
02	3	1.5	10	8	No10
03	1	1.8	7	5	No10
04	2	1.3	9	6.5	No2
05	1	0.7	4	5.7	No2

Name of a feature	Number of a node, in which a feature was measured for the first time
C_1	1
C_2	1
C_6	1
C_{10}	1

2) Matrix T2

Id	C_1	C_2	C_3	C_5	Name of a successor
01	2	0	11	1.5	No3
04	2	1.3	9	1.8	No3
05	1	0.7	30	3.9	No7

Name of a feature	Number of a node, in which a feature was measured for the first time
C_1	1
C_2	1
C_3	2
C_5	2

3) Matrix T3

Id	C_3	C_7	C_9	The class name
01	11	0.7	4.8	1
04	19	0.4	3	2

Name of a feature	Number of a node, in which a feature was measured for the first time
C_1	2
C_7	3
C_9	3

Fig.1. Fragments of segments SNo1, SNo2 and SNo3

III. CONCLUSION

Presented data base will be realized in two versions: integrated data base on SM-4 computer and distributed data base in computer network with SM-4 computers. These both installations will be intended for automatization of experimental research investigations /in chemistry, medicine, mechanics, physics, etc./ where pattern recognition plays main role.

REFERENCES

Kurzyński M.W. /1983a/, The Optimal Strategy of a Tree Classifier, Pattern Recognition, 16, pp.81-87.
Kurzyński M.W. /1983b/, Decision Rules for a Hierarchical Classifier, Pattern Recognition Letters, 1, pp.305-310.

Metainformation as a Tool for Integration

K. Neumann, Berlin

SUMMARY

One of the most important steps to improve the usefulness and the userfriendliness of statistical information systems is a higher degree of integration. As integration is both a goal and a process, it needs a powerful tool for its permanent maintenance. Such a tool is METAINFORMATION which has its own properties, e.g. its capability in respect of data modelling, and requirements, especially the meta-data capture. As a result of the use of metainformation one can expect a new quality of integration and therefore of statistical analysis.

Keywords: Data modelling; integration; metainformation; statistical information system; statistical data analysis.

INTRODUCTION

One of the most important parts of each statistical information system (SIS) is data. Therefore, data organization influences highly both the usefulness and the effectiveness of such systems. Data organization encompasses: data modelling; logical and physical data structuring; data description and storage; access control; ensuring the integrity of data.

Each modern SIS contains
- microdata
- macrodata
- metadata

(Mendelssohn, 1978).

The first level - microdata - represents individual responses to surveys and censuses.

The second level - macrodata - consists of summaries and estimates derived from microdata.

The third level is metadata - data on data:
"Metadata is that data which represents a description of other data, or more precisely the data required to access, interpret and evaluate other data." (Williams, 1978)

Although metadata were a perpetual attribute of each kind of statistical data processing, during the last year a vivacious international discussion has started on its role and functions (see references).

The reason therefore is <u>inter alia</u> the fact that modern information technologies - as e.g. the creation and using of databanks - on the one hand cause a high demand on metainformation and on the other hand offer rather new possibilities to capture, process, store and analyze them.

This paper will focus only on one of the very different functions of metainformation: that of being a tool for integration.

INTEGRATION

Although statistics was in the past unique both in science and in practice, the specialization of labour broke down the former unity of statistics into separate parts. All of these parts have had their own history which led them to special concepts, specialized tools and different methods.

That is true not only of statistics as a very general term, but also of such fields, as social or economic statistics and even of what we call official statistics. And Dörnyei (1983) stated in an excellent essay:

"It was not earlier than the last decade that the dialectic process turned to the reverse and the demand for the internal integration of the statistical system strengthened again in the statistical agencies."

But what do we understand by "integration"? During the abovementioned international discussion very contrary views had been expressed.

Some authors are restricting integration to statistical computing and distinguish between

- integration of data
- integration of functional systems
- integration of control facilities (Williams, 1978).

Very important is the clarification that integration itself can be done in very different ways, namely as subjective integration or as objective integration:

"SUBJECTIVE INTEGRATION stands for the process of unification of the corresponding ranges of indicators in various statistical surveys." ... "The process of subjective integration can be achieved with the help of the semiotic model of a statistical indicator and the registers of the names of indicators."

"By OBJECTIVE INTEGRATION we should understand such an unification of the defining manners of reporting units and survey units that would allow to clearly qualify the objective differences in individual statistical surveys and thus their unification." ... "Objective integration can utilize registers of objects which are reporting units, or objects in which survey units are separated." (Olenski, 1978)

Here we may note the mentioned relation between the way of integration and the possible tools to reach a higher degree of integration. But before we follow this line, we should try to clarify a little more the term "integration" in our context.

Integration is both a goal and a process.

As a goal it defines a situation that enables us to use the SIS as a whole, as a non-conflicting combination of various components, e.g. functions, levels, subject-matter fields, substantial and technical aspects etc. But this situation is artificial, because in reality not only the components of each SIS are changing, but also their interrelationships. Therefore, integration as a goal is no real situation, but an expected ideal, even if existing SIS are often defined as "Integrated Statistical Information Systems".

As a goal integration has changing dimensions so that we have to define its content from time to time in a new way.

So the second side of the medal is that integration is a permanent process. It is a long-term strategy to adapt the SIS to a changing environment, while it itself is changing its elements and its structure.

In a more practice-oriented sense integration is a permanent process to make statistical data "compatible" for a better combination and analysis.

Therefore, integration is a method to bridge the gap between the input into the SIS and the output required by the end users of statistical information. The input is determined by the way of data capture.

During data capture the philosophical principle of universality is to be neglected by the determination of selected attributes as a result of intended concentration of the essential feature of the captured phenomena and processes. Just this way causes necessarily limitations for analysis regarding the complex reflection of the objective reality in its subject-matter oriented, time-oriented and local context.

Metainformation may help to overcome such limitations by

- the improvement of definitions of data or figures (more exact de-

marcation and fitting in of attributes; description of relations to other figures or to their attributes);
- matching of data or groups of figures to models which reflect more realistic complex relations;
- the simulation of complicate analyses on a meta level.

Thus the integration of data from different sources will be ceased.

METAINFORMATION HANDLING

The use of these integrative abilities of metainformation assumes however that the metainformation itself has been captured, processed, stored, maintained and analyzed very carefully.

Similar to the experience gained by the data base management of microdata and macrodata, this carefulness can only be achieved rationally on a high level of automation. Therefore the creation and managing of METADATA BASES is indispensable.

Such a metadata base should contain"the following principal metadata:
- on the information resources stored in different data bases including statistical documents, computer media, publications, etc.,
- explaining the semantic content of data bases, methodological principles concerning particular indicators stored in data bases or inserted in statistical surveys,
- on basic sources of data captured in the form of current reports, full and sample surveys,
- on survey units and statistical population of particular surveys,
- on applied nomenclatures and classifications." (Walczak, 1983)

The main way of using such a metainformation system as a tool for integration is to use it "also for solving the problems of the conceptual level of data modelling" (Soltés, 1984). This seems to be the only way to bring the well known theoretical knowledge of data modelling (e.g. Sundgren, 1973) into practice.

One of the most important obstacles which at present hamper a faster development in this direction are the problems related to the metadata capture, both in terms of an agreement on common standards and definitions and in respect of the heavy burden to the "respondents" which in this case are the statisticians themselves!

The last hindrance could be overcome with the use of new information processing technologies especially in this field which may ensure the formation and maintenance of metadata bases.

They ensure progress in integration not only for already existing data collections, but also and in the first line for the establishment of new data stocks. There efficacy of metainformation in the project phase of new reconstructed SIS will lead to a new quality of integration and therefore of statistical analysis in the future.

REFERENCES

Abeele, R. van den /1978/ Data dictionary/directory project; synthesis of the results of the preparatory study, CES/SEM.10/4.

CRC/1983/ Integration of the statistical information system - Development and present state of the problem area, Report by the Computing Research Centre, Bratislava, CES/WP.9/199.

Dörnyei, J. /1980/ The structure and realization of the data system and data management, CES/SEM.13/R.2.

Dörnyei, J. /1982/ Methods for integrating the statistical information system, CES/WP.9/201.

Dörnyei, J. /1983/ The role of metainformation in statistical integration, ISI-Proceedings of the 44th Session, Madrid, Vol. L, Book 1, 545-560.

GDR /1982/ Metainformation as a method of integration of the German Democratic Republic's statistical information system, Report by the GDR, CES/WP.9/200.

Graves, R.B. /1980/ Semantic modelling in Statistics Canada, ISIS '80 Seminar, Bratislava.

ISIS /1971/ An integrated statistical information system. /Proposal 1971/, Computing Research Centre, UNDP, Bratislava.

Klas, A. /1978/ Structuring of the computerized information system from the aspect of integration. CES/SEM.10/22.

Kühn, J. /1983/ Daten- und Systemdokumentation im Informationssystem des Statistischen Bundesamtes für die Bundesrepublik Deutschland, ISI-Proceedings of the 44th Session, Madrid, Vol. L, Book 1, 561-570.

Mendelssohn, R.C., Juenemann, H.-J. /1977/ An innovation for federal agencies: computer-sharing, Government Data Systems, Sept./Oct. 1977.

METIS /1982/ Functional specifications from the aspects of the management of statistical information systems, SCP/MI/WP-24.

Miller, E.W.W. /1980/ Some aspects of an integrated statistical computing environment, CES/WP.9/180.

Miller, E.W.W. /1982/ The integrated statistical computing environment at the Australian Bureau of Statistics: Progress, Problems and Prognosis, CES/WP.9/198.

Nordbotten /1966/ Automatic files in statistical systems. Conf. Eur.Stats/WG.9/52.

Olenski, J. /1978/ Structure of statistical information systems and methodology of statistical surveys in terms of computerization, CES/SEM.10/2.

Olenski, J. /1980/ Semantic structures of meta-data in the statistical information system, CES/SEM.13/R.14.

Párniczky, G. /1976/ The idea of a nomenclature descriptor language, Proceedings of the ISIS '76 Seminar, Bratislava.

Poulain, C. /1983/ Pour mieux informer, produire autrement, ISI-Proceedings of the 44th Session, Madrid, Vol. L, Book 1, 528-544.

Philips, J. and Podehl, W. /1977/ Contents and structure of a metadata library for a statistical agency, Proceedings of ISIS '77 Seminar, Bratislava.

Rauch, L. /1979/ Meta-data in the statistical information system of the GDR, ISIS '79, CES/SEM.11/R.3.

Soltés, D. /1977/ Description of information system elements from the aspect of metadata base development, CES/SEM.9/7.

Soltés, D. /1978/ The information metasystem in the structure of a computerized statistical information system - definition and functions, CES/SEM.10/24.

Soltés, D. /1980/ Progress report on metadata base development for ASIS and the main spheres of its using, ISIS '80 Proceedings.

Soltés, D. /1984/ The metainformation system and the conceptual level of statistical data modelling, Statistical Journal of the UN Economic Commission for Europe, Vol. 2, No 1, 97-108.

Sundgren, B. /1973/ An infological approach to data bases, Urval No 7. National Central Bureau of Statistics, Sweden.

Sundgren, B. /1977/ Metainformation in statistical agencies, Proceedings of ISIS '77 Seminar, Bratislava.

Walczak, T. /1983/ Meta-data in statistical information system, ISI - 44the Session, Madrid, Contributed Papers, Vol. 1, 79-82.

Statistical Database: An Interactive Language for Logical Schema Definition by Means of a Model Based on Graphs

M. Rafanelli, and F.L. Ricci, Roma

ABSTRACT

The term "Statistical Database (SDB) refers, in scientific literature, to Databases (DBs) that represent statistical or summary informations used for statistical analysis. SDB contain quantitative informations (measured data) and a combination of descriptive informations (parameter data) for each quantitative measure. In this paper we paid particular attention on the logical model (the GRASS model) and on its Data Definition Language (DDL). In the section 1 we introduce selection category and summary attributes, distinguishing between quantitative data (measured numeric data) and qualitative data (or descriptive data). In the section 2 we give a description of the GRASS logical model for SDBs. In the section 3 the DDL of this logical model is explained by means of "syntactical diagrams". Finally in the section 4 current status of work in progress and future plans are given.

KEYWORDS: Statistical Database.Logical Model.Data Definition Language

1. CATEGORY AND SUMMARY ATTRIBUTES.

Statistical Data Bases (S.D.B.s) exhibit, regard to conventional Databases (D.B.s), the following characteristics: a) they recognize a large variety of datatypes (as matrixes, time series, generalized relations, etc.) (Su, 1983); b) they tend to be static, because represent consolidated events in time; c) they are usually large DB, in which some values may be missing and sparse data are very often contained. Moreover they manage two types of data: measured or quantitative data, that refer to numeric (summary) attributes on which statistical analysis is performed, and parameter or descriptive (qualitative) data, that refer to as selection category attributes describe the measured data (Shoshani, 1983). For each summary attribute a "cross product" of category attributes is usually required (Chan, 1981). A further distinction can be by calling the first "data" and the second "metadata" (Mc Carthy, 1982).

2. THE GRASS LOGICAL MODEL.

The GRASS logical model (GRraphical Approach for Statistical Summaries) proposed in this paper is based on the concepts of category and summary attributes (Rafanelli, 1983/A). The GRASS model endowes to the user a tool to know in what manner the SDB is logically organized. The proposed model introduces five types of nodes. These nodes are marked (to distinguish the type) and labelled (to identify each node). The edges are oriented, being the graph a direct, acyclic, partially ordered, connected graph. Such orientation is draw by means of arrows in order to avoid ambiguity in the graph. The semantic description of such nodes is the following: an "S-node" represents the indicator of tables Selection paths, a "T-node" represents the statistical Table physically present in the D.B. (it is the only node able to represent the stored data), a "C-node" and an "A-node" represent the selection category attributes for the statistical tables ; the nodes produce, as result, respectively the sum (1) and the cartesian product (2) of the assumable instances associated to branch nodes:

(1) $$\text{card}(I) = \sum_{1}^{k}{}_{i}\,\text{card}(I_i)$$

(2) $$\text{card}(I) = \prod_{1}^{k}{}_{i}\,\text{card}(I_i)$$

while a "t-node" represents one of the possible values of the definition "domain" of the C-node at the lower level of abstraction. For the T-nodes it is necessary to give the data typology too. Such typology specifies if a datum is an integer or real value, an average, a rate, etc.. Every table, that is represented by only one T-node, can have only one typology. Beacuse often the t_n-nodes are in high number (e.g. several hundred) and can be used as "unique basis" for different C-nodes (fathers), it is convenient to define such nodes defining directly the definition "*domain*" of such t_n-nodes. For this logical model "connection rules" between nodes and "branching rules" as queries to the system have also been defined (Rafanelli, 1983/A). Both the rule types assure the semantic integrity of the model.
An example of a simple SDB graph related to "Data on Railway Transport" is showed in Fig. 1.

3. THE DATA DEFINITION LANGUAGE.

The flexibility of the DDL (Rafanelli, 1983/B) allows to design the logical model without procedural constraints. In fact, it is possible: a) to design the logical model in a sequence of sessions (using the SUSPEND or the END commands); b) "to built" the graph during the same session either going from the root to the domains, or viceversa, or in a causal manner. The phases of building are two: 1) the definition of nodes and domains; 2) the definition of edges between two already

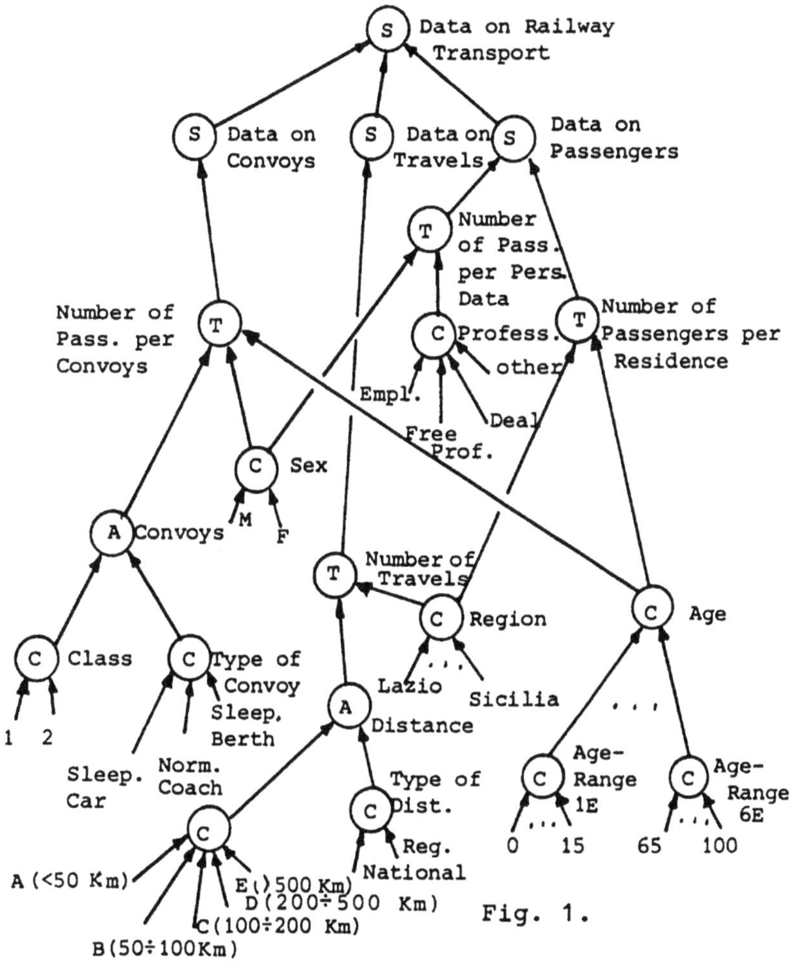

Fig. 1.

defined nodes. That means there are two types of control: the first type refers, for instance, to the "existence" of a node or a domain, the verification of the connection rules, the absence of homonyms, the verification of the syntactical rules, etc..; the second type of control is carried out when the END command is given. In this case is verified that: 1) the graph is acyclic: 2) the graph is connected; 3) the graph is direct; 4) there are not double paths between a T-node and the C-nodes (at lower level of abstraction) connected to it; 5) there is one only root; 6) the lower level of the graph is only formed by domains. The opening and closing instructions are syntactically very sample. There are two ways for closing the session: the command SUSPEND, that "freezes" the building of the graph without carrying out any semantic (or syntactic) check, and the command END, that verifies all the rules above said. The classes of commands are: a) declaration commands; b) deleting commands; c) helping commands; d) listing commands. The declaration commands are of two types: a) definition; b) modify. With the definition commands we can define nodes, domains, edges, synonyms and texts (as remarks). Since we define edges and nodes in two different phases, we can delete only edges or nodes and modify only nodes or domains. An example of a syntactical diagram for the description of a "command" that defines data and category attributes (with values domains) of the DB is showed in Fig. 2.

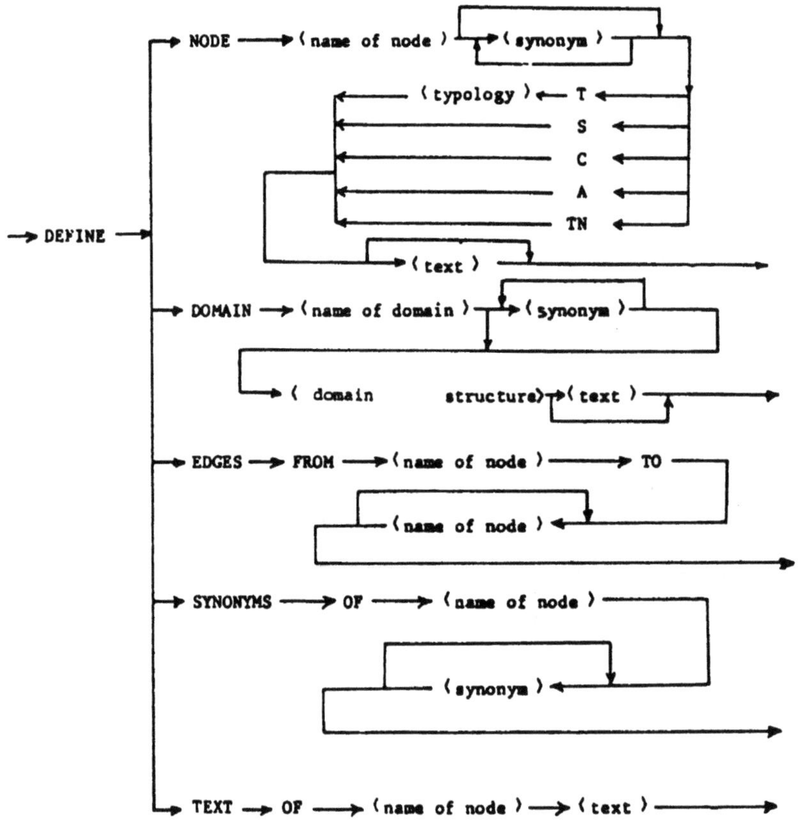

Fig. 2.

The Backus normal form and the complete syntactical diagrams of the D.D.L. is in (Rafanelli, 1983/B). An example of interactive session to define and to modify elements of the GRASS-graph is showed in Fig. 3.

4. CURRENT STATUS AND FUTURE PLANS.

A first simplified version of such language has been implemented on a computer IBM-3033, under MYS/SPI.3 operating system, in ambient TSO. At the present we are working on a statistical friendly-user query language (Rafanelli, 1984), in which is present a sub-schema definition language and a data manipulation language, able to retrieve and to elaborate statistical data. Other problems regarding the future development of this research are: a) the definition of an architecture for a Statistical Database Management Systems, the integration

```
>CALL RAILWAY_TRANSPORT
YOU ARE NOW ENABLE TO BUILD THE SCHEMA OF GRAPH RAILWAY_TRANSPORT.
 PLEASE ENTER ONE OF THE FOLLOWING MAIN COMMANDS:
   DELETE
   DEFINE
   MODIFY
   HELP
   LIST
   END
   SUSPEND
>DEFINE NODE NUMBER_OF_PASSENGERS_PER_CONVOYS  T  ABSOLUTE
>DEFINE NODE SEX
PLEASE SPECIFY THE TYPE OF THE NODE SEX
>C
DEFINE NODE COMMAND COMPLETE
>DEFINE DOMAIN SEX_TYPE  AN  1 2
>MODIFY DOMAIN SEX_TYPE
...DOMAIN SEX_TYPE  AN  1 2
..>DOMAIN SEX_TYPE  AN  M F
DEFINE DOMAIN COMMAND COMPLETE
>DEFINE EDGE FROM SEX_TYPE TO NUMBER_OF_PASSENGERS_PER_CONVOYS
ERROR: IT IS NOT POSSIBLE TO COLLECT A DOMAIN TO A T-NODE
 LINE REFUSED
>DEFINE EDGE FROM SEX TO NUMBER_OF_PASSENGERS_PER_CONVOYS
>LIST,ALL  NODE SEX
.
.   NODE NAME: SEX
    SYNONYMS:  -
    TYPE NODE: C
    TEXT:      -
    ENTERING EDGES: SEX_TYPE
    LEAVING EDGES:  NUMBER_OF_PASSENGERS_PER_CONVOYS
>SUSPEND
.
SUSPENDED SESSION
```

In such manner we have built the part of the following graph:

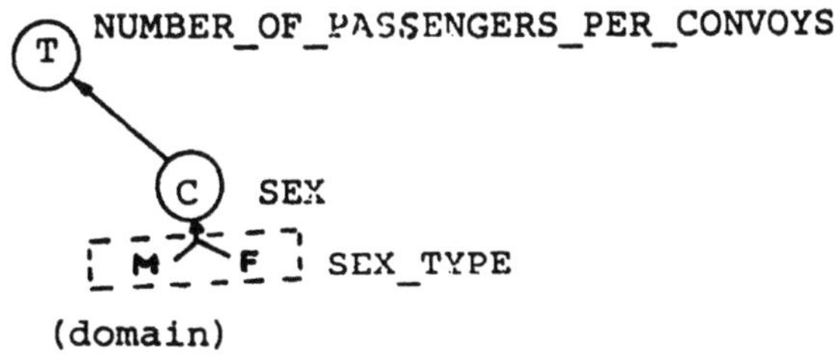

Fig. 3.

between different schema (with regard also to the different units and nodes) and the problem of the summarization (that is not yet completely resolved).

REFERENCES

(Chan, 1981) - P. CHAN, A. SHOSHANI: "SUBJECT: a directory friven system for organizing and accessing large statistical database" VII Int. Conf. on VLDB, Cannes, France, September 1981.

(Mc Carthy, 1982) - J.L. Mc CARTHY: "Metadata Management for Large Statistical Database", VIII International Conference on Very Large Data Bases, 8-10/9/1982, Mexico City, Mexico.

(Rafanelli, 1983/A) - M. RAFANELLI, F.L. RICCI: "Proposal of a logical model for a Statistical Database", II International Workshop on Statistical Database Management, 27-29/9/1983, Los Altos, California.

(Rafanelli, 1983/B) - M. RAFANELLI, F.L. RICCI: "A Data Definition Language for a Statistical Database", Technical Report I.A.S.I. n.R-62, July 1983.

(Rafanelli, 1984) - M. RAFANELLI, F.L. RICCI: "STAQUEL: a statistical friendly-user query language", Technical Report I.A.S.I. (to appear in 1984).

(Shoshani, 1983) - A. SHOSHANI: "Statistical Databases: Characteristics, Problems and Same Solutions", Computer Science and Statistics, Proceedings of 15th Symposium on the Interface, Houston, Texas, March 1983.

(Su, 1983) - S.Y.W. SU: "SAM*: a semantic association model for corporate and scientific-statistical Databases", Information Sciences, Vol. 29 n. 2 and 3, May-June 1983.

Interface Problems Between Software Packages Used in Statistical Information System

L. Rauch, Berlin

Summary

The edp-supported Statistical Information System (SIS) can be considered as a stream of data through different pieces of software, controlled by the system architecture of the SIS and/or by users (endusers, programmers, etc.).
Software packages are in general developed quite independently of each other. They require a well-defined input of data and produce specified output within the frame of their functional characteristics. Often they are black boxes from the user's point of view.
As a rule it cannot be expected that the output of one program will harmonize with the input requirements of subsequent programs.
The presented report will consider interface problems between several software packages used in the SCSO of the GDR.
The problem of metadata in this context is also touched on.
The question of defined standards for interfaces will be raised.

Statistical Information System, Software, System Architecture, Software Packages, Interface

I. Introduction

1. The development and use of software in the Statistical Computing Centre of the State Central Statistical Office (SCSO) of the GDR is increasingly attached with the implementation of outside program products. This situation has raised some new questions compared with the earlier situation using mainly inhouse developed software.
2. The attention is focused on the interfaces between the software components of the whole system.
3. Roughly, the data flow of the data processing within the SIS can be outlined as shown in fig. 1, each box representing not only one program package but several which are jointly connected.

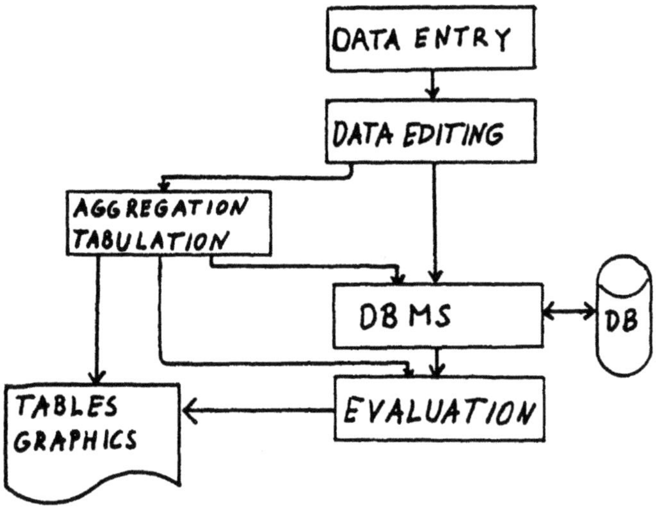

Fig. 1

4. From the fig. 1 we see two levels of interfaces. The first level exists between different phases of the SIS data flow. The second level lies within the boxes in fig. 1 between different programs belonging to one box. In general both types arise similar problems, but sometimes the interfaces between different phases can be solved by suitable standards in the computing centre (e.g. rules for creating files). Serious problems arise from interface conditions which cannot be solved by more or less organisational means.

II. The Main Interface Problems

5. Each program product requires its specific formats for the input data and has specific output formats. To connect two programs by the output/input data is difficult when the requirements do not fit. Often independent data transformations are necessary to provide a suitable link. Extra program development is necessary to solve the problem. You get additional redundant files which must be maintained in your organization. That is very unconvenient, in particular, if the output of one program later is used as input to several other programs with different input formats.
6. In general there are similar problems with the different record formats, but in some cases these problems are smaller because of using flat file structure in many program products.
7. Meta data has an important position in the use of modern software products. The user is able to access data by its name and other kinds of help is provided to make it easier to applicate the program. But again we find different solutions, e.g. data dictio-

naries, code books, meta data memories, etc. At first the user has to provide and store the meta data in the required format of the program product. There are no possibilities to make an interface reference to already existing metadata.

8. A lot of troubles using software come from this situation. Not infrequently the user avoid to use a suitable program for such reasons. In particular data base software has strong demands for meta data. Meta data transformation from one software package to another is much more difficulty as for data formats. It is not only the problem to change the structure of the meta data, but often you also have to change variable names, etc. for the simple reason that the other program requires shorter names, other name conventions, etc.
9. By this way the user is confronted with differing data enviroments which is very difficult to overcome by a kind of umbrella function as a common interface to the end user /2/. At least it requires some substantial efforts to develop a useful interface program.
10. Generalized software, meta data will be more and more important in the future. The costs for an in-house development prevent producing tailor made software for an application. To cut down the software costs you are depending an generalized packages. At the same time the demand grows for more user-friendly interfaces. That means you need flexible interfaces. If the present situation continues, the computing centres 'only' have to replace their capacities from in-house software development from scratch to software development for interfaces. I think this may not be the future.

III. Ways out of the Dilemma

11. In the computing centre of the SCSO the future software architecture will consist of different generalized program packages. A rough picture of this structure is given in fig. 2. From this figure you can see that there are some significant interface problems. One way to improve the situation is the development of specific interface tools which can be used quite flexible.
12. The SCSO of GDR participates in the Statistical Computing Project of the ECE/UNDP (1981-84). In the Joint Group RAPID participating countries are developing so called 'Base Operators' /3/. The Base Operators can be considered as tools to handle data in

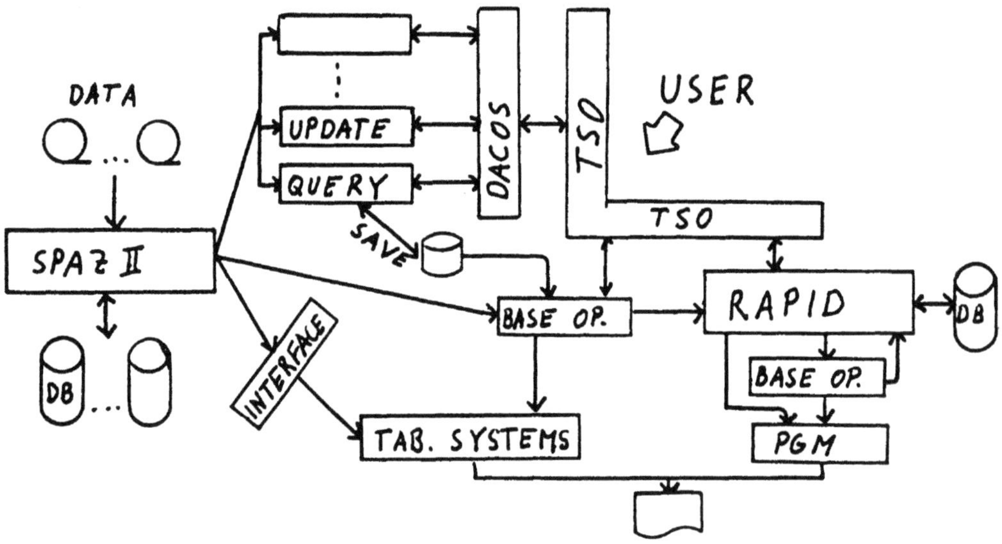

Fig. 2

a relational approach. In their first fashion they will be developed for a RAPID-enviroment. RAPID is a relational DBMS developed by Statistics Canada /4/.

13. The following Base Operators are under development:
    JOIN, PROJECT, SORT, MINUS, UNION, SELECT, MODIFY, AGGREGATE
14. The main principles of the operation of the Base Operators are as follows:
    - The Meta Data Memory (MDM) of RAPID is used as the Meta Data Base.
    - Input: Sequential (flat) files or RAPID-relations, both described in the MDM.
    - Output: Sequential file, automatically described in the MDM.
    - The Base Operators can work in two modes:
      . Each Base Operator in sequence of Base Operators produce a separate file which can be used as input for the next Base Operator
      . There is a pipeline mode between Base Operators which means that no file is created for each Base Operator, only at the end of the sequence of Base Operators a final file is created.
15. The Base Operators in the present fashion provide a better data handling in a RAPID environment to create new files for a later data processing by table generators or other programs. But still it is not possible to provide meta data for an other environment, e.g. table generators. This functional scope is not sufficient to cover the existing and planned software architecture.
16. Therefore the functional characteristics of the Base Operators

will be extended . The Base Operators are programmed very modularly. This allows to include/replace new functions. In fig. 3 possible extensions are indicated. This development will result in more generalized interface tools which quite easy can be adapted to a concrete interface problem. You have to replace some modules by your own instead of programming a whole interface program.

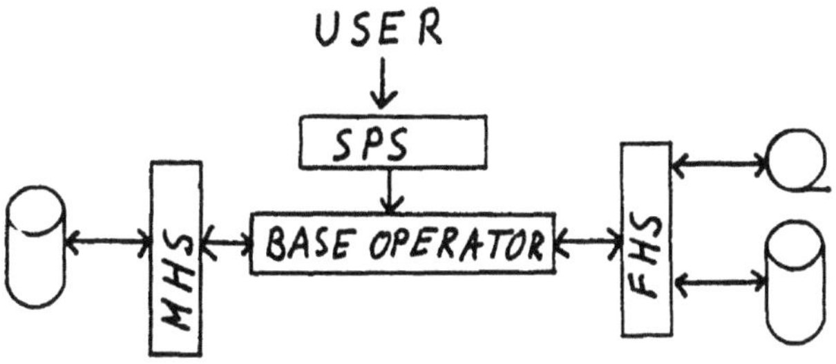

Fig. 3

- a File Handling System (FHS) which allows the adaption of other file formats (e.g. data base)
- a Meta Data Handling System (MHS) to transform meta data from one system to another
- a Statement Processing System which allows to create a user friendly interface to the Base Operators (e.g. on-line dialogoue techniques)
- conversion routines to transform data formats

17. An additional way to avoid complications is to include the possibility to specify references instead of an explicit storing of for instance meta data. In this case the adaption of a software package consists of programming of specific 'call'-routines to retrieve necessary meta data from another source.
This method would also prevent the redundant storing of a lot of meta data which is used by different programs. You will be able to set up centralized meta data bases for your whole system. The maintenance will be easier.

IV. Looking for Standards

18. The way which we try to follow as described in the chapter before is an effort to live with the present situation without changing it. It would be much better to aggree about standards for interfaces. Such standards should get the character of recommendations for software development.
19. From my point of view standards are suitful for the following matters:
    - data formats for input/output
    - record formats for input/putput
    - a standardized DML-macro interface to data bases; such standard

interface can be built as an umbrella to existing DML's or be an additional feature; a relational approach would be suitful (flat file character)

- structure of meta data bases
  definition of entities which may appear in a meta data base ; this need not be a complete specification of standard meta data bases but it should contain the necessary subset of specification for a convenient use of data in data processing
- general software packages should allow to concatenate different meta data sources by a small adaption program (exits). This would make it possible to centralize the meta data base which is of great advantage for computing centres with a more or less homogenuos field of data processing as it is the case in a statistical computing centre.

/1/ Lars Rauch The Software Architecture of the Statistical Information System of the GDR
CES/SEM. 16/R.8 ISIS '84 Geneva 16 - 19 April 84

/2/ Bo Sundgren Statistical Information System Design
Discussion paper CES/SEM.16/2
ISIS '84 Geneva 16 - 19 April 84

/3/ Eyvind Björk Development of the Base Operator System
Working paper Statistical Computing Project
Statistics Sweden, Stockholm 1984

/4/ RAPID-Manuals
Statistics Canada, Ottawa

Software for Parallel Processing in Statistical Data Analysis

Parrallelization of Algorithms in the Practive of Statistical Data Analysis

D. Lafaye de Micheaux, Nice

ABSTRACT

This paper presents a parallelization method, suitable to statistical data analysis algorithms. This method is easy to use, thanks to the simple synchronization tools that it needs. It fits particularly well arising multiprocessor structures, thus making them of great interest for the current practice of statistical processing.

KEYWORDS : parallelism, multiprocessor, MIMD, statistics, computing statistics.

INTRODUCTION

Parallizing a task, in computer science, consists in having it done by several actors (processors, computing units), which are able to work simultaneously and cooperate for this purpose. The aim of such an approach, obviously, is to reduce the global time needed to carry out the task. Algorithms and parallel programming are well-known concepts, and the literature about this topic is particularly rich. For a long time, however, this approach was practically reserved to prototype computers, or to the parallelization of computation with input/output. The present evolution of software (operating systems, programming languages) and of hardware (cost, integration) will probably put the parallelism closer to computer users. This paper proposes a method for getting benefits from parallelism in the current pratice of statistical data analysis (S.D.A.).

1- PARALLEL MACHINE ARCHITECTURES

Many ways to parallelize a task on a computer have been explored; they are distinguished by the choice :

- of the way to divide the task into sub-tasks
- of the capabilities of the various actors
- of the way to assign sub-tasks to the actors
- of the cooperation mode between actors.

This led to a lot of different parallel architectures and to many attempts to classify these architectures. It is impossible, in the limits of this paper to compare the various parallel architectures, and to give a full account of their advantages in S.D.A. . Thus, we shall give only briefly the reasons that led us to interest in a specific architecture, i.e., the multiprocessor.

Pipeline machines (CRAY-1, CDC STAR, ...), as well as parallel SIMD machines (array processors, ICL DAP, MPP...), are especially well-suited to process vectors and matrices, and consequently they are of great interest for S.D.A. (Sy182).

However, these architectures are used in high performance computers, and consequently they may concern only statisticians working on very large amounts of data.

Elementary operations (e.g., addition) or micro-operations (e.g. part of addition) are processed in parallel on these machines. Thus, the hardware or the compiler can manage parallelism, and the users can describe their algorithms in a sequential language. Two reasons, however, reduce this benefit :

- The languages differ from one machine to another (although generally of a Fortran type)
- Programming tricks, specific to the machine, are needed in order to improve run-time performance.

Systolic machines are presently designed for very specific problems, so they are not suitable to all the algorithms used in S.D.A. .

However, the design of such machines for real time statistical processing (e.g., process analysis, sound or image processing) would be an important field for future research.

The parallel architecture that seems the most suitable to a S.D.A. general portable software, is of the multiprocessor MIMD type: several heterogeneous processors share a bus through which they communicate and reach common resources (global memory, disk-drive...). The number of processors is not fixed, and each of them owns private resources (private memory, arithmetics processors, private I/O...). This choice is justified by its generality, because it includes:

- cheap monoprocessor,
- multi-microprocessor : an evolutive structure, where an increased performance can be reached, easily and cheaply by adding some more processors.
- a structure open to the use of high performance specialized processors, (array, systolic processors).

Moreover, algorithms developed on such machines will be also suitable to more advanced computers (e.g., Deneleor-H.E.P., C.I.I.-I.S.I.S.) where several pipeline processors are connected by a communication network.

2 - PARALLELISM ON MULTIPROCESSOR FOR S.D.A.

To reduce the total time needed by a task on a multiprocessor, it is necessary :

- to use the power of each processor at its best, especially by reducing idle periods,
- to minimize the frequency and amount of information transfers between processors,

and consequently

- to parallelize long tasks as much as possible,
- to group variables so that only a few connections between groups will subsist.

Thus, it is possible to improve the usually admitted measure E of parallelization efficiency :

$$E=S_q(n)/q$$

where q = number of processors used

n = complexity parameter of the problem

$T_1(n)$ = run time duration for the sequential algorithm

$T_2(n)$ = run time duration for the parallel algorithm

$S_q(n) = T_1(n)/T_2(n)$ speedup due to parallelism.

Statistics deal with a special type of data : sets of observations or sets of variables. It is thus possible to apply with "efficiency", on a multiprocessor, the same method for parallelizing nearly all algorithms used in S.D.A. : each processor works on the subset of data that has been assigned to it, and partial results are merged, in order to give the global result.

- <u>layout of the data partitioning method</u>

Let $\mathbb{I}$ a set, $\mathbb{P}(\mathbb{I})$ the set of all subsets of $\mathbb{I}$, T a function applying $\mathbb{P}(\mathbb{I})$ to the set $\mathbb{Z}$. $I \subset \mathbb{I}$ represents a data set, $Z = T(I) \in \mathbb{Z}$ is the result of applying T to I.

A binary associative operator F defined on $\mathbb{Z}$ may exist such that T(I) can be split as

$$T(I) = T(I_1)\ F\ T(I_2)\ F \ldots F\ T(I_q) \text{ noted } \mathbb{F}\,(T(I_k))$$

for every partition $\{I_1, I_2, \ldots I_q\}$ of I.

In this case, the evaluation of T(I) can be easily done in parallel, by assigning the processing of different $T(I_k)$ to q distinct processors.

The same method can be used when I is a sequence of **I** and I_k a subsequence of I (made of contignous elements, if **F** is not commutative).

A lot of elementary S.D.A. processes are relevant to this situation, and therefore they can be easily implemented on multi-processors computers : computation of mean, variance, minimum or maximum, histograms, scalar products, correlation matrices, linear or non linear transformations on variables, sorts on a variable, searches of the k nearest neighbours, classifications, etc.

Table 1 shows how well such processes fit the proposed model.

- <u>Algorithmic description of the method</u>

Here are described the sequential and parallel algorithms that compute T(I), using the following notation :

. Zo neutral element of F on **Z** (Zo is only used for easier expression, in fact it decreases efficiency)

. $P_0, \ldots, P_q$ are processors

Sequential algorithm :

```
. Z := Zo
. for i in I loop
      . Y := T({i})
      . Z := F(Z,Y)
      . end loop
```

Parallel algorithm to be implemented on P_0 :

```
. for k in 1..q loop
      . On P_k start process T (in I_k, out Z_k)
      end loop with wait
. Z := Z_1
. for k in 2..q loop
      . Z := F(Z, Z_k)
      end loop
```

- <u>Discussion</u>

The statement " on P_k start process T(in I_k, out Z_k) " means that processor P_0 asks process P_k to calculate $Z_k := T(I_k)$.

The expression "end loop with wait" means that, when processor P_0 gets out of the loop, it must wait for the results of all processes started inside the loop.

Obviously, the computing time of $T(I_k)$ should be the same for k = 1..q, and thus the same size would be chosen for all I_k. Similarly, I_k and perhaps Z_k would be assigned to the local memory of processor P_k, in order to reduce the amount of transfers between processors. It appears that speed up of parallelism is important as long as the merging time of **F** is small compared to the duration of process $T(I_k)$.

Parallelism efficiency is also improved if P_0 and P_q model the same physical processor. The graph on fig 1 represents the different stages of process T(I).

Considering a scalar product with n = 1000 and q = 10, the global processing time is :

$T_1 \simeq$ 1000.tmul + 1000.tadd for the sequential algorithm

$T_2 \simeq$ 100.tmul + 110 tadd + tcom for the parallel one.

In the parallel case, inter-processor communications are only
- 10 requests to start a process
- 10 end of process signals
- 10 accesses by processor P_0 to a number in the memory of an other processor.

If transmission speed between processors is well adapted to processor speed, the duration tcom is short compared to T_2, and consequently the efficiency of parallel algorithm is close to 1.

It may be useful to parallelize the merging operation **F** when merging two results is notably long or when q is large. If q processors are available for computing **F**, it is possible to use the "usual recursive compute/combine" or "associative fan-in" method, yielding a result of order $\log_2(q)$ instead of q. (LiA81), (Hel78).

Fig 2 illustrates this method for q =4 .

- <u>More general applications</u>

Complex processes in S.D.A. (projection, linear regression, factorial analysis, regressograms, hierarchical clustering, dynamic clustering, etc...) are made of successive elementary processes, which can be done in parallel, using the data partitioning method.

However, some special steps also occur : linear system solution, matrix inversion or diagonalization, for example, cannot be processed in parallel so easily. For such computations, different methods are proposed in the literature, which we are beginning to evaluate.

A detailed example of complex S.D.A. processing exceeds the limits of this paper; it will be described in (LdM84)

3- IMPLEMENTATING PARALLEL ALGORITHMS

- <u>Synchronization tools</u>

Two main characterictics of the parallel method just described need to be emphasized :
1) Only one processor (P_0)keeps the initiative, the others are acting as slaves.
2) All slave processors do the same task on different data.

This organization, sometimes called S.P.M.D. (Simple Program Multiple Data), allows to easily express algorithms using simple synchronization tools :

- Slave processors need the description of processes only in the usual form of sequential procedures.
- The only statements absolutely necessary for the master processor are :
  * start a process on a specified processor:
    on P_k **start process** T(**in** I_k, **out** Z_k)
  * wait for a signal of the end of a specific started process :
    wait for process T **on** P_k

  In S.P.M.D. mode, it is generaly more useful to wait for the results of all processes started in a block, so that the wait command is related to the end of block :
    end loop with wait

In (LaN83) and (Ner83), a more complete language is proposed for this situation, including modularity and instantiation.

This mode of extremely simple cooperation between processors could be managed by all users of usual sequential programming languages, ignorant of semaphores, monitors, rendez-vous, etc ... (Ben82),(HLP83).

- Implementation

The simplicity of the synchronisation primitives needed makes these algorithms easily expressed in several parallel programming languages : C-Pascal, C-Euclid, ADA ...

We simulated such parallel processing on a monoprocessor VAX 11-750, using the York ADA workbench compiler. We anticipate their implementation on a multi-MC68000 computer SM90, within a S.D.A. software project, in progress at the University of Nice (LLP83).

CONCLUSION :

In order to make parallelism avaible to stastiticians, in the context of an advanced S.D.A. software, the study related in this paper led us to fix some choices. So we retained :

- a multiprocessor architecture using processing on the SPMD way
- a general method of parallelization by data partition
- a description of parallel SPMD algorithns using some simple synchronisation tools
- a programming language, ADA, chosen not only for its parallel capabilities, but also because it provides modularity, genericity, and overall because it is expected to be widespread.

These choices suggest a solution for reducing computing time in common practice of S.D.A. : a cheap and evolutive architecture based on several micro-processeurs sharing a bus. However,these choices are compatible as well to cheap mono-processors as to high performance computers, like HEP or ISIS.

REFERENCES

(Bar82) J.G.P. Barnes, Programming in ADA, Addison-Wesley, Londres, 1982.

(Ben82) M. Ben Ari, Principles of concurrent programming, Prentice Hall, 1982.

(HLP83) R.W. Hockney, J. Lenfant, R.H. Perrot, notes cours Ecole d'Eté d'Informatique CEA,INRIA,EDF, 08/1983.

(HoJ81) R.W. Hockney, C.R. Jesshope, Parrallel computers, Adam Hilger Ltd, Bristol, 1981.

(LdM84) D. Lafaye de Micheaux, Algorithmes et programmes parallèles en Analyse de Données : des exmples, to appear 1984.

(LdN83) D. Lafaye de Micheaux, C. Neri, Vers une parallélisation des algorithmes en Analyse de données, doc. interne Lab. d'informatique, Un. Nice, 03/1983.

(LiA81) B. Lint, T. Agerwala, Communication Issues in the design and analysis of parallel algorithms, IEEE trans. soft. eng. , V SE-7, No2, 03/1981.

(LLP83) D. Lafaye de Micheaux, J. Lemaire, J. Pouget, Conception et réalisation d'un système d'Analyse de Données, projet de recherche Lab. d'informatique, Un. de Nice, 04/1983.

(Ner83) C. Neri, Parallélisation des algorithmes d' Analyse de données, rap. DEA d'informatique, Un. de Nice, 09/1983.

(Syl82) J.D. Sylwestrowicz, Parallel processing in Statistics, COMPSTAT 1982.

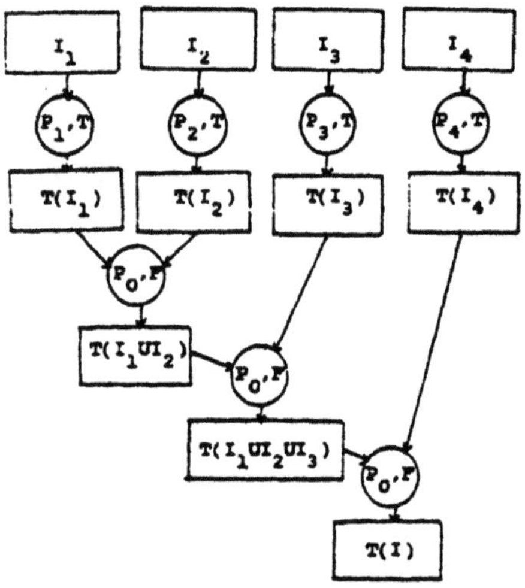

- fig 1 -

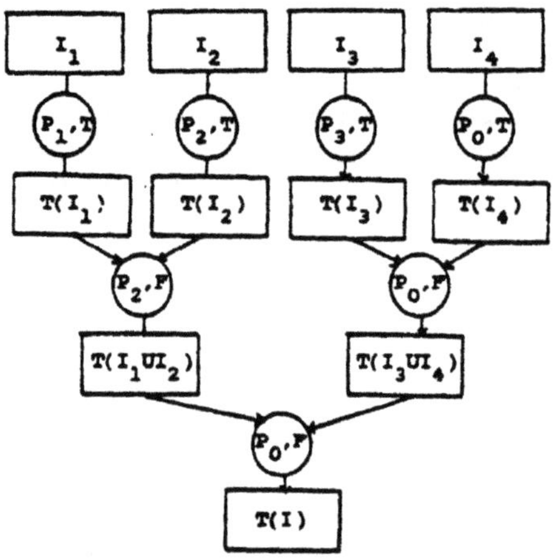

- fig 2 -

I	Z	T({i})	T(I)	Z_1 F Z_2	aplications in S.D.A.
R^2	R	$x1_i \cdot x2_i$	$\underset{i \in I}{SUM}(x1_i . x2_i)$	$Z_1 + Z_2$	mean, variance, ...
O	O	x_i	$\underset{i \in I}{\min}(x_i)$	$\min(Z_1, Z_2)$	min, max
O	{seq.{	x_i	I sorted	Z_1, Z_2 merged	sorting on a variable
R^p	{p,p matrix}	$x_i . x_i^t$	$X^t . X$	$Z_1 + Z_2$	covariance matrix
R^p	{?,p matrix{	$x_i^t . A, A_{p,m}$	X . A	$\left[\frac{Z_1}{Z_2}\right]$	linear and non linear transformations on variables
E	P(E)	x_i	k nearest neighbourgs of x_0 in I	$T(Z_1 U Z_2)$	non linear prediction
E	P(E)	$B(x_0, r) \cap \{x_i\}$	$I \cap B(x_0, r)$	$Z_1 U Z_2$	non linear prediction

Notations : O = ordered space
E = euclidian space
$B(x_0, r)$ = sphere centered on x_0 with radius r
$I = \{\ldots, x_i, \ldots\}$

- table 1 -

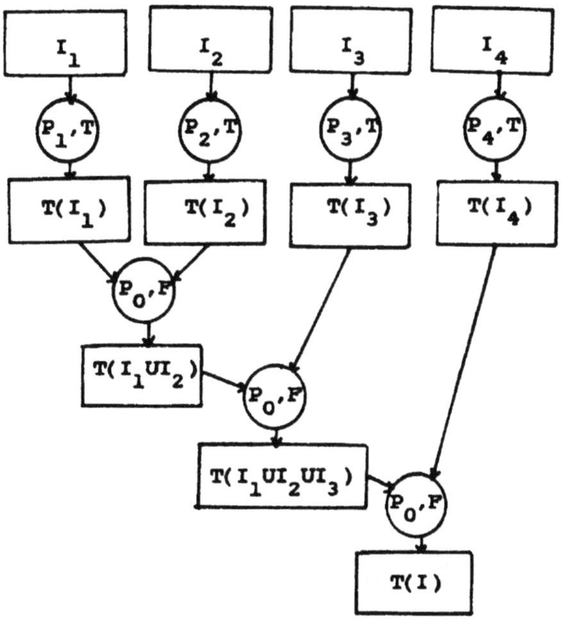
I1
I2
I3
I4
P1,T
P2,T
P3,T
P4,T
T(I1)
T(I2)
T(I3)
T(I4)
P0,F
T(I1UI2)
P0,F
T(I1UI2UI3)
P0,F
T(I)

- fig 1 -

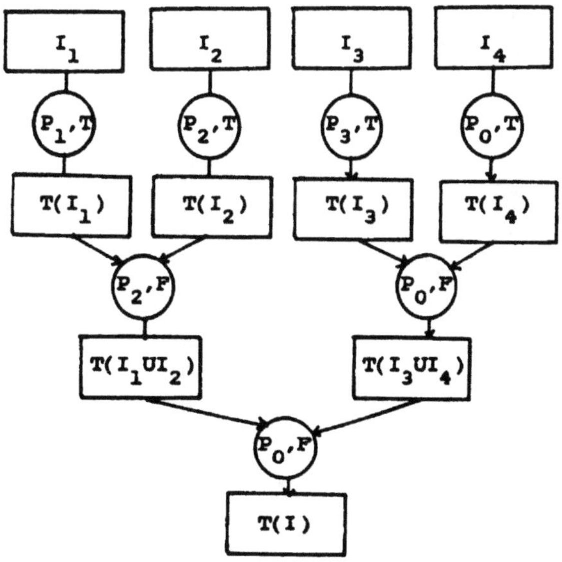
I1
I2
I3
I4
P1,T
P2,T
P3,T
P0,T
T(I1)
T(I2)
T(I3)
T(I4)
P2,F
P0,F
T(I1UI2)
T(I3UI4)
P0,F
T(I)

- fig 2 -

I	Z	$T(\{i\})$	$T(I)$	Z_1 F Z_2	aplications in S.D.A.
R^2	R	$x1_i \cdot x2_i$	$\underset{i \in I}{\mathrm{SUM}}(x1_i \cdot x2_i)$	$Z_1 + Z_2$	mean, variance, ...
O	O	x_i	$\underset{i \in I}{\min}(x_i)$	$\min(Z_1, Z_2)$	min, max
O	{seq.}	x_i	I sorted	Z_1, Z_2 merged	sorting on a variable
R^p	{p,p matrix}	$x_i \cdot x_i^t$	$X^t \cdot X$	$Z_1 + Z_2$	covariance matrix
R^p	{?,p matrix}	$x_i^t \cdot A$, $A_{p,m}$	$X \cdot A$	$\frac{Z_1}{Z_2}$	linear and non linear transformations on variables
E	P(E)	x_i	k nearest neighbourgs of x_0 in I	$T(Z_1 \cup Z_2)$	non linear prediction
E	P(E)	$B(x_0, r) \cap \{x_i\}$	$I \cap B(x_0, r)$	$Z_1 \cup Z_2$	non linear prediction

Notations : O = ordered space
E = euclidian space
$B(x_0, r)$ = sphere centered on x_0 with radius r
$I = \{\ldots, x_i, \ldots\}$

- table 1 -

A Fast Parallel Algorithm for the Recognition of Ionized Particle Tracks

G.A. Ososkov, Moscow, and M. Vajteršic, Bratislava

Summary

The paper presents a fast parallel method for the reconstruction of ionized particle tracks from filmless detection of the streamer chamber information in statistical experiments. To achieve a faster realization of the recognition process, a parallel computer of the SIMD /Single Instruction Multiple Data/ type, working in associative processing mode, is considered. The data to be processed are compressed and given as coordinates of N linear track elements. Each pair of these elements is tested as to whether it belongs to the same approximation circle. In our procedure, the parameters of the circle and the distance relations are computed for N pairs simultaneosly. From resulting pair candidates, the original tracks are re - constructed by using the nearest neighbour - principle. Our modification of this approach allows to realize even this rather complicated phase on the associative parallel computer up to a significant extent. The main advantage of the algorithm lies in parallel execution of the circle fitting procedure which is the most time consuming part of the whole process. Thus, the parallel algorithm developed yields an N-fold reduction of arithmetic operations in comparison to its sequential variant.

Key words:
complexity of statistical algorithms, parallel processing, associative parallel computers

1. Introduction

On grounds of its advantages, the filmlees processing of the streamer chamber images becomes one of frequently used approaches in high energy physics experiments. In order to allow a fast evaluation of these statistical experiments on computers, a reduction of the input data (10^6 - 10^7 bit/sec.) is necessary. Using fast algorithms for track element linking and the piecewise linear approximation of track segments, an on-line reduction method with a considerable data compression ratio has been proposed in Bajla et al. (1982). Results of this reduction process are coordinates of the linear track elements (LTE) from which the tracks of ionized particles have to be re-constructed. The efficiency of the performance of the filmless process depends upon the strategy for recognizing the tracks from the compressed data. The restoration problem is time-consuming for this application and therefore powerful computers are needed to solve it in fast manner. Modern parallel computers are valuable candidates to meet these demands.

The paper presents a fast parallel method for the reconstruction of tracks suited for parallel associative SIMD-type computers working on the content-addressable principle (Yun et al. (1977)). Section 2 brings a brief description of the recognition procedure given in O s o s k o v (1983) which uses the circle fitting principle to find relations between pairs of LTE's. The main portion of the computational work is used to estimate the coordinates of the circle centre and two distance parameters. Those pairs of LTE's are found which belong to the same approximating circle. Using the nearest-neighbour principle, these pairs are connected to yield longer track segments. A parallel fashion of this procedure is presented in the third section of the paper. Some characteristics of the associative SIMD type parallel computer are also given. The circle fitting computations are realized on this machine for all pairs in N parallel itarations by shifting a field of the associative memory, where the coordinates of the LTE's are located, one word upwards in each iteration. This parallel procedure is formulated in terms of vector operations. Further, we formulate also an instruction how to delete, in parallel, track strings which are parts of longer track candidates.

The approach presented is the first one proposed to restore the tracks in filmless processing on parallel machines. Due to its efficiency and considerable reduction in computational time, it seems to be a valuable contribution to the on-line detection of the streamer chamber information.

2. The procedure for recognition of particle tracks

It can be supposed that the particle tracks are piecewise linearized, i.e. they are represented by a set of N linear track elements (LTE's). The problem is to reconstruct the tracks from given coordinates of their end points. The recognition procedure can be assumed to consist of the following stages:

/i/ Testing the end points coordinates for all pairs of LTE's.

/ii/ Computation of parameters of the approximation circle for all pairs which pass /i/.

/iii/ Evaluation of distance values.

/iv/ Finding the track candidates.

/v/ Reconstruction of tracks from track candidates.

The purpose of the first two stages is to find all pairs of LTE's which could belong to one approximation circle. In Ososkov et al. (1984) the procedure CIRCLE is given for effective estimation of the approximation parameters. Before computing the coordinates of the circle, each pair of elements l_i and l_j, $i=1,\dots,N$ with respectively coordinates $(\bar{x}_{i1},\bar{y}_{i1},\bar{x}_{i2},\bar{y}_{i2})$ and $(\bar{x}_{j1},\bar{y}_{j1},\bar{x}_{j2},\bar{y}_{j2})$ of their end points, are tested according to the criterion

$$0 < \bar{x}_{i2} - \bar{x}_{j1} < 1000 \quad . \tag{1}$$

The coordinates X_{ij}, Y_{ij} of the centre of masses are estimated by

$$X_{ij} = (\bar{x}_{i1}+\bar{x}_{i2}+\bar{x}_{j1}+\bar{x}_{j2})/4 \quad , \quad Y_{ij} = (\bar{y}_{i1}+\bar{y}_{i2}+\bar{y}_{j1}+\bar{y}_{j2})/4 \ .$$

The coordinates are translated according to this centre by

$$x_{qk} = \bar{x}_{qk} - X_{ij} \ , \quad y_{qk} = \bar{y}_{qk} - Y_{ij} \qquad q = i,j; \quad k=1,2 \ .$$

In the third phase, parameters σ and δ are evaluated characterizing respectively dispersion and the maximum distance of an examined pair to the approximation circle. Experiments indicate that it is sufficient to evaluate values of these parameters from four end points of the pair under consideration. As given in Ososkov et al. (1984),

$$\sigma^2 = 1/4R^2 \sum_{q=i,j} \sum_{k=1,2} ((x_{qk} - X_c)^2 + (y_{qk} - Y_c)^2 - R^2)^2 \tag{2}$$

where $R^2 = X_c^2 + Y_c^2 + H$, $H = \sum_{q=i,j} \sum_{k=1,2} (x_{qk}^2 + y_{qk}^2)/4$.

The coordinates X_c, Y_c of the centre of the circle are

$$X_c = Y_c(S_4 - S_7 . 2Y_c)/S_1 . 2Y_c$$
$$Y_c = (1/2)(S_1 . S_6 - S_4^2 - (1/4)(S_1 . S_3^2)) . (1/(S_1 . S_5 - S_7 . S_4)), \tag{3}$$

where S_j, $j=1,\dots,7$ denote Gaussian sums

$$S_1=[x^2],\ S_2=[y^2],\ S_3=[x^2+y^2],\ S_4=[x(x^2+y^2)],\ S_5=[y(x^2+y^2)],$$
$$S_6=[(x^2+y^2)^2]\ ,\ S_7=[xy] \tag{4}$$

(e.g. $S_1=[x^2]= x_{i1}^2+x_{i2}^2+x_{j1}^2+x_{j2}^2$).

The parameter δ is computed by

$$\delta = \max_{q,k} |(x_{qk}-X_c)^2 + (y_{qk}-Y_c)^2 - R|/2\sqrt{R}, \quad q=i,j; \quad k=1,2. \tag{5}$$

The values of σ and δ are most significant information on whether a given pair of LTE's belongs to one track or not. Therefore, only such pairs (l_i,l_j) are considered, for which

$$\sigma^2(l_i,l_j) < 700 \quad \text{and} \quad \delta(l_i,l_j) < 40 \ . \tag{6}$$

Let us denote by C a set of all pairs of LTE's for which the criterions (1) and (6) are satisfied and let C_i be that subset of C which contains all pairs with a given element l_i, $i=1,\dots,N$.

To create track candidates, the principle of the nearest neighbour will be used in the fourth stage. Since it often occurs that for a given LTE l_i, $i=1,\dots,N$, C_i contains more than one element, only that pair $(l_i,l_m) \in C_i$ will be considered in this principle, for which σ is minimal on C_i. Let us denote (l_i,l_m) by c_i. (If C_i is an empty set, c_i does not exist .) Thus, for each $i_0=i=1,\dots,N$

(c_i is not empty) strings

$$T_i = \{l_{i_o}, l_{i_1}, \ldots, l_{i_k}\} \qquad (7)$$

can be constructed where $(l_{i_s}, l_{i_{s+1}}) = c_{i_s}$ for $s=0,\ldots,k-1$. The strings (7) will be called the track candidates. As a result of this strategy, there occur among the candidates also branches $T_{i_s} = \{l_{i_s}, l_{i_{s+1}}, \ldots l_{i_k}\}$, $s=1,\ldots,k-1$, $i=1,\ldots,N$. As a last operation of the stage (iv) these branches are deleted from the set of track candidates.

In the fifth stage, the tracks from candidates are to be reconstructed. It must be said that no exact rules can be formulated which would yield absolutely correct recognition results. As one of the ways how to extend our nearest-neighbour principle, the following heuristic rules can be proposed:

- Insertion into a track.
  To insert an LTE l_{i_s} into a track (7) between $l_{i_{s-1}}$ and $l_{i_{s+1}}$ it is necessary that both the pairs $(l_{i_{s-1}}, l_{i_s})$ and $(l_{i_s}, l_{i_{s+1}})$ be from C.
- Decision about the crossing points.
  When an LTE l participates in two tracks T_i and T_j, i.e. when e.g. $l_{i_s} = l_{j_p} = l$, it is necessary to decide which from strings $\{l_{i_o}, \ldots l_{i_{s-1}}\}$ $\{l_{j_o}, \ldots, l_{j_{p-1}}\}$ should by deleted (because parts $\{l_{i_s}, \ldots l_{i_k}\}$ and $\{l_{j_p}, \ldots, l_{j_k}\}$ are respectively in T_i and T_j identical). It is done

by proving neighbourhood connections to l in T_i and T_j from its left and right sides, i.e. to check if pairs $(l_{i_{s-2}}, l_{i_s})$, $(l_{i_s}, l_{i_{s+2}})$ and $(l_{j_{p-2}}, l_{j_p})$, $(l_{j_p}, l_{j_{p+2}})$ are from C. When the number of connections for both track candidates is equal, one has to decide by comparing corresponding values of σ .

3. The parallel recognition algorithm for an associative computer

It can be assumed that $O(N^2)$ executions of stages (i) - (iii) have to be performed in sequential realization of the recognition procedure. Since the stages (ii) and (iii) represent the most time consuming part of the whole process, we turn our attention on their parallel realization. Due to the sort of operations required, a parallel processor of the SIMD type is well amenable to solve this problem.

It consists of a number of modules which are composed of an associative memory, of a permutation network and of a vector of elementary processors. To each word of the memory one processor is assigned.

A more detailed description of the concrete machine under consideration can be found in Richter (1984). Thus, computations can be performed over all words in parallel.

An arithmetic-logical operation $\circledast$ over words of memory fields F1 and F2, with results in a field F3, will be described as

$$F3 \leftarrow F1 \circledast F2(M) \qquad (8)$$

where mask M contains 1's, (0's) to indicate into which words of F3 the result of (8) is (not) to be written. (The word-lenght in operand fields is assumed to be n.) A cyclic shift of a field F k bits upwards (to the right) will be expressed by $F^{k\uparrow}$ ($F^{k\rightarrow}$). The content of i-th word of a field F will be denoted by F_i. The vectors with elements equal to 1(0) are denoted by I ($\emptyset$). Further, let us assume that we have reserved in the memory fields FNR, F1 and F2 of suffieient widths for storing in their i-th words respectively the numbers of LTE's which create the pair candidates with l_i and their corresponding values of σ^2 and γ. First $\bar{n}$ and n bit slices respectively of these fields will be denoted as $(FNR)^{\bar{n}}$ and $(F1)^n$, $(F2)^n$. Then parallel algorithm for computation of stages (i) - (iii) can be formulated as follows:

```
Inputs: FX1, FX2, FY1, FY2  (FX1,FX2 and FY1,FY2 respectively are
                             fields of x and y coordinates of the
                             end points of the LTE's)
      FN                    (field of numbers of LTE's)
       N                    (number of LTE's)
       n̄                    (number of bits for binary representa-
                             tion of N)
       n                    ( width of fields FX1,FX2,FY1,FY2)
   FX1C  FX1,FX2C  FX2,FY1C  FY1,FY2C  FY2
   FNC  FN
   for j=1,...,N
        M←I
        FX1C←FX1C^(1↑), FX2C←FX2C^(1↑), FY1C←FY1C^(1↑), FY2C←FY2C^(1↑)
        FNC←FNC^(1↑)
        FC1←FX1C-FX1
(9)     if 0>FC_i>1000 then
           M_i←0
       {FC1← FX1+FX2+FX1C+FX2C
(10)   {FXT1← FX1-FC1, FXT2← FX2-FC1, FXT1C←FX1C-FC1, FXT2C←FX2C-FC1
       {FC1← FY1+FY2+FY1C+FY2C
       {FYT1← FY1-FC1,FYT2← FY2-FC1,FYT1C← FYC1-FC1,FYT2C←FY2C-FC1
        FC1←FXT1**2,FC2←FXT2**2,FC3← FXT1C**2,FC4←FXT2C**2
(11)    FCX← FC1+FC2+FC3+FC4
        FC5← FYT1**2,FC6←FYT2**2,FC7←FYT1C**2,FC8←FYT2C**2
(12)    FCD← FC1+FC5+FC2+FC6+FC3+FC7+FC4+FC8
(13)    FCY← FC5+FC6+FC7+FC8
        FC1← FC1+FC5,FC2← FC2+FC6,FC3←FC3+FC7,FC4← FC4+FC8
        FC5← FC1*FXT1,FC6← FC2*FXT2,FC7← FC3*FXT1C,FC8←FC4*FXT2C
(14)    FCXQ← FC5+FC6+FC7+FC8
        FC5← FC1*FYT1,FC6← FC2*FYT2,FC7← FC3*FYT1C,FC8← FC4*FYT2C
(15)    FCYQ← FC5+FC6+FC7+FC8
        FC5← FC1**2,FC6← FC2**2,FC7← FC3**2,FC8← FC4**2
```

```
(16)   FC4 ← FC5+FC6+FC7+FC8
       FC1 ← FXT1*FYT1,FC2 ← FXT2*FYT2,FC3 ← FXT1C*FYT1C,
       FCY ← FXT2C*FYT2C
(17)   FCXY ← FC1+FC2+FC3+FC4
(18)   FC2 ← 0.5*(FCX*FCQ-FCXQ**2  -0.25*FCX*FCD**2 /
       /(FCX*FCYQ-FCXY*FCXQ)
(19)   FC1 ← FC2*(FCXQ-2*FCXY*FC2)/(2*FCX*FC2)
       FC3 ← FC1**2+FC2**2+ 0.25*FCD
       FC7 ← 0, FC8 ← 0
       for J = 1,2,10,20
           FC5 ← (FXTJ-FC1)**2, FC6 ← (FYTJ-FC2)**2
           FC4 ← (FC5+FC6+FC3)**2
(20)       FC7 ← FC7+FC4
           if FC4 > FC8 then
              FC8 ← FC4
(21).  FC4 ← ABS(SQRT(FC8)/2*SQRT(FC3)
       ⎧FNR ← FNR^(n̄→)(M)
(22)   ⎨F1 ← F1^(n→)(M)
       ⎩F2 ← F2^(n→)(M)
       (FNR)^n̄ ← FNC(M)
       (F1)^n ← FC7(M)
       (F2)^n ← FC4(M)
       OUTPUT FIELDS: FNR, F1, F2
```

(In bits $(\bar{n}-1)j+1,\ldots,\bar{n}j$ of i-th word in FNR there is a number s of the element l_s which creates the j-th pair candidate with the element l_i.
In bits $(n-1)i+1,\ldots,ni$ of respectively fields F1 and F2 there are located values of $\sigma^2(l_i,l_s)$ and $\delta(l_i,l_s)$).

As seen from the algorithm, all operations are performed on vectors i.e. on all words in parallel. The criterion (1) is realized in (9), the coordinates are transformed in (10). Sums of power series (4) are computed by (11) - (17). The coordinates of the centre result from (18) and (19). In fields FC7 (20) and FC4 (21) respectively values of σ^2 and δ are located. A cyclic shift of fields FNR,F1,F2 to the right is done in (22), in order to make place for new in-coming values.

It is important to note that the number of operations (for the arithmetic and transport of data) is independent on N for each i=1,...,N. Since operations are performed on n bits sequentially, O(n) parallel steps are required for each iteration. Thus, the complexity estimation for obtaining the pair candidates is O(nN) only.

It remains to cut the branches among the candidates. The following procedure can be executed on all words of FNR in parallel:

- set M=I ;
- find the longest non-zero word of FN (with M);
  if this word has all bits from $2\bar{n}$-th position (from right) equal to 0 then end;
- $\bar{M}$=0;

- in M set 0 for this word;
- set its content into comparand register $R=[r_1,\dots,r_s]$;
- compare by R (in a bit by bit manner, from right to left) all words of FNR (set 1 in $\bar{M}$ where words of FNR are of the form $[\underbrace{0,0,\dots,0}_{k-1},r_k,\dots,r_s]$);
- $F1 \leftarrow \emptyset(\bar{M})$, $F2 \leftarrow \emptyset(\bar{M})$, $FNR \leftarrow \emptyset(\bar{M})$.

Again, the procedure compares the tracks almost N times in parallel. Thus, the linear complexity estimation in N remains unaltered.

Bajla I., Ososkov G.A., Prichodko V.I., Turzova M. (1982), On data compression at a filmless detection of the streamer chamber information, P10-82-653, JINR Dubna (in Russian).
Ososkov G.A. (1983), Pattern Recognition Applications in High Energy Physics, P10-83-187, JINR Dubna (in Russian).
Ososkov G.A., Chernov N.I. (1984), Effective Algorithms of Circle Fitting, P5-84-7, JINR Dubna (in Russian).
Richter K. (1984), Parallel computer system SIMD, Proceedings of the AIICSR Conference, (to be published by North-Holland).
Yun S.S., Fung H.S. (1977), Associative processor architecture - a survey, Computing Surveys, 9, 3-27.

Evaluation and Enhancements of Statistical Software

The Entrancement of Fortran for Statistical Analysis

A.J.B. Anderson, Aberdeen

Summary

Fortran has recently been rejuvenated by the appearance of Fortran77 and projected extensions of the language. The next version is expected to provide a basic core of facilities required by the majority of users plus optional but consistent application modules with their own specialised syntax and semantics. The paper suggests some extensions to Fortran that could form the basis of a data analysis module and describes their experimental implementation in a preprocessor-driven system called STATFOR as foreshadowed in Anderson(1980).

KEYWORDS: Fortran; host language; preprocessing.

Introduction.

An unfortunate by-product of the wide availability of statistical software is the gradual decrease in collaboration between researcher and professional statistician. In many fields, attention has been drawn to the deficiencies of the resulting analyses presented in research publications. The change was encouraged by the traditionally poor "response times" of consulting statisticians and, if it is to be reversed, a software environment must be provided that is equally friendly to both the statistical adviser and his client so that neither feels hindered by the other. That is to say, the statistician must be able to recommend an appropriate form for the next stage of analysis unfettered by software limitations and the client must feel able to accomplish such analysis with the minimum of further contact until the results are available for joint inspection. Hence, the software system must be designed so that relatively routine tasks are easily accomplished without obstructing analysis paths for the more ambitious user.

It is clear that requests for simple tasks should be expressible in a reasonably terse way but, to achieve the generality implied in the previous paragraph, Anderson(1980) has suggested that Fortran-like facilities are required as a minimum. Statistical languages such as GENSTAT include their own dialect versions of many such high-level constructs; other systems such as BMD permit incorporation of Fortran segments (albeit in a somewhat disjointed way). Another solution is by means of the subroutine library approach championed by Healy(1981) and exemplified by the specialised RGSP system (Yates,1973). Schemes of this kind are very flexible and therefore powerful but demand programming skills of a fairly high level. Indeed it was experience with an earlier version of RGSP (Anderson, 1966) that led the author towards the work described here.

Fortran is by no means the only candidate as host for statistical constructs and procedures. Languages of the Algol68 type have been advanced on the grounds of inbuilt mechanisms (structured modes, preludes etc.) for declaring new operand types and even (re)defining the meanings of operators. However, the availability of a vast number of published subroutines and of general (e.g. NAG) and specialised (e.g. GINO) libraries of software is a telling argument in favour of Fortran. Whatever might formally be added to the language, access to this software is a vital component of flexibility in analysis.

Furthermore, Fortran has been revitalised by the arrival of Fortran77 and the continuing development of the language by Committee X3J3 of the American National Standards Institute. Fortran8x will almost certainly follow a "core-plus-modules" architecture i.e. a basic repertoire of features essential to most programs together with a number of applications/language extension modules involving specialised syntax and semantics but consistent with the spirit of Fortran. Database maintenance, graphics and multi-tasking have been suggested as examples, but statistical analysis is clearly another such area. This paper presents some extensions to Fortran useful for the data analyst that have been incorporated experimentally in a preprocessor-driven system called STATFOR. Examples of the extensions are interleaved with comments on the philosophy behind them but a full description of the system is not given here.

Statistical requirements.

In considering the extension of Fortran to take account of statistical needs, a balance must be struck between new types of operand and new operators to act on these. Since novice users are unwilling to surmount the psychological barrier of a large number of apparently arbitrary "computer-friendly" constructs, it seems reasonable to minimise the provision of new types of operand in favour of an extended list of procedures that have a more immediate relationship with statistical operations. This is the approach of STATFOR and parallels the subroutine package conception.

(a) Statistical variables.

It is essential, however, that entities representing statistical variables (variates) be made available. If the data are to be held in core or can at least be accessed variate-by-variate (as in the transposed file of database terminology), then such an entity can be thought of as the n-element vector corresponding to one column of the data matrix; if the data matrix can be scanned only row-wise, then a variate value relates only to its (scalar) realisation for the current row.

In addition to the missing data problem discussed by Anderson(1980), variates generally possess a number of attributes that are important for statistical purposes; these could somehow be attached to ordinary Fortran "reals", but, on the whole, a new type "variate" is the better solution. Thus, a statistical variable requires a name suitable for good annotation of output as opposed to use as a program mnemonic. Current proposals to extend Fortran identifiers to, say, thirty one characters could resolve this, but for the time being, it seems better to separate these two functions and allow both mnemonic and annotative names.

Often a variable may have the special function of indexing; in statistical terms, it is a "factor" possessing only discrete (usually integer) coded values. For example, SEX has two levels probably coded 1 and 2 or 'M' and 'F'. Such variates define the shapes/sizes of tables, the values of dummy variables in regression equations etc. The list of (not necessarily consecutive) level codes and associated annotative names of these levels must be supplied. In STATFOR, this is done by declarations such as

VARIATE SEX/1,2:MALE,FEMALE/, CLASS/1...5/

or LEVELS Q1,Q5,Q15...Q27/1:YES,2:NO,9:MAYBE/

Much of the above is related to the specification of schema in database work; for example, factor levels correspond to domains.

It does not seem useful to regard factors as otherwise different from variates, though the distinction parallels that between integers and reals. Indeed, a variate such as AGE could become a derived factor through its raw value being overwritten by an "age group" code. In STATFOR, all variates are held as reals though input formats are provided for reading character data and dates. For example, an unspecified date in April 1963 is held as 19630400.0; use of the Julian day as internal code would not permit representation of the missing "day" field.

(b) Tables.

For some purposes, it is useful to have multiway tables (of frequencies, totals, maxima, minima etc.) as defineable entities. Normally, the declaration can be implied at the point of formation and, if the table is only to be printed when complete, then it need not be further identifiable. If later operations require such identification, a numbering rather than a naming scheme is probably best. STATFOR regards T(I) as referring to a table numbered I rather than the Ith. element of a vector T(). Thus, we can have cell-by-cell assignments such as

T(I)=T(J)/T(J+1)

Division by zero-entry cells and similar impermissible operations must be trapped and default action taken. Such table manipulation facilities are not likely to be used much by beginners but are of great help to more sophisticated users and provide links with, for instance, log-linear modelling.

(c) Additional procedures.

These make up the greatest part of the enhancement provided in STATFOR and the list is open-ended. Subroutine libraries can, of course, be accessed but what is provided is somewhat more user-friendly.

In the first place, many of the Fortran77 generic functions are made to operate on variates (i.e. take account of missing values) and some on tables. A number of special statistical functions are also available, the most important providing Normal equivalent deviates etc. and random deviates from several distributions.

Recoding and class-grouping of variates have been provided as procedures rather than functions because such transformations are often applied simultaneously to several variables. The flavour of these commands is typified by

```
RECODE(1=1,2,5;2=6,3;9=4,7...9) V7,V12,V31

CLASS(0(1)5(5)20(20)80,110) AGEGRP=AGE

RECODE(61024,96347,201124; ...) HCODE=HOSPNO
```

(This last, for example, will cause HCODE to take the value 1(.0) when HOSPNO is 61024,96347 or 201124, and the value 2 otherwise, unless HOSPNO is unknown in which case so is HCODE.)

In general, procedure calls in STATFOR follow the format of Fortran input/output statements. An example might be

```
PLOT(SYMB=SEX;XSCALE=SAME;YSCALE=XSCALE) HEIGHT;FHT,MHT
```

which produces line printer scattergrams of height on father's height and on mother's height; the plotting symbol is determined from the value of the variate SEX; the horizontal scale is to match that of the last diagram produced and the vertical scale is to equal the horizontal. (This last facility is particularly needed for plots of principal component scores.) Scales can be provided by the user or determined automatically to give sensibly annotated axes.

Where possible, data are held in core. If this is not feasible, as with database working, summarisation tasks can be accomplished in a single data pass. The user is provided with a mechanism for controlling the extent and form of this phase. Relational database software is often Fortran-hosted with unit-by-unit access and the FIND directive of STATFOR can readily integrate with such systems.

Conversational working.

Anderson and Lowe (1970) and others have suggested that conversational working is inappropriate for many statistical analyses where some hours between machine interactions gives necessary time for reflection. Certainly the summary stage of a large survey is essentially a batch activity; what might be useful is a facility to switch from batch to conversational working "during" analysis. STATFOR is not yet conversational because the present preprocessor converts source directly into pure Fortran rather than an intermediate interpretable code. However, the system is able to resume an analysis without redefinition of operands since the preprocessor is re-entrant.

In many circumstances, it may be useful to build up at least some part of the analysis request conversationally. At the very least, interactive checking of the user's command sequence should be a separate preliminary to any batch run. Furthermore, a short simulated pseudo-analysis may be incorporated in this phase to confirm that the output is roughly as intended. This is achieved by means of a flag that temporarily switches the input directives to generation of random data. Only sufficient units for adequate testing are produced, the variates having distributions that may be specified but, by default, are taken to be discrete rectangular for factors and otherwise Normal.

Implementation of STATFOR.

Every attempt has been made in the design of STATFOR to maintain a Fortran-style syntax; for instance, blanks remain non-significant and there are no reserved words. Modern program preparation demands column-independence; this has been achieved by using the exclamation mark character to introduce comments and the ampersand character at the beginning or end of lines to indicate continuation. There is a need for good list specification facilities in statistical software and STATFOR uses forms such as A...F and 1(2)11 ; the triple dot syntax seems preferable to using a colon which might arguably be more Fortran-consistent.

STATFOR would clearly be most efficiently implemented using a full-scale compiler. However, the somewhat experimental nature of the extensions and the desire for long-term portability have led to a preprocessor approach akin to that used in RATFOR (Kernighan and Plauger, (1976)) for similar reasons. The fact that other software (e.g. for database access) is likewise preprocessor driven is also of some importance. Gentle(1982) has used the technique in STAT/PROTRAN but certain deficiencies in that product are apparent. For example, pure Fortran and extension commands must be inelegantly distinguished by signal characters and any pure Fortran is not syntax-checked, leading to baffling compiler errors. Furthermore, all data must be held in core, whereas STATFOR has considerable flexibility in data access and storage.

The STATFOR preprocessor follows principles outlined by Anderson(1980). It has recently been converted to Fortran77 to take advantage of the new character handling facilities. Some Fortran facilities (e.g. DOUBLE PRECISION, COMPLEX) are not implemented; EQUIVALENCE will probably be dropped because of the immense amount of code required to check the validity of the declaration and its probable deletion from Fortran8x. Variate identifiers are global as are the names of common blocks, functions and subroutines. Any of the last three could conceivably clash with preprocessor-generated names but this is unavoidable if non-preprocessed libraries are to be accessed; there is no problem with local variables or labels.

Where an analysis is to be continued in a later run, the directory of (main program) identifiers is incorporated into the compiled code and this and selected items are dumped to a scratch file at the end of execution. The preprocessing stage of the later run can then reconstitute this position in an initialisation phase.

Every opportunity is taken by the preprocessor to trap errors so that successful compilation of the resulting Fortran is guaranteed. Each error report pinpoints the position reached and gives a textual message indicating the fault. The source code is available during execution so that run-time error messages can likewise be helpfully informative.

Because STATFOR allows the user control over whether his data are held in core or scratch file or are accessed in a single pass its efficiency in time and core storage compares very favourably with other statistical packages.

Conclusion.

STATFOR has been developed over a number of years and has been confronted with a wide variety of real problems arising from statistical consulting work. As has been implied, the process is no more complete than with other comparable software but further enhancements of STATFOR will cause little increase in resources demanded by the system. Current work relates to the possibility of making STATFOR available on microcomputers and to continuing exploration of database interfaces.

References

Anderson, A.J.B.(1966) A note on the construction of a general survey program in Extended Mercury Autocode. Comp. J., 8, 312-314

Anderson, A.J.B. and Lowe, B.I.(1970) A Multivariate Analysis Computer Program. Appl. Stat., 19, 18-26

Anderson, A.J.B.(1980) The use of preprocessors in statistical software. COMPSTAT 1980, Physica-Verlag, Vienna, 490-496

Gentle,J.E.(1982) A Fortran Preprocessor for Statistical Data Analysis. COMPSTAT 1982, Physica-Verlag, Vienna, 230-235

Healy, M.J.R.(1981) A Subroutine Library for Statisticians. BIAS, 8 (1981) 12-25

Kernighan, B.W. and Plauger, P.J.(1976) Software Tools. Addison Wesley, Reading, Mass.

Yates, F.(1973) The analysis of surveys on computers - features of the Rothamsted Survey Program. Appl. Stat., 22, 161-171

Comparing Statistical Software Packages on Tabulation efficiency

A.L. Dekker, Akasaka

Ten statistical software packages with tabulation capability were installed on a FACOM M360-MP computer system. Each package was then used to produce a set of four simple tabulations which allowed a comparison of the packages on the point of execution efficiency. The results obtained demonstrate considerable variation in CPU-time requirements, with dedicated tabulation packages doing comparatively better than multi-purpose statistical software. Local conditions as well as the benchmark tables chosen influence the outcome of these tests, which, therefore, do not lead to indisputable rankings. However, some guidance is provided, which may assist users in avoiding mistaken choices.

KEYWORDS: Statistical Packages, Tabulation, Efficiency

1. Introduction

Execution efficiency in computer programs is no longer as important a consideration as it used to be. Ever increasing computer power, in terms of instructions executed per unit of time, has moved priorities to other areas, for example, correctness and maintainability of software. However, another effect compensates this tendency, at least partially. Tasks earlier considered too time-consuming for computer processing are now often accepted. This is particularly noticeable in statistical application areas, where increased computer capacity tends to be quickly absorbed by demand for more detailed tabulations, analysis using large microdata files in stead of aggregated data, and so on. Also, microcomputer systems are more and more put to work in statistical processing tasks, and execution efficiency is becoming once again a matter of importance in that context.

Therefore it makes sense today, as before, to include execution efficiency as a criterion in considering tabulation software.

2. Reproducibility

One would expect that identical jobs performing exactly the same computations would require the same number of CPU seconds, but this is not true in a multi-programming environment. The reason is that some of the overheads caused by the operating system, for example for swapping pages of memory, end up being credited to application jobs. CPU-times required to complete a particular job tend to become somewhat longer if the system is under a heavy workload. Ideally, such experiments as described here should be done on a computer temporarily dedicated to the job. This, however, was not possible.

In order to reduce the error margins, we repeated each measurement at least three times, reporting here the mean values. One particular tabulation (Table A through SPSS) was repeated as many as 25 times, the results being reported in Fig. 1. While the variable "CPU-time" could not be described as continuous (due to the distinct time requirements of certain operating system tasks) the distribution of values does not deviate significantly from a normal curve, as confirmed by a Shapiro-Wilk statistic of nearly 1 (in fact: 0.95). Standard deviation is 0.78 sec, or 5.3%. If a sample of three measurements were taken, the standard deviation of the mean would be about 3%.

Repetition of measurements on CPU-time requirement of other jobs gives similar

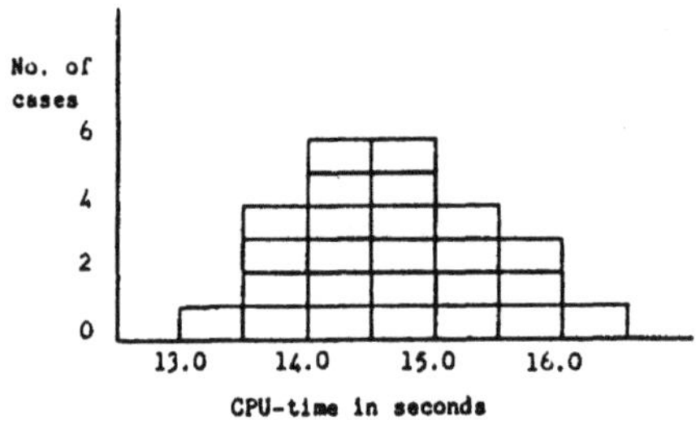

Fig.1 Histogram representing dispersion in CPU-time requirements of identical jobs

percentages. We conclude that the dispersion in observation results should be kept in mind while considering Fig. 3. However, the differences between many individual packages in this table are significant well beyond the noise level.

3. Benchmark tables

The tests were executed with an input file of population census data containing 7,063 household records and 33,752 person records. Table A (Fig.2) is about as simple a two-dimensional tabulation as one could imagine. Tables B introduce the effect of several tables being generated within the same computer run. Efficiency in tabulation would normally mean that additional tables do not add proportionately to the CPU-time required. Tables A and B are for person records only, requiring the household records to be skipped.

Table C is for household records only, a small subset of the sequential input file. This table is slightly more complex in its design, since both rows and columns represent a combination of two variables. Some recoding and adding of elementary variable values is required. Table D (not shown) is identical to Table A but for a ten-times larger input file.

4. Interpretation of the results

In Fig. 3 the final result of the exercise is shown. If two numbers are given, the first is for total CPU-time used, in the second fixed overheads as for compilation of generated programs, preparation of final output, etc., have been deducted to the extent that such tasks were executed in separate job steps.

In considering this table the following points must be kept in mind:

1. Both systems software and hardware have effect on the results, which might not be completely similar on other equipment.

2. Although we exercised great care in programming these test runs, it might be possible to improve the performance of some packages through the use of different control statements.

3. Some packages have newer releases which were not available for our equipment or had not reached us yet.

5. Discussion of individual packages

CENTS4

The successor of COCENTS (see below), similarly a COBOL-based system. An important advantage of using COBOL, and then by preference only the commonly implemented language constructs, is good portability. CENTS4 has improved characteristics in terms of its table lay-out capabilities and better consistency as well as flexibility in its parameter language. These improvements have not resulted in lower

A-TABLE, SAMPLE POPULATION CENSUS

TABLE OF SEX BY AGEGROUP

SEX / FREQUENCY	AGEGROUP: UNDER 25	25 - 49 YRS	50 - 74 YRS	OVER 74	TOTAL
MALE	9800	4816	1813	166	16595
FEMALE	10050	5039	1896	172	17157
TOTAL	19850	9855	3709	338	33752

TABLES-B, SAMPLE POPULATION CENSUS - AGE BY SEX

	MALES	FEMALES	TOTAL
UNDER 25 YRS	9.800	10.050	19.850
25 - 49 YRS	4.816	5.039	9.855
50 - 74 YRS	1.813	1.896	3.709
OVER 74 YRS	166	172	338
TOTAL.......	16.595	17.157	33.752

TABLES-B, SAMPLE POPULATION CENSUS - SEX BY RELATION TO HEAD

	HEAD	WIFE	CHILD	PARENT	OTHER	NONE	UNKNOWN	TOTAL
MALE........	6.545	28	7.215	76	2.441	199	91	16.595
FEMALE......	498	6.953	6.669	464	2.313	221	39	17.157
TOTAL.......	7.043	6.981	13.884	540	4.754	420	130	33.752

TABLES-B, SAMPLE POPULATION CENSUS - RELATION TO HEAD BY AGE

	UNDER 25 YRS	25 - 49 YRS	50 - 74 YRS	OVER 74 YRS	TOTAL
HEAD........	780	4.329	1.803	131	7.043
WIFE........	1.811	4.189	953	28	6.981
CHILD.......	13.642	224	14	4	13.884
PARENT......	5	90	370	75	540
OTHER.......	3.243	902	518	91	4.754
NONE........	261	104	46	9	420
UNKNOWN.....	108	17	5	-	130
TOTAL.......	19.850	9.855	3.709	338	33.752

- 1 -

TABLE-C, SAMPLE POPULATION CENSUS TYPE OF STRUCTURE/NB OF RESIDENTS BY NR OF ROOMS/TYPE OF FLOORS

	<3 ROOMS CEMT/WOOD	<3 ROOMS EARTH	<3 ROOMS OTHER	<5 ROOMS CEMT/WOOD	<5 ROOMS EARTH	<5 ROOMS OTHER	5+ ROOMS CEMT/WOOD	5+ ROOMS EARTH	5+ ROOMS OTHER	TOTAL
HUT <3 RES...........	31	743	8	21	541	2	29	170	-	1.545
HUT <7 RES...........	17	856	11	56	1,006	6	88	617	2	2.659
HUT 7+ RES...........	4	295	1	27	378	4	111	421	3	1.244
SHARED HUT <3 RES...	15	82	-	19	67	-	8	38	-	229
SHARED HUT <7 RES...	12	54	1	37	120	-	18	108	-	350
SHARED HUT 7+ RES...	3	19	-	15	50	-	18	74	-	179
VILLA <3 RES.........	-	-	-	-	-	-	5	1	-	6
VILLA <7 RES.........	-	2	-	2	-	-	6	-	-	10
VILLA 7+ RES.........	-	1	-	4	1	-	2	3	-	11
FLAT <7 RES..........	-	-	-	-	-	-	-	1	-	1
FLAT 7+ RES..........	-	-	-	-	-	-	1	-	-	1
CONCESSION <3 RES...	4	26	-	2	36	-	6	43	-	117
CONCESSION <7 RES...	3	41	-	7	73	-	23	150	-	297
CONCESSION 7+ RES...	1	15	-	4	31	1	35	129	1	217
OTHER <3 RES.........	8	40	2	-	4	-	1	4	-	59
OTHER <7 RES.........	4	78	1	-	3	-	-	3	-	89
OTHER 7+ RES.........	3	38	-	-	4	-	2	2	-	49
TOTAL................	105	2.290	24	194	2.314	13	355	1.764	6	7.063

\- DATA NOT AVAILABLE.

Fig. 2 Test tables. The printouts shown were made through SAS (Table A), CENTS4 (Table B) and TPL (Table C). Table D (not shown) is identical to Table A but for a 10x larger input file.

execution speed for large tabulations.

COCENTS [1]

COCENTS is a program generator. According to the table specifications, a COBOL program is automatically generated, that can be used one or many times to produce tables of this particular design. Execution speeds are comparatively good, especially if time required for overhead tasks such as generating and compiling the COBOL program are deducted.

The package is specifically intended for tabulation purposes and can print publication-quality (camera-ready) reports. Reading the data in binary form provided a small bonus in execution efficiency.

MINI-TAB [2]

MINI-TAB TABLES is a FORTRAN program consisting of only about 400 lines of code. Since it carries an internal FORMAT statement for the data record(s), recompilation is required to process a new data file. The program allows multiple, more-dimensional tables to be generated in one run. Table lay-out facilities are understandably rather limited.

Its small size and straightforward design allow relatively simple user modification or extension of the source program. This was required in order to generate Table C, since the program does not normally allow variables to be added in order to combine them in one table dimension. Simple recoding of variables, however, is possible.

Integer-type numeric variables were used in the unformatted FORTRAN data file.

TAB68 [3]

This is the only pure Assembler language package in our test. It provided superior execution speed. Usual options of algebraic computations on variables

Tabulation packages	Table A	Tables B	Table C	Table D	Pre-processing of data file
CZNTS 4	9.13/3.38	13.87/6.29	8.89/2.57	37.96	-
CZNTS 4 (Binary Input)	8.78/3.03	12.39/5.29	8.76/2.23	-	25.87
COCENTS	7.84/4.05	11.05/7.95	6.52/2.59	40.18	-
COCENTS (Binary Input)	7.10/3.64	9.99/6.27	6.14/2.43	-	25.87
MINI-TAB TABLES	20.28/13.96	24.57/18.45	24.22/18.46	151.54	-
MINI-TAB TABLES (Unf. READ)	13.01/7.96	15.71/10.22	13.10/7.80	76.69	25.87
TAB68	2.43	4.99	2.58	19.20	-
TAB68 (Binary Input)	2.47	4.91	2.55	-	25.87
TPL	13.48/10.90	15.40/12.82	13.53/10.98	97.53	-
TPL (Binary Input)	14.40/11.80	16.40/13.73	14.36/11.83	-	25.87
Multi-purpose packages					
BMDP 1977 (P1F)	32.56	40.52	69.61	326.70	-
BMDP 1977 (P1F) sysfile	27.57	30.73	48.86	-	57.71
RGSP Standard Execution	177.75	186.04	188.31	-	-
RGSP Unformatted Read	14.52/12.30	24.93/22.46	13.42/10.71	123.80	31.66
SAS82.3	15.89	17.30	9.47	156.64	-
SAS82.3 sysfile	13.30	14.68	8.27	-	14.85
SPSS9.1	14.57	25.91	14.59	142.76	-
SPSS9.1 sysfile	10.04	22.00	8.67	-	20.85
STATLIB	18.86	24.04	27.57	177.31	-
STATLIB Unformatted Read	11.29/10.91	13.44/13.20	10.10/9.71	-	31.66

Fig. 3. Execution time of tabulation jobs in CPU seconds

and constants, grouping of variable values into tabulation classes, breaking on area levels, provision of headings and line-stubs, are available. However, in some cases the obtainable table lay-out may be judged inadequate for publishing purposes. This would require post-tabulation editing of table print files.

TPL [1]

The Table Producing Language system has excellent table lay-out capabilities, culminating in the capability to drive photo-typesetting equipment. The package was mostly written in XPL (a PL/I dialect), but uses an Assembler "Submonitor" as interface to the operating system.

Tabulation through TPL requires prior preparation of a "codebook" which defines variable names, formats and acceptable values. This codebook describes variables that are already in a proper form to provide tabulation classifications (control variables) and others which require recoding, or will contribute directly to table cell values (observation variables).

BMDP [1]

It should be noted that we had to use BMDP module P1F, which was superseded together with P2F and P3F by the new module P4F in 1981. The older BMDP release also denied us the use of some other new facilities that could have expedited job execution, such as an inproved data reader and upgraded recode facilities.

RGSP [1]

This package has a wider aim than tabulation alone. It provides many functions important to survey processing, including validation facilities, calculation of standard errors for most sample designs and processing of variable-length records. In the Standard Execution Program these capabilities give rise to a considerable CPU-time overhead, so the use of standard RGSP can not be recommended for simple but voluminous tabulations.

However, it is possible to substantially improve execution speed by creating tailor-made RGSP programs in which the reading of data files is taken care of by simple user-written FORTRAN statements. Required RGSP functions are attached to the program by calling them from the RGSP library.

SAS [1]

With later releases of SPSS not yet available for our equipment, we used procedure CROSSTABS of Release 9.1 in Integer mode. As expected, execution speed was improved is an SPSS system file was first prepared. This, of course, eliminates data conversion from EBCDIC to internal SPSS storage format, which otherwise takes place for each run in which an external file is accessed.

SPSS does not easily provide subtotals within one table frame, since multidimensional tables are printed out in the form of a series of subtables. However, table cell counts may be written to a file and the table can then be reassembled as required.

STATLIB [4]

STATLIB is a statistical computing library containing a large number of FORTRAN-callable routines, and a command language for executing these either in batch or interactively. The cross-tabulation facilities allow several multi-dimensional tables to be generated in one run, and provide many advanced features in data manipulation. The package, however, has only limited support for flexible table lay-out and could not usually produce printout ready for offset reproduction.

STATLIB expects the data to be all in central storage before tabulation for a new table begins. However, if a new cross-tabulation matrix specification is found to be consistent with an existing matrix, the latter is simply added into. This allows for large data sets to be processed in segments.

In order to apply unformatted FORTRAN read statements, the data input and to be taken care of by a small user-written main program which was then linked with STATLIB routines for recoding, selecting, cross-tabulation, etc. The package stores data as double-precision real numbers in this type of processing. In the unformatted read file, therefore, variables were provided in floating-point form (this was also the case for RGSP).

6. Conclusion

Because of the factors mentioned in Section 4, it would not be wise to point out a winner in this contest. Our conclusion is that the existence of significant differences in execution efficiency between tabulation software packages is obvious. If tabulation execution efficiency is a point in a statistical computing operation, it will be worthwhile to have a critical look at available software resources, and possibly install and evaluate other packages. Several systems that did quite well in our tests are usually provided at nominal cost.

References

(1) Francis, I. (1981), Statistical Software A Comparative Review, Elsevier North Holland, New York.

(2) Elkins, H.G. (1971), MINI-TAB EDIT, MINI-TAB FREQUENCIES and MINI-TAB TABLES, Community and Family Study Center, U. of Chicago.

(3) Statistics Sweden (1976), TAB68 Users Manual.

(4) Brelsford, W.M. and D.A. Relles (1981), STATLIB A Statistical Computing Library, Prentice-Hall, Englewood Cliffs, NJ.

A Multilevel Approach to Improve the Flexibility of Statistical Software

V.J. de Jong, Groningen

Summary

In this paper it is investigated whether the flexibility of statistical software can be improved by the introduction of an intermediate level between the application level and the program source level. An implementation of such a system is made. The system is compared with the macro facility in GENSTAT and the MATRIX facility in SAS.

Keywords: multilevel programming, flexibility, statistical software.

1. Introduction

Most statistical software products focus on users who want to apply standard statistical techniques. It is very difficult to modify the statistical techniques in these packages, especially for users with hardly any programming experience and no knowledge of the architecture of the software product. If the user wants to experiment with his own variant of a statistical technique, he must start all over again using programming languages like APL, FORTRAN or PASCAL. Otherwise he depends on the producer of the package to get his software adapted to his needs. This is both money and time consuming.

In this paper I suggest an approach to improve the flexibility of statistical software products for the user. The approach is based on a principle used in computer science to make the actual applications (e.g. problem oriented programming languages) as independent as possible of the hardware by introducing intermediate levels. For statistical software I distinguish three levels. First an application level for users who only want to use statistical techniques. Second a matrix specification level for statisticians with knowledge of the matrix notation of statistical techniques. Third a software level for computer scientists. Each level is made fully independent of the other levels. A user on a particular level does not need knowledge required on the other levels.

An implementation of a threelevel system is developed at our University. This system is compared with systems based on related approaches, like GENSTAT with its macro definitions and the MATRIX facility in SAS.

2. Flexibility of software

In many applications one does not know beforehand which statistical techniques must be used. Software therefore should provide the possibility to make changes in the implemented techniques if required. Though software packages contain quite a few statistical techniques, they do not offer the possibilities to modify these techniques. For a user the implemented techniques are 'black boxes'. The user specifies a data set and the software package returns the answers (the output). This approach, visualized in figure 1, offers a minimum of flexibility to the user.

Figure 1. Black box approach in software packages

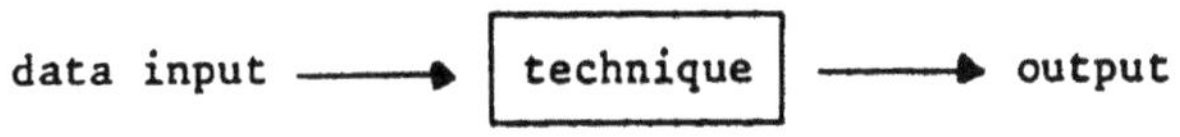

In some of the software packages the situation is slightly different. Besides data input the user can specify control input for the 'black box'. This offers the possibility to choose between different options of the software, containing variants of the implemented technique. Though more flexible than the 'black box' approach, the control approach still offers moderate flexibility. Only those variants that the

Figure 2. Control approach in software packages

developers of the package can think of are implemented. It is without doubt that the user will come up with a lot more. And it would be a great improvement if he could make the required modifications himself.

If flexibility is so much needed, why not use a problem oriented programming language you will ask. Of course these languages offer enough possibilities to implement every statistical technique you can imagine. But the resulting software however has low flexibility due to for statisticians totally irrelevant information in the source text. A statistician most of the time is not interested in problems like for example the exact implementation of input-output routines. Due to the complexity of the source program, he can not adapt his software to new needs. With regard to the statistical software it is useful to distinguish three levels on which changes can be made.

The application level

On this level changes can be made in a fixed set of parameters of a statistical technique. For example (a) changing the number of observations, (b) changing the number of variables included in the analysis.

The matrix specification level

On this level changes can be made in the matrix specification of a statistical technique. For example (a) changing a procedure for ordinary least squares into a procedure for generalized least squares, by changing the matrix notation of OLS, (b) adding a new statistic to a procedure.

The 'source' level

On this level changes in implementation of the software, the source, can be made. For example (a) changes in the memory organisation within a program (hashingtables, symboltables, etc.), (b) changes in the algorithms used for basic mathematical operations such as inversion and the calculation of eigenvalues.

Of course the user must possess the required knowledge to make changes on a certain level. No matter how well designed a software product is, it is only flexible if the user possesses the required knowledge to make the changes. For economic reasons most software produced until now, focussed on flexibility on the application level. The flexibility can be improved if a statistician is able to implement new developments on a matrix specification level as well.

3. An implementation: SIMS

In modern computers many intermediate languages, between the soft-ware packages and the actual hardware, exist (see Tannenbaum [4]). The application programs like TSP [5] and SAS [2,3] are thus just the top of an iceberg. Underneath the surface are a lot of other languages. Using compilation- or interpretation-techniques, the languages at the higher level are transformed to languages on a lower level, which can be executed through the electric circuits of the computer. The introduction of separate levels makes it possible to locate decisions in one particular level of the software/hardware. The resulting flexibility of this approach is impressive. People can, for example, use TSP or SAS without concern for problems like optimality of algorithms or the computer-architecture. In a similar way the introduction of the intermediate matrix notation level greatly improves the possibility to modify or add statistical techniques in statistical software. The implemented techniques are made independent of the program source.

On each of the three distinguished levels there is defined a language: the application language, the matrix language and the programming language in which the actual source is written. The application language should resemble the languages used in most statistical software packages, the programming language is one of the languages like FORTRAN, PASCAL, etc. The matrix language, however, needs to be developed with care. It must contain the mathematical operations and functions used to describe the implemented statistical techniques. Especially matrix operations and matrix functions play an important role in this language. Examples are matrix inversion, eigenvalues, eigenwaarden, kronecker products, etc.

At our University we have implemented a system with the features described above. We have build an interpreter that can handle the matrix language and the application language. This interactive interpreter is called SIMS. At the application level SIMS can be used to select statistical techniques from a file TECFILE. The techniques can be used to analyse data, which the user can select from a datafile DATFILE or which can be typed in on his terminal. An example of the application language in SIMS is given below:

```
SELECT USING OLSQ ON TECFILE
SELECT USING DATA1 ON DATFILE
SHOW BETAHEAD,RSQUARED
RUN (during the run the variables in the SHOW-command are displayed)
SELECT USING DATA2 ON DATFILE
RUN (now the results of the calculations on the second are displayed)
```

In this example a user selects a statistical and analyses two datasets with the use of this technique. The technique, in this case ordinary least squares, is written in the matrix language. A statistician can create or modify these techniques with the use of an editor or in an interactive dialogue with SIMS. The contents of the TECFILE written in the matrix language of SIMS for the example of ordinary least squares is:

```
OLSQ
  VAR [10 by 1] Y,E
  VAR [10 by 3] X
  VAR [3  by 1] BETAHEAD
  VAR RSQUARED
  DEFINE BETAHEAD = INV(X'*X)*X'*X
  DEFINE E = Y - X*BETAHEAD
  DEFINE RSQUARED = 1 - E'*E/(Y'*Y)
```

New statistics and modifications of the existing equations can easily be made by a statistician. For example the adjusted r^2 can be added simply be appending the following lines to the TECFILE:

```
VAR N,K,RSQUAREDADJUSTED
DEFINE RSQUAREDADJUSTED = 1 - ((N-1)/(N-K))(1-RSQUARED)
```

On the lowest level in our implementation we have the programming language in which the program SIMS is written: PASCAL at this level the source level changes can be made. Problems like userfriendlyness, memory organisation, conversion from infix to postfix notation of formulas, etc., are handled on this level. Changes at the source level can only be made by people with fair knowledge of PASCAL and insight in the architecture of SIMS. Because SIMS is developed in a modular fashion it is relatively easy to add new functions and procedures to SIMS. The resulting software package is a package that can be used by application oriented users, statisticians and computer scientists independently. Using their own language they can modify the system to their own needs. The complete system is visualized in figure 3.

4. A short comparison with other software tools

In this section we compare SIMS with other software products, that show some resemblance with SIMS: the macro facility of GENSTAT [1] and the MATRIX language in SAS [2,3]. Of course SAS and GENSTAT are developed with other objectives than discussed in this paper. Both packages focus on the efficient implementation of a

Figure 3. The SIMS system

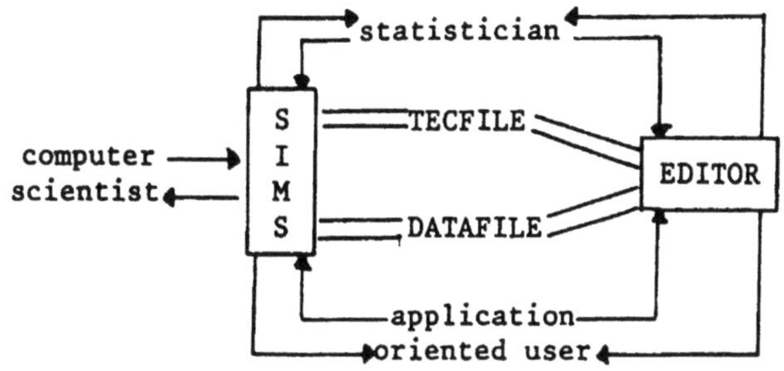

set of standard statistical techniques.
The macro facility in GENSTAT and MATRIX in SAS are added to allow the user to write his own routines. We will indicate the main differences with SIMS.

4.1. GENSTAT

It is possible to regard the macro facility in GENSTAT as a three level system. A user can write and modify macros (the matrix specification level) and the macros can be used by other users through a macro call (the application level). The main difference between SIMS's and GENSTAT's macro facility lies in the implementation of the matrix language and the independence between the three levels. The language on the matrix level does not resemble the mathematical language used in the statistical text book. All operations are performed through procedure calls. The formula of, for example, OLS must be written as

```
'CALCULATE' XACX     = TPDT(X,X)
            XXINV    = INV(XACX)
            XACY     = TPDT(X,Y)
            BETAHEAD = PDT(XXINV,XACY)
```

which is far from resembling original textbook notations. Moreover many source level decisions are made on the matrix notation level. For example the output of the macro and memory organisation commands like 'devalue' (return occupied space) and 'pointers' are controlled on the matrix notation level.

4.2. MATRIX in SAS

The MATRIX language in SAS is a good example of how the matrix language should be implemented. All formulas can be written in readable form and thus can be used and modified easily. Unfortunately MATRIX can not be used as a 'black box' on the application level. It is not possible to call a 'MATRIX-procedure' from a library and assign values to certain parameters. Values are assigned to variables on the same level as the definition of the formulas. As a result the difference between application level and matrix level does not exist.

There remain lots of other differences between the software products not discussed in this section. A full enumeration, however, is outside the scope of this paper, where we only focus on flexibility of the software.

5. Conclusions

On nowadays computers it is possible to implement systems, that are far more flexible than the 'black-box' approaches in most standard packages. Of course there remains always a trade-off between flexibility on the one hand and computing efficiency on the other. Nevertheless I believe that flexible systems will become more and more important, especially in scientific environment. With SIMS I hope to make a small contribution in this direction.

Literature

[1] Nelder, J.A., 'GENSTAT manual', Rothamsted Experimental Station, Harpenden, Herts, U.K., 1977.
[2] SAS, 'User's Guide: Basics', SAS Institute Inc., Cary, NC, 1982.
[3] SAS, 'User's Guide: Statistics', SAS Institute Inc., Cary, NC, 1982.
[4] Tannenbaum, A.S., 'Structural Computer Organisation', Prentice Hall, New Jersey, 1976.
[5] TSP, 'Time Series Processor', User's Manual, Computing Centre of Western Ontario, Canada, 1980.

QUESTOR, a Conversational Preprocessor to BMDP

D.J. Kuik, Amsterdam

Summary.

QUESTOR is a modularly built program, serving as a conversational preprocessor to BMDP programs. It is query-oriented, recognizing two types of questions. The questions and all other relevant texts for communication are stored in one file. Special routines serve to handle the communication with the user. Extensive help facilities, also based on the central text-handling module and on the special tree-like query-structure of QUESTOR, are available. The structure of the text file is discussed. It can be manipulated, independent from the package. The way QUESTOR builds a control language file is discussed, and a sample run is given. Finally, a (short) discussion is given about the usefulness of a conversational statistical package like QUESTOR.

Keywords: hashing, text handling, query structure, conversational statistical package

1 Introduction.

In the medical world the use of computing machinery seems to spread like the proverbial epidemics of yesteryear. Hospital Information Systems, administrative systems, Intensive Care monitoring systems, all seem to spring into use overnight. The physician may consider this as a great help with his day-to-day chores. But there is also a need for support of his clinical or epidemiological research. And here he is left in the lurch. Of course, there are well-designed data-base systems, of which the more recent versions are easy to use. But these are only suitable for simple data analyses. And of course there are also the big statistical packages and program sets, like SPSS and BMDP. But these demand an expert knowledge of some complex control language and therefore are of little use to an overworked clinician.
To bridge this gap, a conversational package, called QUESTOR, has been developed at the department of Medical Informatics of the Free University of Amsterdam. This package seeks to create a link between the clinician and the already mentioned statistical program set BMDP.
In a session of questions and answers, QUESTOR inquires after the wishes of the researcher and translates this into a correct set of control language instructions for one of the programs in BMDP [Dixon, 1981]. We will call such a set of control language instructions a 'set-up'.
Extensive help facilities are available. This makes the package easy to use and easy to learn. There are also facilities to provide links between BMDP and data-base systems.
In this article we will describe some of the technical aspects of how this objective is reached.

2 Structure of the package.

QUESTOR recognizes three main tasks. Each of these tasks consists of one or more modules, as shown in table 1. Initiation consists mainly of opening files, definition of recovery routines and the setting of flags. Communication will be discussed in the following paragraph. Execution consists of two parts: building the set-up, and executing BMDP. The former will be discussed in a later paragraph. Execution is performed by the module XQBMDP. It calls the correct BMDP program; it also takes care that the necessary information is passed on, like the name of the setup file and the output file, as generated by the BMDP program. When it is finished, BMDP returns control to this module. The user then may request another analysis. If he does, XQBMDP will set certain flags and return control to the initiation task.

table 1

initiation	INIT
	BSF
communication	ASKING
	PARSER
	RECALL
	HELP
execution	SETUP
	XQBMDP

To be able to communicate correctly with the user, the package needs to know the contents of the system file, that will be used in the intended BMDP-execution. Therefore the module BSF will retrieve all necessary information, like variable names, categorical information, etc.

Some more information about the way QUESTOR handles variables and the several kinds of files that are in use, can be found in Kuik [1982] and in Kuik [1983].

3 Communication.

QUESTOR is a query-oriented package; i.e. the package decides which information it needs at a certain point and then poses the appropriate questions [AKOS, 1984]. The questions can be rubricized into two types. One kind is concerned with decisions. These questions mainly are trying to determine which analysis the user wants. They often are in menu-form and therefore impose a tree-structure upon the query control. The other kind of questions is concerned about information. For example the name of the BMDP system file to be used, the name(s) of the variable(s) on which the analysis takes place; etc.

When a question is to be asked, the module ASKING is called. Each question is identified by an unique keyword, which is passed to ASKING. This keyword serves several purposes. One of them is to identify the text of the question to be asked, together with all specific information for the question. Two versions of the text of the question are stored on a general textfile, called TEXTS.QST; the user can decide whether he wants the brief or the verbose type. A hashing routine converts the keyword to a range of record numbers, in order that the RECALL module may retrieve the correct text. When the user has given a correct and acceptable answer, this is converted into a standardized form. Together with the keyword for the question, this is written to an answer file ANSWER.usr . Here, the keywords serve to identify the questions, that have been asked; i.e. the path through the query-tree. If the user should realize that he has given a wrong answer somewhere along the line, he may jump back to a previous question. This is done by pressing the escape key ESC or by using the help facility ?<. QUESTOR then jumps back to the previous node in the query tree. Thus the user may go back to any relevant question by typing the correct number of escapes. At each point QUESTOR scans the answer file for the appropriate keyword, reposes the question and replaces the previous answer with the correct one. For each following question, the package takes the previous answer as a default, which can be chosen by just pressing the return key. When the user decides

not to use the default for a decision-asking question, all following information in the answer file will be useless, of course, and therefore can be discarded.
There also exists the possibility for a user, when he starts a session with QUESTOR, to make use of an answer file from a previous session. This may make it easier to do several analogous analyses.

4 Help Facilities.

In the previous paragraph the help facility ?< has been mentioned shortly. In order to be as user-friendly as possible, QUESTOR has extensive facilities in this direction. We will mention a few in short.
Often a new user will not know how to answer certain questions. He then may request help by simply typing the question mark ? . The program will display guide-lines, appropriate for this question. Also, more general information, relevant to this specific question, is available. It can be obtained by typing ?I .
These texts are also found on the textfile and identified by the keyword, associated with the question. When the question was in menu-form, the options can be obtained by typing ?O .
More general facilities include:
?V, ?G, ?S for the names of variables, groups and subsets;
?? to obtain a survey of all help facilities;
?^ to stop the session prematurely, and
?M to send a message to the person or persons responsible for the correct operation of the package (Mailing service).
?:CONCEPT, to get a definition of the meaning of a certain word, like ?:FILE, ?:SUBSET, etc. These definitions are also stored on the textfile and identified by a keyword. This storage takes place in the form of a dictionary.

5 The textfile.

To be of any use to QUESTOR, the texts must be easily available. To this end a file with a well-defined direct access structure is most suitable. This is the file TEXTS.QST , that has been mentioned in paragraph 3.
It consists of two parts, the pointer area and the text area. The pointer area is divided into five sections, each containing pointer-records, pointing to the text area. The sections are: query, parser, errors, notices and the dictionary. The sections are of different lengths. These lengths can be changed by the programmer(s) of QUESTOR. To this end pointers are needed to the five sections; these pointers are stored in the first record of the file. At each point where the package needs some text, the programmers of QUESTOR have defined a unique keyword (i.e. the same that has been mentioned in paragraph 3). By hashing, the appropriate pointer-records for this text are obtained.
In the second part all text is stored. This is done in blocks, for each defined keyword one. When certain (sub)texts are needed at more than one point, they will be stored only once. At the point, where they are needed, a special record will replace the ordinary text line. It is identified by the ASCII BELL character in the first position of the line. In the next positions the pointers to the (sub)text are located, which then may be retrieved. The BELL character may also be discovered in some other position in a text line. This means that some variable text has to be retrieved. This variable text is stored by the program before the RECALL module has been called. This way, messages can be given, containing specific information about the problem encountered.
For easy handling, there exists an other form of the text file. This is the sequential file TEXTS.DAT . On it all relevant texts are writ-

table 2

```
%!NODICW
The dictionary doesn't contain the word '@' .
Type  ?D  for a survey of all available words.
%:SUBSET
A part of the data, that can be
characterized by a joint property;
i.e. Male, Young, Anaemic, Policlinic.
%$OPTKEY
Implemented are:
%.1
Incorrect Option number asked for: @
%.2
Option number @ out of range [0,@].
%??
?^  Stop this session
?<  Jump back one question
?D  List content of Dictionary
?H  HELP: general info about QUESTOR
?M  Send a Message to the Manager
?V  List all variable names
?:N Look for name N in dictionary
%E 54
INCORRECTLY FORMED NUMBER
CALLED FROM ROUTINE 'hstype'
%>1ST USER
Is this the first time you use QUESTOR ?
%.LONG
Is this the first time you use QUESTOR ?
%.QUERY
You can obtain some extensive information,
if you answer  YES  to this question.
You also may obtain this information at
any time during your session by typing  ?H .
%.INFO
If you have trouble that cannot be helped
in this way, you should ask advice from your
Statistician.
He may be able to point out to you,
what went wrong, and why.
You better end this session by typing  ?^ .
Good luck next time !
%]RGROPT
1.  Linear Regression (with groups)
2.  Stepwise Linear Regression
3.  Nonlinear Regression
4.  Regression on Principal Components
5.  Regression on all possible subsets
6.  Derivative-free nonlinear Regression
```

ten in ordinary text records, preceded by specific command lines for defining the keyword belonging to the following text block.
Parts of this file are shown in table 2. This file can be handled by an ordinary text editor and may be changed at any time. Therefore no reprogramming of QUESTOR is needed to change any given piece of the text. Translation into some other language is also easily done !
To convert the sequential file into a direct access file with such a structure, the program TEXTS exists. Several checks are also performed, for example, if a keyword is defined more than once.
We will indicate a few points of the way the conversion is obtained. Each command line is recognizable by the %- character in the first position. The character in the second position determines, for which of the above mentioned 5 sections a text block will follow. If this character is a dot, the following block is linked to the keyword of the preceding block, but it is separately identifiable. If a call is made to such a keyword, the package will choose one of these linked texts, depending on the context.
To indicate places, where the converting program TEXTS has to put the BELL symbol, the character @ is used. If either of these symbols (% or @) has to be used in its own right, it can be preceded by the backslash, to nullify its systemic use.

6 Building the setup.

The setup is built in two stages by the module SETUP. This module has the same tree structure as the module ASKING. It poses the same questions, but gets its answers from the answer file. The information thus obtained is used to construct a temporary setup file. This is a direct access file, which will be converted to a sequential file when it is completed. The two-stage approach is needed, since the order of the setup statements is not necessarily the same as the order in which logically the questions have to be posed.

7 Sample Run.

We present here parts of a fictitious sample run:

```
WHICH KIND OF ANALYSIS DO YOU WANT ?
regression
WHAT TYPE OF REGRESSION ?
?options                                             <----
1. LINEAR REGRESSION WITH GROUPS
2. STEPWISE LINEAR REGRESSION
       :
       :
       :
WHAT TYPE OF REGRESSION ?
linear with groups
PLEASE TYPE THE DEPENDENT VARIABLE(S)
?variables                                           <----
AGE        SMOKER    CHOLEST   SURGERY
SURVIVE    VARSURV
PLEASE TYPE THE DEPENDENT VARIABLE(S)
survive.
TYPE THE INDEPENDENT VARIABLE(S)
age,smoker,cholest.
DO YOU USE WEIGHTS IN YOUR ANALYSIS ?
yes
WHICH VARIABLE CONTAINS THE WEIGHTS ?
varsurv.
WHICH VARIABLE CONTAINS THE GROUPING INFORMATION ?
surgery
DO YOU WANT TO SAVE RESIDUALS ?
yes

QUESTOR HAS STARTED PROGRAM 'P1R' . . .
IT IS FINISHED.
YOU MAY COLLECT YOUR OUTPUT FROM THE PRINTER.

       :
       :
       :
       :
```

In capital letters are shown the texts, produced by QUESTOR: in lower case the answers from the user. The arrows indicate the places, where a help facility was called.

8 Discussion.

QUESTOR is designed for easy use by inexperienced computer users. It brings the following advantages:

1. Data can be efficiently analysed;
2. the user can concentrate on his analysis and is not bothered by programming problems;
3. the package can help by providing necessary information at key points or upon request;
4. and so the user can decide on the spot, which action to take to warrant a correct course for his analysis.

There also are a few disadvantages:

1. default options can be used undiscriminatingly;
2. the user may stumble into the pitfall known as 'Capitalization on Chance', i.e. one may perform so many statistical tests, that the 'significant' results, thus obtained, are not based upon reality, but on chance alone; and
3. the availability of a package like QUESTOR may detract the user from consulting his statistical adviser at critical points in his analysis.

Concerning these disadvantages we may remark the following. Critiqueless use of the data is always possible, whether an easy to use interface is available or not. Scientific responsibility always implies a correct use of the data. So a responsible researcher tries to obtain all information that is necessary to guarantee a correct course of the analysis. This again implies consulting a statistician, when the path of the analysis is not clear to the user. And for the more esoteric statistical techniques most statistical departments have their own programs, so as to be completely up-to-date.

References.

AKOS [1984]; On the Use of Conversational Packages in Statistical Computing; to be published.

Dixon, W.J. et al. [1981]; BMDP Statistical Software

Kuik, D.J. & Hasman, A. [1982] QUESTOR - Een consultatief pakket voor Statistische Analyses; Verslag van het vijfde Informatica Congres; MIC '82; pp 153-158

Kuik, D.J. & Hasman, A. [1983] QUESTOR, a conversational statistical package: Proceedings of the Fourth World Conference on Medical Informatics (MEDINFO '83); pp 939-942.

The System for Multivariate Statistical Data Processing

J. Michálek, M. Němcová, L. Popelinský, and M. Řebíčková, Brno

SUMMARY

The system consists of two relatively independent parts. The first is the general linear model (GLM) which can be especially used for processing of continuous type of data. The second is the general log-linear model (GLLM) and is useful for processing of categorical data.

The GLM works with general model $Y = XB + E_0$ and it enables under normality condition to test general linear hypothesis $CBA = \Gamma$. One can arbitrarily pass from this general form to concrete submodels (e.g. multivariate regression, multivariate profile analysis etc.) by suitable selection of the matrices Y, X, B, G, A, Γ. Further you can work in three processing modes namely in interactive mode in batch mode and in simulation mode.

The GLLM system proceeds from general log-linear model $\log p = \sum_{A \in a} \lambda_A (i_A)$ and includes the procedures for making multidimensional contingency tables and marginal tables, for the work with submodels, for hypotheses testing and for adequate model search. The main idea of the selection of adequate models is in the paper by Havránek (1984).

Keywords: General multivariate Gauss-Markoff model, General linear hypothesis, U distribution, Wilks´Λ criterion, Interactive mode, Batch mode, Simulation mode, Multidimensional Contingency table, Log-linear model, Graphical model, Decomposable model, Adequate models.

1. THE PROCESSING CONTINUOUS TYPE OF DATA

A great part of statistical procedures which are often used is a special case of the general Gauss-Markoff model. There are for example linear regression models, models of analysis of variance or analysis of covariance etc. The processing of these methods by computer is issued usually from some concrete procedures using unique properties each from the above mentioned methods. Each procedure from this great number is programmed. So we have got a great number of programmes for various statistical analysis, each of them is written optimally with respect to performed analysis.

The access which we put forward has got a contrary aim. It is issued from the general linear model and this one we process by a programme. Using common matrix operations this can be executed with a programme of minimal length. Subsequently, the programme of a requirement statistical analysis we obtain by a suitable choice of input parameters of the model. The advantage of this access is that we can process a great number of statistical analyses practically by means of one programme of minimal length. Its disadvantage is the fact that some numerical results are not achieved in optimal way. But this fact has not a material effect upon the time of processing for the data of small or middle range.

1.1 THE GENERAL LINEAR MODEL

We consider the general linear model in the following form

$$\underset{N\times p}{Y} = \underset{N\times q}{X}\ \underset{q\times p}{B} + \underset{N\times p}{E_o}$$

where Y is data matrix, its rows constitute random sample of size N of p-variable vectors

X is a known design matrix, rank X is $r \le q \le N$

B is a matrix of unknown parameters

E_o is a matrix of random errors.

Further we assume that expected value $E(E_o) = 0$ (zero matrix) and variance matrix $V(E_o) = I_N \otimes \Sigma$ (where I_N is identity matrix, Σ is a variance matrix each row vector of Y and $\otimes$ means Kronecker product. Thus $EY = XB$ and $V(Y) = I_N \otimes \Sigma$ which denotes the multivariate Gauss-Markoff setup. A test of general linear hypothesis $CBA = \Gamma$ where C, A and Γ are given matrices is established on the fundamental least-squares theorem (see Timm (1975)).

Theorem: Under the Gauss-Markoff setup consider the $g \times q$ matrix C of rank $g \le r$ such that each element of CB is individually estimable. Let A be any $p \times u$ matrix of rank $u \le p \le N-r$ and

$Q_e = \min\limits_{B} \mathrm{Tr}\left[(YA-XBA)'(YA-XBA)\right]$, $Q_h = \min\limits_{CBA=\Gamma} \mathrm{Tr}\left[(YA-XBA)'(YA-XBA)\right]$

(where Γ is a specified $g \times u$ matrix). Then

(a) $Q_e = A'Y'(I-X(X'X)^-X')YA$, $Q_h = (C\hat{B}A-\Gamma)'\left[C(X'X)^-C\right]^{-1}(C\hat{B}A-\Gamma)$

where $X'X$ is an arbitrary g-inverse matrix of the matrix $X'X$ and $\hat{B} = (X'X)^-X'Y$ is some solution of the normal equations.

(b) if the hypothesis H_o: $CBA = \Gamma$ is true, then the Wilks' criterion

$$\Lambda = \frac{\mathrm{Det}\ Q_e}{\mathrm{Det}(Q_e+Q_h)}$$ has $U(u,g,N-r)$ distribution.

Tests that are established on criterion Λ are derived by using the likelihood-ratio-test criterion.

Further it is well-known (see Gabriel (1968), (1969)) that $100(1-\alpha)\%$ simultaneous confidence intervals for a parametric estimable functions $\gamma = c'Ba$, for arbitrary weight vectors c and a are given by formulae

$$c'\hat{B}a \pm c_0(a'Q_e/(N-r))a\ c'(X'X)^-c)^{\frac{1}{2}}$$

where $c_0^2 = (N-r)\ (1-U^{\alpha})/U^{\alpha}$ and U^{α} is the critical value of $U(1,1,N-r)$ distribution. Alternatively, when interest is focused on estimable functions, Bonferroni method can be used.

1.2 THE DESCRIPTION OF THE GLM WORK

The structure of the GLM is in fig. 1. This system can work in three processing modes namely in an interactive mode, in a batch mode and in a simulation mode. The selection of the mode is carried out in a segment INTRODUCTION.

Now the interactive discipline of the work will be described. First the matrix Y is called (it can already be in a file or it can be in the file WORK which is arised during the work of the GLM and which includes interresults or it can be given from the terminal - for this purpose you can use a special segment which makes easy an entry of some special type of matrix. Further, you can choose a statistical analysis which you need when using a segment WHICH PROBLEM. You can choose for instance multivariate or univariate analysis, profile analysis, analysis of variance etc. or you can choose a general linear problem. In this last case the matrices X, C, A, Γ are called and the matrices $X'X$, $(X'X)^-$, $X'Y$, $\hat{B}$, Q_e, Q_h and criterion Λ are computed and the test of the hypothesis $CBA = \Gamma$ is made (you can control if each element of CB is individually estimable) and then the simultaneous confidence intervals are computed, too. The results of this calculations are written in the work file WORK. Then the control is given to the segment AND NOW for the selection of the next work. The segment AND NOW facilitates to pass to each block of those in fig. 1 to change each matrix of those which were worked out earlier and to change the working discipline.

The work of the system in the discipline BATCH is similar, the departure is through the matrices input. The discipline SIMULATION enables to prepare the requirements for the following processing by BATCH mode.

The reliable work of the GLM depends on the quality of an algorithm for the computation of the percentage points of the U distributions, too. For this purpose a special algorithm was derived.

There were used the relations between U, F, t, χ^2 and normal distributions and the distribution function of the random variable $-m \ln U(u,g,N-r)$ $(m = N-r-\frac{1}{2}(u-g+1))$ and the last which was approximated by Anderson (1958). From this approximation an α percentage point of U distribution is computed by Newton-Raphson method (see Michálek, Němcová (1984)).

2. THE GLLM SYSTEM

The GLLM system is useful for processing of caregorical data. This system proceeds from general linear model (see Havránek (1984)) of the type $\log p = \sum_{A \in a} \lambda_A(i_A)$ and it includes the parts as follows:

- procedures for making multidimensional contigency tables and relevant marginal tables
- procedures for the work with various submodels (e.g. hierarchical, graphical and decomposable). These procedures enable to choose the adequate models which can be accepted. The main idea of this work is written by Havránek (1982), (1984).
- procedures for hypotheses testing. These procedures proceed from the classical method (see Bishop (1975)) and from the multivariate simultaneous methods (see Řebíčková (1984)).

This system works in two modes. The first is for alternative factors, the other is for the nonalternative ones. The first is materially richer in its procedures than the other.

3. SOME FINAL REMARKS

The whole system is written in FORTRAN IV, it is of modular structure which enables to repeat particular steps and to make use of the results of former steps.

The systems is established for statisticians who are not familiar with computer science and it is suitable especially for minicomputers.

REFERENCES

Abramowitz M., Stegun I. A. (1972) Handbook of Mathematical Funcions Dover Publications ,New York

Anděl J. (1972) Kontingenční tabulky, Skripta PGS MFF KU, Praha

Anděl J. (1978) Matematická statistika, SNTL/ALFA, Praha

Anderson T. W. (1958) An introduction to multivariate statistical analysis, John Wiley, New York

Bartlett M. S. (1938) Further aspects of the theory of multiple regression. Proc. Camb. Phil. Soc. 34, 34-40

Bishop Y. M. M., Feinberg S. E., Holland P. W. (1975) Discrete multivariate analysis, MIT Press, Cambridge

Cran G. W., Martin K. J., Thomas G. E. (1977) Inverse of the incomplete beta function ratio, Applied Stat. 26, Algorithm 109, 111-114

Darroch J. N., Speed T. P. (1979) Multiplicative and additive models and interactions, Research rep. No 49, Dept. of Theoretical Statistics. Univ. of Aarhus

Darroch J. N., Lauritzen S. L., Speed T. P. (1980) Markov fields and log-linear interaction models for contingency tables, The Annals of Stat. vol 8, No 3, 522-539

Gabriel K. R. (1968) Simultaneous test procedures in multivariate analysis of variance. Biometrika, 55, 489-504

Gabriel K. R. (1969) A comparison of some methods of simultaneous inference in MANOVA. In P.R. Krishnaiah (Ed), Multivariate analysis II pp. 67-86, Academic Press. New York

Havránek T. (1982) O analýze mnohorozměrných kontingenčních tabulek, Sborník prací ROBUST 82, 11-18, JČMF Praha

Havránek T. (1984) A procedure of model search in multidimensional contingency tables. Biometrics to appear.

Havránek T., Edwards D. (1984) A fast procedure for model search in multidimensional contingency tables, Rapport 84/1 Recku Dokumentation, Kobenhavn

Majumder K. L., Bhattcharjee G. P. (1973) The incomplete Beta Integral. Applied Statistics 22, Algorithm AS 63, 409-411

Michálek J., Němcová M. (1984) Percentage points of U distributions, Not yet published

Olehla M., Věchet V., Olehla J. (1982) Řešení úloh matematické statistiky ve FORTRANU, NADAS, Praha

Rao C. R. (1965) Linear statistical inference and its applications John Wiley, New York

Řebíčková (1984) Využití log-lineárních modelů - soutěžní práce SVOČ KAM PF UJEP, Brno

Sbornik naučnych program na Fortraně (1974) Rukovodstvo dlja programmista. Vypusk 1, 2. Izd. Nauka Moskva.

Timm N. H. (1975) Multivariate analysis with applications in education and psychology, Brooks Cole Publishing Company, Monterey, California

Fig. 1

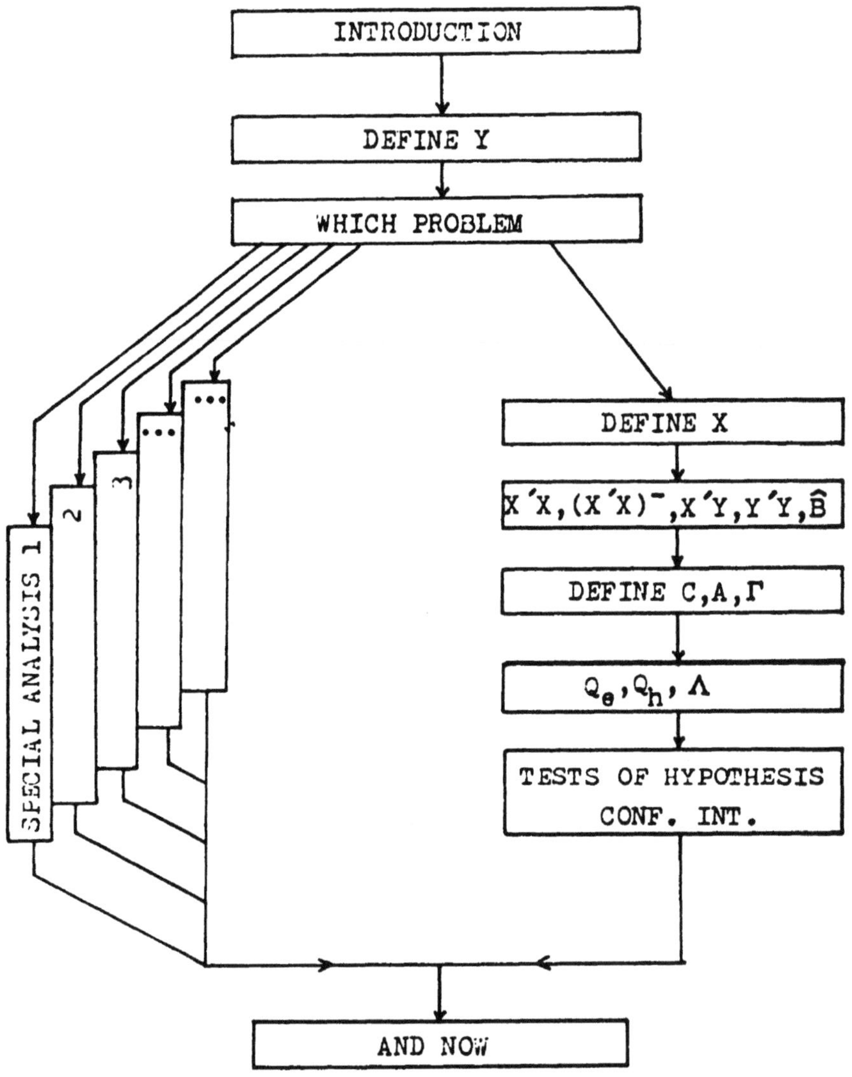

Interpretation of Statistical Software Output: some Behavioral Studies

I.W. Molenaar, and H. Broersma, Groningen

Summary

Systematic observation of statistical software users is seldom reported in the research literature. This paper describes a pilot study in which 33 subjects scanned graphical or numerical output for deviations from the normal distribution. A task analysis is sketched for stem and leaf display, box plot and numerical characteristics (such as moments and quantiles). The promises and pitfalls of this kind of experiment are briefly evaluated. The companion paper Molenaar (1984) discusses more fully how observation of users may contribute to better human computer interaction.

Keywords: behavioral experiments, observation of package users, human computer interaction, exploratory data analysis.

1. Introduction

An effective flow of information between the user and the computer system is of paramount importance for the successfull large scale use of standard software. Nevertheless, there have been surprisingly few experimentally controlled investigations of statistical package users. Somewhat more is known about text editing systems or about the physical characteristics of visual display terminals, where ergonomists and cognitive psychologists have analyzed the various steps involved in the tasks. For text editors, experimental results and cognitive models, including error rates and reaction times,are found e.g. in Card et al. (1983), Scapin (1981) or Roberts and Moran (1983). For the ergonomics of display terminals see e.g. Cakir et al. (1980) and Matula (1981).

In computational statistics, however, debates on user friendliness seem to have largely developed without systematic observation of users under experimentally controlled conditions (Molenaar 1984). The related topic of the use of charts and graphical display methods is also characterized by a paucity of experimental results, as is remarked in the reviews by Fienberg (1979) and Wainer and Thissen (1981). Exceptions are Wainer (1974), Wainer and Francolini (1980) and Naveh-Benjamin and Pachella (1982).

2. Problem formulation for a pilot experiment

The computer is used for a host of widely differing statistical analyses, by a host of users differing widely in their statistical competence. It is obvious, and indeed it was found in similar experiments with text editors or programming languages, that no experiment will lead to general conclusions about the way in which any user interacts with any statistical package: the detailed findings will depend to a large extent on the sophistication of the user, both in statistics and in computer use, and on the specific aspects of the task and the task conditions, which may vary from a simple recoding and counting batch job to the interactive fitting of a complicated loglinear model. For a pilot study, it was decided that the task of correctly interpreting the computer output would be studied in isolation from the input preparation and error correction tasks. Moreover, an output interpretation task was sought which was frequently used and not too complicated.

The scanning of a unidimensional frequency distribution in order to check for drastic deviations from normality is a subtask which looked like a suitable candidate. It is frequently used in a general exploration, in a check for outliers or data entry errors, as a prerequisite for the application of a t-test and in the inspection of the residuals in regression analysis. Both in formal significance tests for normality and in graphical inspection methods, the customary methods differ in the way in which the full information from the data is transformed and/or reduced. One may for instance count the number of observations falling in each of a number of suitably chosen intervals, and compare them with the corresponding normal frequencies. One may also inspect the values of the coefficients for skewness and kurtosis, based on the first four moments of the observed frequency distribution. One could use the information on the quartiles and extreme values contained in a box plot (Tukey, 1977). A comparison of the empirical cumulative distribution to its normal counterpart could take place directly, with a vertical axis linearly divided between 0 and 1, or indirectly via transformation to a QQ-plot.

Note that each inspection method has its natural associated significance test: Pearson chisquare for frequencies, Geary's test, a binomial test on the frequency within so many standard deviations form the mean, Kolmogorov-Smirnov for the maximal cdf distance or the Wilk-Shapiro test for the probability plot. Formal significance tests, however, are somewhat misleading, as their detectional power for drastic deviations depends so heavily on sample size: one hopes to accept, rather than to reject, the null hypothesis.

3. Sketch of a task analysis

A stem and leaf display (Tukey, 1977) is essentially a histogram with natural class boundaries in which the bars are replaced by the next digit of the observations in order to add some more information. Statistical packages show slight variations in their stem and leaf modules. They all use the horizontal axis for the frequencies and the vertical axis for the intervals, since this facilitates the use of digits for the "leaves" that replace the histogram bars. This reversal of roles may complicate the assessment of symmetry: the human eye is more accustomed to vertical than to horizontal symmetry planes, as the human body itself and almost all man-made objects have a vertical plane of symmetry.

Assessment of symmetry is indeed the first check on normality of a given sample that comes to mind. Next, one might check unimodality and look for conspicuous "gaps" in the frequencies, or for outlying observations. In general, frequencies should decrease monotonely on both sides when moving outward from the top, at a rate corresponding to the well known Gaussian curve.

A box plot gives the three empirical quartiles as the box, from which "whiskers" extend to the furthest observations within one box length outward from the quartiles. Observations further away from the median are individually marked. Note that on each side beyond the whiskers there should be about 2.5 percent of the observations when normality holds. This was the reason to adopt this whisker length definition (McNeil, 1977) rather than Tukey's . The internal representation of a normal box plot will be far less dominant in the subject's mind, even after training, than the bell shaped curve. It is easy to check whether the median lies half way between the quartiles and whether the whiskers have the same length. Whiskers also lead to a check on the frequency of extreme observations. Indeed this task is now easier than in the histogram, provided that the subject is aware of the "2.5 percent rule" mentioned above and of the total sample size. The box plot does not lend itself to detection of bimodality or gaps (Lourens, 1984, p.34).

The box plots offered by statistical packages differ among each other (Wesselink, 1983): boxes are vertical or horizontal (which may affect symmetry detectability, see above), and the amount and display method of information on scale values varies. The limitations of a line printer may suggest an asymmetric position of the median in a perfectly symmetric case.

The third presentation form used in the experiment was the list of distribution characteristics used in SPSS: mean, standard error, median, mode, standard deviation, variance, kurtosis, skewness, range, minimum, maximum, sum and number of valid cases (there were no missing data in the experiment). This block of 13 characteristics offers several ways of assessing deviations from normality. One may compare mean, median and mode (with an eye on sample size and standard deviation or range for checking the plausibility of discrepancies among them). Skewness and kurtosis may be compared to the values of 0 and 3 for the normal distribution (taking into account their instability for small sample sizes and the weird habit of SPSS of calling "kurtosis" what everybody else calls "excess", i.e. kurtosis minus 3). The position of minimum and maximum may be mentally expressed in number of standard deviations away from the mean, although sample size interferes and too much weight may be attached to just those two extreme values.

Whereas the internal representation of the histogram or stem and leaf display was purely visual, and a combination of visual and propositional representation of a normal distribution was useful for the box plot, the assessment based on the characteristics is purely based on propositions (knowledge about what the numbers should be for the normal distribution, and how much variability can be due to sampling). Yet it is plausible that the "translation" of asymmetry, bimodality or gaps into their effects on the characteristics is a relatively difficult mental task.

4. Some results

The experimental subjects were 33 students in the social sciences voluntarily taking an additional course in statistics after the compulsory curriculum. This course contained a brief introduction to the display methods used in the experiment.

Subjects were instructed to answer yes or no to the question whether a dataset contained serious deviations from normality, and if yes, to characterize the deviations. They were also asked to indicate whether their conclusion had been reached "at first glance", "after a brief analysis" or "after a careful analysis". Three different datasets were used: approximately normal, skewed (skewness -1.0) and symmetric with outliers (kurtosis 4.8). Six subject groups were formed. Each group received each of the three datasets in mutually different presentation modes (stem and leaf, box plot, characteristics) such that all groups received a different combination of datasets and presentation modes. From the detailed analysis reported in Broersma (1984) it was concluded that the symmetric dataset with outliers was correctly handled by only two of the 33 subjects. Otherwise the stem and leaf display had the highest success rate. The boxplot came second highest in the case of skewed data, but the characteristics came second for approximately normal data. On the ordinal scale for amount of attention paid to the three different presentation modes, the subjects' scores showed significant differences for all three datasets (using the Kruskal-Wallis test at level 0.1). A pairwise analysis of the results (using the Mann-Whitney test at level 0.1) showed no significant differences for the approximately normal dataset. For the other two datasets the subjects' conclusions from the stem and leaf display were most often reached at first glance. However, none of the subjects' conclusions for the dataset with outliers were correct, so we have to be careful in drawing conclusions about the superiority of the stem and leaf presentation mode.

5. Some conclusions and some research projects

Comparison of task analysis and results suggests some "aptitude treatment interaction" as was predicted in Molenaar (1984). A box plot may well be an efficient outlier detector for a user trained in EDA, and a kurtosis value of 4.2 will be very informative for a user who is well familiar with kurtosis values, but still a less trained user may prefer a histogram because (s)he can draw more and better conclusions from the comparison with the bell shaped curve that is so familiar. Given that some display or analysis methods are superior for well trained users only, the question remains which user groups would benefit from such a training. One may conjecture that QQ-plots or box plots are more popular with professional data analysts than with occasional computer users. Indeed one of us answered a student's question "why should I learn about QQ-plots" with "once you know how to interpret them, you have the advantage that the human eye detects deviations from linearity more easily than deviations from a bell shaped curve".

This leads to the - obvious - conclusion that a general statistical package should contain several display methods for a frequency distribution, each of them formatted in a way best suited to the analysis of the task for which they serve. Most packages offer a frequency table (but some try to fool the user by omitting intermediate intervals which have zero frequency). Some allow histograms or stem and leaf displays. Comparison with the corresponding frequencies for the best fitting normal distribution, however, is either impossible or very difficult in most packages.

Molenaar (1984) contains some more proposals for "behavioral studies of the software user", some of which will be carried out in Groningen in the near future. Two examples are the comparison of two frequency distributions and the scanning of the output of modules for t-tests, correlation and regression: what aspects are inspected by most users, what messages are lost and what display format can diminish the time on task and the risk of wrong conclusions? The paper just quoted contains a general evaluation of the promises and pitfalls of behavioral experiments under controlled conditions for the improved performance of package users. It concludes that insights from experiments could help to diminish user frustration (both by package improvements and by improved user training), but that a general understanding of the miracles and the limitations of human information processing is a too ambitious goal for the next decade. It is maintained, however, that the behavioral and cognitive sciences can certainly contribute to improved human computer interaction, by the systematic study of aspects that package designers or computer specialists tend to forget.

References

Broersma, H.J. (1984),Empirical comparison between presentation forms for data (in Dutch), Heymans Bulletin, 84-701-EX, Vakgroep Statistiek & Meettheorie FSW, University of Groningen.

Cakir, A. et al.(1983), Visual display terminals, Wiley, New York.

Card, S.K., et al. (1983), The psychology of human-computer interaction, Lawrence Erlbaum Assoc., Hillsdale (N.J.).

Fienberg, S.E. (1979), Graphical methods in statistics, The American Statistician *33*, 165-178.
Lourens, P.F. (1984), The formalization of knowledge by specification of subjective probability distributions, an experimental approach, Thesis, University of Groningen.
Matula, R.A. (1981), Effects of visual display units on the eyes: a bibliography (1972-1980), Human Factors *23*, 581-586.
McNeil, D.R. (1977), Interactive data analysis, Wiley, New York.
Molenaar, I.W. (1984), Behavioral studies of the software user, Computational Statistics and Data Analysis *2*, 1- ... .
Naveh-Benjamin, M. and Pachella, R.G. (1982), The effect of complexity on interpreting "Chernoff" faces, Human Factors *24*, 11-18.
Roberts, T.L. and Moran, T.P. (1983), The evaluation of text editors: methodology and empirical results, Comm. ACM *26*, 265-283.
Scapin, D.L. (1981), Computer commands in restricted language : some aspects of memory and experience, Human Factors *23*, 365-375.
Tukey, J.W. (1977), Exploratory data analysis, Addison Wesley, Reading (Mass.)
Velleman, P.F. and Hoaglin, D.C. (1981), Applications, basics and computing of exploratory data analysis, Wadsworth, Belmont.
Wainer, H. (1974), The suspended rootogram and other visual displays, The American Statistician *28*, 143-145.
Wainer, H. and Francolini, C.M. (1980), An empirical inquiry concerning human understanding of two-variable color maps, The American Statistician *34*, 81-93.
Wainer, H. and Thissen, D. (1981), Graphical data analysis, Annual Review of Psychology *32*, 191-241.
Wesselink, G. (1983), Exploring in the packages CADA, SPSS and TTWESP: the box plot and the stem and leaf display (in Dutch), Heymans Bulletin 83-679-RP, Vakgroep Statistiek en Meettheorie FSW, University of Groningen.

Design of Eperimental Statistical Packages

P. Nedoma, Praha

SUMMARY: An interactive service software system has been developed suitable to serve as a tool for experimental as well as routine work among others in the field of statistical data analysis. The system is FORTRAN-based emphasizing a transferability feature. It can be employed as a superstructure for FORTRAN subroutine libraries-handling. The project represents a specific step in the effort to create a software base enabling a comfortable design of transferable interactive systems in laboratory or small-group conditions.

1. Introduction

A research and experimental work with computers in the field of statistical data analysis represents very demanding and time-consuming task in which a broad variety of mathematical methods is used. The supporting software must cover among others simulation of stochastic processes and experimental estimation of rounding errors.

The broadly accepted approach to facilitate the work is creation of libraries of computational subroutines expressing the theoretical results in the form of algorithms. However, even with a good library, main control programs must be coded and debugged. It can be very time-consuming especially if a readable and user-friendly presentation of research work's results is expected.

For a routine work in a specific application field, effective programm-products have been developed, e.g. SPSSX, BMDP etc. Those have set a critera for the professional level of software results in the field of computational statistics. Thus an ordinary user is not forced any longer to read library descriptions and make any programming work. The nonprogram user is offered a tempting possibility to lead a dialogue with a computer in a rational and natural way, using a set of commands closely related to the problem solved.

As far as interactive software packages is concerned, a deficiency can be felt that existing software is usually closely linked to a specific computer, operating system or a high-level programming language. A need arises to develop a software base that supports experimental transferable interactive systems design in laboratory or small-group conditions.

An attempt to the solution of the problem is described. It should be considered as a description of required software-system features and as a call for further development in the field.

Programming system under discussion is based upon long-term experience /1/. It was proved that the design of interactive software packages can be solved in accordance with modularity approach similar to the architecture of computational libraries. A set of operations is machine-dependent and must be solved differently for a specific computer. Those machine-dependent parts can be aggregated into a relatively small number of blocks, the rest of the system becoming thus machine-independent and consisting of logical and arithmetical operations only. Then proper selection of a specific hosting language loses most of its importance and FORTRAN was the choice just for the sake of an easy system portability. In this respect minicomputers were at first aimed being a standard equipment of reserch laboratories.

The conversational service system /2/,/3/,/4/ is based upon a construction-library covering problems related to an interactive software package design. A brief account of them is given in following sections. The library makes is possible to incorporate a computational subroutine library into a flexible interactive package and to prepare an "inteligent", easy to use and understand, software tool for practical users. The features of the main supervising program were chosen in such a way so that it can serve as an contents and machine-independent superstructure for any FORTRAN libraries handling.

2. Computational means

a/ Dynamic storage allocation: A data-space is used to store any data. The access to it is not arbitrary, instead, a free space must be asked for and granted with help of "allocation-functions". This facility supports a selection of daty-types allowed by a compiler. Allocations are grouped into "blocks" and deleted if not used any longer. The allocation mechanism itself is very simple one and being slightly machine-dependent follows the way how FORTRAN "EQUIVALENCE" statement is interpreted by the compiler. Emphasis should be given to the fact that all subroutines based upon the access are independent on the way how a compiler allocates data.

b/ System "objects": Data units distinguished by the allocation-procedure, by the data type and by the interpretation are refered to as objects /e.g. numerical vector, matrix etc./. An object can possess a different internal structure /dimensions, a set of relative pointers etc./ described by a supporting "dope" vector. If an object is used inside a computational subroutine, only its type, dope vector and

pointer to the first data unit is to be mentioned.
c/Data-structure unification: was achieved in such a way. The selection of objects supported differs according to an application. E.g. /5/ uses only scalars, vectors and matrices stored FORTRAN-wise. In /6/ the structure of vectors, matrices etc. was generalized in such a way that for example rows and diagonals of existing matrices can be used as vectors in computational subroutines and triangular and symetric matrices are retrieved independently on the way how they are stored in the memory.
d/Controlled precision: During the research work a necessity usually arises to study an algorithm properties with respect to rounding errors. It often means to run the algorithm with different wordlength and precision types. A software tool was designed /6/ that enables dynamic switching between single and double precision /if allowed by the compiler/. A possibility has been included, too, of rounding results after every arithmetical operation, as well as optional counting of operations. Supporting subroutines are written in assembler /IBM 370/ and must be rewritten if used on other computer.

3. Conversational means

a/ External objects: are named objects sharable under the name between different program-units. The approach supposes existence of at least one COMMON block. A kind of garbage collection and data-compress has to be carried out from time to time.
b/ Retrieval of character strings: almost lacks any FORTRAN-support. The sufficient flexibility was achieved using two machine-dependent subroutines that convert character to integer and vice versa. The support can be replaced by the possibility given by a compiler.
c/ File management: Based upon FORTRAN unformatted and formatted statements for input and output operations, a simple file-management system was designed that includes access to basic file characteristics as well as to buffer-allocation mechanism. A binary stream data retrieval and character lines editing are supported by a selection of subroutines. An alphanumerical display unit is considered as a supporting input/output device: on input it can optionally substitute input file operation, on output it can display output lines or to substitute the output operation, too. Subroutines to read one line from terminal-keyboard, to display one character line and to clear a screen must be supplied at any installation.
d/ Control commands: A selection of control commands is incorporated in the input stream and special subroutines take care of their detection, syntax analysis and resulting internal program branching.

e/Special_files: A selection of files is considered as system files. Those have special label and a kind of data-directory at their beginning. Their primary role consist in preserving the content of memory or storing results of repetitive calculations.

f/_Dynamic_module-loading: A selection of subroutines can be called during the program run according to users wishes expressed by means of control commands. Those subroutines are placed into an "on-line" library and brought dynamically into the memory whenever needed. The access was formalized in such a way that it can be simulated within FORTRAN. The loading mechanism was developed to work under operating system OS/VS as well as under DOS /with certain limitations/. A procedure is used to link modules into the on-line library independently of each other and of the whole system.

g/_Errors: A simple and effective error system takes care about error recovering throughout the whole system, but it can hardly recognize such errors as numerical overflow. Errors should not lead to any sort of abnormal program end so that last mentioned errors must be recovered by the means offered by a computer.

4. Supervising program logic

The program run is controlled by means of control commands. Being format-free they enter from terminal keyboard in interactive mode of work or as punched cards in batch mode. Switching between both modes can be accomplished to accelerate processing. Three levels of interactive dialogue are available.

Two sections are to be distinguished within the supevising program: static and dynamic. In the static section, control commands are read, decoded and interpreted. Objects are allocated, filled with data and displayed/printed. A computation can take part - modules are loaded into memory and control is given to them. Conditions are specified for consecutive dynamic section: names of modules and lists of arguments are established and various lists of objects are defined. All of the contents of the data space can be stored in a system file and restored when required. The feature last mentioned allows checkpoints to be taken as well as initial conditions to be restored in the case of repetitive runs. A selection of service commands is interpreted.

The dynamic section carries out the following tasks:

- modules needed are loaded and linked to the program
- those are retitively called in a cycle of "time" /"case number"/ - frequencies of their call are specified in "timer" options
- a selection of system files is read simultaneously /merged/ or files are activated one after one/added/

- data from a binary or character file or in-stream data are read according to specifications made in the static section
- selected objects are periodically displayed - the feature eliminates programming of print points in computational subroutines
- the run can be periodically stopped to allow program-data changes
- when dynamic section is finished, the static one is entered again to prepare conditions for subsequent dynamic section

5. Additional features

a/ Initialization algoriths approach: As an analogy of FORTRAN program writing, names and arguments lists of modules used in the dynamic section can be specified by means of control commands. In order to avoid long list entry a more effective way was developed - it consist in mere naming a single key-word conected with a simple subprogram, refered to as "initialization algorithm". It allocates all objects requiered and fills them with initial values . The argument list and the name of the module used in dynamic section is specified inside it, too. A specific conventions were developed /5/ of initialization algorithms writing so that those can substitute the usual verbal description of the use of algorithm and allows to employ the numerical experience of the designer.

b/ Users commands : Modules can be designed in such a way that they include retrieval of users control commands. Syntactical and semantical analysis is left to designer, but programming is widely supported by the construction-library. No fixed rules are established - it depends only upon the designer's taste how the dialogue related to his problem will look like.

c/ Graphics: Printed/displayed graphs can be prepared from system files by reading and retrieving them in the dynamic section. On-line use of a graphical devise has not been formalized. The output document can be printed and/or displayed. Basic forms of prints are supported by the construction-library, special retrieval is left to users.

6. Applications

The described programming tools are incorporated in a software base developed in the Department of Automatic Control of the Institute of Information Theory and Automation, ČSAV, Prague /see/6//.

The supervising program was designed as a supporting service system for the library for simulation, identification, /adaptive/ control and statistical analysis of /technological/ processes. A broad list of methods can be found there related to the parameter estimation and statistical analysis of regression linear model /see /5/, older work is /1/, minicomputer application: /7/,/8//.

Computer aided documentation is available /2/,/3/,/4/. A wide possibility exist to pre-process source texts according to an application requirements.

REFERENCES:

/1/ Kárný M., Bohm J., Hallousková A., Nedoma P. /1977/
SIC - Program for system simulation, identification and control, Research report ÚTIA ČSAV, Prague

/2/ KOS - Conversational service system: System manual, Research report ÚTIA ČSAV /1984/

/3/ KOS - User's manual, research report ÚTIA /1983/

/4/ KOS - Programmer's guide, research report ÚTIA /1983/

/5/ SIC - Library of subroutines for system simulation, identification and control, research report ÚTIA /1983/

/6/ ÚTIAPACK - Scientific subroutine library, research report ÚTIA /1983/

/7/ SICA - Manual del sistema, research report IMACC, Havana /1982/

/8/ SICA - Manual del usuario, research report IMACC, Havana /1982/

Software in Categorical Data Analysis

A Computer Intensive Approach to the Analysis of Sparse Multidemensional Contingency Tables

D. Edwards, Copenhagen

Summary. An approach to the analysis of sparse multidimensional contingency tables by means of simple exact tests is described. The tests are essentially equivalent to tests for the conditional independence of two factors given the third in a 3-way table, and software for this purpose can be used. The tests are initially applied to the whole table, and then to appropriate marginal tables, using collapsibility properties in provisionally accepted models.

Results proved by Sundberg (1975) imply that the models accessible by such a procedure are precisely the decomposable models. A simple direct proof of this result using graph theory is given here, together with an illustrative example.

Enumeration of the conditional distributions in totality is often infeasible, and monte carlo methods must be used. Hopefully the procedure leads to testing in sufficiently deep marginals at some stage that asymptotic approximations can thenceforth be used.

Keywords: Graphical Models, Interaction Graph, Model Selection, Sparse Contingency Table, Exact Tests.

OUTLINE OF THE METHOD

Exact tests for independence in rxc tables have been treated by many authors: recently Pagano and Halvorsen (1981) and Mehta and Patel (1983) have developed fast algorithms for the enumeration involved, and software packages are available (Verbeek et. al., 1983).

A simple generalisation of these tests can be used to analyse sparse multidimensional tables. Consider first a 3-way table T_{ABC} with classifying factors A,B and C, and the hypothesis that A and B are conditionally independent given C, corresponding to the log linear model [AC][BC]. Exact tests of this hypothesis are evaluated in the conditional distribution of the table given the marginal tables T_{AC} and T_{BC} (see for example Andersen, 1974). Viewing the table as several slices of AB tables, one for each level of C, the conditioning on T_{AC}

and T_{BC} is equivalent to conditioning on the margins in each slice. The conditional probability of the table is just the product of the conditional probabilities in each slice. Further algorithmic aspects are described in Kreiner (1984a).

Consider next an n-dimensional table with factors A,B,C...N, and consider the hypothesis that A and B are conditionally independent given the remaining factors C...N. (Some authors express this by saying that A and B exhibit zero partial correlation. We follow this usage in terming the corresponding test a ZPA test.) Define a new factor X by stacking the factors C...N, i.e. with one level for each combination of levels of C...N. Then the hypothesis is that A and B are conditionally independent given X, i.e. reduced in principle to a 3-way table.

A computer program for performing exact ZPA tests in multi-dimensional contingency tables is available at a nominal charge (Kreiner, 1984b).

Consider finally a 4-way table with factors A, B, C and D and suppose that the model L=[ABC][BCD] has provisionally been accepted. Consider the interaction graph of L, as given in Fig. 1. We use the following result (see Asmussen and Edwards, 1983): under a graphical model L, the removal of an edge *e* can be performed as a ZPA test in a marginal table T_a, if and only if *a* is complete and *e* is not in the boundary of a connected component of the complement of *a*. Thus for example the removal of the edge AB can be performed as a ZPA test in the marginal table T_{ABC}, i.e. as a test for the conditional independence of A and B given C.

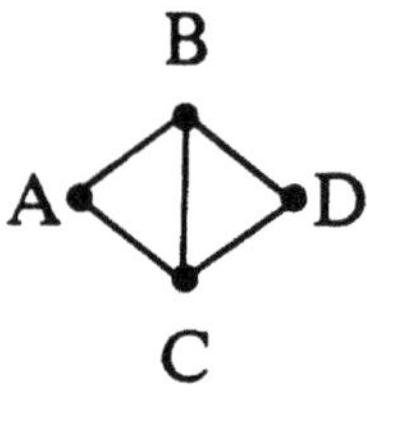

Fig. 1. The interaction graph of [ABC][BCD]

Similarly the removal of the edges AC, BD and CD can be performed as ZPA tests in marginal tables, but not the removal of BC. Note also that the removal of BC gives a non-decomposable model [AB][BD][DC][AC], whereas removal of the other edges gives in each case a decomposable model.

The approach of the present paper to the analysis of multi-dimensional tables by means of exact ZPA tests can now be described. A backwards elimination procedure, as described for example in

Edwards and Kreiner (1983), is adopted. Starting with the complete graph (saturated model), an edge is removed at each step. However here only edges whose removal can be performed as ZPA tests in appropriate marginal tables are considered at each step. The effect of this restriction, in terms of which models are thereby excluded from consideration, is treated in the following section.

THE SCOPE OF THE PROCEDURE

The central result of the present paper is that the models accessible by the procedure just described are precisely the decomposable models. This follows from results given by Sundberg (1975), though in somewhat disguised form. Here a direct proof using simple graph theory is given.

The characterisation of decomposable models given by Darroch, Lauritzen and Speed (1980), as graphical models whose graphs are triangulated, i.e. do not contain a circuit of length ≥ 4 is used. For some theory on triangulated graphs, see Golumbik (1980, ch. 4).

We first prove:

Lemma 1. Removal of an edge from a decomposable model L corresponds to a ZPA test if and only if the resulting model is decomposable.

Write the edge as e and the resulting model as $L\backslash\{e\}$. Suppose that the endpoints of e are v,w and that n generators of L, $g_1 \ldots g_n$ say, contain both v and w. If $n=1$, then the removal of e is clearly a ZPA test since e is not in the boundary of a connected component of g_1^c. Suppose that $L\backslash\{e\}$ contains a circuit of length ≥ 4 without a chord. Since L is decomposable, this circuit must intersect v and w. The circuit cannot intersect g_1^c, since segments in g_1^c have chords in the boundaries, these being complete. But all vertices in g_1 are adjacent except v and w. Hence the circuit cannot be of length ≥ 4 without a chord. Hence $L\backslash\{e\}$ is decomposable.

If $n>1$, then clearly the removal of e cannot be a ZPA test. Let u be a vertex in $g_1\backslash g_2$. There must be a vertex x, say, in $g_2\backslash g_1$ not adjacent to u, for otherwise g_2 could be enlarged to include u. Then [v,u,w,x] is a circuit in $L\backslash\{e\}$ of length 4 without a chord, so $L\backslash\{e\}$ is non-decomposable.

Lemma 1 indicates that the backwards elimination procedure described above can only arrive at decomposable models. To show that all decomposable models are accessible in this manner, we need the following result.

Lemma 2. Given a non-saturated decomposable model L, there can be formed another decomposable model by the addition of an edge.

We use the fact that a graph is triangulated if and only if there exists a perfect elimination scheme, i.e. an ordering $\sigma=(v_1,...,v_n)$ of the vertices such that each set $X_i=\{v_j \text{ adjacent to } v_i{:}j>i\}$ is complete. Moreover, an arbitrary vertex can be chosen as the last vertex v_n in the scheme. See Golumbik (1980, ch. 4). Since L is non-saturated, we can choose a vertex v_n which is not adjacent to at least one of the remaining vertices. Let σ be a perfect elimination scheme with v_n as last vertex, and let v_k be the last vertex non-adjacent to v_n. Consider $L+\{e\}$ where e is the edge (v_k,v_n), and suppose that $L+\{e\}$ contains a circuit of length $\geqslant 4$ without a chord. Clearly the circuit must contain the edge e, since L is decomposable. Furthermore the circuit cannot intersect $s=\{v_1,v_2...v_{k-1}\}$. To see this, note that it cannot intersect v_1 since X_1 is complete, so a segment intersecting v_1 has a chord in X_1. Similarly it cannot intersect v_2 since X_2 is complete. Continuing in this fashion, we obtain that the circuit cannot intersect $\{v_1...v_{k-1}\}$. Thus the circuit can be written $[v_n,v_k,v_\alpha^1...v_\alpha^p,v_n]$ where the $v_\alpha^i \epsilon \{v_{k+1}...v_n\}$. But all v_α^i are adjacent to v_n, so that $p=1$, and the circuit has length 3. Thus $L+\{e\}$ is decomposable.

DISCUSSION

The importance of the results given above is that they show that a quite satisfactory preliminary analysis of sparse multidimensional tables can be made using a simple generalisation of exact tests for rxc tables. Hopefully the procedure leads to testing in marginal tables of sufficiently low dimension that asymptotic tests can be applied: this will of course extend the range of models that can be fitted.

Exact tests for more general log linear hypotheses for multidimensional tables are treated briefly in Andersen (1974): development of an algorithm for the general case would seem a daunting task.

Enumeration of the conditional distributions in totality is often infeasible, and it is in practice quite sufficient to use monte carlo methods, so that the significance levels can be obtained to a given accuracy. This is the approach of Kreiner (1984b).

REFERENCES

ANDERSEN, A.H. (1974). Multidimensional Contingency Tables. *Scand. Jour. Stat.* **1**, 115-27.

ASMUSSEN, S. and EDWARDS, D. (1983). Collapsibility and Response Variables in Contingency Tables. *Biometrika*, **70**, 3, 567-78.

DARROCH, J.N., LAURITZEN, S.L. and SPEED, T.P. (1980). Markov Fields and log linear interaction models for contingency tables. *Ann. Statist*, **8**, 522-39.

EDWARDS, D. and KREINER, S. (1983). The Analysis of Contingency Tables by Graphical Models. *Biometrika*, **70**,3, 553-65.

GOLUMBIK, M.C. (1980). *Algorithmic Graph Theory and Perfect Graphs*, New York: Academic Press.

KREINER, S. (1984a). The Analysis of Multiple Contingency Tables by Exact Conditional Tests for Zero Partial Association. Danish Institute for Educational Research, Hermodsgade 28, 2200 Denmark.

KREINER, S. (1984b). EXABIRCH, A Program for Exact Conditional Analysis of Multiple Contingency Tables. Danish Institute for Educational Research, Hermodsgade 28, 2200 Denmark.

MEHTA, C.R. and PATEL, N.R. (1983). A Network Algorithm for performing Fisher's Exact Test in rxc Contingency Tables, *JASA*, **78**, 382, 427-34.

PAGANO, M. and HALVORSEN, K.T. (1981). An Algorithm for finding the exact Significance Levels in rxc Contingency Tables. *JASA*, **76**, 376, 931-4.

VERBEEK, A., KROONENBERG, P.M. & KROONENBERG, S. (1983). User's Manual to Fisher. Sociological Institute, TMS, Heidelberglaan 2, 3584 CS Utrecht, The Netherlands.

The New Version of the GUHA Procedure ASSOC (Generating Hypotheses on Associations) – Mathematical Foundations

P. Hájek, Žitná

Abstract. Design choices for a new version of the GUHA procedure ASSOC are formulated and mathematical foundations are presented.
Key words: GUHA, hypothesis formation, missing data items.

§ 1. Introduction.

GUHA is a method of mechanized hypothesis formation; GUHA procedures generate hypotheses. Development of GUHA procedures often leads to mathematical (logical and other) problems that may be of independent interest. In the present paper we are going to present such a mathematical theory that has been needed for the construction of a new version of the GUHA procedure ASSOC, which is seemingly the most used GUHA procedure. In its original version, ASSOC accepts as input dichotomous data - a matrix of zeros and ones, possibly with some data items missing - and generates sentences of the form $A \sim S$ (read: A is associated with S) where $\sim$ is a fixed associational quantifier (notion of association, see below) and A, S (antecedent, succedent) vary over some combinations (elementary conjunctions) of properties corresponding to columns of the input matrix. The output consists of a list of those sentences that are true in the data matrix and are strong in a certain logical sense, each together with various additional information. (See Hájek and Havránek 1978a,b, 1977, Hájek, Havránek and Chytil 1983.)

At present, a new generation of the GUHA package of programs is being prepared by Prague GUHA circle, one of first steps being a new version of the procedure ASSOC. Improvements concern two things: implementation (unified user-oriented control language etc.) - this will not be discussed here - and reasonable generalization of the procedure, concerning (i) possibility of processing more general (nominal) data, (ii) possibility of generation of sentences restricted to a subsample, (iii) various kinds of processing of missing information, according to user's choice. This requires some mathematical theory, which will be presented here. Due to space limitation, our presentation will be rather sketchy; proofs and technical details will be published elsewhere.

The procedure will offer a rather rich choice of input parameters, compensated by systematic introduction of default values for a less advanced user.

§ 2. Data and sentences.

Nominal (finitely valued) data may be considered not only as a generalization of dichotomous data (as in Hájek 1973-4, Hájek and Havránek 1978a) but also as their particular case. In the latter view a column with values from $\{1,\dots,h\}$ is understood as coding h dichotomous columns in which each row has exactly one value 1, and this 1 is at i-th place iff the corresponding value in the original column is i. (Thus, instead one column coding colours of objects, say, we have various dichotomous columns coding properties COLOUR:RED,...,COLOUR:GREY (yes-no) etc.) We say that a h-valued nominal quantity Q determines h atomic properties belonging to Q. Atomic properties belonging to input quantities (columns in the input matrix) can be taken for atoms of a logical language; we may form elementary conjunctions from them, e.g. (in numeric form) 4:02 & -6:01 & -7:04 or (in alphanumeric form) COLOUR:GREY AND NOT AGE:YOUNG etc. Given data M with complete information each pair A, S of elementary conjunction determines a four-fold contingency table $T = \begin{pmatrix} a & b \\ c & d \end{pmatrix}$ (written also (a,b,c,d)); each function f associating with each such table zero or one (false or true) defines a quantifier and enables us to evaluate the sentence $A \sim S$: this sentence is true in the data M if for the table T (determined by A, S and M) we have $f(T) = 1$. For simplicity, we shall write $\sim(T)$ instead of $f(T)$.

More generally, given A, S and C (three elementary conjunctions) the truth of $(A \sim S)/C$ (relativized sentence) is determined as follows: first we omit from the matrix M all rows (objects) not satisfying C; the resulting matrix is denoted $M \upharpoonright C$. Then $(A \sim S)/C$ is defined to be true in M iff $A \sim S$ is true in $M \upharpoonright C$.

A quantifier $\sim$ is associational if for each T = (a,b,c,d) and T'= (a',b',c',d') we have the following: $\sim(T) = 1$ and $a' \geq a$, $b' \leq b$, $c' \leq c$, $d' \geq d$ implies $\sim(T') = 1$. (T' can be said to be better than T or to have more coincidences and less differences than T.) For an associational $\sim$, $A \sim S$ can be read "A is associated with S". To obtain the definition of an implicational quantifier delete the inequalities $c' \leq c$, $d' \geq d$ above; thus each implicational quantifier is associatonal. (See Hájek and Havránek 1978a.) $\sim$ is founded with a base $B \geq 1$ if for each T = (a,b,c,d), $\sim(T) = 1$ implies $a \geq B$. (This means: if $\sim$ is founded and $A \sim S$ is true in M then in M there are satisfactory many objects satisfying $A \sim S$.)

All this generalizes for data with incomplete information; there are at least three possibilities. If M is a matrix with incomplete information then a completion of M is any matrix M_o resulting from M by replacing all missing data items by some possible values. Let Φ be $A \sim S$ or $(A \sim S)/C$. Components of $A \sim S$ are A, S, components of $(A \sim S)/C$ are A, S, C.

(1) Φ is true in M in the secured semantics if Φ is true in all completions of M. (Notation: $M \models \Phi$ (sec).)

(2) Φ is true in M in the deleting sematnics if Φ is true in the matrix M' resulting from M by deleting all objects (rows) for which the value of a component of Φ is missing (not determined) (Notation: $M \models \Phi$ (del).)

(3) Φ is true in M in the optimistic sematnics if Φ is true in at least one completion of M and at least one object satisfies the conjunction of all components of Φ. (Notation: $M \models \Phi$ (opt). See Hájek and Havránek 1978a, Havránek 1980.)

Theorem 1. (1) (Havránek) If $\sim$ is associational then $M \models \Phi$ (sec) implies $M \models \Phi$ (del) implies $M \models \Phi$ (opt).

(2) If $\sim$ is founded and $M \models \Phi$ (sec) or $M \models \Phi$ (del) then at least one object satisfies the conjunction of all components of Φ.

§ 3. Design choices.

(DC0) ASSOC will process nominal data as a particular case of dichotomous data. Hence, the control language must be (practically) the same as if only dichotomous data were accepted.

(DC1) In a given run, the quantifier is fixed.

(DC2) It will be possible to have a run generating restricted sentences of the form $(A \sim S)/C$ with fixed C.

(DC3) All admitted quantifiers will be founded.

(DC4) Main attention is paid to associational quantifiers.

(DC5) Only hypotheses with disjoint components are generated.

(DC6) The user can choose any of the three semantics of missing information.

(DC7) Improving literals can be used to compress the results.

(DC8) It will be easy to extend the program by adding new quantifiers.

Remarks. (DC0) leads to a particular choice of control language that refers to input quantities rahter than to dichotomous atoms belonging to them. Details cannot be discussed here.(DC1) is "classical". (DC2) seems to be a reasonable compromise between needs of users and extreme complexity. (DC3 and DC4) makes possible to use the theory presented below and seems to correspond to actual needs. (DC5): Two components are disjoint if no input quantity occurs in both of them. (DC5) is relevant for treating missings as demanded in (DC6). Note that the definition of secured and optimistic semantics is computationally highly unfeasible; see below. Concerning improving literals see § 6. Ad (DC8): six starting quantifiers are used, named SIMPLE, FISHER, CHISQ, FIMPL, UIMPL, LIMPL, some of them being based on tests for statistical hypotheses. For example, the quantifier FISHER with parameters BASE, ALPHA is defined as follows: FISHER(T) = 1 iff $ad > bc$, $a \geq$ BASE and $F(T) \geq$ ALPHA, where F(T) is the Fisher statistic and $T = (a,b,c,d)$. FIMPL has parameters BASE and CPROB; FIMPL(T) = 1 iff $a \geq$ BASE and $a/(a+b) \geq$ CPROB. All six quantifiers are associational, the last three are implicational.

This is approximately what the user has to know (he should also know the rule on improving literals formulated in § 6). In the rest of the paper we present some thoeretical results important for the implementation and theoretical evaluation of ASSOC.

§ 4. Critical completions.

Evaluation of $M \models (A \sim S)$ (del) presents no computational problems; thus we concentrate our attention to (sec) and (opt). Each sentence $A \sim S$ determines in each data matrix (possibly with incomplete information) a nine-fold table (m_{ij}) $(i,j \in \{0,X,1\}$ where X is the sign of missing value). Clearly, given $\sim$, the truth of $A \sim S$ in M with complete/incomplete information depends only on the corresponding 4-f.-t./9-f.-t.; thus we may speak on a valid 4-f.-t. (w.r.t. $\sim$) and on a valid 9-f.-t. (w.r.t. $\sim$ and the chosen semantics SEM = sec, del or opt). Furthermore, if $A \sim S$ has disjoint components then there is an obvious correspondence between completions of M and four-fold reducts of the 9-f.-t. of $A \sim S$. Define the critical symmetric 4-f. reduct of a 9-f.-t. T9 w.r.t sec to be T4 = = (a,b,c,d) which is a 4-f. reduct of T9 with minimal a, d, maximal b + c and minimal $|b - c|$. Thus e.g. if

$$T9 = \begin{pmatrix} 3 & 2 & 4 \\ 1 & 2 & 1 \\ 6 & 2 & 5 \end{pmatrix} \qquad T4 = \begin{pmatrix} 3 & 9 \\ 9 & 5 \end{pmatrix} \qquad T4' = \begin{pmatrix} 7 & 4 \\ 6 & 7 \end{pmatrix}$$ then T4 is the critical symmetric 4-f. reduct of T9 w.r.t. sec and T4' w.r.t opt.

Theorem 2 (Havránek, Rauch). If $\sim$ is one of SIMPLE, FISHER and

CHISQ, $A \sim S$ has disjoint components, T9 is the 9-f.-t. of $A \sim S$ in M and SEM = sec or opt then $M \models A \sim S$ (SEM) iff the critical symmetric 4-f. reduct of T9 (w.r.t. SEM) is valid (w.r.t. $\sim$).

Now consider relativized sentences $(A \sim S)/C$ and observe that each of A, S, C may have missing information . Let $M \restriction (C \neq 0)$ denote the matrix M with all rows omitted for that the value of C is 0 and let $M \restriction (C = 1)$ denote the matrix resulting from M by omitting all rows for that the value of C is not 1 (i.e. is 0 or missing). Let (rij) denote the 9-f.t. of (A,S) on $M \restriction (C \neq 0)$ and (sij) the 9-f.-t. of (A,S) on $M \restriction (C = 1)$. $(i,j \in \{0,X,1\})$.

The <u>critical</u> 9-f.-t. of $(A \sim S)/C$ on M w.r.t. sec results from (rij) by replacing r11 and r00 by s11 and s00. Similarly for opt: replace r10 and r01 by s10 and s01.

<u>Theorem 3.</u> If $\sim$ is associational, $(A \sim S)/C$ has disjoint components and SEM = sec or opt then $M \models (A \sim S)/C$ (SEM) iff the critical 9-f.-t. of $(A \sim S)/C$ on M is valid.

Clearly, the above theorems make the evaluation of $M \models (A \sim S)/C$ computationally feasible. Note that Theorem 2 has an analogon for all implicational quantifiers that is easy to prove.

§ 5. Hopeful cedents.

Generation of antecedents (succedents for a given antecedent) is search in a ceratin tree of formulas. To make this search effective we use the notion of hopeful/hopeless cedents introduced by Rauch. In this section, $\sim$, SEM, M and the condition C are fixed.

A is a <u>hopeless antecedent</u> if there is no A' A and no S such that M (A' S)/C (SEM). Given A, S is a <u>hopeless succedent</u> for A if there is no S' S such that M (A S')/C (SEM).

We need simple sufficient conditions for hopelessness. First we present conditions valid for arbitrary founded quantifiers; then some more powerful conditions for founded associational quantifiers. Let FRQ1(A) be the number of all objects having value 1 for A; similarly for FRQ1X (values 1 or X).

<u>Theorem 4</u>. Let $\sim$ be founded with base BASE. (1) If SEM = sec or del and FRQ1(A & C) < BASE then A is a hopeless antecedent. (2) If SEM = opt and FRQ1(A & C) = 0 or FRQ1X(A & C) < BASE then A is a hopeless antecedent. (3) If SEM = sec or del and FRQ1(A & S & C) < < BASE then S is a hopeless succedent for A. (4) If SEM = opt and FRQ1(A & S & C) = 0 or FRQ1X(A & S & C) < BASE then S is a hopeless succedent for A.

Let $\sim$ be founded with base BASE. Let BA(m) be maximal B such that BASE $\leq$ B $\leq$ m + 1 and for each T4 = (a,b,c,d) such that a + b + + c + d = m and a < B we have $\sim(T) = 0$ (treshold for antecedents). BSU (r,m) is defined for r = 1,...,m as the maximal B such that BASE $\leq$ B < m + 1 and for each T4 = (a,b,c,d) as above which, in addition, satisfies a + b = r, we have $\sim(T) = 0$ (treshold for succedents with given frequency of antecedent).

<u>Theorem 5</u>. Let M, SEM, C and $\sim$ be given, let $\sim$ be founded and associational. Each of the following conditions is sufficient for A to be a hopeless antecedent:

(1) SEM = sec and FRQ1(A & C) < BA(FRQ1X(C))
(2) SEM = del and FRQ1(A & C) < BA(FRQ1(C))
(3) SEM = opt and FRQ1(A & C) = 0 or FRQ1X(A & C) < BA(FRQ1X(C))

<u>Theorem 6</u>. Under the same assumptions, either of the following conditions is sufficient for S to be a hopeless succedent for A:

(1) SEM = sec and FRQ1(A & S & C) < BSU(FRQ1(A & C),FRQ1X(C))
(3) SEM = opt and FRQ1(A & S & C) = 0 or
FRQ1X(A & S & C) < BSU(FRQ1(A & C),FRQ1X(C))

<u>Caution</u>: There is no point (2); for SEM = del we can say in general nothing better than Theorem 4(3). But there is a non-trivial condition for SEM = del if the quantifier is implicational.

Observe that in a run the second argument of BSU in (1), (3) above is fixed, thus one can work with a one-dimensional table precomputed for the given run. Moreover, in a given run, one can explicitly work with M↾(C ≠ 0) instead of M if SEM is sec or opt and with M↾(C = 1) if SEM = del. This gives further computational sim plifications in the above theorems.

§ 6. <u>Improving literals</u>.

Literals are atomic properties and negated atomic properties, i.e. conjuncts in elementary conjunctions. "A literal L improves the antecedent (succedent) of a sentences A S" is a notion satisfying the following <u>rule of improving literals</u>:

If M ⊨ A ∼ S (SEM) and L1, L2, ..., Lk are some literals improving the antecedent of A ∼ S (in M w.r.t. SEM) then M ⊨ A & L1 & L2 ... & Lk ∼ S (SEM).

Thus knowing that A ∼ S is true and that L1, ..., Lk improve the antecedent, we know that $2^k - 1$ other sentences are true. Similarly for the succedent. This is used for reduction in representation of results: (if desired) only <u>prime</u> sentences are printed, i.e. true and not obtainable from simpler true ones through the rule of improving literals. (Each prime sentence is printed together with the list of improving literals.)

There are two notions of improvement: <u>conservative</u> and <u>strict</u>. If K is any elementary conjunction and L is a literal not occuring in K then L conservatively improves K if K & L is three-valuedly equivalent to K, i.e. for each object, the value of K is the same as the value of K & L. Evidently, if L conservatively improves A then the 9-f.-t. of A ∼ S is the same as that of A & L ∼ S. We present a notion of strict improvement by a table giving conditions for L to improve the antecedent of A ∼ S and R to improve its succedent. Each row says: for each object, if the pair of values of A and of S is ... then the value of L (R) must be ....

<table>
<tr><th colspan="3">SEM = sec or del</th><th colspan="3">SEM = opt</th></tr>
<tr><th>A, S</th><th>L</th><th>R</th><th>A, S</th><th>L</th><th>R</th></tr>
<tr><td>11</td><td>1</td><td>1</td><td>11</td><td rowspan="4">1 or X</td><td rowspan="4">1 or X</td></tr>
<tr><td>1X</td><td>1</td><td>1 or X</td><td>1X</td></tr>
<tr><td>X1</td><td>1 or X</td><td>1</td><td>X1</td></tr>
<tr><td></td><td></td><td></td><td>XX</td></tr>
</table>

Theorem 7. If ∼ is associational then strict improvement obeys the rule of improving literals.

Moreover, one can show that if L strictly improves the antecedent of A ∼ S(w.r.t. SEM) and T9, T9' are 9-f.-t.'s of A ∼ S and of A & L ∼ S respectively then T9' is better than T9 in a natural sense (depending on SEM).

<u>Remark</u>. It is impossible that both L and -L (negated L) improve the antecedent of A ∼ S. (Here it is important that ⊨(opt) was defined as it was.)

The case of restricted sentences is solved by the following

Theorem 8. If $\sim$ is associational, if L1, ..., Lk improve strictly the antecedent of $A \sim S$ w.r.t. SEM and if $M \models (A \sim S)/C$ (SEM) then $M \models ((A \& L1 \& \ldots \& Lk) \sim S)/C$ (SEM).

Everything holds analogously for succedent.

The above results are substantial for optimization of computer time needed for a run of ASSOC as well as of the representation of results. They also appear to contribute to the theory (theories) of the processing of missing data items in statistical software.

References.

P. Hájek 1973-4: Automatic listing of important observational statements I-III, Kybernetika 9, 187-205, 251-271 and 10, 95-124

P. Hájek and T. Havránek 1977: On generation of inductive hypotheses, Int. J. Man-Machine Studies 9, 415-438

P. Hájek and T. Havránek 1978a: Mechanizing Hypothesis Formation (Mathematical Foundations for a General Theory), Springer-Verlag Berlin-Heidelberg-New York

P. Hájek and T. Havránek 1978b: The GUHA method - its aims and techniques, Int. J. Man-Machine Studies 10, 3-22

P. Hájek, T. Havránek and M. K. Chytil: Metoda GUHA - automatické generování hypotéz. Academia Prague

T. Havránek 1980: An alternative approach to missing information in the GUHA method, Kybernetika 16, 145-155

J. Rauch 1978: Some remarks on computer realization of the GUHA method, Int. J. Man-Machine Studies 10, 75-86

Methods for Summarizing the Analysis of Categorical Data

P. Lane, Harpenden

Summary

Most models fitted to categorical data involve analysis on a transformed scale. As a result, model parameters are not usually the best form of summary of the effects included in a model. A more informative type of summary is a table of predictions, formed from the parameters, in which the effects of one or two classifications are shown, adjusted for all other effects in the model. The formation of predictions is illustrated here by summarizing the results of two analyses, one using a logit model for proportions, the other a log-linear model for counts. Problems in calculating predictions are discussed with reference to the Genstat system, in which these methods are available.

Keywords

Categorical data; Summary of effects; Prediction; Generalized linear models.

1. Models for categorical data

Categorical data are the result of counting people or things classified by their attributes. Such data are usually presented as a contingency table that shows the numbers of individuals associated with each combination of categories of the classifying attributes or variables. Contingency tables, or marginal summaries of them, can be useful in themselves to summarize the results of some investigation. However, there are often patterns of association between the classifying variables; these can be measured and assessed by calculating statistics of association, or by fitting models to describe the patterns.

A wide range of models for categorical data have been proposed in many applications (Haberman, 1974; Finney, 1971). The type of model that is relevant for a particular contingency table depends on what relationships are to be assessed. One distinction is between models in which all classifying variables have the same status, and models in which one variable is considered as a response variable related to others. In the former, log-linear models are often used to model the counts in terms of the classifying variables. In the latter, it is the proportion of individuals at each category or level of the response variable that is usually modelled; in particular, when a response variable has only two possible levels, a logit or probit model can be used to describe the proportion at one of the levels.

A second criterion affecting the choice of model is the type of variables which classify the table. Models for relationships between variables with purely nominal levels are often restricted to 'independence' models, in which the association between the divisions into categories by each variable is investigated. When the levels of a variable are ordinal, account can be taken of the order of the levels, as in the continuation-ratio model for a multiple response variable. When the levels are cardinal, models may involve polynomial or other quantitative contrasts of the numbers. For example, probit analysis usually involves a linear effect of some explanatory variable such as dose.

Many of the models applied to categorical data are in a class known as generalized linear models (McCullagh & Nelder, 1983). The best-known examples are probably the log-linear model and the model used in probit analysis. Generalized linear models have the advantage of being cast in a form analogous to classical linear models; they also have a common algorithm for the estimation of parameters by maximum likelihood. The models describe dependence of a response variable on explanatory variables; they are characterized by the assumption that the expected value of the response variable is related by a monotonic differentiable function to a linear combination of the explanatory variables. Usually, it is further assumed that observations of the response variable are independent and have a distribution from the exponential family. Log-linear models for counts fit into this framework by specifying as the response variable the variable of counts itself.

2. Summarizing the analysis of generalized linear models

The process of fitting models to data involves several stages, often interrelated:

(1) selection of a suitable model,
(2) estimation of parameters,
(3) checking the adequacy and goodness of fit, and
(4) summary and presentation of the results.

I shall be concerned here only with the final stage, assuming that a suitable model for a contingency table has been found and that estimates of parameters and their precision are available.

The final stage is often neglected in standard texts and articles about statistical method. Clearly, much depends on the nature of the audience for which the analysis is intended, and what is regarded as an appropriate summary is an ad-hoc and subjective decision, but a number of general methods are available. The choice of effects to include in a model can be described by an analysis of deviance - a generalization of the analysis of variance for classical linear models. The fit of the model can be displayed by graphical methods and by standard statistics measuring dispersion or deviation from the model. Both of these methods are also used in earlier stages of the model fitting process, and will not be described further here. Finally, the effects which are included in the model can be summarized by their corresponding parameters and standard errors, or by tables of fitted values derived from the parameters.

The main problem with describing the effects is to find suitable summaries which are at the same time concise and meaningful. The most concise form is the list of parameter estimates themselves, but these are often expressed on some transformed scale, and may not be readily interpretable on the natural scale of the response variable. Indeed the parameters may well have been specified for convenience of parameter estimation rather than of summary of the results. The fullest representation of the effects is the table of fitted values, classified by all the variables in the model; but this table is no simplification of the original data, and does not help to describe the patterns in the data which have been modelled. What is often needed are marginal summaries which describe one or two effects at a time while making allowance for the other effects which have been estimated.

Summaries of individual effects in the model may be formed from the full set of fitted values by standardization with respect to the other effects in the model. Various types of standardization may be employed, corresponding to different requirements of the presentation.

(1) Marginal standardization: the fitted values for each level of a classification which is to be summarized over are averaged, weighting with the numbers of individuals at each level in the whole set of data. Thus each level gives the same proportional contribution at all levels of the variables whose effects are to be described, even though there may have been very different proportions in the data.

(2) Equal-weights standardization: the fitted values for each level of a classification are averaged with equal weights. This method is relevant if it is considered that disproportionate numbers for various levels of the classification have arisen by chance, and that one level should not be given more weight than another just because it happens to occur more often than the other in the data.

(3) Population-weights standardization: the fitted values for each level of a classification are averaged with known weights corresponding to frequencies in some population. This method expresses the results of fitting a model in the context of whatever population is under study.

The values derived from these methods of standardization are called predictions (Lane & Nelder, 1982). They predict what the unweighted marginal fitted values would have been if the numbers of individuals in each subclassification of the data had been exactly in proportion according to the frequencies corresponding to the type of standardization employed.

3. Analysis of proportions

To illustrate the methods described above, I shall first look at an analysis of proportions. Yates (1961) gave a summary of the incidence of milk fever in cows, derived from a survey of diseases of dairy cattle. The numbers of occurrences and non-occurrences of fever associated with calving are given below, classified by season and by the lactation number of the calving.

	Lactation									
Season	1-2		3		4		5		6+	
Jan-Apr	16	3790	31	2106	55	1689	66	1118	122	1888
May-July	4	2348	31	975	52	654	46	442	92	729
Aug-Sept	21	3148	49	944	58	559	40	326	73	649
Oct-Dec	18	5099	51	1690	76	1176	76	715	127	915

The incidence of fever is higher when cows have had several lactations, and higher in summer than winter. The size of these effects can be estimated by relating the proportion of occurrences to the levels of the classifications. I shall use a generalized linear model assuming that the numbers are binomially distributed and that the effects of the classifications are additive on the logit scale. It is clear from the data that the effects cannot be additive on the natural scale: they appear to be approximately multiplicative, or additive on the log scale. In fact, there is little difference between log and logit transformations for the range of proportions in these data (0.002 to 0.14), but the log is never suitable for the analysis of proportions which range both above and below 0.5.

A simple 'independence' model for the effect of the two classifications will serve to illustrate the presentation of results; I shall not attempt to take account of the ordinal nature of the lactation classification nor of the cyclic nature of the season classification. Using the program Genstat (Alvey et al. 1977) to fit the model, the parameter estimates are as follows (with approximate standard errors):

		Constant	-6.10 (0.14)			
Lactation 1-2	0.00	(----)		Season Jan-Apr	0.000	(-----)
3	2.01	(0.15)		May-July	0.639	(0.092)
4	2.78	(0.15)		Aug-Sept	0.985	(0.091)
5	3.19	(0.15)		Oct-Dec	0.649	(0.082)
6	3.39	(0.14)				

(Note: Genstat parameterizes a classification by differences with the first level.)

The residual deviance from this model (deviance = -2×log(likelihood ratio)) is 17.6 with 12 degrees of freedom. The deviance is an approximate χ^2 statistic for goodness of fit, so there is no evidence here of an interaction between the effects on the logit scale, nor of any heterogeneity in the distributions of the counts.

The parameters of this logit model are not easy to interpret. For example, the parameter 'Lactation 3' has the following interpretation: "the log odds ratio of a cow in its third lactation having milk fever is 2.01 higher than a cow in its first or second lactation." By transforming to remove the logarithms, the odds for Lactation 3 can be described as 7.5 times the odds for Lactations 1 and 2. (The odds ratio is the expected probability of occurrence divided by the complementary probability: E[prob] / (1 - E[prob]).)

Simpler summaries of the effect are the predictions for lactation adjusted for the effect of season. Weighting by the observed proportion of calvings in each season (10881, 5393, 5677, 9942) the predicted percentages of disease incidence are:

1-2	3	4	5	6+
0.39 (0.05)	2.79 (0.22)	5.85 (0.36)	8.55 (0.54)	10.18 (0.48)

This table effectively summarizes the change in disease incidence with lactation number in this survey.

Alternatively, it may be preferable to adjust for some other proportion of calvings in each season. To get predictions relevant for assessing the chance of disease for cows with a given number of previous lactations, but when any season is thought equally likely for calving, we can give the seasons equal weights:

1-2	3	4	5	6+
0.42 (0.05)	3.01 (0.23)	6.28 (0.40)	9.17 (0.58)	10.91 (0.52)

These are higher than the previous predictions because in the survey there were fewer calvings recorded in the summer, when disease was more prevalent.

4. Analysis of counts

To illustrate further the use of predictions I shall consider an analysis in which the counts are not naturally represented in the form of proportions. McCullagh & Nelder (1983) gave a summary of damage to certain types of cargo-carrying ships. The numbers of damage incidents are given below with the aggregate numbers of months of service, classified by the ship type, year of building and period of operation.

		Type									
Operated	Built	A		B		C		D		E	
1960-74	1960-64	0	127	39	44882	1	1179	0	251	0	45
	1965-69	3	1095	58	28609	0	781	0	288	7	789
	1970-74	6	1512	12	7064	6	783	2	349	5	1157
1975-79	1960-64	0	63	29	17176	1	552	0	105	0	0
	1965-69	4	1095	53	20370	1	676	0	192	7	437
	1970-74	18	3353	44	13099	2	1948	11	1208	12	2161
	1975-79	11	2244	18	7117	1	274	4	2051	1	542

McCullagh and Nelder show that a log-linear model can describe these counts well, including main effects of each classification and an offset term to take account of the length of service. (An offset variable is one which is related to the response variable in a predetermined way, i.e. without a parameter.) This model can be written as follows:

$$E(\text{number of accidents}_{ijk}) = \text{Exp}(\text{constant}+\text{Log}(\text{service})+\text{type}_i+\text{built}_j+\text{operated}_k)$$

The parameter estimates for this model are given below.

Constant -6.41 (0.27)

Type		Built		Operated	
A	0.00 (----)	1960-64	0.00 (----)	1960-74	0.00 (----)
B	-0.54 (0.22)	1965-69	0.70 (0.19)	1975-79	0.38 (0.15)
C	-0.69 (0.41)	1970-74	0.82 (0.21)		
D	-0.08 (0.36)	1975-79	0.45 (0.29)		
E	0.33 (0.29)				

These parameters may be interpreted by transforming them with the exponential function to the natural scale. For example, the effect of ship type may be summarized directly in terms of the parameters as follows:

Type	A	B	C	D	E
% change in accidents compared to Type A	0 (--)	-42 (13)	-50 (21)	-8 (33)	39 (40)

Such a summary may be sufficient, but relative effects are not as evocative as absolute values. By summarizing the effect in terms of the numbers of accidents, the results are immediately interpretable. Predictions for ship type are given below, adjusting with equal weights for the effects of year of building and of operation, and using 1000 months service as a basis for comparison.

Type	A	B	C	D	E
Predicted accidents per 1000 months	3.7 (0.7)	2.1 (0.2)	1.8 (0.7)	3.4 (1.1)	5.1 (1.2)

5. Computational problems

The computation of predictions is usually straightforward once the type of standardization has been specified. Predictions may now be formed after the analysis of any generalized linear model by the Genstat system (Alvey et al., 1977). In the simplest, default mode the program carries out marginal standardization with respect to classifying variables and will adjust for quantitative explanatory variables using their mean values. A range of options are available to request equal weighting or population weighting, or to specify particular values of explanatory variables at which to form the predictions.

Two problems remain to be solved in the current implementation of Genstat. Firstly, the calculation of standard errors for the predictions may involve a very large intermediate covariance matrix. For example, if there are three variables with 3, 6 and 10 levels respectively the matrix will have 180 rows and columns and its lower triangle will require over 16000 storage locations. It seems unlikely that a general solution to this problem will be found, because of the non-linearity introduced by the link function. With a linear function, as in classical regression models, the full matrix does not have to be formed and the problem can be avoided.

The second problem occurs when some parameters in a model are not estimatable. In the example of the previous section, ships cannot be operated in 1960-74 and built in 1975-79. Thus, a model cannot be fitted if it includes a parameter for this combination, for example including the interaction between building date and operation date: Genstat would report that a parameter cannot be estimated and proceed to fit a model excluding the parameter. A naive attempt to form predictions from this model would result in meaningless averages. Such attempts are inhibited in Genstat at present, but it is hoped to provide in a future version of the system more comprehensive checks to enable such models to be summarized adequately.

References

Alvey N.G. et al. (1977) Genstat: a general statistical program. NAG, Oxford.
Finney D.J. (1971) Probit analysis. University Press, Cambridge.
Haberman S.J. (1974) The analysis of frequency data. University Press, Chicago.
Lane P.W. & Nelder J.A. (1982) Analysis of covariance and standardization as instances of prediction. Biometrics **38**, 613-621.
McCullagh P. & Nelder J.A. (1983) Generalized linear models. Chapman & Hall, London.
Yates F. (1961) Marginal percentages in multiway tables of quantal data with disproportionate frequencies. Biometrics **17**, 1-32.

Computational and Statistical Methods for Exploratory Analysis of Textual Data

A. Morineau, Paris

Abstract : Methods for analysing textual data (responses to open-ended questions, interview transcripts, etc.) are described, involving mainly Correspondence Analysis. A suitable software makes the methods utilizable on a computer.

Keywords : Textual and Qualitative Data, Correspondence Analysis, Data Analysis Software.

WHAT KIND OF TEXTUAL DATE ?

1. Responses to open-ended questions

This is the main domain of application of the methods presented here. People are asked to answer freely a question : "What do you think of ...?" The text of the responses will be submitted to analysis, without any preliminary coding. The emphasis is put on the connection between textual and quantitative or qualitative data in the questionnaire.

2. Analysis of more literary texts

The same automatic processing can be used for any text in natural language : marketing announcements for competitive products, summaries of scientific journals, poems, and so on. To illustrate our purpose let us consider 13 articles from english economic newspapers, with about 1000 words in each article (extract from a study by MOULD, Paris).

WHAT KIND OF PROBLEMS ?

1. To summarize complex information

Statistical criteria can select the most typical words in each text (according to their frequencies), and the most racteristic sentences in order to insert words in their context. Figure 1 illustrates the choice of words in the two first articles of the example. This output will be discussed below.

2. To visualize associations between words

Figure 2 is a synthetic display of relationships between the hundred words most frequently encountered in the 13 articles. Such graphical display will be the main output of the method (the rules to read it are connected with Correspondence Analysis, which is assumed to be known).

Figure 1

Selection of the 12 most typical words

in the first two articles of the example

SELECTION DES FORMES LEXICALES CARACTERISTIQUES

	LIBELLE DE LA FORME LEXICALE	CRITERE DE CLASSEMENT	FREQUENCE GLOBALE	FREQUENCE INTERNE	POURCENTAGE GLOBAL	POURCENTAGE INTERNE
TEXTE NUMERO 1	*01* - HOW MANY MIDDLE MEN					
1	DOES	4.908	12.	6.	0.18	0.98
2	THOSE	4.581	17.	7.	0.25	1.15
3	DATA	4.176	11.	5.	0.16	0.82
4	WHAT	4.011	20.	7.	0.30	1.15
5	MAKING	3.906	12.	5.	0.18	0.82
6	PEOPLE	3.906	12.	5.	0.18	0.82
7	ONE	3.578	39.	10.	0.58	1.64
8	BETWEEN	2.905	12.	4.	0.18	0.65
9	DO	2.905	12.	4.	0.18	0.65
10	WE	2.896	17.	5.	0.25	0.82
11	MORE	2.796	55.	11.	0.82	1.80
12	NOT	2.686	71.	13.	1.06	2.13
TEXTE NUMERO 2	*02* - PERFECT LAUNCH OF AR					
1	SPACE	12.614	17.	15.	0.25	2.98
2	ARIANE	11.621	11.	11.	0.16	2.19
3	SATELLITES	10.476	11.	10.	0.16	1.99
4	LAUNCH	9.193	19.	12.	0.28	2.39
5	EUROPEAN	8.049	20.	11.	0.30	2.19
6	SATELLITE	4.657	20.	7.	0.30	1.39
7	FOR	3.482	164.	24.	2.45	4.77
8	WAS	3.443	56.	11.	0.84	2.19
9	EUROPE	3.386	12.	4.	0.18	0.80
10	WERE	2.827	21.	5.	0.31	0.99
11	PUT	2.808	15.	4.	0.22	0.80
12	YESTERDAY	2.480	11.	3.	0.16	0.60

3. To connect textual and numerical data

When textual data are responses to open-ended questions, it is of importance to know how to connect the vocabulary and the respondent characteristics : sex, age, number of children, etc.. With literary information, we may have at our disposal many numerical and nominal variables characterizing each text.

WHAT KIND OF PROCESSING ?

1. Data handling and elementary statistics

The input is the text in natural language. All distinct words are recognized and ranked according to their frequencies (specific parameters allow us to eliminate words with less than k letters, or words occuring less than p times). It is not necessary to know beforehand the number of texts, the number of sentences, nor the number of words.

This preliminary step produces also elementary statistics as shown in Figure 3 relating to the previous example. Figure 4 gives a truncated list of the frequencies (for words with more than 2 letters, and appearing at least 11 times).

Figure 2

Example of graphical display by Correspondence Analysis

(axes 1 and 2). Articles appear as numbers

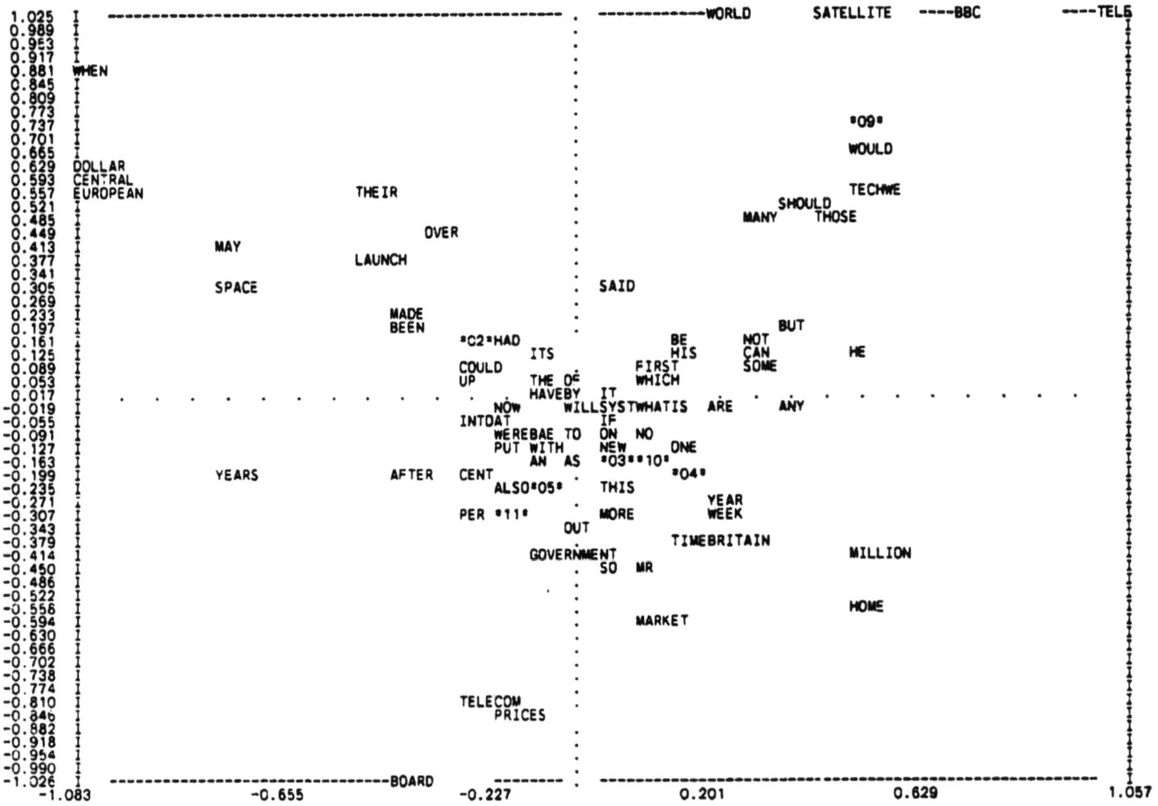

2. Typical words of a text

Let m_{ik} be the frequency of the word i within the text k. Let m_i be the global frequency of the word i, m_k the number of words in text k, and m the total number of words. The word i characterizes the text k if its frequency within this text is significantly high. Under the null hypothesis H_0 of random (without replacement) drawing of the m_k words among the m words, the frequency X_{ik} of word i in text k has a hypergeometric distribution with parameters m_i, m_k and m :

$$[1] \quad p_k(i) = \mathrm{Prob}_{H_0}(X_{ik} \geq m_{ik}) = \sum_{x \geq m_{ik}} \binom{m_k}{x}\binom{m - m_k}{m_i - x} \Big/ \binom{m}{m_i}$$

The higher the absolute frequency m_{ik} is, the weaker the probability $p_k(i)$ is, and the more typical the word is. A normal approximation is used to sort the words ; an example of result is provided by Figure 1.

Figure 3

Elementary statistics related to the 13 articles.

Frequencies and percentages concerning all the words,

and all the distinct words (minimal frequency : 10 ;

minimal lenght : 2 letters)

EFFECTIFS SELON LE NOMBRE DE LETTRES

NOMBRES DE LETTRES	1	2	3	4	5	6	7	8	9	10	11	12	13	14	15	16
EFFECTIFS OBSERVES	0	81	169	353	404	442	438	381	307	206	143	77	44	22	8	3
EFF.MAXIMAUX PREVUS	40	110	220	430	550	550	600	530	450	300	210	110	75	60	50	25

BILAN DU TRAITEMENT

NOMBRE TOTAL DE REPONSES = 531

NOMBRE TOTAL DE MOTS = 12599
NOMBRE DE MOTS DISTINCTS = 3078
POURCENT. MOTS DISTINCTS = 24.4

SELECTION DES MOTS

SEUIL DE FREQUENCE = 10

TOTAL DES MOTS RETENUS = 6669
MOTS DISTINCTS RETENUS = 149

REPARTITION DES MOTS DANS LES TEXTES

NUMERO DU TEXTE	IDENTIFICATEUR	NOMBRE DE MOTS	/1000 DU TOTAL	MOYENNE PAR REPONSE	NOMBRE DE MOTS DISTINCTS	/100 MOTS DU TEXTE	MOYENNE PAR REPONSE	NOMBRE DE MOTS RETENUS
1	*01* HOW MANY MIDDLE MEN	1106	87.8	22.1	476	43.0	9.5	611
2	*02* PERFECT LAUNCH OF AR	931	73.9	21.2	423	45.4	9.6	503
3	*03* EMPLOYMENT . TO INTE	929	73.7	24.4	431	46.4	11.3	443
4	*04* WIZARDRY FROM THE AR	1221	96.9	19.4	502	41.1	8.0	699
5	*05* TELECOM ISSUE FACES	1115	88.5	27.9	514	46.1	12.8	592
6	*06* AS NATO FALLS UNDER	1015	80.6	23.6	510	50.2	11.9	472
7	*07* STATE FUNDING FOR AI	1088	86.4	29.4	442	40.6	11.9	612
8	*08* RETAILERS AUTOMATE S	939	74.5	24.7	468	49.8	12.3	474
9	*09* CASH CLOUDS THE BBC'	1074	85.2	21.1	455	42.4	8.9	631
10	*10* FARM MINISTERS AVOID	590	46.8	22.7	290	49.2	11.2	311
11	*11* BULKING UP THE POTAT	1097	87.1	26.8	479	43.7	11.7	592
12	*12* THE LONG-TERM DOLLAR	679	53.9	30.9	344	50.7	15.6	354
13	*13* CUTS DILEMMA FOR THE	815	64.7	21.4	433	53.7	11.5	375
	G L O B A L	12599	1000.0	23.7			5.8	6669

3. Typical sentences of a text

The typical sentences in a text will contain as many typical words as possible. No universal criterion exists since there is competition between the number of words and their ability to discriminate between the texts.

(a) We may characterize each sentence by the mean value of the criteria of its words, and sort the sentences according to these mean values.

(b) We may as well search for the sentences whose lexical profiles are nearest to the average profile of the text.

Using the above notations, the lexical profile of the text k reads : (m_{ik}/m_k) for $i = 1,2,\ldots,m$. If m_{ij} designates the frequency of word i in sentence j, containing m_j words, its lexical profile will similarly be : (m_{ij}/m_j) for $i = 1,2,\ldots,m$. The CHI-2 distance between these profiles is written :

$$[2] \qquad d^2(j,k) = \sum_{i=1}^{m} \frac{1}{m_{ik}/m_k} \left\{ \frac{m_{ij}}{m_j} - \frac{m_{ik}}{m_k} \right\}^2$$

Figure 4

Truncated list of distinct words

(ranked according to global frequency)

TABLEAU DES PROFILS COLONNES (TOTAL/COLONNE=100)

FREQUENCES	POURCENTAGES	LIBELLES	1	2	3	4	5	6	7	8	9	10	11	12	13
948	14.22	THE	11.3	15.9	10.6	10.6	16.4	12.8	16.0	15.0	14.7	11.9	17.4	19.2	12.3
420	6.30	TO	5.9	5.2	7.9	3.9	7.9	9.1	5.6	4.4	4.1	11.6	7.4	6.2	6.1
379	5.68	OF	7.2	6.0	5.0	4.3	6.4	7.0	5.2	6.1	6.0	1.9	5.9	6.2	5.3
294	4.41	AND	5.9	3.6	4.5	5.4	3.2	5.9	3.8	4.6	3.8	2.3	2.4	3.7	8.5
268	4.02	IN	3.4	4.8	4.7	5.3	3.5	3.4	5.2	5.9	3.6	0.6	4.1	2.8	2.4
193	2.89	IS	4.6	1.2	0.7	3.7	2.9	6.1	2.3	4.0	3.3	3.2	2.0	1.4	0.8
164	2.46	FOR	1.6	4.8	2.3	2.4	2.7	3.0	3.9	2.5	1.3	2.3	1.2	1.1	2.9
154	2.31	BE	2.3	1.6	1.6	1.3	2.2	3.4	2.5	1.5	4.3	2.6	3.4	1.7	1.1
147	2.20	THAT	2.6	1.2	3.4	1.3	1.9	2.1	2.5	4.2	1.9	2.9	2.9	1.4	0.5
136	2.04	IT	2.0	2.2	0.7	3.3	1.2	2.8	2.1	1.3	2.2	2.6	2.2	2.0	1.6
110	1.65	ON	0.8	1.6	3.6	1.4	1.5	2.1	1.1	3.4	1.3	1.9	0.8	1.1	1.6
94	1.41	HAS	0.8	0.6	0.7	2.1	2.0	0.6	2.0	2.5	0.6	0.6	1.4	1.7	2.1
76	1.14	ARE	2.1	0.2	0.9	1.1	0.8	1.1	0.5	1.7	1.4	1.6	0.8	0.3	2.4
74	1.11	BY	0.3	0.4	1.6	0.4	1.2	2.1	1.6	1.1	1.3	0.3	1.9	1.1	1.1
73	1.09	AS	2.0	1.0	1.1	1.0	2.5	1.3	0.8	0.4	0.5	1.0	0.5	0.8	1.1
72	1.08	WILL	0.3	1.2	0.2	0.4	2.2	1.1	0.5	1.3	1.6	1.0	2.7	0.8	0.3
71	1.06	NOT	2.1	0.2	0.9	1.4	0.8	2.1	1.1	0.2	1.9	0.3	0.5	0.6	0.5
69	1.03	ITS	0.5	0.6	0.5	0.7	2.0	0.8	2.3	0.0	1.3	0.3	0.5	1.4	2.4
69	1.03	HAVE	1.5	1.2	0.0	1.4	1.4	1.3	0.5	0.2	0.6	0.6	0.7	1.4	2.9
69	1.03	WITH	0.7	1.4	1.6	0.9	0.7	1.1	0.8	2.1	0.5	0.3	1.0	0.6	2.4
68	1.02	AT	1.3	1.0	1.8	0.9	0.7	0.4	0.7	1.7	0.5	0.0	1.0	1.4	2.4
67	1.00	WHICH	1.5	1.0	0.7	0.7	0.6	1.7	0.2	0.6	1.3	2.6	1.2	0.8	0.5
60	0.90	THIS	0.7	0.6	2.0	1.9	1.2	0.2	0.8	0.6	0.3	1.9	0.2	0.8	0.8
56	0.84	WAS	0.5	2.2	2.7	0.9	0.5	0.0	0.2	0.4	0.3	1.6	0.8	0.6	1.1
55	0.82	MORE	1.8	0.6	1.6	1.1	1.2	0.2	0.2	0.6	0.2	1.6	0.7	0.6	0.3
55	0.82	WOULD	0.3	0.4	1.1	0.3	0.0	1.1	1.8	0.0	3.0	1.0	0.5	0.3	0.5
54	0.81	FROM	0.2	1.0	0.9	0.7	0.2	0.6	1.6	0.8	0.8	0.3	1.0	0.6	1.9
52	0.78	BUT	1.0	0.4	1.6	0.9	0.5	0.2	0.7	0.2	1.7	1.3	0.7	0.3	0.5
49	0.73	GOVERNMENT	0.3	0.2	0.7	0.0	1.7	0.0	2.1	0.0	0.3	0.0	2.4	0.0	1.1
45	0.67	BEEN	0.6	0.4	0.2	0.3	0.8	0.6	1.6	0.4	0.5	0.0	0.3	1.7	1.1
43	0.64	MARKET	0.5	0.4	1.4	2.1	1.2	0.0	0.0	0.8	0.0	0.0	1.0	0.0	0.2
39	0.58	ONE	1.6	0.4	0.7	1.3	0.2	0.4	0.0	1.3	0.3	0.0	0.0	0.3	0.8
38	0.57	LAST	0.3	1.0	0.7	1.6	1.2	0.0	0.3	0.0	0.3	0.3	0.2	0.8	0.3
38	0.57	AN	1.0	1.2	0.9	0.1	0.7	1.1	0.0	0.0	0.2	0.6	0.8	0.3	0.8
36	0.54	ABOUT	0.3	0.6	0.0	1.3	0.7	0.4	1.0	0.6	0.5	0.0	0.3	0.6	0.0
36	0.54	THERE	0.7	0.2	0.5	0.9	0.2	1.1	0.2	1.1	0.3	1.3	0.2	0.6	0.5
35	0.52	IF	1.0	0.0	0.3	0.3	0.2	0.6	0.7	0.8	0.5	1.0	0.6	0.6	0.0
34	0.51	YEAR	0.2	0.0	0.0	2.0	0.3	0.6	0.0	0.2	0.3	1.3	0.3	0.3	1.1
32	0.48	THEY	0.3	0.6	0.5	0.7	0.5	0.8	0.2	0.2	0.5	0.3	0.3	0.3	0.8
32	0.48	OR	1.0	0.0	0.0	0.6	0.3	1.1	0.5	0.2	0.3	1.0	0.3	0.6	0.5
31	0.46	UP	0.2	0.6	1.1	0.1	0.0	0.6	0.5	0.2	0.3	1.0	0.7	0.6	0.3
30	0.45	NOW	0.5	0.6	0.0	0.6	0.3	0.0	0.5	0.4	0.2	1.0	0.2	0.6	1.3
30	0.45	INTO	0.2	1.0	1.4	0.1	0.3	0.6	0.2	0.6	0.2	0.3	0.5	0.6	0.3

The characterization is improved by dividing this distance by the average distance to all the other texts.

4. Multivariate Analysis of Textual Data

The multivariate analysis leads to synthetical displays as shown in Figure 2. Various types of data arrays issued from the same textual data set can be processed:

a) The "words-texts" contingency table : the general entry (i,k) is the frequency m_{ik}. In our example this table cross-tabulates 149 lexical forms with the 13 articles.

b) The incidence matrix "words-sentences" : the entry cell (i,j) contains the frequency m_{ij}. This matrix, which cross-tabulates 149 lexical forms with 531 sentences, is a sparse matrix, since the average number of words in a sentence is about 10.

c) The compressed matrices of lexical profiles : responses of open-ended questions can be gathered into "texts" in as many ways as there are ways of partitioning respondents (sex, age, category of revenue, etc.). The matrix of all the profiles of these texts characterizes the groups' vocabulary.

The statistical method used to describe associations between words (inside the sentences or the texts) will mainly be Correspondence Analysis (Benzécri, 1973 and 1980 ; Lebart *et al.*, 1984). Justification essentially lies in the

"Distributional Equivalence Property" of the CHI-2 distance (see : formula [2]) : *If two words i and i' have the same distributional profiles through the texts (which is a definition of synonyms : these words have the same context), the distances between any pair of texts $d^2(k,k')$ remain unaltered when considering that the two words are identical.*

IMPLEMENTATION OF A TEXTUAL DATA ANALYSIS SOFTWARE

The software consists of a FORTRAN program of about 5000 lines. It is divided into 11 steps, controlled by key words and parameters in free format. This section contains the list of steps with a brief commentary on their functions.

1. AFLEX : Creation of the archive textual data file. There is no limitation on the number of texts and their lengths.
2. DOTEX : Preparatory step that organizes the textual responses to a battery of open-ended questions.
3. DISCO : Selection of the responses to a particular open-ended question in the questionnaire.
4. COLEX : Creation of the glossary of all the distinct lexical forms ; computation of frequencies and percentages. This step creates the contingency matrix cross-tabulating words and texts, then the incidence matric of the words in the sentences (coded in "reduced form").
5. TALEX : Creation of the table of lexical profiles corresponding to any partitions of the respondents.
6. ASPAR : Correspondence analysis of the sparse matrix cross-tabulating words and sentences.We use a special algorithm adapted to sparse matrices (Lebart, 1982).
7. APLUM : Correspondence analysis of the contingency tables created by COLEX and TALEX.
8. MOCAR : Selection of the most typical words of a text. Selection of the characteristic sentences based upon the mean criterion of the typical words.
9. RECAR : Selection of the characteristic sentences based upon the CHI-2 distance between the lexical profiles.
10. POLEX : This step permits to display the m words of the glossary as illustrative points onto any graphical configuration.
11. TALEX : This steps allows the projection of nominal variables as illustrative points onto a graphical display of words issued from textual analysis.

The program is part of a more general software called SPAD, which contains 41 steps for performing exploratory data analysis of large sets of nominal and numerical variables (Lebart *et al.*, 1982 and 1984 ; Morineau, 1982).

REFERENCES

BENZECRI J.P. (1973) : L'Analyse des Données : L'Analyse des Correspondances (tome 2) Dunod, Paris.

BENZECRI J.P. et Coll. (1980) : Linguistique et lexicologie, Pratique de l'Analyse des Données (tome 3). Dunod, Paris.

LEBART L. (1981) : Une procédure d'Analyse Lexicale Ecrite en Langage FORTRAN. Les Cahiers de l'Analyse des Données, vol. 6, n° 2, pp 229-243.

LEBART L. (1982) : Exploratory Analysis of Large Sparse Matrices with Application to Textual Data. COMPSTAT-1982, pp 67-76.

LEBART L., MORINEAU A. (1982,1984) : SPAD, Système Portable pour l'Analyse des Données (tome 1, 1982). Analyses des Données Textuelles (tome 3, 1984). CESIA, 82, rue de Sèvres, 75007 Paris.

LEBART L., MORINEAU A., WARWICK K.W.(1984) : Multivariate Descriptive Statistical Analysis, Correspondence Analysis and Related Techniques for Large Matrices. Wiley, New-York.

MORINEAU A. (1982) : Choice of Methods and Algorithms for Statistical Treatment of Large Arrays of Data. COMPSTAT-1982, pp 342-347.

Similarities and Differences Between Linear, Logistic and Log-Linear Models for Survival Analysis; some Practical Observations.

B.P. Murphy, Nedlands

We review briefly some modelling procedures currently available in the computer literature. The question of consistency of results is examined through some examples from the experience of our epidemiological colleagues.

KEYWORDS: Linear Discriminant Function, Logistic, Loglinear, Survival, Branch and Bound Searching.

1. Background.

The history of classical Survival modelling exhibits a microcosm of statistical and computational development. The problem is to predict the binary 'Event' (Survival/ Death) from a set of possible explanatory variables x1 x2 ... . With powerful computers, the extra gloss of searching for suitable models becomes a desiderata, so that the problem becomes that of finding not only θ but also X, and even f from a class or classes of models, of the form

$$s = f(X,\theta) + e$$

where s is the event variable and e is some error term.

Social interests have led to a proliferation of studies of the incidence and risks of heart disease death. The entire issue of the final Journal of Chronic Disease of 1979 was given over to the parallel analyses of 21 such studies being undertaken in various countries. Potential 'risk' factors were substantially agreed (e.g. blood pressure, smoking habits, family history, blood lipids concentrations, etc.) and linear and logistic linear models fitted by each participating research group. In our case we were also able to fit log-linear contingency table models, and this paper reports on these experiences.

2. Some Specific Models and Searching Techniques

The oldest model form still (alas) in practical use is the linear model

[i] $$s = \beta_0 + \beta_1 x_1 + \beta_2 x_2 + .. + \varepsilon$$

which in effect yields the discriminant function for 'best' separation of the two classes. The model is open to serious objections on many counts, but has the advantages of easy calculation and long familiarity. Further, it has long been amenable to stepwise and 'all-regression branch-and-bound' search procedures, as arose from the algorithms of Efromyson (1960) and Furnival (1971).

The logistic model has also been long available with approximate solutions, but proper calculation is a relatively recent demand by journals. Here

[2] $$\log (p/(1-p) = \beta_0 + \beta_1 x_1 + \beta_2 x_2 + \quad ..$$

where p is the probability of survival. The method was widely promoted in the medical literature as a result of the work of Truett et al (1967) on the Framingham heart disease data. Full maximum likelihood solution is not trivial computationally, and this impeded its acceptance for some while. The program GLIM (Baker and Nelder, 1978) does however yield accurate solutions; indeed this program alone accounts for much of the interest in the technique.

There is no search algorithm for logistic models as efficient as the branch and bound method of linear modelling, and little work seems available on such searches. We have developed an inefficient one, in that all interesting models are actually fitted, up to the bound of assuming submodels of nonfitting models are irrelevant and supermodels of fitting models also fit. This program PLMM is really a general non-linear least-squares estimation procedure, of great efficiency within a single model (Miller, 1981). It is important only in that it runs on relatively inexpensive microcomputers, so the inefficiency of the bounding principle is largely unnoticed; it is being incorporated in our general microcomputer package MASS (Henstridge et al., 1984). Microcomputers will make a great difference to the work of the statistician in this way - what is needed will be computed without financial constraint, and new methodologies and theories will be generated. This is the exciting challenge.

Extending beyond logistic models are the wider class of 'hazard' models (Cox, 1972), also now attracting great interest and work. We are not yet able to report model searching experience in these, nor are we able to talk on covariance models (Dempster, 1972), which seem even more attractive, being a richer class in terms of possible parameters and practical interpretation. Some discussion of their potential is given in Wermuth (1976) and Knuiman (1978).

The covariance selection models lead on to the Log-Linear models of contingency tables. Though this theory has been long studied, practical interest on a large scale stems from the tome of Bishop et al (1974), and the implementation of the technique in GLIM and the more efficient iterative proportional algorithm of Habermann (1971). The model can be expressed in the form

[3] $$\log m_{ijk} = u_i^{(0)} + u_i^{(1)} + u_j^{(k)} + u_k^{(3)} + u_l^{(4)} + u_{ij}^{(12)} + u_{ik}^{(13)} + u_{jk}^{(23)} + u_{ijk}^{(23)} + \ldots + u_{ijkl}^{(1234)}$$

The general loglinear model study is hardly yet started, the computational problems being time consuming if not particularly difficult, though many theoretical difficulties remain. Current work centres on interesting subclasses of models.

Wermuth (1976) identified the subclass of 'direct' models with conditional probability interpretations and simple calculation needs, and gave a simple stepwise search algorithm. Darroch et al (1979) described fully the extension of the direct models to the class of 'graphical' models, and Murphy (1982) provided stepwise and branch and bound 'all models' searches in the class. Havranek (e.g. 1984) has provided more efficient algorithms, and more work will be found elsewhere in the proceedings of this Conference. Again, the microcomputer will give great thrust to this type of research and practice.

Some work has been done in searching amongst the hierarchical class of log-linear models by Benedetti and Brown (1976) and Murphy (1982), and more will be reported below. In this, the availability of inexpensive computing is a sine qua non.

3. Comparison of some models.

Data from the Busselton Mass Health Surveys of 1966-81 were available for study. Readings for over 50 suspected 'risk variables' and other concomitants were available in 1966 for 1433 men, and in 1981 the heart disease status (CHD) for 1426 could be accurately determined. In the analysis, eventually only 12 variables were studied in depth, including blood cholesterol and glucose level (BCHOL & BGLU), bloodpressure (DBP), Agegroup (AGG), obesity index (O.I.), Insulin and phosphate levels (IN & APO) and a biochemical ratio (CH/HDL). The event variable was Survival (i.e. no CHD).

Table 1 summarises the results of all model searches using linear and logistic linear models ([1] & [2]) above. Each line of the table shows the finally selected variables amongst models of the given size. The 5 variable models in each case accounted for about 40% of the variance.

TABLE 1. Variables selected by two 'all-models' procedures.

Model Size	AGG	CH/HDL	BCHOL	BGLU	DBP	APO	OI
1	*						
2	*	*					
3	*	+	+			-	-
4	*	+	+	+	-	-	-
5	*	+	+	*	*	-	-

(* indicates selected by both procedures, + indicates linear model selection only, - indicates logistic model selection only)

The result of these analyses, and all that we have even seen in practice is that the linear model, though theoretically objectionable, is always very close to the logistic; in fact here the 'best' three models of each type were usually permutations of one another. This behaviour has been widely noted we believe, but not often documented in print. It also leads to speculation as to the real worth of even more expensive, more sophisticated modelling.

In this new age of inexpensive computing (all of the calculations here recorded were done on microcomputer using the programs LOLITA (Murphy, 1980), PLMM (from Miller, 1981) and MASS (Henstridge et al, 1984), in which all are now being embedded) and we can consider all such modelling. The appropriate next models to consider for the data would be the Covariance Selection models of Dempster (1972), but as algorithms have not yet been developed, we were forced to pass over to the corresponding discrete versions, the log-linear contingency table models.

In fitting contingency table models to such data major problems arise. Firstly, it is well known that choice of cut points can grossly distort conclusions. We do feel that this can be checked, so have had little hesitation in investigating poly-chotomised data, which after all is the usual form of displaying much information. Table 2 arises from the data and is the form of input to the LOLITA program for searching models. It is only one of many tables formed from the data, and does well represent it. Note that the data is entered in "FORTRAN ORDER'.

The second problem concerns the 'explanatory' variables - should these be fixed in the search so that all main effects and their interactions are forced into the model, and thus the search now only needs to search the interactions of survival with each explanatory variable. We deduce that this is not appropriate and thus the whole concept of explanatory variable term fixing, even in 'standard' situations, is dubious, and leads to spurious results reflecting overfitting, which we are documenting elsewhere.

TABLE 2. Polychotomised form of CHD data, as LOLITA program input.

FACTOR-NAMES CHD BGLUcode BCHOLesterol CH/HDL AGGroup DBPressure
FACTOR-LEVELS 2 · 2 2 3 2 2
DATA

534	133	22	6	3	0	0	0	2	1	0	0	1	1	0	0
135	59	17	6	0	0	0	0	1	0	0	0	2	1	0	0
220	134	24	24	4	1	1	0	0	5	0	1	4	3	0	0
14	6	0	0	0	0	0	0	0	0	0	0	0	0	0	0
14	6	1	2	0	0	0	0	0	0	0	0	0	0	0	0
19	12	2	1	0	0	0	0	0	0	0	0	0	0	0	0

Thirdly, one has a choice of graphical, hierarchical and general log-linear models. We cannot pursue arguments here, but merely state some results comparing graphical and hierarchical models for searches in which explanatory variable terms were held fixed in the data of Table 2.

The hierarchical graphical model search with all environmental factors (numbers 2 to 6 in Table 2) fixed gave a minimal fitting model which we denote by (14/23456). This is interpreted as CHD - age dependency only [14] given the other factors fixed [23456]. If the fixed factors condition is dropped, the model searches yield (14/24/345/46) as minimal good fit model. This is a fortunate circumstance in which the environment-fixing solution agrees with the more general one; we have often found that the fixed terms demand a larger model involving the survival variable, and which seems to be caused by the fact that the sum of survived and not survived must remain fixed in table of expected values, is a powerful condition for a model to satisfy.

Further, the similarity between graphical and hierarchical models found here reflects the simplicity of the interaction structure, a situation one would not often expect, but we have not found practical instances of striking discrepancies between the two procedures.

Comparing these results with those of the logistic and linear models, we again see reasonable cohesion - age is the most dominant factor and cholesterol and the biochemical ratio next. It has been a major interest to look for discrepancies between linear and log-linear modelling here, where the effect of a single high order interaction between a few of the concomitant variables in the contingency table (and covariance selection) models could be masked in the general miasma of main effects of the linear models. So far we have found no examples of this potential difficulty in actual data, but it can be manufactured in data where the predictive power of the loglinear model is high; clearly in practical circumstances in which such modelling is done, predictive power of the models is generally poor.

One mechanical difficulty worth noting in searching is the difficulty of organising the voluminous output in meaningful ways, since simple criteria for 'best fit' and bounds are not available when a monotone objective function cannot be defined.

4. Acknowledgements.

The author wishes to thank colleagues who helped in this over the years - K.L.Wearne, J.D.Henstridge, J.C.Poulsen of the MASS project, and J.D. Bovey and C.A.Boom on the LOLITA project.

5. References.

Baker,R.J., & Nelder,J.A. (1978). Generalised Linear Interactive Modelling, Release 3. Numerical Algorithms Group, Oxford.

Cox,D.R. (1972). Regression Models & Life tables. J.Roy.Stat.Soc., B34, 187-220.

Dempster,A.P. (1972). Covariance Selection. Biometrics, 28, 157-175.

Efroymson, M.A. (1960). Multiple Regression Analysis in Mathematical Methods for Digital Computers. A. Ralson & H. Wilf (eds), Wiley, N.Y.

Furnival,G.S. (1971). All possible Regressions with less Computation. Technometrics, 13, 403-408

Havranek T., (1984). A Procedure for Model Search in Multidimensional Contingency tables. Biometrics (in press)

Henstridge,J.D., Murphy,B.P., Poulsen,J.C., & Wearne,K.L. (1984). MASS 3.0 User Manual. Western Australian Regional Computer Centre, Perth.

Knuiman,M. (1978). Covariance Selection. Suppl. Adv. Appl. Proby, 10, 123 - 130.

Miller,A.J. (1981). LMM - A Subroutine for Unconstrained Non-Linear Least-Squares Fitting. Report No. VT.81/23. CSIRO Division of Mathematics & Statistics, Melbourne.

Murphy,B. (1982). Fitting & Searching Log-linear Models. Short Communications, COMPSTAT '82, Physica-Verlag, Vienna.

Truett,J., Cornfield,J., & Kannel,W. (1967). A multivariate Analysis of the Risk of Coronary Heart Disease. J. Chronic Dis., 20, 511-524.

Optimal collapsing of Two-Way Contingency Tables Including Structural Zeroes

D. Pokorný, Praha

SUMMARY.

In the previous papers procedures were suggested for explanation of the dependency structure in two-way contingency tables by means of optimal collapsings of optimal subtables. Here we shall propose their possible generalization to the case of tables with missing entries (structural zeroes). Besides the general case, particular cases of one missing entry and missing diagonal are discussed.

Keywords: two-way contingency tables, optimal collapsing, optimal subtables, structural zeroes, missing entries, COLLAPS, GUHA.

I. THE ANALYSIS OF COMPLETE TWO-WAY TABLES.

We shall briefly recall previously introduced GUHA procedures for two-way contingency tables, restricting a description to those features neccessary for understanding of the present text.

I.1. OPTIMAL SUBTABLES.

The procedure, based on Gabriel's results was proposed by Havránek (1981). Let an R*C contingency table T and parameters r,c,N (where 2≤r≤R, 2≤c≤C, N natural) be given. The output of the procedure is a set of N subtables of the table T having greatest values of MLCHISQ, maximum likelihood chi-squre statistic. The most significant r*c subtable leads our attention to the most important categories of both variables considered. Next subtables offer alternative solutions. Unlike the Pearsonian CHISQ, the MLCHISQ statistic has a property of monotonicity: MLCHISQ of a subtable is less than or equal to the value of MLCHISQ of the original table. Thus, subtables can be organized in a tree, where the original R*C table is the top and r*c tables are leaves. Although results are exhaustive, it means really best r*c tables are found, the search need not be exhaustive and considerable time savings are possible. This is one of typical aspects of procedures based on the GUHA approach, which remains theoretically hidden in the present paper.

I.2. OPTIMAL COLLAPSED TABLES.

Let us consider R*C contingency table T and parameters r,c,N again. An r*c table t collapsed from T is a table, where some rows and/or columns are joined (collapsed) together. The output of a procedure is now a set of N tables collapsed from T having greatest values of statistic. "Similar" rows and columns are joined in the optimal collapsed table. Thus (1) rows (and columns) are reasonably clustered and (2) the dependency structure is simplified to that of the collapsed table. Next collapsed tables represent alternative solutions. Both the CHISQ and MLCHISQ are monotonic w.r.t. collapsing. Thus tree-search techniques can be applied.

Let us remark that most of theoretical results, especially quick algorithms as well as practical experience was gained for the particular case of collapsing to 2*2 tables (cf. Pokorný, Havránek (1978), Pokorný (1980), Pokorný (1982), Enke, Zientz (1983)). Algorithms for iterative r*c collapsing were considered also by Diday (1982).

I.3. PROCEDURES FOR SQUARE TABLES.

Procedures searching optimal subtables (resp. optimal collapsed tables) can be easily modified for a R*R square table, where one desires to perform simultaneously in the same way for both variables the search for subsets (resp. partitionings) of the set of categories (1,...,R). From the point of view of computational complexity it represents a great simplification.

Modified procedures are applicable to social mobility tables or to transition matrices of behavioral sequences. However, complete two-way tables are not always satisfactory in this connection.

II. NOTES ABOUT DECOMPOSITION OF MLCHISQ.

II.1. DECOMPOSITION A.

A value of MLCHISQ statistic can be decomposed in various ways (cf. Gokhale, Kullback (1978)). The one described below will be particularly useful in the context of collapsing R*C tables to r*c ones.

Consider some partitioning P of row indices $(1,\dots,R) = F_1 \cup \dots \cup F_r$, resp. partitioning Q of column indices $(1,\dots,C) = G_1 \cup \dots \cup G_c$.

Let for i,j $(i=1,\dots,r\ ,\ j=1,\dots,c)$ x_i (resp. y_j) be the cardinality of an index subset F_i (resp. G_j).

Consider following tables (cf. Fig. A):

T the original R*C table,

S an r*c table collapsed w.r.t. partitionings P and Q,

H_i an $x_i * c$ collapsed subtable: it contains rows having indices in F_i , columns are collapsed w.r.t. Q,

V_j an $r * y_j$ collapsed subtable: it contains columns having indices in G_j , rows are collapsed w.r.t. P,

B_{ij} an $x_i * y_j$ subtable: it contains rows from F_i , columns from G_j .

Total value of Y=MLCHISQ can be now decomposed as follows:

$$Y(T) = Y(S) + \\ + \Sigma_{i=1..r} Y(H_i) + \Sigma_{j=1..c} Y(V_j) + \Sigma_{i=1..r} \Sigma_{j=1..c} Y(B_{ij}) \qquad \text{(A)}$$

The equation is true for degrees of freedom, too.
Monotonicity of MLCHISQ w.r.t. collapsing follows from the equality (A) immediately.

II.2. DECOMPOSITION B.

Let h_i be a subtable with rows from F_i , unlike H_i columns are not collapsed ($x_i * C$ subtable),

let v_j be a subtable with columns from G_j , unlike V_j rows are not collapsed ($R * v_j$ subtable). Then (cf. Fig. B):

$$Y(T) = Y(S) + \\ + \Sigma_{i=1..r} Y(h_i) + \Sigma_{j=1..c} Y(v_j) - \Sigma_{i=1..r} \Sigma_{j=1..c} Y(B_{ij}) \qquad \text{(B)}$$

Again, the equation is true for corresponding numbers of degrees of freedom.

Fig. A. Decomposition A of MLCHISQ.

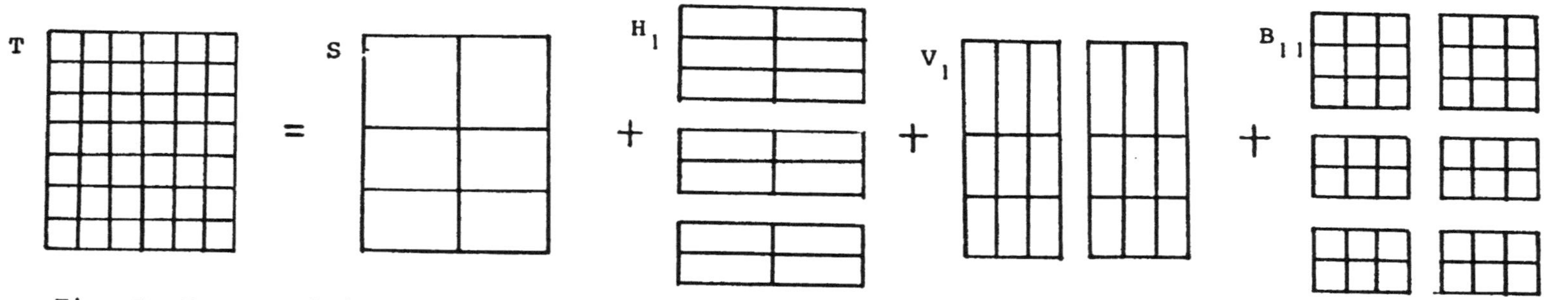

Fig. B. Decomposition B of MLCHISQ.

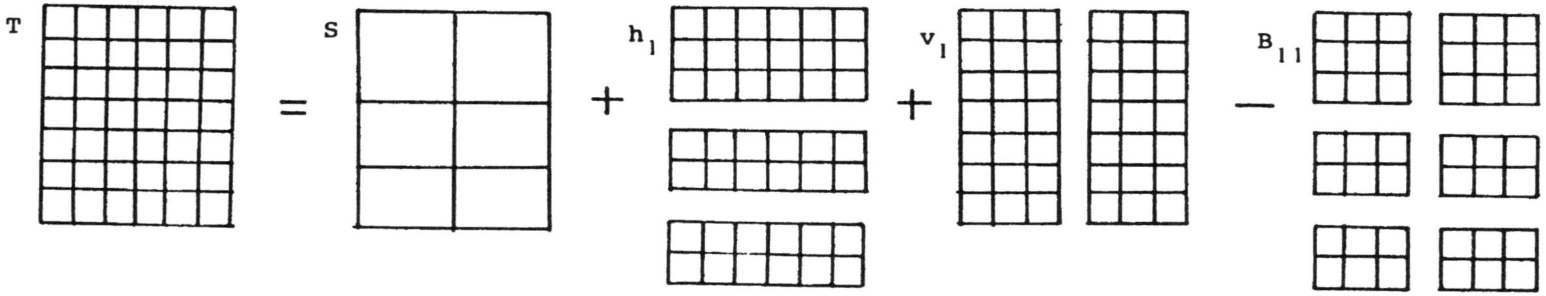

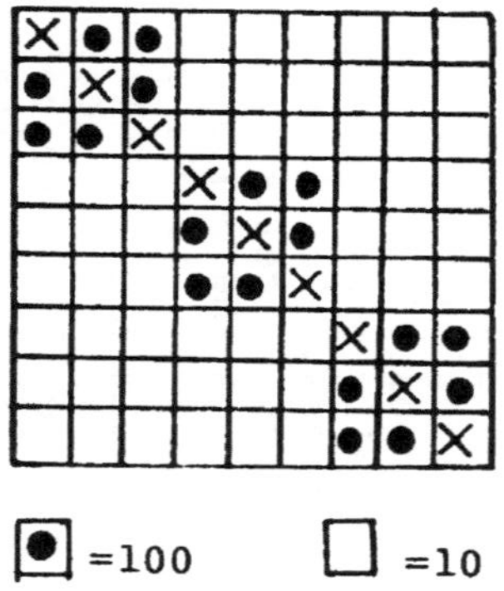

Fig. C. Square table with missing diagonal.

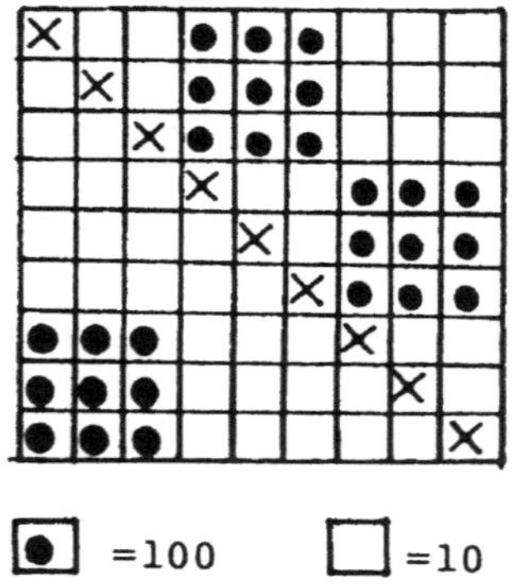

Fig. D. Another square table with missing diagonal.

III. OPTIMAL SUBTABLES WITH STRUCTURAL ZEROES.

III.1. MONOTONICITY OF MLCHISQ.

(a) One empty cell. Let the cell (1,1) be empty, denote $a=a_{11}$, $b=a_{1.} - a$, $c=a_{.1} - a$, $d=a_{..} - a - b - c$. Under the model of quasi-independence the value a in the cell (1,1) can be replaced by the value $e = b\, c / d$. For such a table T* , where an expected value in the empty cell of T is considered, the following inequality holds:

$$Y(T^*) \le Y(T) \qquad (C)$$

Analogous inequality for Pearsonian CHISQ is not true.
The difference $Y(T)-Y(T^*)$ is equal to MLCHISQ in the fourfold table a,b,c,d ; this can be seen from equality (A).

(b) Incomplete two-way table. For more empty cells in a two-way table T, the inequality (C) is preserved. For inseparable tables it follows directly from the continuity of the function Y=MLCHISQ and from the behavior of Brown's algorithm for iterative fitting of expected values, which in each step deletes one cell in the way described above.

(c) Incomplete subtables. Let S be a subtable of an (incomplete) table T. Let $Y(S^*)$, $Y(T^*)$ be MLCHISQs computed under the model of quasiindependence. Then $Y(S^*) \le Y(T^*)$.

III.2. PROCEDURE.

Havrånek's procedure for optimal subtables can be easily generalized for the search for optimal r*c subtables of an incomplete R*C table, since the monotonicity of MLCHISQ is preserved. Moreover, before fitting expected values for a given subtable, we can ask an auxiliary question: Is MLCHISQ large enough with some arbitrary values in empty cells? If the answer is negative we can leave this node. It is good to use values already estimated in the nearest supertable as "arbitrary" values. It is advanteous to avoid unneccessary fittings for these patterns of missing entries, where expected values must be computed iteratively. Unfortunately, an important case of missing diagoal belongs to them (Enke (1983)).

IV. OPTIMAL COLLAPSING OF TABLES WITH STRUCTURAL ZEROES.

IV.1. THE BASIC TROUBLE.

One plus zero is one.
One plus nothing is ... ?

IV.2. CRITERIA.

The criterion for optimal collapsing of a complete table was the maximality of chi-square in a collapsed table. A simple generalization of this criterion to the case of an incomplete table meets troubles which were discussed in a great detail in the paragraph IV.1: How to collapse an actually observed frequency and a structural zero together.
Definitions. Consider a cell of the collapsed table, which corresponds to a block of cells (=subtable) in the original table.
The cell of the collapsed table will be called:
(a) zero cell if all cells in the corresponding block are empty,
(b) mixed cell if some cells in the corresponding are empty and some are not,
(c) full cell if no cell in the corresponding block is empty.
We shall give some proposals for collapsing criteria in the following.

Criterion I. Maximize MLCHISQ in a collapsed table, where all zero or mixed cells are considered empty.
Discussion: Probably the most conservative and secure approach. Criterion is monotonic w.r.t. collapsing. However, it may cause a great loss of information. For instance, the expectation that an optimal collapsing of the squre table in Fig. C as well as in Fig. D will lead in both cases to the partitioning ((1 2 3) (4 5 6) (7 8 9)) is not satisfied by the criterion I.
Criterion II. Expected frequencies are estimated for empty cells only once in the original table.Then the procedure for collapsing of complete tables, maximizing MLCHISQ, is used.
Discussion: No loss of information is caused by a cell deletion. Criterion is monotonic w.r.t. collapsing. Iterative fitting must be performed only once during the computation.
We can avoid (maybe doubtful) summing up observed and fitted frequncies by the following reformulation of the criterion: expected frequencies for empty cells are fitted in the original table and substituted into appropriate cells of tables h_i, v_j, B_{ij} in the equation B. (Cf. Fig. B).
The criterion is now to minimize the expression $\Sigma Y(h_i)+\Sigma Y(v_j)-\Sigma\Sigma Y(B_{ij})$.
Criterion III. For given partitiongs of row, resp. column, indices consider empty all zero or mixed cells in the equation A (cf. Fig.A) and fit expected values in tables H_i, V_j, B_{ij} separately.(Denote tables with fitted values by the * symbol.)
The criterion is to minimize the expression $\Sigma_i Y(H_i^*)+\Sigma_j Y(V_j^*)+\Sigma_i\Sigma_j Y(B_{ij}^*)$.
Discussion: The criterion is monotonic w.r.t. collapsing. It represents an indirect solution of the collapsing problem. Instead of maximizing an association between blocks, we minimize following associations:
between rows within "row clusters" and
between columns within "column clusters" and
within blocks.
A well interpretable pattern is obtained particularly for the case of the missing diagonal.

Non-criterion IV. Consider analogously the equation B (cf. Fig. B).
Minize the expression $\Sigma_i Y(h_i^*) + \Sigma_j Y(v_j^*) - \Sigma_i\Sigma_j Y(B_{ij}^*)$.
Discussion: The criteion is not monotonic w.r.t. collapsing. Moreover, the expression considered can be negative. Hence, the criterion is not recommended.

IV.3. DEGREES OF FREEDOM.

We propose below described degrees of freedom (d.f) for the consideration. Clearly, both the MLCHISQ and d.f. do not play their usual statistical role here. Usual rules of computation of d.f. for incomplete tables (c.f. Goodman (1968)) are utilized below.
Criterion I. D.f.for the MLCHISQ statistic, considered in the Criterion I, are computed from the collapsed table, where both zero and mixed cells are considered empty.

Criterion II. D.f. computed from the collapsed table, where only zero cells are considered empty.

Criterion III. Sum of d.f. from H_i^*, V_j^*, B_{ij}^*, where both zero and mixed cells are considered empty.

There are theoretically more possibilities how to include d.f. in a corresponding criterion:
(a) Ignore d.f. completely: the (much too) simple way. Roughly said, the general effect is the penalization for partitionings with some singleton subset. (E.g. the partitioning ((1)(2)(3 4 5 6 7 8 9)) is penalized, the partitioning ((1 2 3)(4 5 6)(7 8 9)) not.)
(b) For each distinct d.f. find optimal N collapsings. This is practically possible if the number of various d.f., which can occur in a

collapsed table, is relatively small, as for instance in the case of one empty cell or a square table with missing diagonal.
(c) Consider the criterion MLCHISQ / d.f. in the computation.
(d) Consider the significancy, corresponding to MLCHISQ and d.f., as the criterion. This possibility is more reasonable than (c). Some algorithmic problems, which have not been fully solved yet, occur here: for instance, often so small significances occur, that they are represented by computer zero;hence they are not directly comparable.

The monotonicity of criteria w.r.t. collapsing are not preserved when various d.f. are considered. This fact is reflected in algorithms.

IV.4. PROCEDURES.

For the development of quick and accurate algorithms, two things are neccessary:
(a) various sophisticated hints,
(b) (for advanced programmer:) when computing the MLCHISQ statistic not to forget to multiply the expression ... + x ln x + ... by two.

CONCLUSION.

We have developed programs for the particular case of a square table with the missing diagonal in Fortran-77 under RSX-11M compatible operating system. The analysis of well-known Pearson's 14*14 table (father /son occupation) led to reasonable results by time savings up to 99.6 percent.

On the other hand, it is easy to generalize procedures for a set of isomorphic two-way tables which are to be collapsed in the same way. The criteria are then based on the sum of MLCHISQ values in particular tables. These procedures can be useful e.g. if we have to analyse more transition matrices coming from a repeated behavioral experiment.

Our next plan include: (a) to develope more general procedures under PDP-11 and IBM/370 standards, (b) to investigate algorithmic possibilities of futher time savings and (c) after gaining more practical experience to include procedures into the GUHA package.

LITERATURE

Brown M. B. (1974), The identification of sources of significance in two-way contingency tables, Appl. Statist. 23, 405-413.
Diday E. (1982), personal communication.
Enke H. (1977), On the analysis of incomplete two-dimensional contingency tables, Biom. J. 19, 561-544.
Enke H. (1983), personal communication.
Enke H., Zientz R. (1983), Zu einigen Möglichkeiten der Verwendung der Daten des Krebsregister für Untersuchungen über Verzögerungszeiten bei der Diagnose und Behandlung von Bronchialkarzinome, Z. ges. Hyg. 29, 541-544.
Gokhale D.V., Kullback S. (1978), The information in contingency tables, Dekker, New York and Basel.
Goodman L.A. (1968), The analysis of cross-classified data:independence, quasiindependence and interactions in contingency tables with or missing entries, JASA 63, 1091-1131.
Havránek T. (1981),Formal systems for mechanized statistical inference, Int. J. Man-Machine Studies 15, 333-350.
Pokorný D. (1980), Knowledge acquisition by the GUHA method, Int. J. Policy Analysis and Information Systems 4, 379-399.
Pokorný D. (1982), Procedure for optimal collapsing of two-way contingency tables, in: COMPSTAT 82, Physica Verlag, Wien, 96-102.
Pokorný D., Havránek T. (1978), On some procedures for identifying sources of dependence..,in:COMPSTAT 78, Physica Verlag, Wien, 221-227.
Victor N. (1983), A note on contingency tables with one stuctural zero, Biom. J. 25, 283-289.

Stepwise Discriminant Analysis Procedure for Categorical Variable

T. Rudas, Budapest

The majority of multidimensional statistical methods is applicable only to continouos variables. There is, however, a growing demand to apply these methods to categorical data. In this paper a stepwise discriminant analysis procedure is described for categorical variables. Having chosen a set of classificatory variables, instead of discriminant functions, a set of classifying rules is used to decide which group the observation belongs to. Starting from the empty set that variable is entered the predictor set which yields the greatest increase in the probability of correct classification. The probability of correct classification is shown to be a monotonic increasing function of the number of variables in the classificatory set. A test of the hypothesis that either of two variables increases the probability of correct classification equally is suggested. A computer program realizing this procedure is described and an application to a social psychological problem is given.

Keywords: contingency tables, discriminant analysis

Introduction

In the usual discriminant analysis procedure (cf. Morrison (1976), pp 230-246) the problem is to find those (linear)functions of given classificatory or explanatory variables that discriminate or separate the known groups of the data the most. The stepwise version of this procedure first choses the "best" classificatory variable then the "second best" (depending on the first) etc. Appropriateness of the actual set of classificatory variables is usually measured by the percent of correct classification which can be achieved (cf. Dixon (1981), p 520).

Let $V = \{v_1, v_2, \ldots, v_p\}$ denote the set of classificatory variables w the variable which represents the classification to be explained by the variables V. The categories of w will be denoted by the natural numbers $I = \{1,2,\ldots,c\}$. In this paper the classificatory variables will be categorical i.e. $v_i \in V$ is a grouping of individuals into c_i exhaustive and mutually exclusive categories denoted by $\{1,2,\ldots,c_i\}$.

If $H \subseteq V$ then the common categories of the variables H are

$$I(H) = \mathop{X}_{v_i \in H} \{1,2,\ldots,c_i\}$$

$I(H)$ is said to contain the indices of the variables H.

Each observation can be characterized by

$$(a,b) \in I \times I(V).$$

Here b is a vector of the p dimensional Euclidean space i.e. $I(V) \subset R^p$. If $G \subseteq H \subseteq V$ and there are r variables in H and s variables in G then $I(H) \subset R^r$ and $I(G) \subset R^s$ where $s \leq r \leq p$. Let $(h-g)$ denote that coordinate projection from the r dimensional Euclidean space to the s dimensional that picks out coordinates of $b \in I(H)$ constituting an index of the variables in G i.e. by definition if $b \in I(H)$ then

$$(h-g)(b) \in I(G) .$$

Suppose that there is a probability distribution over $I \times I(V)$ i.e. there are numbers $P(a,b)$ such that

$$P(a,b) \geq 0 \quad \text{and} \quad \sum_{a \in I, b \in I(V)} P(a,b) = 1.$$

This distribution is usually not known but we have a sample of N independent observations. The observations are recorded in a $p+1$ dimensional contingency table. Let the observed frequency in the cell of the table having the index (a,b) be denoted by

$$F(a,b) \qquad a \in I, \quad b \in I(V) .$$

Then these random variables are distributed according to the multinomial law with parameters N and $P(a,b)$, $a \in I$, $b \in I(V)$.
It is clear that if $H \subseteq V$ and $b \in I(V)$ then

$$P(a,c) = \sum^{*} P(a,b) \quad \text{and} \quad F(a,c) = \sum^{*} F(a,b)$$

for all $c \in I(H)$, where $\sum^{*}$ denotes summation over those $b \in I(V)$ for which

$$(v-h)(b) = c .$$

Classifying rules

Let $H \subseteq V$ be a given subset. If an observation is in the $b \in I(H)$ category the following forecast of its w category is given:

(1) $a(b) = \min(a\colon a \in I,\ P(a,b) \geq P(c,b) \text{ for all } c \in I)$

i.e. the observation is being classified in that category of w that has the greatest conditional probability given $b \in I(H)$. The idea is similar to that used by Goodman and Kruskal (1954) in their lambda measure of association.

If the probability distribution is not known then the classifying rules are given by replacing the probabilities in (1) by their maximum likelihood estimates, namely the relative frequencies, or equivalently

$$\hat{a}(b) = \min(a\colon a \in I,\ F(a,b) \geq F(c,b) \text{ for all } c \in I) .$$

Since the relative frequencies converge to the probabilities (in several senses) as N increases, for large samples these estimated classifying rules can be regarded as being correct.

Accuracy of classification

For a given $H \subseteq V$ the achievable accuracy is defined as the probability of correct classification. This is

$$A(H) = \sum_{b \in I(H)} P(a(b),b)$$

Theorem 1. For $G \subseteq H \subseteq V$

$$A(G) \leqslant A(H)$$

with equality if and only if

$$P(a(b),c) = P(a(c),c)$$

for all $c \in I(H)$ and $b = (h-g)(c)$, where $(h-g)$ is the projection from $I(H)$ to $I(G)$.

Proof:

$$A(G) = \sum_{b \in I(G)} P(a(b),b) = \sum_{b \in I(G)} \sum_{\substack{c \in I(H) \\ (h-g)(c)=b}} P(a(b),c) \leqslant$$

$$\leqslant \sum_{b \in I(G)} \sum_{\substack{c \in I(H) \\ (h-g)(c)=b}} P(a(c),c) = \sum_{c \in I(H)} P(a(c),c) = A(H)$$

Since

$$P(a(b),c) \leqslant P(a(c),c)$$

for all $c \in I(H)$, $b = (h-g)(c)$ and these numbers are non-negative their sums can be equal if and only if

$$P(a(b),c) = P(a(c),c)$$

holds for all $c \in I(H)$, $b = (h-g)(c)$.

Stepwise discriminant analysis procedure

In the k-th step of the procedure the set of classificatory variables is denoted by E_k, $k = 0,1,\dots,p$. E_0 is the empty set. E_k $(k>0)$ is $E_{k-1} \cup \{v_r\}$, where $v_r \in V-E_{k-1}$ if

(2) $A(E_{k-1} \cup \{v_r\}) \geqslant A(E_{k-1} \cup \{v_s\})$ for all $v_s \in V-E_{k-1}$

In some cases the above rule may not yield a unique E_k. In such cases a prior preference list of the variables can be used to decide which variable is to enter the classificatory set.

As the underlying probability distribution is not known the accuracy in (2) needs to be replaced by its estimate, namely

$$\hat{A}(H) = N^{-1} \sum_{b \in I(H)} F(\hat{a}(b),b) \quad .$$

Theorem 1'. For any $G \subseteq H \subseteq V$

$$\hat{A}(G) \leqslant \hat{A}(H)$$

with equality if and only if

$$F(a(b),c) = F(a(c),c) \quad \text{for all} \quad c \in I(H) \quad \text{and} \quad b=(h-g)(c).$$

Proof: Omitted since is similar to the proof of Th 1.

A test for variable selection

Theorem 2. Let

$$E = E_{k-1} \cup \{v_r\} \qquad F = E_{k-1} \cup \{v_s\}$$

Let us suppose that

$$A(E) = A(F)$$

and the classifying rules are correct. Then

$$(3) \quad P(\hat{A}(E)-\hat{A}(F) \geq \tfrac{m}{N}) = \sum_{i=0}^{N-m} \sum_{j=0}^{\left[\frac{N-i-m}{2}\right]} \sum_{k=m}^{N-i-2j} \frac{N!}{i!j!(j+k)!} P(H^1)^{(j+k)} P(H^2)^j P(H^3)^i$$

where H^1 , H^2 , H^3 are defined under (4), (5), (6), below.

Proof: The following subsets of the $I \times I(V)$ index set are to be defined:

For given $H \subseteq V$ and $b \subseteq I(H)$

$$H_b = \{(a,c): (a,c) \in I \times (V),\ a=a((v-h)(c))\}$$

$$H_1 = \bigcup_{b \in I(E)} E_b$$

$$H_2 = \bigcup_{b \in I(F)} F_b$$

Now it is clear that

$$A(E) = \sum_{(a,d) \in H_1} P(a,d)$$

$$A(F) = \sum_{(a,d) \in H_2} P(a,d)$$

The sets H^1 , H^2 , H^3 are defined as follows

(4) $$H^1 = H_1 - H_2$$

(5) $$H^2 = H_2 - H_1$$

(6) $$H^3 = I \times I(V) - H^1 \cup H^2 \quad .$$

With this definitions

$$A(E) - A(F) = \sum_{(a,d) \in H^1} P(a,d) - \sum_{(a,d) \in H^2} P(a,d) \quad .$$

Similarly

$$\hat{A}(E) - \hat{A}(F) = N^{-1}\Big(\sum_{(a,d) \in H^1} F(a,d) - \sum_{(a,d) \in H^2} F(a,d)\Big) \quad .$$

Since the random variables $F(a,d)$ are distributed according to the multinomial law, the variables

$$F(H^1) = \sum_{(a,d) \in H^1} F(a,d),\ F(H^2) = \sum_{(a,d) \in H^2} F(a,d),\ F(H^3) = \sum_{(a,d) \in H^3} F(a,d)$$

are distributed according to the trinomial law with parameters N and

$$P(H^1)=\sum_{(a,d)\in H^1} P(a,d), \quad P(H^2)=\sum_{(a,d)\in H^2} P(a,d), \quad P(H^3)=\sum_{(a,d)\in H^3} P(a,d)$$

(cf. Bishop, Fienberg, Holland (1975), p 444).

Thus

$$P(\grave{A}(E)-\hat{A}(F)\geq \frac{m}{N}) = P(F(H^1)-F(H^2)\geq m) = \sum_{i=0}^{N-m} \sum_{j=0}^{\left[\frac{N-i-m}{2}\right]} \sum_{k=m}^{N-i-2j} \frac{N!}{i!j!(j+k)!} P(H^1)^{(j+k)} P(H^2)^{j} P(H^3)^{i}$$

Supposed that the sample is large enough for the classifying rules to be correct the hypothesis that for a given E_{k-1} the entering of either variable v_p or v_r yields the same increase in accuracy can be tested using (3) so that the probabilities are replaced by their estimates.

Stopping rules

The procedure can be continued untill $E_p=V$ is reached or can be stopped if any of the following criteria is fulfilled.

a.) $A(E_k)\geq u$, where u is a preassigned constant.

b.) $A(E_{k+1})-A(E_k)\leq u$, where u is a preassigned constant.

c.) $(A(E_{k+1})-A(E_k))(A(E_k))^{-1}\leq u$, where u is a preassigned constant.

d.) $A(E_{k+1})/A(E_k)\leq u$, where u is a preassigned constant.

The program and an application

The input of the computer program for stepwise discriminant analysis for categorical variables can be either a data matrix or a contingency table. A preference list of the variables can be given to be used in variable selection if necessary. A set of pairs of the variables can be specified. If in a step the variable to enter the classificatory set is one of these variables, the increase in accuracy is compared to the increase in accuracy that could be achieved by entering, instead of this variable, its pair in the set using the test suggested above. The stopping rule can be chosen from among a.)-d.) above. Applying this procedure data on 500 divorse cases at a Budapest court (cf. Lengyel (1984)) were analysed. In 389 of these suits the married couple had common children. This 389 cases were taken as a simple random sample. In such cases the children should be adjudicated to one of the parents. A naive approach to this problem would suppose that in such cases the judges take several factors of the personalities of the parents into consideration. Our hypothesis states, to the contrary, that the two groups of plaintiffes, those who the children were adjudicated to and those who were not, can be discriminated by the use of the following variables:

v_1: sex of the plaintiff (2 categories).

v_2: educational level of the plaintiff (5 categories).

v_3: educational level of the respondent (5 categories).

v_4: whether the plaintiff wants the children or not (2 categories).

We applied stopping rule a.) with $u=0.8$.

Our hypothesis proved to be true to such an extent that the procedure stopped after one step with $E_1 = \{v_4\}$, $\dot{A}(E_1) = 0.92$.

Acknowledgement

The author wishes to express his indebtedness to Mr Gy. Kóbor for his programming work and to Dr Zs. Lengyel for the data analysed.

References

Bishop,Y.,M.,M., Fienberg,S.,E., Holland,P.,W. (1975) Discrete Multivariate Analysis. MIT Press

Dixon,W.,J. (1981) ed. BMDP Statistical Software. University of Calif. Press

Goodman,L.,A., Kruskal,W.,H. (1954) Measures of association for cross classifications. J. Amer. Statist. Assoc. 49 732-764

Lengyel,Zs. (1984) The Divorce. manuscript in Hungarian

Morrison,D.,F. (1976) Multivariate Statistical Methods. 2nd ed. McGraw-Hill

A Recursiv Algorithm for the Determination of the Uniformly Most Powerful Unbiased Test in Fourfold Tables

H. Scherb, and G. Welzl, München–Neuherberg

SUMMARY

An algorithm to determine the uniformly most powerful unbiased test (UMPUT) in 2x2 tables is introduced and shown to provide a useful tool in power computations. Sample sizes based on exact power functions are obtained and compared with existing sample size estimations.

Key words: Fourfold table, uniformly most powerful test (UMPUT), algorithm for the determination of critical values, exact power functions, sample size estimation.

INTRODUCTION

One of the classical problems in statistical theory is that of testing for independence in fourfold tables. For quite some time a uniformly most powerful unbiased test has been known to exist, comparing two binomials or testing independence in 2x2 contingency tables. Despite this knowledge, little effort was obviously invested in deriving well-understood and efficient algorithms that determine the critical values of these tests. This may be due to the prevailing prejudice that critical values can only be determined by 'trial and error' as was stated by Tocher (1950, p. 136) and supported by similar statements in other papers and books (e.g. Blyth and Hutchinson (1960), Lehmann (1959), Witting (1966)). Slight simplifications of the problem and some trial algorithms are known: Witting (1966), Zenga (1974), Welzl and Siegerstetter (1975). With such algorithms tabulations of the most important exact tests have been compiled by Zenga (1974), Siegerstetter and Welzl (1975). In this paper we shall consider Welzl's and Siegerstetter's approach to then present an alternative method demonstrating that the determination of unbiased two-sided exact tests shows, in fact, a surprizingly close analogy to the determination of one-sided tests. In addition to mere testing, two important application areas of the algorithm do exist. We shall, first, compute exact power functions which offer a standard for assessing the great variety of power approximations and, second, present exact sample sizes and compare these with various published approximations. Although authors of publications on sample sizes often claim to have improved existing sample size estimates, their improvements criteria remain unclear. Here, too, exact sample sizes derived from uniformly most powerful tests can serve as standards for comparisons and assessments.

STATING THE PROBLEM

Consider a fourfold table:

n_{11}	n_{12}	$n_{1.}$
n_{21}	n_{22}	$n_{2.}$
$n_{.1}$	$n_{.2}$	$n_{..}=n$

Let $x = n_{11}$. It is well known that the UMPUT applied for testing independence, or, comparing two binomials is given by

$$\varphi(x) = \begin{cases} 1 & x < c_1,\ x > c_2 \\ \gamma_i & x = c_i,\ i = 1,2 \\ 0 & c_1 < x < c_2. \end{cases}$$

The critical values c_i, γ_i, $i = 1,2$ are determinded by

$$\text{(I)}\ \alpha = \sum_{\substack{x<c_1 \\ x>c_2}} h(x) + \gamma_1 h(c_1) + \gamma_2 h(c_2) \quad \text{and}$$

$$\text{(II)}\ \alpha e = \sum_{\substack{x<c_1 \\ x>c_2}} x h(x) + \gamma_1 c_1 h(c_1) + \gamma_2 c_2 h(c_2)$$

where $\alpha \in]0,1[$ and where $h(x)$ and e are the density and expected value of the hypergeometric distribution with parameters $n_{1.}$, $n_{.1}$ and n. This problem is known to be difficult as the critical values c_1 and c_2 define regions where the summations have to take place, and where, at the same time, c_1 and c_2 are factors and arguments of h in (I) and (II).

WELZL'S AND SIEGERSTETTER'S ALGORITHM

The first possible step toward determining c_i and γ_i, $i = 1,2$ is to rearrange (I) and (II) and solve for γ_1 and γ_2. This is possible since $c_1 \leq c_2$ is immediately obvious from (I), $c_1 = c_2$ occurs rarely and will provisionally be excluded. We are then given the following expressions

$$\gamma_1(c_1,c_2) = \frac{\alpha(c_2-e) - \sum\limits_{\substack{x<c_1 \\ x>c_2}} (c_2-x)h(x)}{h(c_1)(c_2-c_1)}$$

$$\gamma_2(c_1,c_2) = \frac{\alpha(e-c_1) - \sum\limits_{\substack{x<c_1\\x>c_2}} (x-c_1)h(x)}{h(c_2)(c_2-c_1)}$$

The problem now is to determine c_1 and c_2 such that $0 \leq \gamma_1(c_1,c_2) \leq 1$ and $0 \leq \gamma_2(c_1,c_2) \leq 1$. This problem can be solved - as has been shown by Welzl and Siegerstetter (1975) - by choosing an arbitrary initial value $c_1^{(0)}$ (e.g. max $\{0,n_{1.}+n_{.1}-n\}$ or [e]) to then proceed in computing a $c_2^{(1)}$ such that $0 \leq \gamma_2(c_1^{(0)}, c_2^{(1)}) \leq 1$. Analogously to $c_2^{(1)}$ a $c_1^{(1)}$ will be found with $0 \leq \gamma_1(c_1^{(1)}, c_2^{(1)}) \leq 1$, followed by a $c_2^{(2)}$, etc. It has been shown that the final solution is reached if in two successive steps of the iteration $c_2^{(i+1)} = c_2^{(i)}$ or $c_1^{(i+1)} = c_1^{(i)}$. Normally, it needs only two iterations and three iterations are required solely in rare situations.

SOLVING THE PROBLEM

To solve (I) and (II) for the critical values, we suggest as an important expedient the following

Definition. Let $M = \{x_1,\ldots,x_r\}$ be the support of the hypergeometric distribution. On MxM we define functions $s,\bar{s},t,\bar{t}$:

In the case $c_1 = c_2 = e$

$$s(c_1,c_2) = 1-h(e) \qquad \bar{s}(c_1,c_2) = 1$$

$$t(c_1,c_1) = 1-h(e) \qquad \bar{t}(c_1,c_2) = 1$$

In the case $c_1 \neq e$ and $c_2 \neq e$

$$s(c_1,c_2) = \sum_{\substack{x<c_1\\x>c_2}} \frac{(c_2-x)}{(c_2-e)} h(x) \qquad \bar{s}(c_1,c_2) = \sum_{\substack{x\leq c_1\\x>c_2}} \frac{(c_2-x)}{(c_2-e)} h(x)$$

$$t(c_1,c_2) = \sum_{\substack{x<c_1\\x>c_2}} \frac{(x-c_1)}{(e-c_1)} h(x) \qquad \bar{t}(c_1,c_2) = \sum_{\substack{x<c_1\\x\geq c_2}} \frac{(c_2-x)}{(c_2-e)} h(x)$$

The significance of the functions $s,\bar{s},t,\bar{t}$ can best be understood by considering an

Example. Let $n_{1.} = n_{.1} = 7$ and $n = 19$. According to an immediate obvious pattern we compute values for $s,\bar{s},t,\bar{t}$:

(c_1,c_2)	$s(c_1,c_2)$ / $t(c_1,c_2)$		$\le \alpha <$	$\bar{s}(c_1,c_2)$ / $\bar{t}(c_1,c_2)$
(0,7)	.000000 .000000		$\le \alpha <$	.024887 .000054
(0.6)	-.000006 .000054		$\le \alpha <$	.027561 .003932
(0,5)	-.000705 .003932		$\le \alpha <$	.031756 .057261
(1,5)	.031756 -.004600		$\le \alpha <$	.243836 .065083
(1,4)	.022499 .065083		$\le \alpha <$	.293489 .355430
(2,4)	.293489 -.121795		$\le \alpha <$	.758044 .406109
(2,3)	.216063 .406109		$\le \alpha <$	1.000000 1.000000

We move from c_1 to c_1+1 or from c_2 to c_2-1 resp. iff $\bar{s}(c_1,c_2)$ or $\bar{t}(c_1,c_2)$ resp. is the minimum of these two values. With the new pair (c_1+1,c_2) or (c_1,c_2-1) resp., the value $s(c_1+1,c_2)$ or $t(c_1,c_2-1)$ resp. is automatically the maximum of these two values and equals the former minimum. Proceeding in this manner we obtain a complete disjoint partition of $[0,1]$, which contains all α. Proofs and generalizations of these relations are given by Scherb (1983). For a fixed $\alpha\in]0,1[$, according to the above scheme, we recursively compute $\bar{s}(c_1,c_2)$ and $\bar{t}(c_1,c_2)$ until their minimum is greater than α for the first time. The corresponding c_1, c_2 are the required solutions of (I) and (II). The computation of γ_1,γ_2 turns then out to be trivial. With the abreviation

$$d = \sum_{\substack{x<c_1 \\ x>c_2}} (e-x)h(x)$$

the following important recursion formulae do hold:

$$s(c_1+1,c_2) = s(c_1,c_2) + \frac{(c_2-c_1)}{(c_2-e)} h(c_1),$$

$$t(c_1+1,c_2) = t(c_1,c_2) + \frac{d}{(e-c_1-1)(e-c_1)}$$

$$s(c_1,c_2-1) = s(c_1,c_2) + \frac{d}{(c_2-1-e)(c_2-e)}$$

$$t(c_1,c_2-1) = t(c_1,c_2) + \frac{(c_2-c_1)}{(e-c_1)} h(c_2)$$

Together with the simple relations $\bar{s}(c_1,c_2) = s(c_1+1,c_2)$ and $\bar{t}(c_1,c_2) = t(c_1,c_2-1)$ we obtain recursiveness for functions $\bar{s}$ and $\bar{t}$, too. The programming of the described procedure offers no numerical problems, whereas, if we invert the algorithm in the sense that we determine c_1 and c_2, starting at the expected value e, elimination of leading digits will take place. However, this can be neglected in most applications.

POWER FUNCTIONS OF THE EXACT RANDOMIZED FISHER TESTS

The unconditional power function of the test φ for comparing two binomial distributions $B(1,p_1)$ and $B(1,p_2)$ has the form

$$\text{(III)} \quad \beta(p_1,p_2) = \sum_{n_{.1}=0}^{n} \sum_{x} \varphi(x) b(n_{1.},p_1;x) b(n_{2.},p_2;n_{.1}-x),$$

where b denotes the probability function of the binomial distribution. The unconditional power function of the test φ for the stochastic indepdendence of two events A and B, with the probabilities p_A and p_B resp., has the form

$$\text{(IV)} \quad \beta(\lambda,p_A,p_B) \sum_{n_{.1}=0}^{n} \sum_{n_{.1}=0}^{n} \sum_{x} \varphi(x) f(x,n_{1.},n_{.1}).$$

with $\lambda = p_{11}/p_A p_B, f(x,n_{1.},n_{.1}) = b(n,p_A;n_{1.}) b(n_{1.},p_1;n_{11}) b(n_{2.},p_2;n_{21}), p_1 = p_{11}/p_A, p_2 = p_{21}/(1-p_A)$, b as above and, p_{11} the probability that A and B will occur simultaneously. Programs utilizing the introduced algorithm to compute (III) and (IV) have been written and applied to determine exact sample sizes for given β.

COMPARISON OF EXACT SAMPLE SIZES WITH APPROXIMATIVE SAMPLE SIZES

We shall restrict attention to the more simple case of comparing two binomials, when additionally $n_{1.}=n_{2.}$ holds. The following table shows the uncertainty in the literature (e.g. Fleiss (1973), Fleiss (1981)) as to what correct sample sizes are. The table also shows that a considerable reduction in required sample size results in utilizing exact randomized tests and exact power functions.

Required sample sizes to detect the difference $p_1 = .3$ and $p_2 = .85$ with power β at level α according to different methods.

α \ β	.9	.9	.9	.8	.8	.8	.5	.5	.5
.01	28	25	23	24	21	18	17	14	11
.02	25	22	20	21	18	16	15	12	9
.05	21	18	16	18	15	13	12	9	7
.10	18	15	13	15	12	10	10	8	6
.20	15	12	10	13	10	8	.8	6	4
Method	A	B	C	A	B	C	A	B	C

A: corrected CHI^2 - approximation by Casagrande-Pike-Smith,

B: corrected CHI^2 - approximation by Kramer-Greenhouse,

C: UMPUT.

REFERENCES

Blyth, C.R., Hutchinson, P.W. (1960), Table of Neyman-shortest unbiased confidence intervals for the binomial parameter, Biometrika, 47, 381-383.

Fleiss, J.L. (1973), Statistical methods for rates and proportions, Wiley, New York-London-Sidney.

Fleiss, J.L. (1981), Statistical methods for rates and proportions, Wiley, New York-London-Sydney.

Lehmann, E.L. (1959), Testing statistical hypothesis, Wiley, New York-London-Sidney.

Scherb, H. (1983), Die Bestimmung der kritischen Werte exakter Tests, Dissertation, Universität des Saarlands, Saarbrücken.

Siegerstetter, J., Welzl, G. (1975), Tafeln zur Durchführung des zweiseitigen exakten Tests nach Fisher für Vierfeldertafeln, GSF-Bericht MD 145, München.

Tocher, K.D. (1950), Extension of the Neyman-Pearson theory of tests to discontinuous variates, Biometrika, 37, 130-144.

Welzl, G. Siegerstetter, J. (1975), An Algorithm for determining the critical values of Fisher's exact test with randomization, Unpublished paper, München.

Witting, H. (1966), Mathematische Statistik, Teubner, Stuttgart

Zenga, M. (1974), Tavole per alcune ipotesi bilaterali riguardanti distribuzioni di Poisson, binomiali, ipergeometriche e per l'indipendenza in una tavola tetracorica, Pubblicazione della Universita Cattolica del Sacro Cuore, Milano.

Fitting All Possible Decomposable and Graphical Models to Multiway Contingency Tables

J. Whittaker, Lancaster

SUMMARY
Fitting heirarchical loglinear models to contingency tables classified by k response factors is simplified, both theoretically and computationally by restricting attention to conditional independence (graphical) models. In an exploratory study there is a need to assess all such models. A procedure is outlined here by which most, though not quite all, of the relevant information can be rapidly computed and processed. This procedure is illustrated with a five way table.

1. INTRODUCTION

A list of the many objectives that confronts the practioner faced ab initio with a multi-way contingency table, includes those
a) of determining which dependencies (or effects) are present,
b) of deciding which of these are major,
c) of deciding if the table can be collapsed onto some of its margins without loss of information and, finally,
d) of selecting one (or perhaps a few) parsimonious models that adequately represent these effects.
These considerations fuel the desire to examine everything, a desire that is intensified by the high costs of data collection and the diminishing costs of data processing; but implementation of these objectives faces the problems caused by the inherent lack of balance in the table and the multitude of contending loglinear models.

Balance

When the effects are balanced, which is a combined consequence of both the probability model and the experimental design, there is relatively little difficulty: an estimate of a dependency between two factors cross classifying the table will not vary according to which, if any, of the other effects have been adjusted for. That this is almost never the case with multinomial response data is a consequence of the following theoretical consideration: if A, B and C are three events then the independence of A and B neither implies nor is implied by the condtional independence of A and B given C.

The number of models

The formidable number of heirarchical log-linear models for k-way contingency tables (Edwards and Kreiner, 1983) quickly dispels the possibility of fitting all such models. As independence and conditional independence are comparatively fundamental notions a sensible contraction is acheived by focussing attention on the $2^{\binom{k}{2}}$ graphical or conditional independence models first discussed by Darroch, Lauritzen and Speed (1980).

Whittaker(1982) used a deviance addition theorem implicit in the work of Andersen (1974) and Sundberg (1975) to simplify calculation of the deviances of some of these graphical models. Extensive application of deviance addition shows that the deviances of all decomposable models can be obtained from those of an irreducible set of equi-probability models. So by restricting the class of graphical models to decomposable models the number can be reduced to 2^k, the cardinality of the set. A coherent way to digest the information in this set of deviances is by transforming to the additive elements, Whittaker(1984).

Additive elements

Additive elements generalise the regression elements of Newton and Spurrell (1967). They are one way of extending the technique of the analysis of variance to cope with unbalanced models. In the regression context the primary elements are the sums of squares uniquely attributable to each regressor, the secondary elements are the amounts that can be attributed to exactly two of the regressors, and so on through the higher order elements.

In the analysis of a k-way contingency table the additive elements of the irreducible equi-probability models have a different interpretation. The primary elements are test statistics for tests of conditional equi-probability of one factor given the rest and the secondary elements are test statistics for the conditional independence of two factors given the rest. The higher order elements are differences between these tests and tests based on varying the conditioning set. An implication of additivity is that the test statistics for marginal independence can be derived by adding certain elements together. In this sense, the additive elements contain the necessary information to simultaneously evaluate dependencies in the marginal tables.

Computing

The deviances for the irreducible equi-probability models can be computed directly without fitting a full log-linear model and so do not require any iterative fitting procedures or matrix inversions. Nor do they impose any large storage demand. The transformation to the additive elements is linear with an upper triangular matrix and so can be cheaply calculated. The additive elements plot provides a visual summary of the analysis.

The procedure is fairly easily adapted to cope with tables that are classified by treatment factors as well as response factors and can extend to cover more general recursive structures. On the other hand there are some drawbacks: it cannot give the deviances for irreducible non-decomposable graphical models, for example, one that contains a four-cycle in its graph; it is an exploratory rather than confirmatory data analytic tool; the sampling distributions of higher order elements appear extremely intractable. Here we exemplify the technique by considering an artificial 3-way table and discuss a more practical application in a 5-way table.

2. A THREE WAY TABLE

In 3 dimensions the three way table AxBxC of counts n(i,j,k) is cross classified by factors A, B and C. A numerical example where A, B and C each take two levels is the following cross classification of 280 units.

Table 1. An artificial 3-way table of counts.

A=1	C 1	C 2	A=2	C 1	C 2
B 1	40	40		10	30
2	40	40		20	60

The table has been constructed so that the BxC table exhibits independence at each level of A. The deviances for testing the conditional independence of any two factors given the third and the for testing marginal independence are

model	dev	df	model	dev	df
(B+C).A	0.0	2	B+C	.50	1
(A+C).B	17.94	2	A+C	18.44	1
(A+B).C	7.36	2	A+B	7.86	1

They have been computed in the usual way with the deviances for the tests of marginal independence being computed in the relevant margin.

Graphical models

There are eight graphical models in 3 dimensions:

A.B.C, A.(B+C), B.(A+C), C.(A+B), A.B+C, A.C+B, B.C+A, A+B+C

Except for A.B.C all are reducible, examples of the application of deviance additivity are

dev(A.B+A.C) = dev(A.B)+dev(A.C)-dev(A)

and

dev(A+B+C) = dev(A)+dev(B)+dev(C)-2dev(1).

Each model is reducible to a linear combination of the models A.B.C, A.B, A.C, A, B.C, B, C, 1. These are the irreducible models of equi-probability. For example, dev(A.B) is the test statistic for testing if the levels of C are equally represented at each of the levels of the AxB classification.

There is an explicit expression for the deviance of any of these equi-probability models; an example is

dev(A.B) = Σ n(i,j,k)log{n(i,j,k)/n(i,j,+)/K}

Each deviance is of the form Σ n(t)logn(t) where t represents one of the index sets ijk,ij+,...,+++. These quantiti.es are directly computable from the original table of counts.

Table 2. Deviances of the irreducible equi-probability models.

model	dev	df
A.B.C	0.0	0
A.B	31.39	4
A.C	13.59	4
A	44.99	6
B.C	31.53	4
B	44.99	6
C	37.76	6
1	50.72	7

Additive elements of irreducible equi-probability models

Though this set of models contains the necessary information the best way of extracting it is not immediately obvious. A systematic method of transforming to the additive elements is described in Whittaker(1984a); applied here this leads to the following transformation:

Put

G(:ABC) = dev(A.B.C)
G(A:BC) = dev(B.C)-dev(A.B.C)
G(AB:C) = dev(C)-dev(A.C)-dev(B.C)+dev(A.B.C)
G(ABC:) = dev(1)-dev(A)-dev(B)+dev(A.B)-dev(C)+dev(A.C)+dev(B.C)-dev(A.B.C)

and similarly define G for each permutation of A, B and C. The additive elements of deviances in table 2 are the outcome and are presented in the next table together with the result of applying the same transfrom to the degrees of freedom.

Table 3. Additive Elements of Equi-probability Models.

order	element	G	df
residual			
	:ABC	0.0	0
primaries			
	A:BC	31.53	4
	B:AC	13.59	4
	C:AB	31.39	4
secondaries			
	AB:C	-7.36	-2
	AC:B	-17.94	-2
	BC:A	0.0	-2
third order			
	ABC:	-0.50	1

The important remarks to make are that

(i) the transformation is linear, invertible and triangular;

(ii) the test statistics for conditional independence above are recovered in the secondary elements, for example G(AB:C) = -7.36 = -dev(B.C+A.C) ;

(iii) the test statistics for marginal independence are obtained by adding the third order element to the appropiate secondary element. For example, G(AB:C)+G(ABC:) = -7.36-.50 = -7.86 = -dev(A+B). In a two way table the latter term is just the secondary element G(AB:); this provides an illustration of the essential additivity of the elements: G(AB:) = G(AB:C)+G(ABC:).

(iv) The sum of the elements has to equal the deviance of the minimal model in the set of irreducible equi-probability models, some 50.72.

3.A FIVE WAY TABLE

The theory extends to higher dimensions without any great difficulty but in practise more effort is required as there is more information to process. To illustrate we have chosen an example of a five dimensional table that has previously been analysed by several authors, including Goodman in his fundamental 1970 paper.

Table 4. Knowledge of cancer and media exposure.

	radio				no radio			
	read		no read		read		no read	
	takes newspapers							
knowledge	good	poor	good	poor	good	poor	good	poor
lectures	23	8	8	4	27	18	7	6
no lectures	102	67	35	59	201	177	75	156
	takes no newspapers							
knowledge	good	poor	good	poor	good	poor	good	poor
lectures	1	3	4	3	3	8	2	10
no lectures	16	16	13	50	67	83	84	393

Let A denote the factor corresponding to reading newspapers, B to attending lectures, C to radio listening, D to solid reading and finally E to a good or poor knowledge of cancer. Rather arbitrarily we have chosen to regard each factor as a response factor. The deviance of the model of complete independence is 596.8 on 26 degrees of freedom. The deviances of the 32 irreducible equi-probability models were calculated; they range from dev(A.B.C.D.E)=0 on 0df to dev(1)=2666 on 31df. The transform to the additive elements results in the following table.

Table 5. Additive elements of equi-probability for the five way table.

residual		3rd order	
:ABCDE	0.0		
primaries		ABC:	-0.4
A:	384.8	ABD:	-3.3
B:	1513.0	ACD:	-7.7
C:	600.7	BCD:	2.5
D:	365.8	ABE:	4.1
E:	306.4	ACE:	-4.2
secondaries		BCE:	1.6
AB:	-10.6	ADE:	-47.8
AC:	-49.8	BDE:	4.3
BC:	-16.6	CDE:	0.1
AD:	-178.8	4th order	
BD:	-17.2	ABCD:	-1.1
CD:	-5.3	ABCE:	-2.9
AE:	-40.0	ABDE:	-6.9
BE:	-15.3	ACDE:	-7.6
CE:	-9.2	BCDE:	-1.6
DE:	-90.5	5th order	
		ABCDE:	-0.6

The labels behind the colon have been suppressed for clarity e.g. AB: is AB:CDE.

The residual element is necessarily zero. The primary elements are extremely large reflecting the inequalities of the number of students represented at each of the factor levels; this is especially so with factor B (lectures). The secondary elements are necessarily negative; they vary in size and the larger dependencies are AD, DE, AC and AE. With the exception of the third order element ADE most of the higher order elements are comparatively small. Their relative disposition is made apparent in a rough plot.

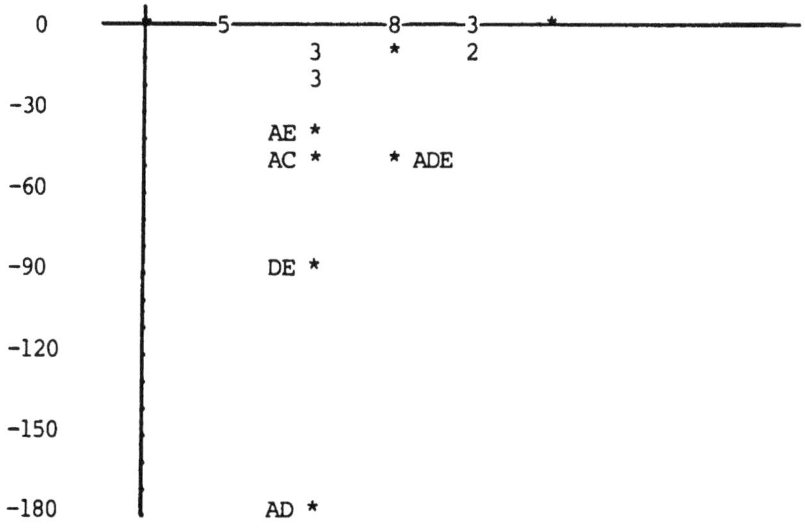

Figure. Additive Elements Plot with suppressed primaries.

A model that does not include A.D.E+A.C+B cannot hope to explain the data. The deviance for this model is 75.5 on 21df, so though it does not provide an adequate representation it accounts for the vast proportion of the deviance of 596.8 against the model of complete independence.

Including these terms in the model leaves the remaining secondary elements: AB:CDE -10.6, BC:ADE -16.6, BD:ACE -17.2, CD:ABE -5.2, BE:CDE -15.3 and CE:ABD -9.2; together with the small third order elements BCD:AE and BCE:AD. That the sum of these secondary elements, -74.1, is very nearly equal to the unexplained deviance of 75.5 strongly suggests that these effects are balanced; a view which is reinforced by the small higher order elements. As the 5% point of chi-squared on 8df is 15.5 the dependencies between Band C, B and D and B and E should be included in the model. This leads to the model A.D.E+B.D.E+A.C+B.C; its deviance is an acceptable 24.2 on 17df. Interestingly this model is graphical but not decomposable and so its deviance has to be calculated iteratively, though the near balance in this table allows a reasonable approximation.

REFERENCES

Andersen, A.H. (1974). Multidimensional Contingency Tables. Scand J. Statist 1, 115-127.

Darroch, J., Lauritzen, S. and Speed, T. (1980). Markov fields and log linear interaction models for contingency tables. Ann. Stat. 8. 522-539.

Edwards, D. and Kreiner, S. (1983). The analysis of contingency tables by graphical models. Biometrika,70,3.

Goodman, L.A. (1970). The multivariate analysis of qualitative data : interaction among multiple classifications. J.Amer.Stat.Soc. 65, 226-256.

Newton, R.G. and Spurrell, D.J. (1967) A development of multiple regression for the analysis of routine data. Appl. Stat. 16, 51-64.

Sundberg, R. (1975). Some results about decomposable (or Markov-type) models for multidimensional contingency tables : distribution of marginals and partitioning of test. Scandinavian J. of Statistics 2, 71-79.

Wermuth, N. and Lauritzen, S.L. (1983). Graphical and recursive models for contingency tables. Biometrika,70,3.

Whittaker, J. (1982). GLIM syntax and simultaneous tests for graphical log linear models. in Gilchrist R. (Ed). GLIM 82 Springer Verlag, p.98-108.

Whittaker, J. (1984) Model interpretation from the additive elements of the likelihood function. Appl. Statist. 33, 1.

Optimization Techniques in Statistics

A Computer Program Package for Solving Spline-Regression Problems

V.K. Kaishev, Sofia

The problem of fitting spline-regression models of one and more dimensions to response surfaces of complex form (posessing peaks, discontinuities, etc.) is discussed. Special attention is payed to computational aspects of optimal design of experiments in the case of spline-functions. For the solution of spline-regression problems a computer programs package named SADP is developed. In the present paper the basic features and programs of SADP are discussed. Examples are given of applications of SADP in real problems.

KEYWORDS: spline, regression, D-optimal design, B-spline.

I. INTRODUCTION

In many regression problems it is often necessary to fit multidimensional response surfaces, posessing peaks, discontinuities in derivatives, etc. In order to construct an adequate approximation in this case the polynomial tensor product spline-regression can be successfully applied provided that the independent variables $x_1, \ldots, x_n$ take values from a rectangular domain $X \subset R^n$. The univariate spline-function $S_{m,k}(x)$ in $x \in [a,b] \subset R$ with breakpoints $a=\beta_0 < \ldots < \beta_{k+1}=b$ is a piecewise polynomial function coinciding with a polynomial of degree less than or equal to m on $[\beta_j, \beta_{j+1}]$, $j=0,\ldots,k$. If we define the knot set as $\Delta_k = \{t_{-m} \leq t_{-m+1} \leq \ldots \leq t_0 = \beta_0 = a < \beta_1 \leq t_1 \leq \ldots \leq t_{l_1} < \ldots < \beta_{k+1} = b = t_{l_1+\ldots+l_k+1} \leq \ldots \leq t_{l_1+\ldots+l_k+m+1}\}$, where $1 \leq l_j \leq m+1$ are knot multiplicities, then $S_{m,k}(x)$ has $m-l_j$ continuous derivatives at β_j, $j=1,\ldots,k$. For representing $S_{m,k}(x)$ on Δ_k the B-spline basis $B_{m+1,i}(x)$, $i=1,\ldots,p$, $p=m+1+l_1+\ldots+l_k$ can be used (c.f. Curry & Schoenberg (1966)); Such bases are well conditioned, have a local support, i.e. $B_{m+1,i}(x)=0$, for $x \notin [t_{i-m-1}, t_i]$, hence for $x \in [a,b]$ only m+1 B-splines are nonzero. B-splines are evaluated by the recurrence relation of De Boor - Cox (1972). A multidimensional spline-regression on a fixed set of knots $\Delta = \Delta_k \times \ldots \times \Delta_u$, forming a rectangular grid in X, is defined as

$$(1) \qquad E(y/x) = \sum_{i=1}^{p_1} \ldots \sum_{j=1}^{p_n} \theta_{i,\ldots,j} B_i(x_1) \ldots B_j(x_n) ,$$

where $\theta = (\theta_{1,\ldots,1}, \ldots, \theta_{p_1,\ldots,p_n})'$ is a vector of unknown parameters. and $E(y/x)$ is the expectation of the response variable y, including a noise component.

In certain applications the investigator might be interested only in computing the estimates of θ on the basis of N existing data values of y at some points in X. In other cases such a set of data might not be initially available and a special experiment, often expensive and continuous, should be carried out for its collection. The problem of optimal experimental design arises, namely to find a minimum number of appropriate values of $x_1,\dots,x_n$ at which measurements of y should be taken in order to achieve a spline-regression model of highest possible quality. For the solution of the above mentioned problems a computer programs package named SADP was developed. In Section II the basic methods implemented in SADP are discussed. In Section III the main programs of SADP are shortly described. Examples of applications of SADP are given.

II. BASIC METHODS

The construction of spline-regression of one and more dimensions is carried out following the sequential procedure of Kaishev (1977). It includes the stages:

(i) fix the initial degree, number and values of knots in Δ.

(ii) determine the optimal experimental design and take the measurements y_i, $i=1,\dots,N$ at the design points.

(iii) find the least squares (LS) estimates of θ.

(iv) make a statistical analysis of the spline-regression. If the result is acceptable, go to the end, else make a correction of the set Δ and return to (ii).

Obviously, only stages (ii), (iii) and partially (iv) can be strictly formalized. In stage (i) it is recommended to take as breakpoints the values of independent variables corresponding to the extremums, jumps, etc. in the response surface if such information exists. Let us first consider the optimal design problem for the univariate B-spline regression $E(y/x) = \sum_{i=1}^{p} \theta_i B_{m+1,i}(x)$. Since Δ_k is fixed (stage (1)) the above equation is linear in $\theta_i, i=1,\dots,p$ and G- (D-) optimality criteria can be used to generate optimal designs. Recall that the G-optimal design minimizes the maximum of the variance of predicted value $\hat{y}$ on $[a,b]$ and is at the same time, due to Kiefer-Wolfowitz equivalence theorem, D-optimal. We define a normalized continuous design $\xi \subset \begin{Bmatrix} z_1,\dots,z_q \\ \xi_1,\dots,\xi_q \end{Bmatrix}$ as a probability measure $\xi(x)$, concentrating mass $\xi_1,\dots,\xi_q$ at some values $z_1,\dots,z_q$ of the independent variable x, $0 \le \xi_i \le 1$, $\sum_{i=1}^{q} \xi_i = 1$. Then the D-optimal design ξ^* is a design for which the equation $\det M^{-1}(\xi^*) = \min_{\xi} \det M^{-1}(\xi)$ holds. Here $M(\xi) = \sum_{i=1}^{q} \xi_i \, B_{m+1}(z_i) \, B'_{m+1}(z_i)$ is the information matrix with $B_{m+1}(\cdot) = (B_{m+1,1}(\cdot),\ldots,B_{m+1,p}(\cdot))'$. The set z_i^*, $i=1,\ldots,q$ is the spectrum of the optimal design and ξ_i^*, $i=1,\ldots,q$ are the frequences. Thus a D-optimal design can be found through a direct maximization of $\det M(\xi)$ in 2q variables z_i, ξ_i, $i=1,\ldots,q$, $q \geq p$, under the restriction $0 \leq \xi_i \leq 1$, $\sum \xi_i = 1$. However this approach leads to serious computational difficulties when q is large. A way out is the use of a design easily computed and at the same time close enough to the D-optimal. Such a design is the "B-spline maximum" one (BSM), introduced in Kaishev (1977), maximizing $\mathrm{tr}\,\| B_{m+1,i}(z_j) \|$, $i=1,\ldots,p$, $j=1,\ldots,q$, for the saturated case p=q. BSM-design concentrates equal mass $\xi_1=\ldots=\xi_q$ at the spectrum points $z_1^{max},\ldots,z_q^{max}$, corresponding to the maxima of $B_{m+1,1}(x),\ldots,B_{m+1,p}(x)$, respectively. The spectrum values are found by solving the equations $B^{(')}_{m+1,i}(x)=0$, $i=1,\ldots,p$. In order to estimate the deviation of a BSM-design from a G-optimal one, the G-efficiency criterium can be used, $G_{eff}=1$ for the G-optimal design. Spectra, frequencies of BSM- and D-optimal designs and G_{eff}, computed for several fixed sets of knots Δ_k are given in the Table below. As seen from the Table, G_{eff} for each BSM-design is greater than 0.94 and its spectrum is very close to that of the corresponding D-optimal design. The spectra No 2 and

TABLE. Saturated BSM- and D-optimal Designs for Cubic Splines on $[-1,1]$

No	k	Design	Position of knots in Δ_k	Spectrum	G_{eff}
1	4	BSM	-0.35,0.05,0.05,0.70	-0.730,-0.370,0.020,0.420, 0.690, 0.870	0.957
2	4	D-opt	- " -	-0.780,-0.366,0.024,0.420, 0.686, 0.875	0.995
3	8	BSM	-0.38,-0.15,0.07,0.36, 0.50,0.50,0.70,0.70	-0.760,-0.460,-0.170,0.080,0.290 0.450,0.550,0.670,0.770,0.900	0.940
4	8	D-opt	- " -	-0.811,-0.463,-0.170,0.060,0.288 0.450,0.540,0.650,0.770,0.907	0.991
5	3	BSM	-0.80,0.00,0.40	-0.90,-0.60,-0.05,0.40,0.75	0.945
6	4	BSM	-0.80,0.00,0.40,0.60	-0.90,-0.60,-0.10,0.30,0.60,0.85	.940

4 of the Table are computed by the use of a minimax sequential designing algorithm for splines (Kaishev (1977)) implemented at present in SADP. Another algorithm of direct maximization of $\det\| B_{m+1,i}(z_j)\|$ starting from a BSM-design is currently included in the package. Designs No 5 and 6 of the Table illustrate the fact that the addition of a new knot to the set of knots brings only to a local change in the spectrum. This fact can be explained by the small support of

B-splines. The property of local change is an argument in favor of the application of optimal designs in cases when variation of the number and position of knots is necessary, as prescribed in stage iv.

In the multivariate case (eq.(1)) the D-optimal design is easily obtained as a direct product of the univariate D-optimal designs for each set $\Delta_k, \ldots, \Delta_u$ respectively (Krug & Kaishev (1977)).

According to stage (iii) the LS estimates of θ in (1) are computed in SADP as the tensor product

(2) $$\theta = (F_1'F_1)^{-1} \otimes \ldots \otimes (F_n'F_n)^{-1} (F_1' \otimes \ldots \otimes F_n') Y$$

assuming that experimental design points form a rectangular configuration, as in BSM- and D-optimal designs. $F_1, \ldots, F_n$ in (2) are matrices with elements $B_{m+1,i}(x_{1,j})$, $j = 1, \ldots, N_1$, $i = 1, \ldots, p_1$, ... , $B_{s+1,v}(x_{n,w})$, $w = 1, \ldots, N_n$, $v = 1, \ldots, p_n$. $Y = (y_1, y_2, \ldots, y_{N_1 \ldots N_n})$ is a vector of experimental observations in the design points. LS estimates can be computed in SADP also for arbitrary set of data.

Statistical analysis of spline-regression equation carried out in stage (iv) includes testing of the goodness of spline fit by an F-test, determination of confidence intervals about $\hat{y}$. Hypothesis testing about linear combinations of the spline regression coefficients in the B-spline representation on the basis of F-test, using the idea of P. Smith (1982) can also be performed.

If the spline is not adequate, an additional knot is included in Δ . It should be placed in the subdomain of X where a worst fit of the regression function to the response surface, after residual analysis, is observed. If the design is saturated, as is in the D-optimal case, some additional measurements should be carried out for checking the adequacy.

III.SADP PROGRAMS

The package SADP includes 4 main programs, namely SADP S1, S2, S3 and DO. Programs SADP S1, S2 and S3 allow computation of spline-regression of one two- and three independent variables respectively, of degree less than or equal 20 in each variable. The program SADP DO is dedicared to generation of univariate BSM-designs and of D-optimal designs of desired accuracy (stage ii). A common set of input instructions is developed for communication with the above programs. The data is entered in a free format. The values of knots and the degrees of spline in each variable are a necessary information for all the four programs. A set of experimental measurements of the dependent variable y in the points of the experimental design should be entered as well if SADP S1, S2 or S3 are to be used. Each of the programs is designed to follow the sequence: initializing parameters in common areas, reading input instructions, checking for consistency and constraints violation of input data, precise report about errors in the input instructions and data, report about input information read by the programs, basic computations (LS estimates and D-optimal designs), statistical analysis, listing of output results, plotting by means of a graph-plotter. Facilities exist in SADP, allowing minimum amount of storage used for current computations. LS estimates are computed by a direct information matrix inversion or using Choletski factorization algorithm. Statistical analysis includes computation of mean and variance at each experimental design point, etc. SADP instructions are English-based and are used to describe the data input, name the variables and specify optional features and results to be printed and plotted. The following example of application of SADP S1 demonstrates the SADP instructions language.

EXAMPLE

```
TITLE                Surface Tension Isotherm
ARGUMENT FIRST       Log of Concentration
FUNCTION             Surface Tension Difference
INDEPENDENT FIRST -6.097 -5.848 -5.602 -4.600 -3.600 * *
DEPENDENT            0.420  2.087  9.646 17.000 37.717 * *
KNOTS FIRST          -5.602 -5.602 * *
PARAMETERS FIRST DEGR    2 * *
END
```

The above example is illustrated by Fig.1. The data is taken from A. Nikolov et. all. (1980). The curve in Fig.1 represents the spline-regression model of degree 2 with a double knot at -5.602, computed by SADP S1 on the basis of y-values (black circles) measured at the points of a D-optimal design. The latter was generated by help of SADP DO. The empty circles in Fig.1 correspond to additional measurements taken in order to perform subsequent statistical analysis. They illustrate the goodness of fit obtained by the spline. In Fig.2 a quadratic spline (the continuous curve 1) and a fourth order polynomial (dotted curve 3) with equal number of regressors are plotted on the basis of experimental data shown in Fig.1. Curves 2 and 4 in Fig.2 represent 0.99% confidence intervals about the regression curves 1 and 3 correspondingly. As seen from Fig.2 and 1, spline fits the data much better than a polynomial with equal degrees of freedom.

SADP has been used as well for solving multivariate

regression problems in electrochemical investigations (M. Kaisheva et. all. (1984)), metallurgy, etc.

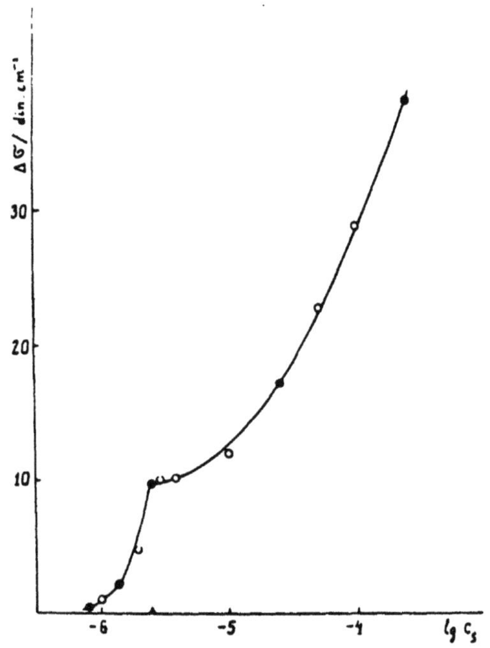

Fig. 1

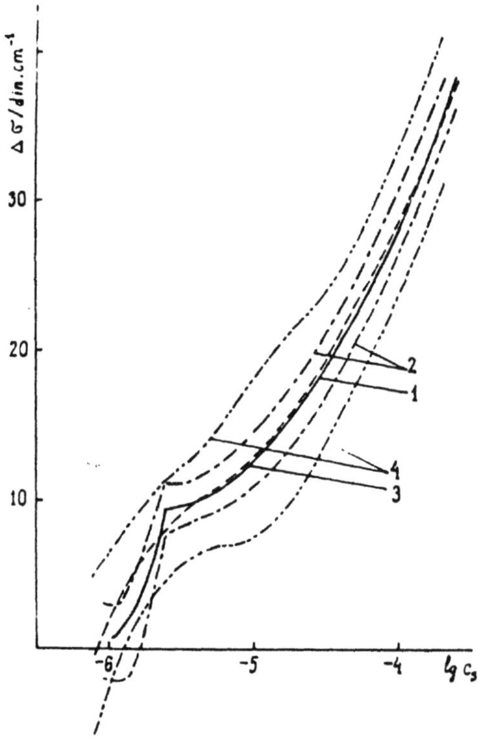

Fig. 2

REFERENCES

Curry H.B., Scoenberg I.J. (1966), On Polya Frequency Functions IV: The Fundamental Spline Functions and their Limits, J. Analyse Math., 17, 71-107.

De Boor C. (1972), On Calculating with B-Splines, J. Approximation Theory, 6,50-62.

Kaishev V.K. (1977), Design of Experiments in Construction of Spline-Regression Models, Ph.D. Thesis, Moscow.

Kaisheva M.K., Kaishev V.K., Matsumoto M.(1984), Adsorption of Dodecylhexaoxyethylene Glycol Monoether at a Stationary Mercury Electrode. A Spline Regression Model, J.Electroanalyt. Chem., in press.

Krug G.K., Kaishev V.K. (1977), D-optimal Design for a Class of Regression Functions, Zavodskaya Laboratoriya, 7, 858-860.

Nikolov A., Martinov G., Exerova D., Kaishev V. (1980), Micellization and Properties of Adsorption Layers of Ionic Surfactants, Kolloidnii Zhurnal, XLII, 672-679.

Smith P. (1982), Hypothesis Testing in B-Spline-Regression, Commun. Stat.-Simula. Computa., 11, 143-157.

A New Method for Solving Nonlinear Least Squares Problem

T. Kutas, Kende

A new method for solving the nonlinear curvefitting problem has been developed in order to construct such a method which is as stable in matrix inversion as the Marquardt method, its convergence rate is near to that of Gauss-Newton method and it calculates the goal function only once in each iteration step. The new method has been compared to other methods /Marquardt, Gauss-Newton/ on several test problems. Based upon previous experience an algorithm has been constructed which automatically chooses among Gauss-Newton type methods.

KEY WORDS: NONLINEAR CURVEFITTING PROBLEM, GAUSS-NEWTON AND MARQUARDT METHODS, NUMERICAL COMPARISON

1. INTRODUCTION

One of the most frequent problems of mathematical statistics is the least squares estimation of parameters. This paper deals only with its numerical solution. The problem is as follows

$$\min H(\underline{\vartheta})$$

where
$$H(\underline{\vartheta}) = \sum_{i=1}^{n} [f(\underline{x}_i, \underline{\vartheta}) - y_i]^2$$
$$\underline{\vartheta} \in \mathbb{R}^l$$
$$\underline{x}_i \in \mathbb{R}^k \qquad i = 1, \dots, n$$
$$y_i \in \mathbb{R} \qquad i = 1, \dots, n$$

the function f is twice differentiable and the following equations hold

$$y_i = f(\underline{x}_i, \underline{\vartheta}) + \varepsilon_i \qquad i = 1, \dots, n$$

where ε_i-s are normally distributed with zero mean and with covariance matrix $\sigma^2 I$. In linear case, when the f function is linear in parameters, these latter assumptions guarantee some good features of the optimal solution. In nonlinear case these assumptions can be neglected, and sometimes the interpolation between the measurement points is more important than the estimation of parameters.

Except for linear case the H function is not a convex one, so only local minima can be reached. This is the reason why it is so important to find new algorithms. Developing a new algorithm a special point of view was to construct such a one which can be used for large systems, in case a function evaluation needs too much computer time, the function to be minimized mustn't be evaluated more than once in one iteration step. Another point of view was - as there is no unique method for solving the problem - to find an algorithm which automatically chooses among methods. All methods described by the author are modifications of Gauss-Newton method.

2. GAUSS-NEWTON TYPE METHODS

The Gauss-Newton method applies the principle of Newton method for the minimization of a general function using a special approximation of the Hessian of sum of squares:

$$\nabla H(\vartheta) = 2 \cdot \nabla^T \underline{f}(\vartheta) \cdot [\underline{f}(\vartheta) - \underline{Y}]$$

$$\nabla^2 H(\vartheta) = 2 \cdot \sum_{i=1}^{n} [f(x_i, \vartheta) - y_i] \cdot \nabla^2 f(x_i, \vartheta)] + 2 \cdot \nabla \underline{f}(\vartheta) \cdot \nabla^T \underline{f}(\vartheta)$$

$$\underline{f}(\vartheta) = [f(x_1, \vartheta), \dots, f(x_n, \vartheta)]^T \in \mathbb{R}^n$$

$$\underline{Y} = (y_1, \dots, y_n)^T \in \mathbb{R}^n$$

The special approximation means that the first term in the Hessian is removed. The remaining matrix is always positive semi-definite, if its inverse exists /i.e. the matrix is positive definite/ it will be positive definite as well, so the Gauss-Newton direction which is the multiplication of the inverse and the gradient is an effective direction /Bard /1974//. The Gauss-Newton method gives the following direction:

$$\delta_{GN} = A_{GN}^{-1} \cdot \nabla^T \underline{f}(\vartheta) \cdot [\underline{f}(\vartheta) - \underline{Y}]$$

$$A_{GN} = \nabla \underline{f}(\vartheta) \nabla^T \underline{f}(\vartheta)$$

if the inverse exists.

This method has two disadvantages: firstly it is not monotonous, secondly the inverse doesn't exist always or it is near a singular matrix, the smallest eigenvalue is near zerp. The second disadvantage leads to the divergence of the algorithm or in computer realization to overflow. The method has a great advantage as well: if it converges it does it very quickly. All the algorithms described later try to improve the Gauss-Newton method.

The monotonity can be reached by using line search. This method is cited in the literature as Hartley method. This is not a real solution as the matrix to be inverted can be singular,and the line search is not an effective method in nonlinear curve-fitting as the goal function is not convex and the evaluation of it needs too much computer time /the n number can be very large/.

Marquardt's idea eliminates both disadvantages of Gauss-Newton method: a positive number (λ) is added to the diagonal of the matrix A_{GN} . In any case when the λ number is positive the matrix $(A_{GN} + \lambda I)$ will be positive definite so it can be inverted. It can be proved /see e.g. Marquardt /1963// that if the λ number tends to infinity the direction

$$\underline{\delta}(\lambda) = (A_{GN} + \lambda I)^{-1} \nabla^T \underline{f}(\underline{\vartheta}) \cdot [\underline{f}(\underline{\vartheta}) - \underline{Y}] \quad (1)$$

tends to zero, and the angle between $\underline{\delta}(\lambda)$ and the gradient of H function tends to zero as well. This fact results that a better iteration point can be reached by increasing the λ number.

The method proposed by R.R. Meyer /Meyer /197o// combines Marquardt's and Hartley's ideas: a positive number is added to the diagonal of matrix A_{GN} and a line search is fulfilled in that direction.

Contrary to the most nonlinear programming problems the evaluation of goal function needs more computer time than the calculation of the direction. So a demand has arisen for such an algorithm which calculates the goal function only once in every iteration steps, and the matrices to be inverted are positive definite, the smallest eigenvalues are greater than a given positive number. The new method tries to combine the high convergence rate of Gauss-Newton method and the robustness of Marquardt method, and it calculates only once the goal function in each iteration step. The root of this algorithm is the following: let us consider $\underline{\delta}(\lambda)$ in /1/ to be a curve, and instead of finding a new iteration point on the curve /this is the Marquardt method/ let us find the new point on the tangent line near the starting point of the curve, the Gauss-Newton point. Formulating this sentence we get:

$$\underline{\delta}_{MM}(\lambda) = \{(A_{GN} + \lambda I)^{-1} + \alpha \cdot (A_{GN} + \lambda I)^{-1} \cdot (A_{GN} + \lambda I)^{-1}\} \nabla^T \underline{f}(\underline{\vartheta}) \cdot [\underline{f}(\underline{\vartheta}) - \underline{Y}]$$

Similar to Meyer method the λ number tends to zero decreasing from iteration step to iteration step. Now two steps have to be taken: first to find appropriate value for α parameter and second to prove that the goal function /at least in linear case/ decreases during iteration steps. Minimizing $\|\nabla f \cdot \underline{\vartheta} - \underline{Y}\|^2$ with respect to α where $\underline{\vartheta} = \underline{\vartheta}_r - \underline{\delta}_{MM}$ in the r-th iteration step and introducing

$$\underline{g} = (A_{GN} + \lambda I)^{-1} \cdot (A_{GN} + \lambda I)^{-1} \cdot \nabla^T \underline{f} \cdot (\nabla \underline{f} \cdot \underline{\vartheta}_r - \underline{Y})$$

we get

$$\alpha = \lambda + \lambda^2 \cdot \frac{\|\underline{g}\|^2}{\underline{g}^T \cdot A_{GN}\, \underline{g}}$$

Instead of choosing this an approximation of it has been used:

$$\alpha = \lambda$$

<u>Theorem</u>: The algorithm is monotonous in the linear case.

Let

$$B = (A_{GN} + \lambda I)^{-1} + \alpha \cdot (A_{GN} + \lambda I)^{-1} \cdot (A_{GN} + \lambda I)^{-1}$$

In linear case the value of the goal function is the following

$$H(\underline{\vartheta}_{v+1}) = \| \nabla \underline{f} \cdot \underline{\vartheta}_v - \underline{Y} \|^2 -$$

$$- [\nabla^T \underline{f} \cdot (\nabla \underline{f} \cdot \underline{\vartheta}_v - \underline{Y})]^T \cdot [2 \cdot B - B \cdot A_{GN} \cdot B] \cdot [\nabla^T \underline{f} \cdot (\nabla \underline{f}\, \underline{\vartheta}_v - \underline{Y})]$$

$$\underline{\vartheta}_{v+1} = \underline{\vartheta}_v - B \cdot \nabla^T \underline{f} \cdot (\nabla \underline{f}\, \underline{\vartheta}_v - \underline{Y})$$

The theorem is equivalent with tho following: the matrix $[2B - BA_{GN}B]$ is positive definit, so the goal function will decrease except the case when $\nabla^T \underline{f} \cdot (\nabla \underline{f}\, \underline{\vartheta}_v - \underline{Y}) = \underline{0}$ that is the optimum has been reached. We can write the following $A_{GN} = U^T D U$ where $U^T U = I$ and D is diagonal matrix with elements d_i. The d_i values are the eigenvalues of matrix A_{GN}, as the matrix was positive semidefinite the values are positive or zero. Using this notation the matrix $[2B - BA_{GN}B]$ can be written in the form $U^T S U$ where S is a diagonal matrix with elements s_i:

$$s_i = \frac{d_i + \lambda + \alpha}{(d_i + \lambda)^4} \cdot [d_i^2 + (3\lambda - \alpha) \cdot d_i + 2\lambda^2]$$

As all the d_i values are nonnegative, the λ number is positive and $\alpha = \lambda$, the s_i values are positive, so the $[2B - BA_{GN}B]$ matrix is positive definit. q.e.d.

3. <u>NUMERICAL COMPARISON</u>

The new method has been compared to other methods which calculate the goal function once or only a few times in every iteration step. The methods were the following: Gauss-Newton, Marquardt, a modification of the Gauss-Newton method, where a positive number has been added to the diagonal of the matrix to be inverted. The basis of comparison was the number of iterations. The problems arose in the practice in the field of biology and agriculture. This results that the optimum values are far from zero. The test problems were as follows:

1. Empirical relation between diversity and cell number of algae. Four function with three parameters have been tried to fit on data, the number of points was 225. The results can be seen in Table 1. /The star means that the algorithm has diverged./
2. Logistic function with 4 parameters has been fitted on cattle growth data, 108 data series were avaiable, the numbers of points were between 14 and 17. Table 2. shows the average

and the standard deviation of numbers of iteration and the numbers of diverging cases.

3. Empirical relation between chemical compounds and discharge rate of a river. A logarithmic function with 2 parameters has been fitted on 3 data series, the number of points was 1280. The results can be seen in Table 3.

These results show that the new method is between the Marquardt and Gauss-Newton methods. It is less stable but it needs less goal function evaulations than the Marquardt method, and it is more stable but slower than the Gauss-Newton method.

Based on previous curve fits an algorithm has been developed for automatic choosing among curvefitting methods. This algorithm decides in every iteration step which method will be carried out. The test problems were the following: the first and the third ones are the same as the first and third problems in the numerical comparison of the new method, the results can be seen in Tables 4 and 6. In the second problem a function with 5 parameters has been fitted in two data series, the number of points were 302 and 241.

The numerical results showed that in most cases the automatic method was the quickest except that the choice itself needed some computer time. Here the basis of comparison was the computer time, as a common basis for direction determination and function evaluation.

REFERENCES

Bard Y. /1974/, Nonlinear parameter estimation, Academic Press, New York and London.

Marquardt D.W. /1963/, An algorithm for loast squares estimation of nonlinear parameters, J. Soc. Indust. Appl. Math. Vol.11 No.2. 431-441.

Meyer R.R. /1970/, Theoretical and computational aspects of nonlinear regression, in: Nonlinear programming ed. J.B.Rosen, O.L. Mangasarian, K.Ritter, Academic Presss, New York and London, 465-486.

method / function	Gauss-Newton	Marquardt	new meth.	mod. G-N
$b_1 \cdot x^2 + b_2 \cdot x + b_3$	2	3	3	3
$\frac{b_1}{x+b_2} + b_3$	25	42	25	*
$-e^{b_1 \cdot x + b_2} + b_3$	22	22	22	22
$\sqrt{b_1 - b_2 \cdot x^2} + b_3$	*	4	4	4

Table 1.

method	num. of div.	average	stand. dev.
Gauss-Newton	7	4.95	0.89
Marquardt	0	7.02	1.64
new meth.	5	5.39	0.70
mod. G-N	4	6.05	0.74

Table 2.

method	I.	II.	III.
Gauss-Newton	2	2	2
Marquardt	3	3	3
new.meth.	2	2	3
mod. G-N	3	3	4

Table 3.

method / function	Gauss-Newton	Marquardt	Meyer	Hartley	Autom.
$b_1 \cdot x^2 + b_2 \cdot x + b_3$	6.8	6.5	18.7	24.9	16.1
$\frac{b_1}{x+b_2} + b_3$	*	12.1	28.4	28.7	14.6
$-e^{b_1 \cdot x + b_2} + b_3$	97.4	15.0	53.6	68.8	16.9
$\sqrt{b_1 - b_2 \cdot x^2} + b_3$	*	15.6	45.1	*	19.7

Table 4.

method	I.	II.
Gauss-Newton	*	*
Marquardt	60.0	60.0
Meyer	71.5	138.6
Hartley	71.7	*
Autom.	47.6	125.6

Table 5.

method	I.	II.	III.
Gauss Newton	33.5	33.8	29.3
Marquardt	94.0	50.5	42.8
Meyer	*	*	*
Hartley	*	140.9	156.0
Autom.	122.6	61.2	71.2

Table 6.

Optimization in Non-Parametric Regression

M. Lejeune, Lausanne

SUMMARY : Non parametric regression is approached through linear estimation, a less restrictive view than the kernel approach since the solution can be adaptative for any pattern of distribution of the abscissae. Local polynomial regression happens to be optimal in the sense of minimum variance for a given order of biais reduction. This last notion is in fact the main issue. Links with kernel theory and computational aspects are given.

KEYWORDS : Non parametric estimation, regression, kernel.

Following the kernel approach introduced by Rosenblatt (1956) and extended by Parzen (1962) for non parametric estimation of a density one can derive an estimator for a regression curve r(x) of Y on X in the most natural way, via conditional expectation, as :

$$\hat{r}(x) = \sum_{i=1}^{n} Y_i K((x-X_i)/h) \;/\; \sum_{i=1}^{n} K((x-X_i)/h)$$

where $\{(X_i,Y_i) \mid i=1,\dots,n\}$ denotes the bivariate sample.

This estimator has been first proposed by Watson (1964) and then studied for asymptotic properties by various authors among which Rosenblatt (1969) himself. In that framework the concern has been primarily the optimal choice of h, the window width, and secondarily the choice of the kernel function K. In fact this last aspect has been rather neglected because of intuitive reluctance towards kernels outside the non-negative class of functions. And, truly, there is not much of an issue within this mere class.

In this paper we will consider the problem of optimality from a more general point of view and we will see that it leads to the natural solution of local polynomial regression. Before all, let us note that practically we could just be interested in the purely descriptive problem of fitting a smooth curve through a set of points (x_i,y_i) of $\mathbb{R}^2$, the finality of which would be restoration or prediction of y-values given x-values. In order to overcome the ambiguities of smoothness of the curve and goodness of prediction we find convenient, as a statistician, to integrate our data into a probabilistic model that we want to keep as crude as possible.

BIAIS ORDER AND MINIMUM VARIANCE

Let us assume the existence of a family of real-valued random variables $\{Y(x) \mid x \in I \subset \mathbb{R}\}$, where I is an interval, with expectations r(x) and common variance σ^2. By $\{Y_i \mid i=1,\dots,n\}$ we denote in short a finite sample of the family, corresponding to x-values $\{x_i \mid i=1,\dots,n\}$ and we pose the problem of estimating r(x). Note that our view is a conditional one with respect to the x-values. Let x be fixed, we will first look at optimality for an estimator within the class of all linear estimators written briefly as :

$$\Sigma w_i Y_i \,/\, \Sigma w_i \tag{1}$$

and we shall set some limitations later.

For an estimator of the form (1) biais and variance are respectively equal to :

$$\Sigma w_i [r(x_i)-r(x)] \;/\; \Sigma w_i \tag{2}$$

$$\sigma^2 . \Sigma w_i^2 \,/\, (\Sigma w_i)^2 \tag{3}$$

It has been usual in the literature to consider a quadratic loss function leading to minimizing least square error. Our approach will be less restrictive since we will try to control separately for biais and for variance following thus Gasser and Muller (1979). Whereas we can easily express minimization of the variance (3) we are only able to consider reduction of biais up to a certain order of polynomial approximation. Let us assume derivability of r(x) up to any desired order, the biais (2) can then be expanded as :

$$(\Sigma w_i)^{-1} \, [(\Sigma w_i x_i) r'(x) + (\Sigma w_i x_i^2) r''(x)/2 + \dots] \tag{4}$$

where we also assume for convenience that the x_i's are centered on x.

<u>Definition</u> : An estimator is said to be of biais order k if it has zero biais with respect to the Taylor approximation of r(x) of degree k-1 and non zero biais for degree k.

We will now search for optimality as minimum variance within the class of linear estimators of a given biais order, say k, and postpone overall comparison. From (3) and (4) the problem is to find w_i's such that :

$$\Sigma w_i^2 = \min!$$

with k constraints
$$\begin{cases} \Sigma w_i = 1 \\ \Sigma w_i^s = 0 \quad \text{for } s=1,\dots,k-1 . \end{cases}$$

<u>Theorem</u> : The minimum variance linear estimator of biais order k for r(x), x being

fixed, is given by the polynomial fit of degree k-1 evaluated at x.

Proof : Let Y and w denote the nx1 vectors of components Y_i's and w_i's respectively and let

$$\underset{n\times k}{A} = \begin{pmatrix} 1 & x_1 & x_1^2 & \dots & x_1^{k-1} \\ 1 & x_2 & x_2^2 & \dots & x_2^{k-1} \\ \vdots & & & & \vdots \\ 1 & x_n & x_n^2 & \dots & x_n^{k-1} \end{pmatrix}$$

and
$$\underset{1\times k}{u'} = (1\ 0\ \dots\ 0) \quad .$$

Then, using a kx1 vector of Lagrange multipliers $\lambda=(\lambda_i)$ we have to minimize

$$w'w - \lambda'(A'w-u)$$

which implies
$$\begin{cases} 2w - A\lambda = 0 \\ A'w = u \end{cases}$$

Multiplying the first equation by A' we have $2A'w=A'A\lambda$ or $2u=A'A\lambda$ and, assuming that A'A is of rank k (i.e. there are at least k distinct values among the x's in order to yield uniqueness), we can solve for λ and substitute to obtain

$$\hat{w} = A(A'A)^{-1}u \quad .$$

The minimum variance estimator will be :

$$\hat{r}(x) = \hat{w}'Y = u'(A'A)^{-1}A'Y$$

in which we recognize the constant coefficient, through least square estimation, of the polynomial regression model for Y. Since the x_i's are centered on x the theorem is proved.

Let us now look at limitations for the above solution. First note that if we let the estimation rest on the whole sample all we have accomplished is a polynomial fit of degree k-1 across the data. Obviously this will keep the biais high because reduction of biais has been designed through local approximation only, via Taylor series. So we should limit the range of values x_i's to be taken into account and be ready to loose on the variance, which means that we cannot escape the usual window width problem encountered in kernel estimation as well as in other approaches. Nevertheless, the window width being fixed, the minimum variance solution is the polynomial solution above where n stands for the number of points within the window.

In the presence now of a limited window we have to deal with the discontinuity of $\hat{r}(x)$, which is a drawback in relation to descriptive aspects. This can be easily overcome in the following manner. Let us impose weights $\{v_i \mid i=1,\dots,n\}$ on the x_i's and note that the weighted polynomial solution still yields an estimator of biais order k. Namely, defining V as the diagonal matrix of the v_i's, that solution is :

$$\hat{w}_V'Y = u'(A'VA)^{-1}A'VY$$

and obviously $\hat{w}_V$ verifies $A'w_V = u$. Thus to get a differentiable estimation along x we have to choose v_i according to a differentiable function. A convenient choice is the biquadratic weights :

$$v_i = \begin{cases} [1-((x_i-x)/h)^2]^2 & \text{when } |x_i-x| \leq h \\ 0 & \text{else} \end{cases}$$

As we shall see this will be done at low cost for the variance.

EVALUATION AND COMPARISONS

In our approach we cannot speak literally of kernel estimation since the coefficients w_i for the Y_i's are specifically tailored to the pattern of locations of the x_i's within the window. More precisely, in the simple case V=I, w_i is the value at x for the polynomial fit on the ordinates 0 for x_j, when $j \neq i$, and 1 for x_i. However if we assume that the x_i's are evenly distributed on the interval I we are in classical kernel theory since the pattern is invariable and comparisons are possible. This has been done in an other study by Lejeune (1983) and its main results are briefly mentionned.

Polynomial regression of degree k-1 yields yet the minimum variance kernel of biais order k. Its expression can be derived from the Legendre polynomial of degree k-1 and k is necessarily even when there are no boundary effects, i.e. when the window is entirely contained in I. For k equal to 2 and 4 one finds the kernels 1/2 and $(3/8)(3-5x^2)$. If we want to impose continuity at both ends of the support of the kernel we can use weights v_i's according to the quadratic function $1-u^2$ in the polynomial regression. Then we get the kernel of Epanechnikov (1969) $(3/4)(1-x^2)$ for k=2 and the kernel $(15/32)(1-x^2)(3-7x^2)$ for k=4. It can be proven that these kernels have minimum variance among kernels with minimal number of changes of sign (equal to k-2) and this result is likely to extend to any value for k (see Gasser and Muller, 1979). These kernels can also be derived from Legendre polynomials. Using biquadratic weights to attain derivability as suggested preceedingly, one will obtain, for biais order 2k, a kernel of a polynomial form of degree 2k+2.

Let us see now how much we loose in variance when we use non constant weights. From (3) we get, up to a constant equal to $\sigma^2(b-a)/(nh)$ where I=[a,b], the following values for the variance :

			k=2	k=4	k=6
constant weights	(optimal)	:	0.50	1.13	1.76
weights as $1-x^2$	(continuous)	:	0.60	1.25	1.89
weights as $(1-x^2)^2$	(differentiable)	:	0.71	1.41	--

These figures show that it is true that the shape of the kernel is not crucial only if one remains in the same biais order class, a fact that has been overlooked in the past. Consequently the main issue is biais order, not minimum variance. Since the variance is increasing quite substantially with k we must compromise with a moderate value of k. In practice we have experienced that polynomial regression of degree 2 is very satisfying. Finally let us remark that the comparison of unweighted regression and weighted regression is not quite fair to the latter because, for a same value of h, the slight loss in variance will be compensated by a lower biais.

COMPUTATIONAL ASPECTS

Suppose we want to estimate r(x) at some fixed point x through local polynomial regression of degree r and let J be the set of indices for the x_i's inside the window centered on x. The solution to the corresponding normal equations can be formally expressed in terms of a ratio of two determinants and further in terms of the sums :

$$S_s(x) = \sum_J (x_i-x)^s \qquad s=0,\dots,2r$$

$$S'_s(x) = \sum_J (x_i-x)^s y_i \qquad s=0,\dots,r \quad .$$

Suppose we increment now x to x+t and let J' be the set of indices for the window centered on x+t. Let us simply denote the sums corresponding to the new window by $S_s(x+t)$ and $S'_s(x+t)$. With $L=J\wedge\bar{J}'$ and $R=\bar{J}\wedge J'$ we can break $S_s(x+t)$ into three terms :

$$S_s(x+t) = \sum_{J'}(x_i-x-t)^s = \sum_J (x_i-x-t)^s - \sum_L (x_i-x-t)^s + \sum_R (x_i-x-t)^s \quad .$$

The first term can be derived directly from the sums $S_k(x)$, with $k \leq s$, by expanding $[(x_i-x)-t]^s$ and thus only the two last terms have to be newly computed. This will be quite time-saving when incrementing x along the whole range of values of interest. If we now consider weighted regression with biquadratic weights the solution to the normal equations is identical except that we have to substitute $T_s(x)$ to $S_s(x)$ where:

$$T_s(x) = S_{s+4}(x) - 2S_{s+2}(x) + S_s(x) \quad .$$

Note that the algorithm involves a number of operations proportional to the sample size, say n. However, unless n is extremely large, it will still be advantageous to first sort the data for increasing x_i's, a process with a number of steps proportional to nlogn. A subroutine consisting of about 100 fortran statements, non including the sorting routine, has been written with the following arguments :

```
SUBROUTINE REGLISSE(X,Y,N,XT,YT,M,D)
```

where : X and Y are N-dimensional vectors containing the observations,
XT is a M-dimensional ($M \geq 1$) vector of abscissae of interest,
YT is the corresponding M-dimensional vector of estimations,
D is the window width.

DEVELOPMENTS AND CONCLUSIONS

By the direct approach of linear estimation we have seen that we encompass a larger class of estimators than kernel estimators which appear to be too restrictive for not being adaptative to the data. In particular, local polynomial regression will take care of boundary effects, which makes it quite efficient for time series analysis, i.e. trend smoothing and periodogram smoothing, since no information is lost on the edges. Note that this method has been selected empirically for a long time in time series and more recently in EDA by Cleveland (1979).

By formally differentiating the estimator of the empirical distribution function one obtains an estimate of density with biais order k-1. Preliminary results and trials in that direction are quite promising.

The problem of chosing h has been left aside in our presentation and we think that the approach of cross validation would be suitable here given the low cost in computation. It can be shown, though, that a higher biais order also means more stability of the estimation with respect to varying h (Lejeune, 1983). From an academic point of view we finally remark that the problem of optimality in the continuous or differentiable class of estimators is still open.

References :

Cleveland W.S. (1979), Robust locally weighted regression and smoothing scatterplots,JASA,74,829-836.

Epanechnikov V.A. (1969), Nonparametric estimation of a multidimensional probability density,Theory Prob. Appl.,14,153-158.

Gasser T.,Muller H.G. (1979), Kernel estimation of regression functions, in Smoothing techniques for curve estimation,ed. Gasser & Rosenblatt, Springer, Heidelberg,23-68.

Lejeune M. (1983), Régression non-paramétrique par noyaux, Manuscript submitted for publication.

Parzen E. (1962), On estimation of a probability density function and mode, Ann. Math. Stat.,33,1065-1076.

Rosenblatt M. (1956), Remarks on some nonparametric estimates of a density function, Ann. Math. Stat.,27,832-837.

Rosenblatt M. (1969), Conditional probability density and regression estimators, in Multivariate Analysis, ed. Krishnaiah,25-31,Acad. Press,N.Y.

Watson G.S. (1964), Smooth regression analysis, Sankhya A,26,359-372.

Maximum Likelihood Estimation for Mixtures of Distributions

L.E. Meester, Delft

Summary

This paper describes a simulation study of (small sample) maximum likelihood estimation for mixtures of distributions. Maximum likelihood (ML) leads to a multi-dimensional maximization problem for which a modified Newton method and the EM algorithm are used. The moment and Kabir method are applied to provide initial points. For twenty-seven different mixture distributions a number of samples is generated by means of a pseudo-random number generator. Frequently the supplied initial points are not admissible. Although an ad hoc method for initial points behaves well, still no estimates are obtained for a number of samples.

Keywords: mixtures of distributions, maximum likelihood, Newton method, EM algorithm, initial point dependence

1. Introduction

The project which is reported here is a feasibility study, performed at the Philips Research Laboratories (Eindhoven, the Netherlands), with the aim to assess the applicability of mixture parameter estimation methods, having in mind application to (large) industrial cases. Part of the study is theoretical and consists of selecting and combining methods. The remainder is practical and deals with application of the methods to a number of "pseudo-real" cases, which are chosen with some care. The samples are chosen as small as seems reasonably possible, in order to study the performance of the estimation methods at their limit.

2. Problem definition

In this paper univariate mixtures of distributions are considered, having a probability density function (PDF) of the type $g(x;\alpha)=\Sigma_{j=1}^{k} p_j f(x;\theta_j)$ with $x\in\mathbb{R}$, $\alpha=(\theta_1,\dots,\theta_k,p_1,\dots,p_k)$, $k\in\mathbb{N}$, $k\geq 2$, $p_j\in(0,1)$, $p_1+p_2+\dots+p_k=1$ and $f(x;\theta)$ a member of the exponential class of probability density functions, with $\theta\in\Omega\subset\mathbb{R}$; $\theta_1<\theta_2<\dots<\theta_k$ is assumed. The densities $f(x;\theta_1),\dots, f(x;\theta_k)$ are called the components, the scalars $p_1,\dots, p_k$ the mixing proportions.

A necessary condition for mixture parameter estimation is that the considered mixtures be identifiable, i.e. the representation of the mixture density by its parameters must be unique. For the cases considered this condition is satisfied (proofs by Teicher, 1963).

Assuming identifiability, the estimation problem is:

Given a sample $x_1, x_2,\dots, x_n$ of independent observations from a mixture having PDF $g(x;\alpha)$, with k and the type $f(x;\theta)$ known, determine an estimate $\hat{\alpha}=(\hat{\theta}_1,\dots, \hat{\theta}_k, \hat{p}_1,\dots,\hat{p}_k)$.

The complexity of the estimation depends on a number of factors:

1. The type of componentdistribution.
2. The number of components. With k components there are 2k-1 parameters to be estimated.
3. The mixing proportions. A small mixing proportion results in few observations on the corresponding component, and thus the estimate of its parameter will be less accurate.
4. The degree of separation of components. With small differences among component parameters it is more difficult to discern the components than when differences are large. (The notion "degree of separation" is difficult to quantify, but it appeals intuitively).
5. The sample size. The accuracy of the estimates depends on the number of observations, and through this the number of discernible components is influenced by the sample size.

In literature concerning mixture parameter estimation often cases with normal components are treated, other distributions appear less frequently. Theoretical papers treat the general case $k \in \mathbb{N}$, but in papers dealing with applications usually two or three component cases are treated, four component cases being exceptional. For a review of literature on the subject the reader is referred to the book by Everitt and Hand (1981).

3. Maximum likelihood estimation

This estimation method is to be preferred to other estimation methods. Several authors (Hasselblad,1969; Holgersson and Jorner,1978; Everitt and Hand,1981) give evidence which supports this choice with respect to the moment method. Another motive is the asymptotic property of ML estimates (asymptotic unbiasedness and minimum variance; estimates of the asymptotic variances can be obtained). This property may to some extent be preserved for non-asymptotic cases.

The likelihood function is defined by $\mathcal{L}(x_1, \dots, x_n;\alpha) = \Pi_{i=1}^{n} g(x_i;\alpha)$. Let $L = \ln \mathcal{L}$, the log-likelihood function (LLF). To obtain a ML estimate the following maximization has to be performed:

$$\max_{\alpha} \ L(x_1, \dots, x_n;\alpha)$$

$$\text{sub} \quad p_j \in (0,1), \ j=1,\dots,k$$
$$p_1+p_2+\dots+p_k=1$$
$$\theta_j \in \Omega, \ j=1,\dots,k$$

Two methods for the maximization are considered:

1. A modified Newton method, which includes a line search and modifications by Fiacco and McCormick (1968); in contrast with the original Newton method this version provides a sequence of iteration points with increasing log-likelihood.
2. The EM algorithm (see Dempster et. al., 1977; and Hasselblad, 1969).

It is known that with the second method convergence may be extremely slow, a disadvantage which is not present with the first method. A reliable convergence rule for the EM algorithm is difficult to formulate. Everitt and Hand (1981) suggest that for the EM algorithm the choice of initial points is not critical. However, this hypothesis seems to lack a solid basis, so that the Newton method is to be preferred.

4. A simulation

Intending to assess which cases can be treated using ML estimation, a simulation is performed. Samples are generated using a pseudo-random number generator. Choices with respect to the aspects 1 to 5 mentioned in section 2 are:

1. Component distribution types: a. normal, with mean θ and variance fixed at 1; b. exponential, with parameter θ (mean θ^{-1}); c. binomial, with parameters 10 and θ.
2. Two, three and four components.
3. Mixing proportions: $p_j = k^{-1}$, $j=1,\ldots, k$.
4. Component separation is based on the differences of the means of adjacent components for the normal case. A difference of 1 is considered small, 2 medium and 3 large. For the other two types parameters are computed using the following (ad hoc) criterion: $\left|\frac{\Pr(\underline{x}_1>\underline{x}_2)}{\Pr(\underline{x}_1\neq\underline{x}_2)} - \frac{1}{2}\right|$, for $\underline{x}_1$, $\underline{x}_2$ being drawn from adjacent component distributions, should be equal to the value in the corresponding normal case. Using this criterion, which originates from the field of distribution-free methods, the degrees of separation for different types of distributions can be made "equal". The parameter values resulting from this computation are given in Table 1. For the normal and exponential mixtures the values in column $1,\ldots, k$ are used (for k components); in the binomial case column 2 and 3 for two components, 2, 3 and 4 for three components and all for four components.
5. Sample sizes are chosen deliberately small: 100, 150 and 200 for two, three and four components, respectively.

Table 1 Component parameter values used for the simulation.

distribution	separation	1	2	3	4
normal	small	0	1	2	3
	medium	0	2	4	6
	large	0	3	6	9
exponential	small	1	3.16	10	31.6
	medium	1	12	140	1600
	large	1	58	3300	193000
binomial	small	.249	.369	.500	.631
	medium	.068	.244	.500	.756
	large	.0055	.135	.500	.865

So, there are twenty-seven different mixture distributions for each of which ten samples are generated. The moment method (Blischke, 1964) and the Kabir method (Kabir, 1968) are used to supply initial points for the Newton iteration. However, other initial points (among others, the true parameter vectors) are used as well, but these are based on foreknowledge of the mixture samples. Hence, there is an impression how many times estimates would be found if it were "real-life" cases, and a lower bound for the total number of solvable cases can be given. (To observe differences between Newton and EM, 50 EM-steps are performed for each of the mentioned initial points, afterwards switching to Newton; differences in the limit-points are rare, so the simulation is treated as if EM is not used at all).

5. Obstacles

There are some problems to overcome in applying the Newton method to the generated samples of which we mention a few:

1. Numerical problems. Numerical second derivatives have been used but these are too inaccurate, so they are replaced by analytical derivatives. Further, over- and underflow have to be avoided.
2. Degenerations. A mixture distribution is called degenerate if $p_j=0$ for some j or $\theta_i=\theta_j$ for some $i\neq j$; that is, the mixture can be represented with (at least) one component less. It appears that degenerations may constitute points of attraction for the Newton method, also in cases where the LLF possesses a non-degenerate local maximum. Degenerations constitute subspaces of the parameter space which, once entered, the Newton iteration cannot leave. So, for a number of samples no (non-degenerate) solutions are found. No means to overcome this, beside trying as many initial points as possible, are found.
3. Initial points. The methods to supply initial points prove to be insufficient, because often the computed values are inadmissible (e.g. complex). Therefore, another way to generate initial points is devised. For this (ad hoc) method the observations are ordered and k-1 separation points are computed (depending on the type of distribution, the minimum and maximum observation). These points divide the sample into k groups of which each is treated as consisting of observations from one component. Using this division values of the mixture parameters are computed. The method shall be referred to as "the interval method".
4. Multiple maxima. Some cases with two local maxima of the LLF are found. It is remarkable that in some of these the parameter values corresponding with the highest maximum do not resemble the true values most. Furthermore, the values of the estimated asymptotic variances for these estimates are relatively high.

6. Results

Of the vast amount of data only a small part can be presented. Rather than giving detailed information on a few of the twenty-seven cases, part of the information concerning all the cases is given. It is recalled that for each of the twenty-seven cases ten samples are drawn; in how many of these samples actually estimates are obtained is tabulated in Table 2. It shows that the frequency with which solutions are found decreases with increasing number of components and decreasing component separation. An exception to this is found with the four component binomial mixtures; here the small parameter values .068 and .0055 (see Table 1) cause numerical difficulties, resulting in frequent non-convergence of the Newton iteration, which explains the small number of solutions.

Table 2 Number of times ML estimates are obtained in ten samples, for each of the twenty-seven mixture distributions.

distribution	two components			three components			four components		
	(s)	(m)	(l)	(s)	(m)	(l)	(s)	(m)	(l)
normal	10	10	10	7	10	10	2	10	10
exponential	10	10	10	9	10	10	8	10	10
binomial	6	10	10	4	10	10	0	4	3

Table 3 shows the number of estimates found a. using the moment and Kabir method for initial points (KM); b. using the interval method (Int); c. using the three methods (KMI). In the fourth column the total number of ML estimates is tabulated. The numbers concern 90 samples each for two, three and four components. It is obvious that a considerable number of solvable cases is added due to the use of the interval initial points. Of these, a large number (56 out of 66) concern the exponential mixtures, for which the other two methods often yield inadmissible initial points. The differences between the columns KMI and ML are due to the use of additional initial points, such as the true parameter vectors (see also section 4).

Table 3 Number of cases (out of 90) for which estimates are found for various (groups of) initial point methods; and the total number of ML estimates found.

k	KM	Int	KMI	ML
2	76	86	86	86
3	41	67	69	77
4	22	50	50	57

Table 4 concerns the bias of the estimates. For each group of ten or less samples for which an estimate is found, the normalized bias (the difference between the mean of the estimated values and the true value divided by the standard deviation of the mean) is computed for each parameter. Let these values be t_i, $i = 1,\ldots,2k$ and let $t = \max_i |t_i|$. The values in Table 4 are the percentiles corresponding with these values: $\Pr(|\underline{t}| > t)$, $\underline{t}$ having a t-distribution with appropriate degrees of freedom.

Table 4 Percentiles corresponding with the maximal biases of the estimated parameters for each of the twenty-seven cases.

distribution	two components (s)	(m)	(l)	three components (s)	(m)	(l)	four components (s)	(m)	(l)
normal	.06	.46	.08	.008	.013	.07	-	.05	.28
exponential	.13	.46	.64	.001	.009	.15	.004	.11	.17
binomial	.39	.37	.57	.12	.17	.64	-	.02	-

These values are somewhat too small, because testing against a t-distribution is too severe. With independent estimates for the parameters $\theta_1,\ldots,\theta_k, p_1,\ldots,p_{k-1}$ the test should be against the maximum of 2k-1 t-distributions and a percentile β in Table 4 would be replaced by $1-(1-\beta)^{2k-1}$. Since assuming independence is not justified the results on bias are summarized in this way. A more extensive treatment of all the results of the simulation is to be found with Meester (1983).

7. Conclusions drawn from the simulation

1. It is advisable to use several methods to supply initial points for the Newton iteration. Where estimation methods, like the moment and Kabir method, prove to be insufficient, the ad hoc interval method supplies adequate initial points.
2. Only for a part of the treated small sample cases the ML estimates appear to be unbiased, that is for all the two component cases and some of the three and four component cases (see Table 4).
3. The estimated asymptotic variances of the ML estimates are approximately equal to their samples variances, at least for the two component case.
4. The Newton and EM method seem to be equally dependent on the initial point. The Newton method is faster than EM for mixtures with badly separated components, for very widely separated mixtures EM is faster. The steps of the EM algorithm may be enlarged by using a line search, which might speed up convergence.
5. The results of this project suggest that, to obtain highly accurate estimates for the parameters of a mixture with many or badly separated components, a very large sample has to be taken.

Acknowledgement

The author is grateful to H.J. Prins, Philips Research Laboratories, and to Prof. Dr. M.S. Keane and Dr. F.M. Dekking, Delft University of Technology, for their contributions to the project.

References

Blischke W.R. (1964), Estimating the parameters of mixtures of binomial distributions, J. Amer. Statist. Assoc., 59, 510-528.

Dempster A.P., Laird N. and Rubin D.B. (1977), Maximum likelihood from incomplete data via the EM algorithm, J. R. Statist. Soc., series B, 39, 1-38.

Everitt B.S. and Hand D.J. (1981), Finite mixture distributions, Chapman and Hall, London.

Fiacco A.V. and McCormick G.P. (1968), Non-linear programming, sequential unconstrained minimization techniques, Wiley, New York.

Hasselblad V. (1969), Estimation of finite mixtures of distributions from the exponential family, J. Amer. Statist. Assoc., 64, 1459-1471.

Holgersson M. and Jorner U. (1978), Decomposition of a mixture into normal components: a review, Int. J. Bio-Med. Computing, 9, 367-392.

Kabir A.B.M.L. (1968), Estimation of parameters of a finite mixture of distributions, J. R. Statist. Soc., series B, 30, 472-482.

Meester L.E. (1983), A simulation study of the small sample behaviour of parameter estimation methods for mixtures of distributions, Philips Research Laboratories Report no. 5885.

Teicher H. (1963), Identifiability of finite mixtures, Ann. Math. Statist., 34, 1265-1269.

HAUS 82: a New Tool for Nonlinear Fitting and the Study of Nonlinear Models

A. Messean, and O. Nicole, Jouyen–Josas

SUMMARY : We developped a general nonlinear least squares software characterized by :
- a special attention given to the naive user through interactive definition of the model and clear options setting
- more sophisticated facilities oriented towards the study of nonlinearity effects
- a logical structure which allows the use of the fitting algorithm in a higher level environment as simulation or P.D.E. integration softwares (CSMP, FORSIM), general purpose statistical softwares (Consistent System, ...).

Key-words : nonlinear least squares, conversational driver, nonlinearity measures

1. INTRODUCTION

At the COMPSTAT 82 meeting, we presented the first step of our nonlinear least squares software project (LE ROUX et al., 1982), through a comparison of nonlinear model fitting algorithms and the implementation of an orthogonality convergence criterion. Our project developped and gave birth to the HAUS82 program.

A threefold objective was aimed at : at a first level an easy-to-use software, at a second level , the possibility of studying nonlinearity effects on statistical characteristics of estimations and last but not least, affording facilities of use through general purpose softwares.

Each of the three next paragraphs is devoted to each of these characteristics in turn.

2. METHODOLOGICAL AND STATISTICAL ASPECTS

The main features of the HAUS82 software are the following ones:

1) The model to be fitted can be defined by a system of algebraic or differential equations.
2) The algorithm used to solve the nonlinear least squares problem is the wellknown Levenberg-Marquardt algorithm. It was chosen after comparison of the most reknown fitting algorithms (LE ROUX et al., 1982). The orthogonality convergence criterion and its combined use with classical "relative change" criteria has also been previously presented in the same paper.
3) Several kinds of weights can be defined, fixed or varying with parameters, allowing parametrized error variances.
4) Bounds values can be introduced for the parameters.

The NONLIN (METZLER, 1975) output frame has been retained, as one of the best complete and wellknown: parameters and expected responses estimations with their standard errors, residuals,..., 95 % confidence intervals and extreme points of the 95 % confidence ellipsoid, variances-covariances and correlation matrices of the parameters.

The residual sum of squares and mean squares are complemented by the orthogonality criterion value. Graphical representations of observed and fitted responses are displayed.

With the other more sophisticated types of facilities, the statistician can use HAUS82 to study nonlinearity effects :

1) Graphical displays of likelihood contours which were shown to realize a good compromise between linearized ellipsoidal regions and more complex lack of fit confidence regions (MESSEAN, 1982).

2) Extreme points on these likelihood contours in each parametrical direction, in order to get conservative confidence intervals for parameters.

3) Nonlinearity measures, structuring nonlinearity in an intrinsic component (model + data) and a parametric component showing the effect of a specific parametrization. In addition to the classic Beale's theoretic measures (BEALE, 1960), HAUS82 computes more expressive measures which are the relative curvatures of the solution locus at the least squares point (BATES and WATTS, 1980) :

- the intrinsic ones : the quadratic mean and the maximum of the normal curvature at the least squares solution.
- the parametric ones : the quadratic mean and the maximum of the tangential curvature at the same point. As a by-product, second order biases approximations (Box's biases) are computed for the parameters.

HAUS82 also allows computing all these nonlinearity measures for any experimental design and theoretical error variance independently of any previous fit.

3. A LOGICAL STRUCTURE FOR A GENERAL USE OF THE FITTING ALGORITHM

In the usual nonlinear software, the user has to write a subroutine describing the model to be fitted. This subroutine is called by the fitting algorithm every time theoretical responses values for specific parameters are needed, in particular for the optimal updating of the parameters.

We call this classical operating way the "autonomous mode". It is satisfactory when the theoretical responses can be easily calculated as in the case of algebraic equations for the model, or even differential equations if an external integration routine is available.

But there are cases when the theoretical responses evaluation has to be done at a higher logical level than the fitting algorithm itself in a "carrier program". For example, for some complex differential problems as partial differential equations , high level specific softwares are available: FORSIM, CSMP, ... But they generally cannot be used through a simple CALL statement : at the contrary, they have to be the master program of a run and then have to control the external iterative fitting process by passing it updated values, each time the fitting routine returns to the carrier program with new parameter values. We call this operating way the "inverted mode".

We illustrate this mecanism by the following two diagrams of the classical and inverted structures.

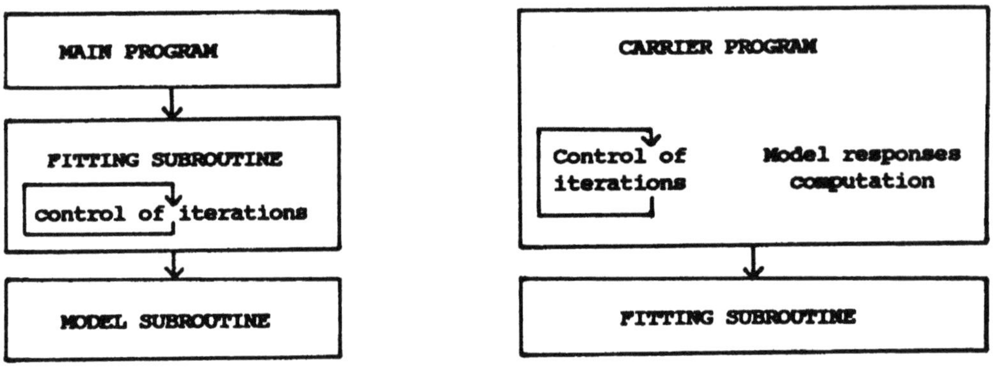

Autonomous Mode — Inverted Mode

The carrier program can be an ordinary FORTRAN program or a specific software as CSMP, FORSIM, Consistent System ...

As an example, let us consider the following sample of instructions where the carrier program is the wellknown Continuous System Modelling Program (CSMP).

```
INITIAL

*Initial conditions, initial values and space allocation
INCON Y0=1
      I=1
PARAMETER ALPHA=1.2
STORAGE YCALC(100),YOBS(100),X(100)

*Reading the different data values YOBS,X,...

DYNAMIC

*Model differential equations
.
DY = ALPHA * Y
Y = INTGRL (Y0,DY)
.
NOSORT

*Sampling the theoretical model responses YCALC

       IF (SAMPLE (0., 100., 1.).EQ. 0.0) GO TO 10
       I = I+1
       YCALC (I) = Y
10     CONTINUE

TERMINAL

C CALL OF THE HAUS82 INTERFACE
       CALL PREPHAUS82 (YCALC, YOBS, X, 100, ALPHA, I, ISTOP)
       IF (ISTOP.EQ.1) GO TO 10
       CALL RERUN
10     CONTINUE

END
```

In order to implement the inverted structure of the program, we dealt with two kinds of problems.

The first one is to save all the context of the fitting routine when returning to the carrier program for new response values : as a matter of fact after these calculations the process has to be restarted from the point (the restart point) of the subroutine which needed the new responses values.

This implies to save all the local variables of all the intermediary subroutines which led to this restart point and, secondly, to store the trace of the chaining of these subroutines in order to retrieve it for restarting the fitting process.

The first point was solved by SAVE statements in the FORTRAN 77 code and the second one through the use of a storage stack. To each routine of the chaining a rank number r_i is associated : the rank of the corresponding CALL statement in the routine which calls it. An index, "the stack value", is set in the following way :

if $r_1, r_2, \ldots, r_k$ is the chaining, the stack value will be

$$r_1 * 9^0 + r_2 * 9^1 + \ldots + r_k * 9^{k-1}$$

in which each level of subroutine is represented by a power of 9 (the maximal number of CALL statements in a routine).

The stack value is then used to come back to the restart point by a decoding process. It is also used to indicate the end of the fitting process to the carrier program.

The other problem we had to solve was the correspondence between real and dummy parameters in the routines calling sequences. To do it we used a systematic control of the parameters with arrays'copies when necessary.

Of course, as this inverted structure determines an increase in the computing time we kept the classical one for the autonomous mode.

4. THE CONVERSATIONAL DRIVER PROGRAM

The second major point of the software is the presence of a conversational driver program in the autonomous mode.

This driver can be used in two ways :

- a naive mode by which all the process options are set by a clear dialogue with the user.
- a free mode by which the statistician can compose his own menu through a set of mnemonics.

In both modes an "Help" assistance is available :

. before the start of the process : a complete on-line information on the capabilities of the program (nonlinear fitting, study of nonlinearity)
. at each option setting : meaning and default values if any.

The driver is responsible for the Fortran coding of the model to be fitted and its derivatives when necessary. The user model definition consists in the introduction of the right member of the algebraic or differential equations. It is made easy by a line editor which allows character addition, correction or deletion. This editor also checks the parenthesis balancing. Model definition obeys Fortran syntax with specific conventions :

$p_1, p_2, \ldots$ for the parameters,
$x_1, x_2, \ldots$ for the independent variables.

An original feature is the possibility of storing fractions of algebraic expressions occurring several times in the algebra of the model, as "intermediary" variables ($e_1, e_2, \ldots$). These variables are then used for the concise definition of the model and its derivatives.

For example, let the model $y = p_1 * exp(p_2 * x_1)$. The expression $exp(p_2 * x_1)$ also occurs in the parameter derivatives and it would be worth to define it as intermediary variable e_1.

Finally, the user has to set the options which are about thirty. Let us consider the following sample session (naive mode) :

```
Number of likelihood contours to be displayed [0] ?
1     <------------------------------ user specification of the number of contours
Projection axis of the likelihood contours [1,[illegible] ?
1 3
Critical contour level [0.95] ?
      <--------------------------- default value 0.95 specified by carriage return
Size of the contour grid [11, 11] ?
```

After the model definition and the options setting the user can submit his job in interactive or batch mode.

5. NEXT AND FUTURE DEVELOPMENTS

Our "nonlinear project" is still in progress. The next version of the program will present some logical and statistical improvements :

- the model will be interpreted and the options stored, allowing quick further rerunning with partially modified entries.

- statistical outputs will be complemented :
  - . residuals analyses
  - . model tests
  - . displays of sensitivity functions

- the study of parametrization effects will be enhanced :
  - . second order correction for the asymptotic variances-covariances of the parameters.
  - . automatic determination of the "best" parameter transformation in a class, with quick updating of parametric nonlinearity measures.

- in the future, problems with complex variance-covariance error structure will be dealt with, and new functions added, as optimal designs of experiments and powerful graphical displays of nonlinearity efects.

The code is FORTRAN 77 and about five thousand instructions long.

BIBLIOGRAPHY

BATES, D.M. ; WATTS, D.G. 1980. Relative curvature measures of nonlinearity. J.R.S.S. B40, 1-25.

BEALE, E.M.L. 1960. Confidence regions in nonlinear estimation. J.R.S.S. B22, 41-88.

CARVER, M.B. ; STEWART, D.G. ; BLAIR, J.M. ; SELANDER, W.N. 1978. **The FORSIM VI Simulation package for the automated solution of arbitrarily defined partial and/or ordinary differential equation systems.** Clark River Nuclear Laboratories, Ontario - Canada.

IBM, 1975. Continous system modeling. Program III (CSMP III). **Program Reference Manual, number 5734-X69.**

KLENSIN, J.C. 1980. **Consistent System : Handbook of programs and data.** Laboratory of Architecture and Planning - Massachussetts Institute of Technology, Cambridge - USA.

LE ROUX, S. ; MESSEAN, A. ; VILA, J.P. 1982. Standardized comparison of Nonlinear model fitting algorithms. In Caussinus, H. Ettinger, P. Tomassone, R. (Ed), **Compstat 82 : Proceedings in Computational Statistics, Part I,** 306-311.

MESSEAN, A. 1982. Régions de confiance dans le modèle non-linéaire. **Journal de la Société de Statistique de Paris,** 123(2), 134-143.

METZLER, C.M. ; ELFRING, G.L. ; MC EWEN, J.A. 1976. **A user's manual for NONLIN and associated programs.** Research Biostatistics, The Upjohn Company, Kalamazoo - Michigan.

On Special Purpose Algorithms for the MSAE and the MMAE Regression

S.C. Narula, Richmond, and J.F. Wellington, Erie

In the last quarter century the minimum sum of absolute errors MSAE regression and the minimization of the maximum absolute error MMAE regression have attracted much attention as alternatives to the popular least squares regression. The MSAE and the MMAE regression problems can be formulated and solved as linear programming problems. Several efficient special purpose algorithms have been developed for each problem. Thus one needs two algorithms and two computer codes to find the MSAE and the MMAE regression equations. Our objective in this paper is to highlight the special structure of and the similarities between the two problems; thus making it possible to develop an efficient algorithm (and a computer code) to solve both problems.

Keywords: Chebychev Regression, L_1 Regression, L_∞ Regression, Least Absolute Error, Linear Programming, Minimax Regression

INTRODUCTION

The least squares regression has dominated the statistical literature for a long time. This dominance and popularity of the least squares regression can be ascribed, at least partially, to the fact that the theory is simple, well developed and documented. The least squares regression is optimal and results in the maximum likelihood estimators of the unknown parameters of the model if the errors are independent and follow a normal distribution with mean zero and a common (though unknown) variance σ^2. The least squares regression is very far from optimal in many non-Gaussian situations.

In many practical problems, the errors may not follow a normal distribution, as pointed out by Box (1967) and Cox (1967), in the discussion of Anscombe's (1967) paper.

> Box (1967) stated, "I feel certain that platykurtic (*short-tailed)* distributions do occur in practice for one reason because of deliberate or unconscious truncation and these ought not to be ruled out."
>
> On examining several sets of data, Cox (1967) observed that, "the surprising conclusion was that while there were frequent departures from normality these were about equally often towards long-tailed and towards short-tailed distributions."

The results of Monte Carlo studies of Bourdon (1974) and Forth (1974) on the relative performance of the minimum sum of absolute errors MSAE regression and the minimization of the maximum absolute error, MMAE regression for various symmetric (error) distributions indicate that the MSAE regression performs best for the long-tailed distributions and the MMAE regression for the short-tailed distributions.

The MSAE and the MMAE regression problems can be formulated and solved as linear programming LP problems, Wagner (1959). As such they can be solved by any linear programming algorithm. However, a number of very efficient special purpose algorithms have been proposed for each problem. These algorithms exploit the special structure of the problems. At present, the algorithms of Barrodale and Roberts (1973) and Barrodale and Phillips (1974) appear to be the most efficient for the MSAE and the MMAE regression problems, respectively. Interested reader may refer to Narula (1982) and Narula and Wellington (1982, 1984a) for other special purpose algorithms for the problems.

Until now these two problems have been studied separately which leaves one with the impression that one needs two different algorithms and thus two computer codes to solve for the MSAE and the MMAE regression equations. Our objective in this paper is to show that because of the special structure of and the similarities between the two problems, it is possible to develop an efficient algorithm to solve both problems. The rest of the paper is organized as follows: In Section 2, we give the relevant linear programming formulations of the problems. In Section 3, we highlight the special structure of and the similarities between the problems. We conclude the paper with a few remarks in Section 4.

2. *PROBLEM FORMULATION*

Let $\underset{\sim}{y}$ be a nxl vector of observations on a response variable corresponding to $\underset{\sim}{X}$, a nxk matrix of predictor (regressor) variables. Then

$$\underset{\sim}{y} = \underset{\sim}{X}\underset{\sim}{\beta} + \underset{\sim}{\varepsilon}$$

is the multiple linear regression model, where $\underset{\sim}{\beta}$ is a kxl vector of unknown parameters and $\underset{\sim}{\varepsilon}$ is a nxl vector of unobservable random errors.

Let $\underset{\sim}{b}$, a kxl vector, denote an estimator of $\underset{\sim}{\beta}$ and let $\underset{\sim}{e} = \underset{\sim}{y} - \hat{\underset{\sim}{y}}$, nxl vector represent the observed residuals, where $\hat{\underset{\sim}{y}} = \underset{\sim}{X}\underset{\sim}{b}$. We shall use $\underset{\sim}{b}$ to denote the MMAE and the MSAE estimator of $\underset{\sim}{\beta}$.

For the MMAE regression, our objective is to find $\underset{\sim}{b}$ that minimizes (over all observations) the $\underset{i=1,\ldots,n}{\text{maximum}}\ e_i$. Let $e = \underset{i=1,\ldots,n}{\text{maximum}}\ e_i$. Arthanari and Dodge (1981, pg. 96-97) have shown that the MMAE estimate $\underset{\sim}{b}$ is the solution of the following LP problem.

(PL_∞) Maximize e^*

Subject to

$$\underset{\sim}{y}e^* + \underset{\sim}{X}\underset{\sim}{b}^* + \underset{\sim}{S} = \underset{\sim}{1} ,$$

$$\underset{\sim}{0} \leq \underset{\sim}{S} \leq 2.\underset{\sim}{1} ,$$

where $\underset{\sim}{b} = -\underset{\sim}{b}^*/e^*$ and $e = 1/e^*$. The dual LP problem to (PL_∞) is

(DL_∞) Minimize $\underset{\sim}{1}'\underset{\sim}{v}^+ + \underset{\sim}{1}'\underset{\sim}{v}^-$

Subject to

$$\underset{\sim}{X}'\underset{\sim}{v}^+ - \underset{\sim}{X}'\underset{\sim}{v}^- = \underset{\sim}{0} ,$$

$$\underset{\sim}{y}'\underset{\sim}{v}^+ - \underset{\sim}{y}'\underset{\sim}{v}^- = 1 ,$$

$$\underset{\sim}{v}^+ , \quad \underset{\sim}{v}^- \geq \underset{\sim}{0} ,$$

where $\underset{\sim}{v} = \underset{\sim}{v}^+ - \underset{\sim}{v}^-$ represents the dual variables.

For the MSAE regression problem, our objective is to find $\underset{\sim}{b}$ that minimizes the sum of absolute errors. Let $\underset{\sim}{b} = \underset{\sim}{b}^+ - \underset{\sim}{b}^-$ and $\underset{\sim}{e} = \underset{\sim}{e}^+ - \underset{\sim}{e}^-$. The MSAE estimate $\underset{\sim}{b}$ is the solution of the following problem:

(PL_1) Minimize $\underset{\sim}{1}'\underset{\sim}{e}^+ + \underset{\sim}{1}'\underset{\sim}{e}^-$

Subject to

$$\underset{\sim}{X}\underset{\sim}{b}^+ - \underset{\sim}{X}\underset{\sim}{b}^- + \underset{\sim}{e}^+ - \underset{\sim}{e}^- = \underset{\sim}{y} ,$$

$$\underset{\sim}{b}^+ , \underset{\sim}{b}^- \geq \underset{\sim}{0} ,$$

$$\underset{\sim}{e}^+ , \underset{\sim}{e}^- \geq \underset{\sim}{0} .$$

The dual problem to (PL_1) is

(DL_1) Maximize $\underset{\sim}{y}'\underset{\sim}{f} - \underset{\sim}{1}'\underset{\sim}{y}$

Subject to

$$\underset{\sim}{X}'\underset{\sim}{f} \leq \underset{\sim}{X}'\underset{\sim}{1} ,$$

$$\underset{\sim}{0} \leq \underset{\sim}{f} \leq 2.\underset{\sim}{1} ,$$

where f denotes the dual variables.

We shall solve (DL_∞) and (PL_1) to determine $\underset{\sim}{b}$ for the MMAE and the MSAE regression problems, respectively.

3. *SOME OBSERVATIONS*

Before we develop an algorithm, we observe the following similarities between (DL_∞) and (PL_1):

(i) Both problems are minimization problems with equality constraints

(ii) By definition, $\underset{\sim}{v} = \underset{\sim}{v}^+ - \underset{\sim}{v}^-$ for (DL_∞) and $\underset{\sim}{e} = \underset{\sim}{e}^+ - \underset{\sim}{e}^-$ and $\underset{\sim}{b} = \underset{\sim}{b}^+ - \underset{\sim}{b}^-$ for (PL_1). Consequently, in any solution $\underset{\sim}{I}\underset{\sim}{v}^+\underset{\sim}{I}\underset{\sim}{v}^- = \underset{\sim}{0}$ $\underset{\sim}{I}\underset{\sim}{e}^+\underset{\sim}{I}\underset{\sim}{e}^- = \underset{\sim}{0}$ and $\underset{\sim}{I}\underset{\sim}{b}^+\underset{\sim}{I}\underset{\sim}{b}^- = \underset{\sim}{0}$. This allows us to improve the solution by replacing a basic variable with its complement before an ordinary simplex iteration, thus passing several corner points in one simplex iteration, Barrodale and Roberts (1973).

(iii) The slack variables $\underset{\sim}{S}$ for the dual to (DL_∞) and the structural variables $\underset{\sim}{f}$ for the dual to (PL_1) are bounded between zero and two. This implies that the reduced costs of $\underset{\sim}{v}^+$ and $\underset{\sim}{v}^-$ in (DL_∞), and $\underset{\sim}{e}^+$ and $\underset{\sim}{e}^-$ in (PL_1) are bounded between zero and two.

(iv) From (ii) and (iii), it follows that the reduced costs of the pair (v_i^+, v_i^-) in (DL_∞) and the pair (e_i^+, e_i^-) in (PL_1), $i=1,\ldots,n$, sum to two. Further, the reduced costs of the pair (b_j^+, b_j^-), $j=1,\ldots,k$ in (PL_1) sum to zero.

To further appreciate the similarities between the two problems, we present the initial tableau for the MMAE and the MSAE regression problems in Tables 1 and 2, respectively.

TABLE 1

Initial Tableau for the MMAE Problem

Basic Variables	$\underset{\sim}{v}^{+'}$	$\underset{\sim}{v}^{-'}$	$\underset{\sim}{w}^{+'}$	$\underset{\sim}{w}^{-'}$	RHS
$\underset{\sim}{w}^+$	$\underset{\sim}{X}'$	$-\underset{\sim}{X}'$	$\underset{\sim}{I}$	$-\underset{\sim}{I}$	$\underset{\sim}{0}$
	$-\underset{\sim}{y}'$	$-\underset{\sim}{y}'$			1
Reduced Cost	$\underset{\sim}{1}'$	$\underset{\sim}{1}'$	$\underset{\sim}{0}'$	$\underset{\sim}{0}'$	$-z=0$

TABLE 2

Initial Tableau for the MSAE Problem

Basic Variables	$b^{+'}$	$b^{-'}$	$e^{+'}$	$e^{-'}$	RHS
$\underset{\sim}{e}^+$	$\underset{\sim}{X}$	$-\underset{\sim}{X}$	$\underset{\sim}{I}$	$-\underset{\sim}{I}$	$\underset{\sim}{y}$
Reduced Cost	$-\underset{\sim}{1}'\underset{\sim}{X}$	$\underset{\sim}{1}'\underset{\sim}{X}$	$\underset{\sim}{0}'$	$2.\underset{\sim}{1}'$	$-z=-\underset{\sim}{1}'\underset{\sim}{y}$

In Table 1, $\underset{\sim}{w} = \underset{\sim}{w}^+ - \underset{\sim}{w}^-$ represent (artificial) variables and $\underset{\sim}{w}^+$ the vector of initial basic variables. As such to obtain the basic feasible solution to the problem, we restrict the first k+1 ordinary simplex iterations to remove $\underset{\sim}{w}^+$ from the basis. Further, once $\underset{\sim}{w}^+$ leaves the basis, neither $\underset{\sim}{w}^+$ nor $\underset{\sim}{w}^-$ ever re-enter the basis.

In Table 2, $\underset{\sim}{e}^+$ provides the initial starting solution. Since the objective function coefficient of $\underset{\sim}{b}$ in (PL_1) is $\underset{\sim}{0}$, we restrict the first k simplex iterations to make $\underset{\sim}{b}$ basic. Further, once $\underset{\sim}{b}$ becomes basic, it never leaves the basis.

Now we make certain observations (from Tables 1 and 2) which together with our earlier observations help reduce the storage space requirements and the computational effort.

From Table 1, we observe that due to linear dependencies among $\underset{\sim}{v}^+$ and $\underset{\sim}{v}^-$ as well as $\underset{\sim}{w}^+$ and $\underset{\sim}{w}^-$, we need to maintain no more than n+k+2 columns explicitly. Further, at each iteration k+1 variables will be basic. We can further reduce the number of columns by k+1. Since the column of w_{k+1}^+ is the same as the right hand side (RHS) column, we need to maintain only one of the two columns explicitly. Thus, we need to maintain no more than n columns explicitly in the working tableau. We also note that the solution to our original problem is contained in the reduced costs of $\underset{\sim}{w}$.

Similarly, from Table 2, we observe the linear dependencies between $\underset{\sim}{b}^+$ and $\underset{\sim}{b}^-$ as well as $\underset{\sim}{e}^+$ and $\underset{\sim}{e}^-$. Thus, we need to maintain no more than n+k+1 columns explicitly. Further, at any iteration, n variables are basic. Therefore, we need to maintain only k+1 columns explicitly in the working tableau.

4. A FEW REMARKS

In this paper we have shown that the MSAE and the MMAE regression problems have a number of similarities as well as special structure that can be exploited to develop an efficient algorithm to solve both problems. One such algorithm has been developed by Narula and Wellington (1984b). This makes it possible to solve both problems using one algorithm and one computer code.

REFERENCES

Anscombe, F. J. (1967). Topics in the investigation of linear relations fitted by the method of least squares. J. Royal Statist. Soc. Series B., *29*, 1-52.

Arthanari, T. S. and Dodge, Y. (1981). Mathematical Programming in Statistics. New York: Wiley.

Barrodale, I. and Phillips, C. (1974). An improved algorithm for discrete Chebychev linear approximation. Proc. 4th Manitoba Conf. on Numer. Math. (Univ. of Manitoba, Winnipeg, Manitoba), 177-190.

Barrodale, I. and Roberts, R. D. K. (1973). An improved algorithm for discrete L_1 linear approximation. SIAM J. Numer. Anal., *10*, 839-848.

Bourdon, G. A. (1974). A Monte Carlo sampling study for further testing of the robust regression procedures based upon the kurtosis of the least squares regression residuals. Unpublished M.S. Thesis, Air Force Institute of Technology, Wright-Patterson AFB, Ohio.

Box, G.E. P. (1967). Discussion. J. Royal Statist. Soc. Series B, *29*, 42-43.

Cox, D. R. (1967). Discussion. J. Royal Statist. Soc. Series B, *29*, 39.

Forth, C. R. (1974). Robust estimation techniques for population parameters and regression coefficients. Unpublished M.S. Thesis. Air Force Institute of Technology, Wright-Patterson AFB, Ohio.

Narula, S. C. (1982). Optimization techniques in linear regression: A review. TIMS Studies in the Management Sciences, *19*, 11-29.

Narula, S. C. and Wellington, J. F. (1982). The minimum sum of absolute errors regression: A state of the art survey. Int. Statist. Rev., *50*, 317-326.

Narula S. C. and Wellington, J. F. (1984a). A branch and bound procedure for selection of variables in minimax regression. SIAM J. Scientific and Statist. Comput., to appear.

Narula, S. C. and Wellington, J. F. (1984b). An efficient algorithm for the MSAE and the MMAE regression problems. Working paper No. 8, Institute of Statistics, VCU, Richmond, Virginia.

Wagner, W. H. (1959). Linear programming techniques for regression analysis. J. Amer. Statist. Assoc., *54*, 206-212.

Optimization Under Probabilistic Constraints and its Applications in Statistics

A. Prékopa, Budapest

In this paper we present solution techniques to problems of the following kind

$$\text{minimize } h(x)$$
$$\text{subject to}$$
$$(1) \quad \begin{aligned} & h_o(x) = P\left(g_1(x,\xi) \geqq 0,\dots, g_r(x,\xi) \geqq 0\right) \geqq p \\ & h_1(x) \geqq p_1,\dots, h_m(x) \geqq p_m , \end{aligned}$$

where for the sake of simplicity we assume that the functions $h, h_1,\dots, h_m$ are defined on the whole n-dimensional space. Similarly, the functions $g_1(x,y),\dots, g_r(x,y)$ are supposed to be defined in the whole $n+q$ -dimensional space, $x \in R^n$, $y \in R^q$.

Various engineering and economic problems can be cast into this form. Now we do not intend to survey the applicational models belonging to this category. We only refer to a few papers [7-14], where the interested reader may find model formulations and references to applications. Our purpose is to summarize the available solution techniques and then show how we can apply these in classical statistical problems. As regards the definition of the science of Statistics we accept that one motivated by the works of Abraham Wald saying that "Statistics is the science of decision making under uncertainty". In this sense Problem (1) itself is a statistical problem and the above-mentioned classical statistical problems will be: construction of tests and tolerance regions. For further statistical applications of Problem (1) see [11].

The most important special case of Problem (1) is obtained by specializing the functions $g_i(x,y)$, $i=1,\dots,r$, so that

$$g_i(x,y) = T_i x - x_i \ , \ i=1,\dots,r$$

where $T_1,\dots, T_r$ are rows of an $r \times n$ matrix T. In this case the probabilistic constraint in Problem (1) takes the form

$$(2) \quad P(Tx \geqq \xi) \geqq p \ .$$

COMPSTAT 1984 © Physica-Verlag, Vienna for IASC (International Association for Statistical Computing), 1984

Introducing the notation $F(z)$ for the joint probability distribution function of the components of the random vector ξ, i. e. $F(z) = P(\xi \leqq z)$, the constraint (2) can be written in the following manner

$$(3) \qquad F(Tx) \geqq p \; .$$

Before proceeding to describe the numerical solution techniques to Problem (1) we mention the following theorem that serves as a basis of the convergence theory in many special cases.

Theorem [15]. If $g_1(x,y),\dots, g_r(x,y)$ are concave functions in R^{n+q} and ξ has a continuous probability distribution with logarithmically concave probability density function f, i. e. for every $x_1, x_2 \in R^n$ and $0 < \lambda < 1$ we have

$$f(\lambda x_1 + (1-\lambda)x_2) \geqq [f(x_1)]^{\lambda} [f(x_2)]^{1-\lambda}$$

then the function h_0 is also logarithmically concave in R^n.

This theorem implies that if has the required property then the function standing on the left hand side of (2) is logarithmically concave.

Maximization of probability and the method of two phases

Together with problem (1) we also formulate the problem

$$(4) \quad \begin{array}{l} \text{maximize } h_0(x) = P(g_1(x,\xi) \geqq 0,\dots, g_r(x,\xi) \geqq 0) \\ \text{subject to} \\ h_1(x) \geqq p_1,\dots, h_m(x) \geqq p_m \; . \end{array}$$

This problem has practical importance too. Many reliability problems belong to this category. For one practical application we refer to the paper [13] where a sequential decision process consists of a sequence of problems of the type (4).

Another importance of problem (4) is that when solving problem (1) a two phase method can be applied where in the first phase we seek a feasible solution and in the second phase we solve the original problem. Assuming that we possess a method to find a feasible solution to the system of inequalities $h_1(x) \geqq p_1,\dots, h_m(x) \geqq p_m$, a feasible solution to problem (1) can be found in such a way that we start to solve problem (4) and stop the procedure when we reach an x satisfying $h_0(x) \geqq p$. This x is a feasible solution to problem (1).

For the solution of problem (1) we propose the application of suitable nonlinear programming methods supplied by Monte Carlo simulation procedures to find function values and gradients of the function h_0. There exist other proposals too to solve stochastic programming problems among which the stochastic quasi gradient method of Yu. Ermolev and his collaborators should be mentioned. There is, however, little experience regarding how this method works in case of problem (1) and (4).

Solution by the SUMT method with logarithmic penalty function

Assumptions:

- h is convex in R^n ,
- $h_1,\dots, h_r$ are continuous logconcave functions in R^n ,
- $g_1,\dots, g_r$ are concave functions in R^{n+q} ,
- there exists an x satisfying $h_i(x) > p_i$, $i=0,\dots, r$,
- ξ has a continuous probability distribution with logarithmically concave density.

The Sequential Unconstrained Minimization Technique applied to our problem works in the following manner. We define the penalty function

$$(5) \quad T(x,r) = h(x) - r \sum_{i=0}^{r} \ln\left(h_i(x)-p_i\right)$$

for every x satisfying $h_i(x) > p_i$, $i=0,\dots, r$ and for every $r > 0$. Take a sequence $r^1 > r^2 > \dots$ with the property that $\lim_{k\to\infty} r^k = 0$ and minimize the function $T(x,r^k)$. If the set of feasible solutions is compact then the minimum of $T(x,r^k)$ exists. Let x^k be an optimal solution to this problem. Then

$$\lim_{k\to\infty} T(x^k,r^k) = \lim_{k\to\infty} h(x^k) = \min_{x\in D} h(x) ,$$

where D denotes the set of feasible solutions. It is remarkable that under the mentioned assumptions the function $T(x,r)$ is a convex function for every fixed r thus various unconstrained optimization techniques work effectively. To compute the values and the gradients of h_o remain difficult problems to which we return later. The sequence $r^1, r^2, \ldots$ in practice is chosen as a geometric sequence and the procedure frequently stops after a few number of steps, where the optimality criterion usually is formulated so that the $T(x,r)$, $h(x)$ do not change significantly in about three subsequent steps.

Solution by the method of feasible directions

Assumptions:

- the probabilistic constraint has the form (3),
- h is convex and has continuous gradient in R^n ,
- $h_1, \ldots, h_r$ are quasi-concave and have continuous gradients in R^n ,
- the constraints in which the constraining functions are linear determine a bounded set,
- there exists an x satisfying $h_i(x) > p_i$, $i=0,\ldots, r$,
- ξ has a continuous probability distribution with logarithmically concave density.

The method uses subsequent linearization of the constraints and the objective function. We start from an arbitrary feasible vector x^1 and if $x^1, \ldots, x^k$ are already fixed then first we solve the following direction finding problem:

$$\text{minimize } y$$
$$\text{subject to}$$
$$(6) \quad \begin{aligned} & h(x^k) + \nabla h(x^k)(x - x^k) \leqq y \\ & h_i(x^k) + \nabla h_i(x^k)(x - x^k) + \vartheta_i y \geqq p_i , \quad i=0,\ldots, m . \end{aligned}$$

If x^{k*} is an optimal solution of this problem then we solve the following step length finding problem:

$$(7) \quad \min_{\lambda} \; h(x^k + \lambda(x^{k*} - x^k))$$

where the minimization is extended over such λ values for which $x^k + \lambda(x^{k*} - x^k)$ is feasible. If λ^k is an optimal solution of problem (7) then we define

$$x^{k+1} = x^k + \lambda^k(x^{k*} - x^k) \quad .$$

The above procedure was published by Zoutendijk in 1960. In the practical computation we frequently disregard the antizigzagging precaution that is expressed by the use of y. Then the direction finding problem (6) reduces to the following:

$$\text{(8)} \qquad \begin{aligned} &\text{minimize} \quad h(x) \\ &\text{subject to} \\ &h_0(x^k) + \nabla h_0(x^k)(x - x^k) \geqq p_1 \quad , \\ &h_i(x) \geqq p_i \quad , \quad i=1,\dots,m \ . \end{aligned}$$

Solution by the supporting hyperplane method

Assumptions:

- there exists a bounded convex polyhedron K^1 such that the set of feasible solutions is contained in K^1,
- the functions $-h$, $h_1,\dots, h_m$ are quasi-concave and have continuous gradient on K^1,
- there exists an x such that $h_i(x) > p_i$, $i=0,\dots,m$,
- ξ has continuous probability distribution and logconcave density in R^n furthermore h_0 has continuous gradient in R^n,

We assume that we have an initial feasible vector x^1. Then we perform subsequent iterations where the rth iteration in this method consists of two subsequent steps.

Step 1. Solve the problem

$$\begin{aligned} &\text{minimize } f(x) \\ &\text{subject to } \ x \in K^r \ , \end{aligned}$$

where K^r is a convex polyhedron. Let x^r be an optimal solution to this problem. If $h_i(x^r) \geqq p_i$, $i=0,\dots,m$, then x^r is an optimal solution to problem (3.1). Otherwise go to step 2.

Step 2. Let λ^r be the largest λ $(0 \leqq \lambda \leqq 1)$ for which the following inequality holds

$$h_i(x^1 + \lambda(x^r - x^1)) \geqq p_i \ , \quad i=0,\dots,m \ .$$

Various one-dimensional methods can be applied to solve this problem.

Let

$$y^r = x^1 + \lambda^r(x^r - x^1) .$$

If $f(y^r) - f(x^r) \leqq \varepsilon$ where ε is a previously chosen small positive number then we stop and accept y^r as an approximate solution to the optimization problem. Otherwise choose a subscript i_r for which $h_{i_r}(y^r) = 0$ and define

$$K^{r+1} = \{x \mid x \in K^r , \nabla h_{i_r}(y^r)(x - y^r) \geqq 0\}$$

and go to Step 1 using $r+1$ instead of r .

<u>Solution by a variant of the general reduced gradient method</u>

A variant of the GRG method suitably adapted to problem (1) where the stochastic constraint reduces to (2) and the other constraints are linear has been reported in [6]. It differs from the GRG method primarily in the formulation of the direction finding problem. Here we generate always feasible solutions and thus we avoid the application of intermediate methods to return to the feasible set which is very important because our function values are noisy due to the evaluation technique although low dimensional random vectors are allowed only and probabilities are evaluated by numerical integration and not by simulation.

<u>Solution by a primal-dual type algorithm</u>

The problem to be solved has the following form:

$$\begin{aligned} &\text{minimize } c'x \\ &\text{subject to} \\ &F(y) \geqq p \\ &Ax = y , \quad Dx \geqq d , \end{aligned}$$

where $x \in R^n$ and $y \in R^m$. We assume that the multivariate probability distribution function F is strictly logarithmically concave and has continuous gradient in R^n. We will shortly describe the method proposed in [3].

To this problem we assign a problem that we will call dual problem although it is not dual in the classical sense. This dual problem is the following:

$$A'u + D'v = c$$
$$u \geqq 0, \; v \geqq 0$$

$$\max \left[\min_{F(y) \geqq p} u'y + v'd \right] .$$

The procedure works in the following manner. First we assume that a pair of vectors (u^1, v^1) is available for which

$$(u^1, v^1) \in V = \{u, v | A'u + D'v = c \; , \; v \geqq 0\} .$$

Suppose that (u^k, v^k) has already been chosen, where $u^k = 0$. Then the following steps have to be performed

Step 1. Solve the problem

$$\begin{aligned} &\text{minimize } u^{k'}y \\ &\text{subject to} \\ &F(y) \geqq p . \end{aligned}$$

Let $y(u^k)$ denote the optimal solution to this problem. Then we solve the following direction finding problem

$$\begin{aligned} &\text{maximize } u'y(u^k) + d'v \\ &\text{subject to} \\ &(u, v) \in V . \end{aligned}$$

Let (u_k^*, v_k^*) be an optimal solution to this problem. If $u_k^* = \varrho u^k$ for same $\varrho \geqq 0$ then (u_k^*, v_k^*) is an optimal solution of the dual problem and the pair $y(u^k)$ is an optimal solution of the primal problem where $\hat{x}$ is an optimal solution of the linear programming problem:

$$\begin{aligned} &\text{minimize } c'x \\ &\text{subject to} \\ &Ax \geqq y(u^k) \; , \; Dx \geqq d . \end{aligned}$$

Otherwise go to

Step 2. Find λ_k $(0 < \lambda_k < 1)$ satisfying

$$u_k^{*'} y \left(\frac{\lambda_k}{1 - \lambda_k} u^k + u_k^* \right) > u^{k'} y(u^k) + v^{k'} d .$$

Then we define

$$u^{k+1} = \lambda_k u^k + (1 - \lambda_k) u_k^* \quad ,$$

$$v^{k+1} = \lambda_k v^k + (1 - \lambda_k) v_k^* \quad .$$

If the procedure is infinite then it can be proved that the sequence (u^k, v^k) converges and the limiting pair has the same property as (u_k^*, v_k^*) in Step 1.

The case of discrete probability distributions

The following problem will be considered

$$\text{(9)} \qquad \begin{aligned} &\text{minimize } c'x \\ &\text{subject to} \\ &F(z) \geqq p \ , \\ &Tx = z \ , \ Bx \geqq b \ , \end{aligned}$$

where F is the probability distribution function of the random vector ξ. If ξ has possible values $z_1, \dots, z_N$ then the above problem is equivalent to the following mixed variable problem

$$\text{(10)} \qquad \begin{aligned} &\text{minimize } c'x \\ &\text{subject to} \\ &y_1 F(z_1) + \dots + y_N F(z_N) \geqq p \ , \\ &y_1 + \dots + y_N = 1, \ y_1, \dots, y_N \geqq 0 \ , \ \text{INTEGERS} \\ &Tx = y_1 z_1 + \dots + y_N z_N \\ &Bx \geqq b \end{aligned}$$

Another mixed variable formulation will be illustrated in the case when ξ is a two-dimensional random vector the possible values of which are nonnegative lattice points with coordinates $\leqq N, M$, respectively. The mixed variable reformulation of the problem is the following

minimize $c'x$
subject to

$$
(11)\quad
\begin{aligned}
&p_{oo}y_{oo}+\dots+p_{NO}y_{NO}+p_{o1}y_{o1}+\dots+p_{N1}y_{N1}+\dots+p_{OM}y_{OM}+\dots+p_{NM}y_{NM} \geqq p\\
&y_{oo}+\dots+y_{NO} = z_1\\
&y_{oo} \qquad +y_{o1}+ \qquad \dots \qquad +y_{OM} = z_2\\
&Ax = z\\
&Bx \geqq b\\
&y_{ik} \leqq y_{i-1,k}\ , \quad i=1,\dots,N\ ; \quad k=0,\dots,M\ ,\\
&y_{ik} \leqq y_{i,k-1}\ , \quad i=0,\dots,N\ ; \quad k=1,\dots,M\ ,\\
&y_{ik} = 0 \quad \text{OR} \quad 1 \quad \text{FOR ALL} \quad i,\ k
\end{aligned}
$$

This record model can be used in connection with continuously distributed random vector ξ when approximating its distribution by a discrete distribution. In the higher dimensional case, however, the number of 0, 1 variables becomes too large.

Construction of statistical tests

The problem of construction of statistical tests can be formulated as an optimization problem. This fact has been known for a long time. In the book of Lehmann [4] a simple test construction problem is formulated as an optimization problem which can easily be seen to be equivalent to a mathematical programming problem known as the knapsack problem. A general form of the test construction problem was given in [11]. Suppose we want to test the hypothesis that the true distribution belongs to the family H and the alternative hypothesis is represented by another family K. The test construction problem can be formulated in the following manner:

$$
(12)\quad
\begin{aligned}
&\text{maximize } \beta\\
&\text{subject to}\\
&P(X \in S|F) = \alpha \qquad \text{for every } F \in H\ ,\\
&P(X \in S|F) = \beta \qquad \text{for every } F \in K\ ,
\end{aligned}
$$

where X is the outcome of the experiment performed concerning testing the hypothesis, S is the critical region and α is the prescribed upper bound for the first kind error. The probability standing on the left hand side in the second constraint is the power function depending on the variable $F \in K$. Thus in problem (12) we maximize the minimum power. Although the maximization of the power could be formulated as a multiobjective optimization problem and relevant methods could be proposed to apply, now we do not intend to treat the problem that way.

We assume that $\overline{S}$, the complementary set of S, depends on the n-dimensional parameter x so that $\overline{S}(x)$ is a concave family of sets i. e.

$$\overline{S}(\lambda x_1 + (1-\lambda)x_2) \supset \lambda \overline{S}(x_1) + (1-\lambda)\overline{S}(x_2) .$$

(We remark here that the family of sets

$$A(x) = \{ y \mid g_i(x,y) \geqq 0 \ , \ i=1,\dots,r \}$$

is concave provided $g_1,\dots,g_r$ are concave functions; this fact is the background to prove that if ξ has a logconcave density then h_0 is a logconcave function of the variable x.) We write $\beta = e^{-\gamma}$ and reformulate problem (12) in the following manner:

$$(13) \qquad \begin{aligned} &\text{minimize } [k(\gamma) + h(x)] \\ &\text{subject to } \gamma \geqq 0 \text{ and} \\ &P(X \in \overline{S}(x) \mid F) \geqq 1-\alpha \ , \quad \text{for every } F \in H \ , \\ &e^{\gamma} P(X \in \overline{S}(x) \mid F) \leqq 1 \ , \quad \text{for every } F \in K \ , \end{aligned}$$

where k, h are some functions of γ and x, respectively; k is supposed to be nondecreasing.

In case of a single alternative i. e. K has only one element F_1, say, we may formulate instead of (13) the following problem:

$$(14) \qquad \begin{aligned} &\text{maximize } P(X \in \overline{S}(x) \mid F_1) \\ &\text{subject to} \\ &P(X \in \overline{S}(x) \mid F) \geqq 1-\alpha, \quad \text{for every } F \in H \ . \end{aligned}$$

Similar reformulation can be given when taking the average with respect to the alternatives belonging to the set K.

Now if $\overline{S}(x)$ is a convex family of sets and the sets H, K contain logconcave probability distributions then the function

$$P(X \in \overline{S}(x) \mid F)$$

of the variable x is logconcave for $x \in H \cup K$. Thus the first constraint in (13) determines a convex set of x vectors but the second constraint spoils it. Similarly, problem (14) is not a convex programming problem because we maximize and not minimize a convex function although the set of feasible solutions is convex.

In problem (13) the objective function can be unpleasant too, from the mathematical programming point of view.

For the solution of problem (14) we propose a method based on a procedure by Majthay and Whinston [5] supplied by an original vertex finding algorithm of Fülöp [1] that can be used in general concerning cutting plane methods.

Various applications of the above mentioned test construction method can be mentioned. The most significant are those for the structure of random vectors having large number of components we have some assumptions and we want to test one special structure against the remaining ones as alternatives. Finding the probability distributions of statistics that can be defined in such cases is frequently impossible. A test of the above mentioned type, however, can be constructed.

Construction of tolerance regions

The tolerance region is a random set constructed only on the basis of a sample and containing a large proportion out of the true probability distribution. More exactly, the random set S is a β-content tolerance region of confidence level γ if the following inequality holds:

$$(15) \qquad P\left(\int_S dF \geqq \beta\right) = \gamma .$$

Here β and γ are fixed probabilities chosen arbitrarily by ourselves, F is the true probability distribution and P is the probability of the event determined by the relation within the parantheses.

The type of the tolerance region construction is illustrated by the exemple given below. Suppose we want to construct a tolerance region for the n-dimensional normal distribution with independent components where the region is assumed to have the form:

$$S(k_1,\ldots,k_n)=(\bar{x}_1-k_1s_1^*,\ \bar{x}_1+k_1s_1^*)x \ \ldots\ x(\bar{x}_n-k_ns_n^*,\ \bar{x}_n+k_ns_n^*)\ ,$$

where $\bar{x}_1,\ldots,\bar{x}_n$ and $s_1^*,\ldots,s_n^*$ are the empirical means and standard deviations of the n independent samples further $k_1,\ldots,k_n$ are unknown decision variables. It is easy to see that the probability content of $S\ k_1,\ldots,k_n$ does not depend on the true expectations and standard deviations of the individual (marginal) distributions. To find $k_1,\dots,k_n$ a nonlinear programming problem (a special case of problem (15)) is formulated and analysed from the point of view of solution technique.

References

[1] J. Fülöp, Finding Original Vertex and its Application to Minimizing a Concave Function Subject to Linear Inequalities, Alkalmazott Matematikai Lapok 9 (1983) 51-72 (in Hungarian).

[2] J. Hájek and Z. Šidak, Theory of Rank Tests, Acad. Publ. House of the Czechoslovak Acad. Sci., Prague, 1967.

[3] É. Komáromi, A Dual Approach to Probabilistic Constrained Programming Problem, Mathematical Programming Study, to appear.

[4] E. L. Lehmann, Testing Statistical Hypotheses, Wiley, New York, 1959.

[5] A. Majthay and A. Whinston, Quasi-Concave Minimization subject to Linear Constraints, Discrete Mathematics 9 (1974) 35-59.

[6] J. Mayer, A Nonlinear Programming Method for the Solution of a Stochastic Programming Model of A. Prékopa, Survey of Mathematical Programming (Proc. 9th Int. Math. Prog. Symp.), Vol. 2, 129-139.

[7] A. Prékopa, Contributions to the Theory of Stochastic Programming, Mathematical Programming 4(1973) 202-221.

[8] A. Prékopa, I. Deák, J. Ganczer and K. Patyi, The STABIL Stochastic Programming Model and its Experimental Application to the Electrical Energy Sector of the Hungarian Economy, in: Stochastic Programming Proceedings of the International Conference on Stochastic Programming, Oxford, England, 1974.

[9] A. Prékopa and T. Szántai, Flood Control Reservoir System Design, Mathematical Programming Study 9(1978) 138-151.

[10] A. Prékopa and P. Kelle, Reliability Type Inventory Models Based on Stochastic Programming, Mathematical Programming Study 9(1978) 43-58.

[11] A. Prékopa, The Use of Stochastic Programming for the Solution of Some Problems in Probability and Statistics, in: Extremal Methods and Systems Analysis Proceedings of an International Conference in honour of A. Charnes' sixtieth birthday, Austin, Texas 1977, Lecture Notes in Economics and Mathematical Systems, 174, Springer Verlag 198o, 522-536.

[12] A. Prékopa and T. Szántai, A New Multivariate Gamma Distribution and its Fitting to Empirical Streamflow Data, Water Resources Research 14(1978) 19-24.

[13] A. Prékopa and T. Szántai, On Optimal Regulation of a Storage Level with Application to the Water Level Regulation of a Lake, European Journal of Operations Research 3(1979) 175-189.

[14] A. Prékopa, Network Planning Using Two-Stage Programming under Uncertainty, in: Recent Results in Stochastic Programming (Proceedings of the International Conference on Stochastic Programming, Oberwolfach, Germany, 1979), Academic Press 1980, 63-82.

[15] A. Prékopa, Logarithmic Concave Measures and Related Topics, in: Stochastic Programming Proceedings of the International Conference on Stochastic Progr.,Oxford,England, 1974, Academic Press 1980,63-82

Parallel Model Analysis: Fitting Non-Linear Models to Several Sets of Data

G.J.S. Ross, Harpenden

Keywords: Parallel Model Analysis, non-linear models, optimisation, analysis of parallelism.

Summary

The parallel linear regression model may be generalised in many ways. When the same model is fitted to several sets of data, some or all of the parameters may be constrained to be common to all data sets. For non-linear models the computations then may be organised as a sequence of optimisations of low dimensionality.

1.1. Introduction

When a non-linear model is fitted to more than one set of data the estimated parameters cannot readily be compared without recourse to further computation. The statistical framework for comparing such estimates is the likelihood-ratio test (Neyman and Pearson (1928)) which compares the likelihood obtained by fitting all p parameters of a model with the likelihood obtained for the constrained model in which $q<p$ parameters have fixed values. Theory shows that under very general assumptions the distribution of twice the log-likelihood ratio criterion is asymptotically χ^2 on (p-q) degrees of freedom.

The term Parallel Model Analysis, derived by analogy from parallel linear regression, is applied in this paper to the following procedure: Given K data sets, $\underset{\sim}{y}_1 \dots \underset{\sim}{y}_k$ with sample sizes $n_1 \dots n_k$, let the same form of model with p parameters $\underset{\sim}{\theta}$ be fitted to each. The parallel model M_q is then the model in which the first q parameters are common to all data-sets, and the remaining p-q parameters are specific to each data set. The number of parameters to be fitted in M_q is thus $q+k(p-q)$. The deviance(which for Normal errors becomes the residual sum of squares twice the minimised negative log-likelihood) for model M_q is D_q. A sequence of parallel models from M_0 (all sets separately) to M_p (all sets pooled) with monotone increasing deviances gives rise to an analysis of deviance (or analysis of parallelism) as follows:

Source	Deviance	Degrees of Freedom
Between specific parameters	$D_p - D_q$	$(k-1)(p-q)$
Between common parameters	$D_q - D_o$	$(k-1)q$
Within subsets	D_o	$\Sigma n_i - kp$

If the model has Normal errors the variance is estimated from the pooled within subsets residual mean squares, and the table is an analysis of variance. For models assuming Binomial or Poisson distributions each deviance may be tested against the chi-squared distribution, but if the within subsets deviance is significantly large there is evidence of heterogeneity and the analysis of variance procedure should be used.

Additional subdivisions of the table may be achieved by fitting intermediate models M_r where $0 < r < q$ or $q < r < p$.

1.2. Applications of Parallel Model Analysis

(1) Fitting parallel straight lines

The standard analysis for testing regression lines for common slope is decribed in some (but not all) textbooks (e.g. Snedecor and Cochran). The slope is the common parameter and the intercept is the specific parameter. The variance of the common slope is less than that of individual slopes, and the model with fewer parameters, if it fits adequately, is then more readily applicable to other data sets where only the specific parameters need to be re-estimated.

(2) Parallel line assays

In biological assay different substances are tested at a range of concentrations on groups of randomly selected subjects. If the dose-response curves are similar in shape and can be transformed to approximate straight lines (for example by plotting probit of response against log-concentration) parallel lines may be fitted. In this case the horizontal difference is of interest (relative log-concentration), the parallelism being necessary if the simple conclusion is to be reached that substance A is r times as effective as substance B. The analysis of parallelism for binomial responses is described by Finney (1971).

(3) Covariance analysis

In factorial experiments with concomitant observations the common parameters are the covariance regression coefficients and the specific parameters describe the block and treatment structure of the model. Interest is in the specific parameters rather than the common parameters.

(4) Parallel non-linear curves

Non-linear curves such as exponentials, logistics or hyperbolae, may be fitted to several sets of data with the constraint that the curves are identical apart from a change in origin, so that the curve

$$y = f(x, \underset{\sim}{\theta})$$

becomes $$y - y_0 = f(x - x_0, \underset{\sim}{\theta}).$$

The algebraic formulation would in most cases have to be reparameterised to allow the parallel model to be expressed in the form of a constraint on the first (p-2) or (p-1) parameters, but the analysis of deviance is unaffected. It is usually straightforward to estimate the parameters x_0 and y_0 directly in all except the first data set. Slopes of the curves are only parallel at corresponding points. Estimation of x_0 alone is a calibration problem (when curves are known to be identical apart from x_0).

(5) Other types of constraint

The word 'parallel' no longer applies literally when we postulate models such as the following: curves constrained to pass through a common point, curves with a common maximum, splines with common nodes, periodic curves with common frequency, growth curves with common shape but different scale.

(6) Multinomial samples

When several samples are classified in the same way, a two-way contingency table results, with k rows and r columns; if a p-parameter model such as a frequency distribution is fitted to each row, there are (r-p-1) degrees of freedom for goodness of fit, for each row. If the same model is fitted to the marginal totals the increase in the deviance has (k-1)(r-p-1) degrees of freedom. If no model is fitted, but the rows are tested for homogeneity, the deviance has (k-1)(r-1) degrees of freedom. The difference between these deviances gives a test for homogeneity of the fitted models, with (k-1)p degrees of freedom.

2. Algebraic formulation

The least squares solution for parallel linear models is as follows: If A is the (n×p) design matrix the model

$$\underset{\sim}{y} = A\underset{\sim}{\theta} + \underset{\sim}{\varepsilon}$$

has solution $$\hat{\underset{\sim}{\theta}} = (A'A)^{-1}A'\underset{\sim}{y} \qquad (2.1)$$

and the dispersion matrix of $\hat{\underset{\sim}{\theta}}$ is $(A'A)^{-1}\sigma^2$, where σ^2 is the variance of $\underset{\sim}{\varepsilon}$. For the parallel model M_q of section 1.1. the submatrices A_j are partitioned into a $(n\times q)$ common part B_j and a $(n\times(p-q))$ specific part C_j, and the composite design matrix has the following structure:

$$\begin{matrix} B_1 & C_1 & 0 & 0 & \dots \\ B_2 & 0 & C_2 & 0 & \dots \\ B_3 & 0 & 0 & C_3 & \dots \\ \dots & \dots & \dots & \dots & \dots \end{matrix}$$

The matrix A'A is then

$$\begin{matrix} \Sigma B_j'B_j & B_1'C_1 & B_2'C_2 & \dots \\ C_1'B_1 & C_1'C_1 & 0 & \\ C_2'B_2 & 0 & C_2'C_2 & \dots \end{matrix} \qquad (2.2)$$

Since the elements of A'A and $A'\underset{\sim}{y}$ have already been computed for the individual analyses, the parameters of M_q and their dispersion matrix may be estimated without reanalysing the data (using the inverse of A'A).

When the model is <u>non-linear</u>, an approximate analysis may be computed as follows:

Given parameter estimates $\hat{\underset{\sim}{\theta}}_1 \dots \hat{\underset{\sim}{\theta}}_k$ and dispersion matrices $\underset{\sim}{V}_1 \dots \underset{\sim}{V}_k$, reconstruct equivalent matrices to

$$(A'A)_j = V_j^{-1}/\sigma^2, \quad (A'\underset{\sim}{y})_j = V_j^{-1}\,\theta_j/\sigma^2.$$

Then form A'A and A'y as required, and solve the equations (2.1). The solutions obtained in this way are often close to the exact solutions.

2.1. <u>Separable linear parameters</u>

An important special case occurs when the (p-q) specific parameters are all linear, for example the parameters β_0 and β_1 in the model

$$\underset{\sim}{y} = \beta_0 + \beta_1 \exp(-k\underset{\sim}{x}) + \underset{\sim}{\varepsilon}$$

Each set of specific parameters may be estimated by multiple regression given particular values of the common parameters. The objective function for the q-parameter optimisation is therefore the sum of the within-subset residual sums of squares. This function is readily optimised and the common parameters have smaller variance than for the within-subset estimates of these parameters. Unfortunately the model specified in this way is not always of great practical interest compared with some of the models discussed in 1.2.

3. Computing schemes for Parallel Model Analysis

Data for parallel model analysis may be stored either as a series of separate data matrices, or as a single data matrix with one or more grouping factors. A table of parameter estimates, log likelihoods and degrees of freedom is needed.

Given a sequence of instructions for the expectation function for a single data-set there is no difficulty in defining the common model M_p; the data sets, must simply be amalgamated by removing the grouping factor.

To fit intermediate models M_q there are three possible procedures:

(1) Simultaneous estimation of all q+k(p-q) parameters. The model must be redefined so that the specific parameters differ for each group. For large k there may be difficulties in controlling the convergence of so many parameters, but for small k this method is usually the quickest to compute.

(2) Nested estimation of the specific parameters for trial values of the common parameters. If the specific parameters are linear this is a simple procedure (section 2.1) in which the total within-subset residual sum of squares is minimised. If the specific parameters are non-linear the procedure is probably uneconomic except when q=1, a single common parameter estimated by optimising the sum of the optimised likelihoods for each subset, with the first parameter fixed.

(3) Alternate estimation of common and specific parameters. There are various choices of algorithm for each stage, and it may be simpler to program two separate algorithms for common and specific parameters rather than redefine the model each time for a single algorithm. To save time it is only necessary to show descent at each stage rather than show convergence. The problem is that if the parameters are highly correlated the process will converge very slowly, in the manner of simple steepest-descent.

This type of algorithm is very similar to the EM algorithms of Dempster et al (1977) in which 'estimation' and 'minimisation' are alternated.

To accelerate convergence it is worth using Aitken's d^2 procedure, when it is safe to do so, using results of two successive iterations to predict the final values of the specific parameters.

The particular case of a calibration model in which there is only one specific parameter, the unknown origin of one of the variables, is especially suited to alternate optimisation. To fit α in the model $y = f(x-\alpha,\underset{\sim}{\theta})$ with $\underset{\sim}{\theta}$ fixed, the Gauss-Newton adjustment

$$d\alpha = \Sigma(y-f)\frac{\partial f}{\partial x} \Big/ \Sigma\left(\frac{\partial f}{\partial x}\right)^2$$

uses the slope of the curve and the residual only.

Alternate estimation is easy to specify, but may not converge very quickly. If it fails to converge, however, it may still give useful results.

4. Numerical example

The following artificial data illustrate the methods described in this paper.

Set 1:	X =	1	2	3	4	5	6	Set 2:	X =	3	4	5	6	7
	y =	5	15	18	23	24	25		y =	10	17	21	23	26

Model: $E(y) = a+br^x$ by least squares (Normal errors)

Individual estimates of parameters:

Set	r	b	a	Residual S.S.	d.f.
1	.5775	-36.71	26.39	2.9874	3
2	.6542	-67.43	29.01	0.8632	2

Fitting a common value of r (the single non-linear parameter) by nested minimisation of the pooled within-set residual sum of squares:

Set	r	b	a	Residual S.S.	d.f.
1	.5980	-35.97	26.84	3.0968	} 6
2	.5980	-82.19	27.48	1.1574	}

Fitting a common curve, with a variable origin α for x in set 2. Convergence in 8 iterations of alternate estimation of (r,a,b) and α

Set 1 parameters:	.6000	-36.55	27.14
x-origin for Set 2:	1.5078		
Residual S.S.	4.6342	on 7 d.f.	

Fitting common r and b but separate a (vertically parallel curves) by redefining the model for 4-parameter optimisation.

Set	r	b	a	Residual S.S.	d.f.
1	.7871	-39.18	36.74	} 26.6668	7
2	.7871	-39.18	31.93	}	

The method of alternate estimation converged very slowly, even with Aitken's acceleration, and the residual S.S. was reduced to 27.02 after 15 iterations. However this was sufficient to show that the fit was extremely poor.

5. Practical Implementation

Early releases of MLP (Ross (1980)) contain analyses of parallelism for curve fitting, probit analysis and linear regression. In version 3.08 of MLP, released in 1984, the methods described in this paper may be applied to general user-defined models.

References

Dempster, A.P., Laird, N.M and Rubin, D.B. (1977). Maximum Likelihood from Incomplete Data via the EM Algorithm. J.R. Stat.Soc.B 39 1-38.

Finney, D.J. (1971). Probit Analysis (3rd Edition). Cambridge University Press.

Neyman, J. and Pearson, E.S. (1928). Biometrika, 20 175 and 263.

Ross, G.J.S. (1980). Maximum Likelihood Program (MLP). Rothamsted Experimental Station, Harpenden.

Software for Data Preprocessing (Checking, Handling of Missing Values Reduction)

Is data preprocessing a computational process only?

M.K. Chytil, Madrid

SUMMARY: The aim of data preprocessing is to furnish the data analyst with appropriate data, i.e. data which are, at least, correct, semantic representative and context dependent. Data preprocessing includes the following procedures: data and problem planning, data acquisition, data input, data defects checking and deleting, data retrieval, data transformation, reduction and description. Data preprocessing involves both the preventive and additional care for the quality of data. It is demonstrated in this paper that only some of the above kinds of procedures can be written as algorithms, and, therefore, programmed. They can be embedded either in the usual general purpose statistical program packages or designed as specialized program systems. The remaining part of procedures must be performed by man and that either in a computer--aided manner or without this aid. Various advanced software forms are remembered here for this purpose such as expert systems, question-answerers, data cleaner, data planner, etc.

Data preprocessing, data quality, data planning, data acquisition, data cleaning, symbolic data, knowledge based systems, catalogues

DATA PREPROCESSING AS A NECESSARY PART OF DATA ANALYSIS.

We speak about data preprocessing in the connection with the multivariate exploratory data analysis rather than with the remaining kinds of statistical data analyses. MEDA is, in fact, a systematic and continuous cognitive process, leading to the solving of some global research problem (Bailyn, 1977, Chytil, 1982, Chytil - Havránek, 1983) stated implicitly only by the help of a body of certain carefully planned and collected data. MEDA cannot be regarded, consequently, as an analysis of *any* data but as an analysis of data *appropriate* to the solving of the problem in question. We must conceive it, therefore, as a complete, nevertheless iterative process consisting of the following three main parts:

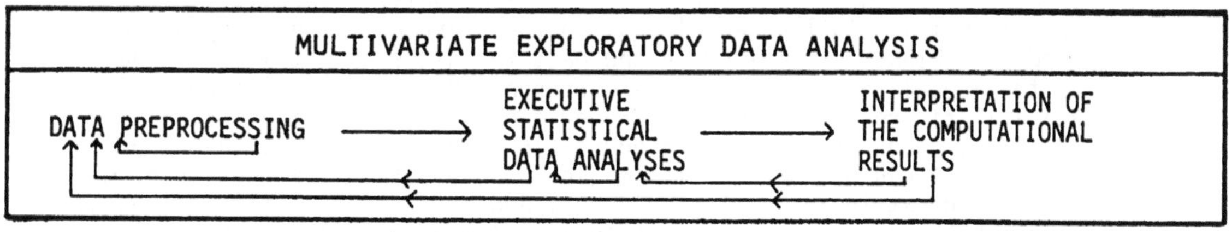

Data preprocessing is conceived here as one part of the whole process of data analysis. Its aim is to furnish in a planned manner the data analyst with appropriate data, i.e. data which fulfil, at least, the requirement of correctness, the requirement of semantic representativeness and the requirement of contextualness. The process of data preprocessing preceeds the proper executive statistical data analysis and runs iteratively in accordance with an open, unified and structured system of special rules, modes, methods or heuristics which are developed not only on the grounds of statistics but, moreover, on the grounds of logic, linguistics, general methodology of science, informatics, measurement theory, cybernetics or similar cognitive sciences (Chytil (1983)). Data preprocessing can be classified as follows:

THE PROCESS DATA PREPROCESSING

	PARTIAL PROCEDURES		THE RESULTS OF THE PROCEDURES
PREPARATORY PROCEDURES	PLANNING	Problem formulation Scheme of the data model	A well stated problem
	DATA ACQUISITION	Measurement Observation Scaling Questionning Interviewing Monitoring Text and graphical data capturing	Rough data
DATA FILTERING	DATA INPUT	Punching Data input Data storage and structuring (DBS)	Input data
	DATA CLEANING	Data defect checking and deleting	Correct data
	DATA ORGANIZAT-ION	Data retrieval Data transformation Data description	Data appropriate to the analysis

In comparison with the hitherto existing spontaneous rather than planned and exhaustive care for data the above scheme presents a systematized and well theoretically backed process, supplying data appropriate to its following analysis.

Obviously, one of the main goals of data preprocessing consists in the care for the *quality of data* which is considered to be analyzed. This care can be *preventive*, i.e. made before the data arizes and/or *additional*, i.e. made after the data was acquired. The quality of data cannot be expressed only by the help of the number of the usual kinds of errors and shortcomings such as of missing values, outliers or clerical errors. In the broader sense of the word defective data is any data which does not fulfil some of its possible roles and which biases, therefore, the results of its analysis. There exists an unlimited number of kinds of defects of the data whether occurring in general or under some specific conditions only. The quality of data is, for instance, essentially influenced already in the planning stage. Many writers in the foreground with FINNEY criticized in the past the hitherto situation in data quality with the appeal to systematize the corresponding processes (Finney, 1974). Clearly, data analysis restricted on the proper executive statistical procedures and lacking, therefore, on preprocessing procedures is incomplete and curtained on the exploratory power. It can become even misleading by neglecting the quality of data and the semantic context in which it should run. And in this connection the question arises how shall be this process accomplished at all and in which measure can be computerized. It can be seen on the first glance that the most part of procedures of which the preprocessing process consists can be scarcely written as algorithms. We analyze the reasons below.

IS DATA PREPROCESSING A COMPUTATIONAL PROCEDURE ONLY?

We start the following consideration with the assumption (Chytil, 1982):

It holds in general for the procedures with information: some of them can be written as algorithms and, consequently, programmed and processed on a computer. Let us call them, therefore, routine. The remaining part of the procedures with information has a non-routine, i.e. a creative character and can be processed by man only. The processing of some of such creative procedures can run, nevertheless, in a computer-assisted mode.

The class of procedures with information includes, therefore

(i) programmable procedures
(ii) computer-assisted procedures
(iii) creative procedures performable by man only.

Examples of such procedures: Rounding of the values of a variable can be programmed planning of an experiment can be computer-assisted (cf. Friedland, 1979), a decision making which of two variables is more relevant to the solving of the problem in question is a creative procedure and can be accomplished by man only. Whereas programmable procedures can be completely performed by a machine, computer-assisted procedures are made by man with a partial support of a machine. Data preprocessing requires procedures of all three kinds. Programmable procedures occur in this process less frequently than in the proper executive data analysis. The reasons are elucidated bellow:

a) The key-problem consists in the character of the *data*. For we cannot restrict the analysis on well-stated numerical data and, therefore, on computational procedures only. We must handle, in addition, various symbolic, text and narrative data or data occurring in a graphical form. This kinds of data which are considered often to be "soft" are rich in the most important domains of application. For instance, they are quite predominant in medicine and social sciences. The elimination of such kinds of data from the data analyses were, therefore, neither warrantable, nor rightful, if possible at all. But symbolic data requires an essentially deeper care, than the pure numerical one. The proportion between the effort expended on the preprocessing of some data and on the analysis of such data is often discussed. SCHNEIDER notes in this connection that in most application the problem of data handling, data manipulation and data description is even predominant over the problems of its proper analysis (Schneider, 1980). Similarly, in a historic study data preparation tooks many years whereas its computational analysis is very short (Thaller, 1980). A general rule can be derived on this place: *The softer the data is, the more requires to be preprocessed.*

b) A further reason presents the fact that the most part of the *methods* for data preprocessing are of such a kind that comprehensive mathematical formulations do not exist or even cannot exist (Freiher, 1980). The use of such methods requires, therefore, a creative reasoning of man rather than some routine action. Many other methodological tools such as rules, advices, relevant knowledge bases can be offered for this purpose by special computer software and intended to serve as a consultant, not as a replacement of a man-solver. Such tools can have different theoretical background. Some are derived from mathematics, while the remaining ones originate in other cognitive sciences (see above). Some are based on the hitherto successful real-world experience and judgement and they are called *heuristics*. Thus the chance of the computer in data preprocessing consists besides the performance of the programmable procedures in easy operating if it is to bridge the gap between what can be offered from the relevant background theoretical knowledge to the performance of the non-routine preprocessing procedures and what a human solver is able to remember, know or apprehend. No one can possibly memorize and respect, for instance, all the rules and advices necessary to a successful problem formulation or to a successful prevention of the defective data.

c) Data preprocessing due to its character described above must be conceived as a *knowledge based system* (Davis-Lenat, 1982). An essential part of the suitable programs differ from the usual computational programs in two key ways:

- they emphasize manipulation of symbolic rather than numerical data and
- they are largely informal and nonroutine decision making rules, advices, hints, methods or heuristics gained from experience or from the relevant theories of the individual cognitive sciences or from the general methodology of science rather than some mathematically proved algorithms (Freiher, 1980) and must be performed rather in a man-machine mode.

Hence, data preprocessing is not only a computational process.

SOFTWARE FOR DATA PREPROCESSING AND ITS FORMS.

1. What is done until now. The contemporary situation in the development of the software for data preprocessing is influenced as by the fact that some of the procedures cannot be written as algorithms as by an insufficient stating, systemizing and unification of the special procedures, of which the preprocessing consists. Data planning runs, for instance, spontaneously only whereas the procedures for defect checking are sufficiently elaborated. In more details:

a) The contemporary opinions on the matter can be represented by numerous suggestions which can be found either in medical, or in statistical or in other kinds of special literature. For instance, FINNEY when speaking about the quality of data, errors and statistical software packages concludes that: "We need to ensure that protection is built into the more popular programs perhaps by incorporating into each an interval monitor that can identify some of the more common flaws in the quality of data and in the appropriateness of standard models." (Finney, 1974). SCHNEIDER appeals to the establishing of "banks of methods" for the description, analysis and interpretation of medical data. These method banks should include
- the relevant data handling methods,
- numerical and statistical methods as well as
- data presentation methods

which should be incorporated in computer networks and have access in dialogue and batch mode (Schneider, 1980).

b) The situation in the general purpose statistical program systems. The programs for some routine preprocessing procedures are embedded still now in the greater part of the contemporary general purpose packages. They are restricted, unfortunately, usually on traditional procedures of this kind as, for instance, on the range checking, management of missing values and outliers, the creation of newly derived variables, simple arithmetic transformations of the data sets, data reduction, simple data description including frequencies, tabulation, and graphical display, and some other simple error detections and correction procedures which can be programmed (cf. Dixon-Brown (1981), Lauro-Serio (1982)). Supposing that the programmable procedures make only a certain part of all kinds of procedures which are necessary to the perfect preprocessing of data we must conclude that the general purpose statistical program systems are in this respect unsufficient and do not secure wholly the user against various defects arizing thanks to the lack of well-preprocessed data.

c) Packages specialized on data preprocessing programs were developed in medicine and social sciences. They involve especially "non-traditional" preprocessing procedures such as the domain of data preparation and data acquisition. The program system DUSP for data collection and storage and for handling coded and/or free text medical records acquired by the help of questionnaires and permitting an effective search for input errors, i.e. not "allowed" contents can be mentioned as an example. The system is accompanyied by another one called DUTAP enabling decoding and manipulation the data acquired by DUSP. The mentioned systems provide the user with a sort of shorthand for the reporting the data and offer, however, also program means to decode the shorthand to an understandable narrative text (Bogdanski-Gassinger-Giere, 1981). A similar system is APHAS (Automated Patient History Acquisition System) which can provide a method for obtaining a standardized history from a large segment of the population and offers access, storage, and retrieval (Simmons-Miller, 1970). The preprocessing of historical sources securs CLIO, a databank oriented system for historians. CLIO can be used to administrate a card file of gigantic dimensions on the one hand and on the other provides access to statistical programs. It involves the following modules: preparing the data, plausibility checks, retrieving the data, transforming the data for statistical analysis and listing the data (Thaller, 1980). As a third example let us mention a program package designed by ABEL and BERGER for handling of clinical follow-up data which is compatible with the statistical package ADAM (Abel-Berger, 1983). As software for the procedure "problem formulation" can be exemplified PSL, SSA, SADL and similar methods as described in (Wasserman 1980). Further valuable ideas and attempts how to design software for data preprocessing are described, for instance, in (Jesdinsky, 1978, Pietrzyk, 1981, Wolf, 1980) and in many others.

d) Manuals and conventional paper catalogues of methods. The performance of procedures which cannot be programmed and must be, therefore, made by man can be facilitated and at least significantly supported by various paper forms such as handbooks, manuals or catalogues of relevant rules, hints, advices, heuristics or methods how

to preprocess the data. *Handbooks* contain the total available knowledge and experience that can be brought to bear on a data preprocessing process. They are equipped with indices facilitating linear or even relational references and are capable to be conventionally handled. As an example let us mention the Manual for the Planning and Performance of Therapy Studies (Biefang, 1979). A suitable form of a written dialog of the type "teacher-scholar" can be found, e.g. in (Slotnicky, 1981).
Catalogues of methods for data preprocessing are complete enumerations of relevant hints, advices, rules, heuristics or methods arranged systematically with the necessary descriptive details and equipped with various pointers or indicators making possible references or crossreferences. They can be specialized on the main areas of data preprocessing as, for instance, on problem formulation, data model planning, causes of errors, kinds of errors, defects checking, etc.. Let us emphasize that such and similar conventional paper form tools are in no way oldfashioned. Because, first, they are irreplaceable in cases when they contain methods the performance of which is the exclusive domain of man, second, they can serve as an alternative form to a computerized preprocessing process, third, they are a prerequisite when such catalogues shall be put in a computer-aided form to run in a man-machine dialogue. The design of such paper catalogues should fulfil the same advanced demands which are lay down on the modern software today (Zelkowitz, 1978), at least the:
- top down design,
- stepwise refinement modularity,
- clear cut points and interfaces to other catalogues, to data banks or to relevant program systems.

2. *What could be done.* As it concerns the *routine procedures:*

a) the general purpose statistical program systems should be enriched of all remaining programmable preprocessing procedures

b) the constructing of special program packages should be supported. Such packages should be oriented on a certain kind of procedures or on a certain discipline.

Nonroutine procedures for data preprocessing can be facilitated:

a) by designing and writing of various useful and on a high-level structured conventional paper tools such as guides, manuals, handbooks, dictionnaries and catalogues of methods,

b) by implementing such tools on computers using some of the known software forms: question-answering systems, consultants, banks of rules and methods, or expert systems runing, as a rule, in an interactive mode. Thus systems like a data planner, data cleaner or data transformer could be set up (Friedland, 1979, Freiherr, 1980, Chytil, 1982, Davis-Lenat, 1982). The success of the introducing of such data, preprocessing software depends on the measure of the respecting:

- the compatibility with the hitherto used statistical program systems,
- the modularity of their design and on a coordination in the development of separate models,
- the respecting of all other advanced forms of software engineering (Zelkowitz, 1978, Wasserman, 1980)
- the techniques of symbolic computation (Freiherr, 1980),
- the contemporary research in AI (Davis-Lenat, 1982).

REFERENCES

Abel U., Berger J. (1983), Ein Dialogsystem zur Aufbereitung klinischer Verlaufsdaten, *Statistical Software Newsletter*, 9, 66-70.

Bailyn L. (1977), Research as a cognitive process: Implications for data analysis. *Quantity and Quality*, 11, 97-117.

Biefang S., Köpcke W., Schreiber M. (1979), Manual für die Planung und Durchführung von Therapiestudien. *Medizinische Informatik und Statistik*, 13, 1-92, Berlin-Heidelberg-New York: Springer.

Bogdanski K., Gassinger C., Giere W. (1981). Gesicherte Datenqualität durch Datentypisierung und Dialogprüfung bei Befunderfassung durch DUSP, in Victor N,, Dudeck J., Broszio E.P. (eds.), *Therapiestudien 26. Jahrestagung der GMDS*, Giessen 1981, Berlin-Heidelberg-New York, Springer, pp. 369-377.

Chytil M. K. (1981), Mathematical method: Its form and role in factual cognitive problem solving, *Nature and System*, 3, 209-222.

Chytil M. K. (1983), Data preprocessing and computational data analysis, *Statistical Software Newsletter*, 9, 3-16.

Chytil M. K., Havranek T. (1983), Mechanizing hypothesis formation - A way for computerized exploratory data analysis, *Bull. of ISI*, 44th Meeting, Madrid, 194-119.

Davis R., Lenat D.B. (1982), *Knowledge-based systems in artificial intelligence*, McGraw-Hill, New York.

Dixon W.J., Brown M.B. (1981), *BMDP Statistical Software*, Dept. of Biomathematics UCLA, Univ. of Calif. Press.

Finney D.J. (1974), Problems, data and inference. *J. R. Statist. Soc. A* 137, 1-23.

Freiherr G. (1980), *The seeds of artificial intelligence*, U.S. Dept. of Health, Education and Welfare, NIH, Bethesda, pp. 6-74.

Friedland P.E. (1979), Knowledge-based experiment design in molecular genetics, Stanford Heuristic Programming Project, *MEMO HPP-79-79*, pp. 1-128.

Jezdinsky H. J. et al. (1978), *Memorandum zur Planung und Durchführung kontrollierter klinischer Therapiestudien*, F.K.Schattauer Verlag, Stuttgart-New York.

Lauro N., Serio G. (1982), Criteria for evaluating and comparing statistical software: a multidimensional data analysis approach, *Statistical Software Newsletter* 8, 102-119.

Pietrzyk P., Klar R. (1981), *EDV-Methoden zur Qualitätssicherung der Erfassung medizinischer Daten*, Universität Göttingen.

Schneider B. (1980), Statistical packages of the analysis of clinical studies, *MEDINFO ´80*, Lindberg-Kaihara (eds.), IFIP, North-Holland Publ. Comp., pp. 1011-1013.

Simmons E.M., Miller O. W. (1970), A new concept in automated patient history, in Anderson J., Forsythe J. (eds.), *Information Processing of Medical Recors*, Proc. of the IFIP-T64 Conf. Lyon, North-Holland, pp. 116-132.

Slotnick H.B. (1981), An approach to mathematical literacy for medical students, *Medical Education*, 15, 11-16.

Thaller M. (1980), *Automation on Parnassus. CLIO - a databank oriented system for historians*, Max Planck Institut für Geschichte, Göttingen, pp. 40-65.

Wasserman A.I. (1980), Information system design methodology, *J. Amer. Soc. Infor. Sci.*, January, 5-24.

Wolf G.F. (1980), Auforderungen zur Daten anbereitung aus der Sicht der Epidemiologie, *Statistical Software Newsletter*, 6.

Zelkowitz M.V. (1978), Perspectives on software engineering, *Computing Surveys*, 10, 197-216.

Data Capture and Validation Using Portable Terminals

J.M. Dickson, Edinburgh

SUMMARY

A computing system has been developed for the capture and validation of data on microcomputers. The aim has been to replace pen-and-paper recording with a system which is as simple to use, but in addition provides automatic data validation at the time of recording.

KEYWORDS: microcomputers; data validation; data capture.

1. Introduction

Microcomputers are becoming increasingly cheap and portable. Systems with a microcomputer providing facilities for storage and some elementary processing of the data; possibly linked directly to an electronic measuring device; are replacing pen-and-paper for data recording in many experimental areas. While such a system speeds operations and eliminates errors involved in transcription, the removal of the human element also takes away some of the checks that the individual would intuitively apply. Once identified, however, many of these checks can be incorporated into a computer-based system.

This paper describes such a system for data capture and validation using a portable microcomputer which can be linked to an automatic measuring device such as an electronic balance in a laboratory.

2. Outline of the system

2.1 Program structure

The system is written in BASIC and consists of a number of modules, each being responsible for a specific function, e.g. data input, validation, editing, storage, printing, summarising. All modules can be used independently and new modules can be added easily.

2.2 Program control

The operator controls the program by means of commands. On entry, and whenever a response is required of the operator, a list of options is displayed on the screen. The operator enters the chosen two-character command name at the keyboard followed by "return".

For standard procedures, the appropriate command parameter values can be set within the program, thus simplifying the keyboard input required of the operator

2.3 Data input

Numeric data can be entered at the keyboard or transferred directly from an electronic device via the serial port of the microcomputer. In the latter case, data are transferred continuously from the device until a stable reading is sent.

Data can be input in random order, or in one of several standard orders based on the experimental layout. For the former, the operator enters the order number of the observation to be recorded; for the latter,the order number of the current unit to be recorded is displayed on the screen.The operator then, either presses a key to indicate that a reading is to be transferred from the measuring device, or inputs the value at the keyboard. Once an observation has been validated, it is displayed on the screen and listed on a printer, if attached.

Data input can be varied by the use of control codes, eg:

- to return to command level;
- to indicate the end of the current experiment;
- to cancel input for the current unit;
- to input a data value at the keyboard;
- to move to a particular unit.

On completion of recording, the operator can:

- edit data values for that experiment;
- print the data in derandomised order;
- key in a textual comment associated with the experiment;
- store the data on backing store.

2.4 Calculation of new variates

Once recorded, data can be transformed, variates combined and new variates derived.

2.5 Summarising data

Information from an experimental plan can be input at the keyboard in the same manner as quantitative data. The recorded data can then be derandomised and treatment means and standard errors calculated. Facilities are provided for simple complete block analysis of variance.

2.6 Printing and transmitting data

Data for a single variate or a combination of variates can be displayed on the screen or printer in derandomised order.

Data can be transmitted in character format from the microcomputer to a mainframe or other computer for further processing.

2.7 Data storage

Data for several experiments can be held in computer memory simultaneously. The data are stored in locations which are identified by experiment and by variate. By default, commands operate on all experiments for which there are data in memory. However the operator has the option of specifying the experiments and variates on which the commands are to operate.

For the purposes of recording data, no information is required about the experiment plan; although the number of observations and the number of replicates and treatments must be known. Details of the experimental design are only needed if summaries are required.

Data can be transfered from computer memory to files in character format on cassette tape or diskette. A file contains data for a particular experiment and variate and may be accessed by the experiment and variate name.

3. Data Validation

3.1 Error prevention

Data integrity is best achieved by eliminating, as far as possible, potential causes of error. Much of the information required by the system is program defined; thus a minimum of key input is required of the operator. All responses are checked for errors and inconsistencies; invalid characters are rejected on input. Escape sequences are available to allow return to command level. Much use is made of operator prompts to confirm actions and provide reminders.

3.2 Error recognition and reporting

Successful error prevention depends to a large extent on speed of error detection. In this system error recognition is achieved by the application of both range checks and credibility checks as the data are recorded. Warnings of possible errors can be given on the screen and aurally, if the microcomputer has a sound box. Appropriate corrections can be made immediately.

Gross errors, such as failure to tare the balance, are detected by applying immediate range checks to all directly recorded variates; the limits are program defined for standard measures, but can be user defined if required. Similar validation can be applied to any derived variate calculated as data are recorded. When a data value falls outwith the defined limits, a message to that effect is displayed on the screen and an aural warning given. The operator has the option of accepting the value or repeating the recording.

There is much statistical literature on the subject of the detection of outliers. However the conditions of data acquisition by microcomputer differ markedly from those described in much of this literature. With a microcomputer-based system, data values can be examined in the sequence recorded; also constraints of time and equipment limit computational complexity and the use of graphical techniques.

The approach adopted in the data capture system described, has been to assign a degree of credibility to each new observation from experience of previous recording. This credibility check is based on a predicted value for the new observation and an estimate of the distribution of recorded values around predicted values. In the simple case the predicted value is the mean of all values previously recorded for the current experiment, and the distribution is assumed to be normal with variance based on long term experience with the measure. Warnings, at increasing levels of urgency depending on the deviation of the actual from the expected value, are displayed on the screen. An aural warning is also sounded.

Other data validation techniques which are being considered are:

a) Print out of the mean, variance and coefficient of variation on completion of the experiment;
b) Use of previously recorded data to provide limits for range checks or to compare values directly;
c) Comparison of values with those of the same treatment in another replicate, if details of the experimental layout are known.
d) Use of one of the standard techniques for detecting outliers, e.g. the maximum normed residual (Stefansky, 1971), the Daniel test statistic (Daniel, 1960) or the variance ratio statistic proposed by Goldsmith and Boddy (1973).

3.3 Error correction

Data errors are corrected simply by the operator repeating the recording. The new data value will automatically replace that previously recorded. On completion of recording, facilities are available for correcting values which may, for example, have had their order numbers transposed or recorded incorrectly.

4. Applications

The system described here was designed for the Scottish Agricultural Colleges to record data from crop variety trials. Grass fresh and dry sample weights are recorded directly from a Mettler balance into an Epson HX-20 microcomputer. Data are stored temporarily on microcassette tapes before being transferred directly to a mainframe computer for standard processing. Samples from up to 35 individual trials varying in size from 18 to 120 plots, are recorded on up to nine occasions each year.

A similar system is also in use for recording yield data from horticultural trials. Here, two or three small trials are recorded, via the keyboard, on perhaps 60 occasions during the year; cumulative data only being stored at the the end of each harvest.

It is anticipated that the system will in future be used to record data in the field with crop yields recorded directly from the balance on the harvester.

5. Conclusions

Data validation is an important statistical technique. The advent of the portable microcomputer has meant that extensive data validation can take place at the time of recording. Errors from which, later, it would be impossible to recover, can be spotted and corrected immediately. Direct data recording from electronic devices avoids human transcription errors and measurements are transferred onto a magnetic medium faster and more accurately. However the microcomputer must also attempt to mimic what is best of the human and his pen-and-paper methods; otherwise automatic data capture will replace human error with errors which are more difficult to detect.

In designing the data validation system, we have attempted to incorporate the expertise of a human recorder with some statistical training. It seems likely that systems of this kind will have an increasing role to play in computer-based data acquisition.

References

Daniel, C. (1960), Locating outliers in factorial experiments, Technometrics, 2, 149-156

Goldsmith, P.L. and Boddy, R. (1973), Critical analysis of factorial experiments and orthogonal fractions, Appl. Statist., 22, 141-160.

Stefansky, W. (1971), Rejecting outliers by maximum normed residual, Ann. Math. Statist., 42, 35-45

Missing Data Problems in Growth Curves

E.P. Liski, Tampere

Summary: The paper considers estimation and testing problems in growth curve models, especially when data are incomplete. Two approaches to growth curve estimation and testing for missing data are presented: A technique based on Kleinbaum's model, and a missing data estimation procedure using a Bayesian approach. Also computer programs for growth curve analysis are introduced.

Keywords: Growth curves, missing data, MANOVA, Bayesian linear model.

1. INTRODUCTION

Potthoff and Roy (1964) proposed a generalized analysis of variance model, which provides a general format for a variety of multivariate situations, for example clinical studies of the effect of a drug over time, learning curves and growth curve analysis. Their model accepts, however, no missing data. It is nevertheless often the case in longitudinal analyses such as growth studies that the data are incomplete. Kleinbaum (1973) formulated a generalization for the Potthoff - Roy model which also admits incomplete longitudinal data. Unfortunately the applicability of Kleinbaum's model is impaired by the fact that estimates of the dispersion matrix from incomplete data may not be good.

Here we first briefly describe growth curve analysis when the data are complete. In this case we have several alternative ways of carrying out the necessary calculations. Thereafter our GCM programs for incomplete longitudinal analysis are introduced. These programs are written in FORTRAN and can be divided into three main parts: Estimation and testing programs based on Kleinbaum's model and missing data estimation using a Bayesian approach. A missing data estimation program using the method of maximum likelihood is in preparation. In this approach the principle is to derive maximum likelihood estimates for missing data from the likelihood of the complete data. Dempster, Laird and Rubin (1977) refer to this kind of procedure as the EM algorithm. All the programs and computer systems to be discussed in the sequel have been implemented on the DEC 2060 in the University of Tampere.

2. A GMANOVA MODEL

Our general format is the generalized multivariate analysis of variance (GMANOVA) model put forward by Potthoff and Roy. Their modification was the addition to the MANOVA model of a within-subject design matrix T. The model becomes

$$E(Y) = \underset{n\times m}{X}\ \underset{m\times p}{B}\ \underset{p\times q}{T'}\ , \qquad (2.1)$$

where each row of Y has a multivariate normal distribution with mean vector $\underset{\sim}{\mu}_i$ and variance-covariance matrix Σ, and X and T are known $n \times m$ and $q \times p$ matrices of rank m and p, $m < n$ and $p \leqq q$. B is an $m \times p$ matrix of unknown parameters.

To reduce model (2.1) to the usual multivariate analysis of variance model, Potthoff and Roy suggested the following transformation of Y to Y_1:

$$Y_1 = YG^{-1}T(T'G^{-1}T)^{-1}, \qquad (2.2)$$

where G is any symmetric positive definite $q \times q$ matrix. Then model (2.1) yields $E(Y_1) = XB$, so that the usual multivariate analysis of variance can be applied, resulting in a solution

$$\tilde{B} = (X'X)^{-1}X'YG^{-1}T(T'G^{-1}T)^{-1}. \qquad (2.3)$$

If Σ were known and $G = \Sigma$, $\tilde{B}$ would be the minimum variance unbiased estimator of B under the model (2.1). Potthoff and Roy propose using an estimate of Σ obtained from an independent experiment or choosing $G = I$.

C.R. Rao (1965, 1966) and Khatri (1966) independently gave an alternative reduction of the model (2.1) leading to a conditional model of the form

$$E(\underset{n\times p}{Y_1} \mid \underset{n\times s}{Y_2}) = XB + Y_2\Theta \qquad (2.4)$$

where Y_1 is given by (2.2) and $Y_2 = YH_2$, H_2 being a matrix of order $q \times s$ $(0 \leqq s \leqq q - p)$ and rank s such that $T'H_2 = \theta$. Θ denotes the matrix of unknown regression coefficients on the s concomitant variables contained in Y_2. If we denote $H_1 = G^{-1}T(T'G^{-1}T)^{-1}$ and $H = (H_1, H_2)$, we easily see that $YH = (Y_1, Y_2)$ and $E(Y_1, Y_2) = (XB, \theta)$. Thus the expected value of Y_1 given Y_2 can be shown to be of the form specified by (2.4).

3. COMPUTATION USING MANOVA

Calculation of the estimates and tests of hypotheses for the Potthoff - Roy model and for the model (2.4) can be effected by a conventional multivariate general linear model program that can test a hypothesis of the form $CBD = \theta$. Grizzle and Allen (1969 p. 366) explain how the form (2.4) of the growth curve model can be analyzed by means of a standard multivariate linear model program. It may be pointed out that the classes of Potthoff - Roy estimators and Rao - Khatri estimators coincide (Baksalary, Corsten and Kala 1978). Khatri (1966) showed that the maximum likelihood estimator of B is $\hat{B} = (X'X)^{-1}X'YS^{-1}T(T'S^{-1}T)^{-1}$, which is obtained via the Potthoff - Roy model by substituting $S = Y'[I-X(X'X)^{-1}X']Y$ in place of G in (2.2). The same estimate is obtained from (2.4) by using all q - p concomitant variables. However, testing the hypothesis H_0: CBD = Θ is not the same under the Potthoff - Roy and Rao - Khatri reductions.

There are available a number of computer programs for MANOVA (See Bock and Brandt 1980), which can be used for complete data growth curve analysis. An interactive computer system MATRIX (cf. Pukkila and Puntanen 1980) developed in the Computer Centre of the University of Tampere has also proved useful in these calculations. MATRIX is easily applicable in all statistical operations expressible in matrix language. For example the MATRIX command

```
Y1: = Y*INV(G)*T*INV(TRANS(T)*INV(G)*T);
```

performs the Potthoff - Roy transformation (2.2). The system has a collection of standard functions for matrices like trans (the transpose of a matrix) and inv (the inverse of a matrix), and also functions for scalars. The user can define new functions and store them for later use. As an example we define a function for the Potthoff - Roy transformation:

```
FUNC PR(Y,G,T): = Y*INV(G)*T*INV(TRANS(T)*INV(G)*T);
PR(Y,G,T);
OUTPUT PR: PROY;
```

The function PR can now be used for doing the Potthoff - Roy transformation, and it is stored in the file PROY. Also elementwise functions can be defined.

In fact we can do MATRIX programs simply by converting the matrix formulas of our analysis into MATRIX expressions. These two forms of presentation are rather near to each other. Naturally these programs are not computationally very efficient. MATRIX is also especially useful in co-operation with special purpose programs like MANOVA. Preprocessing of data can be conveniently done by MATRIX, for example the Potthoff - Roy transformation for the usual MANOVA program.

Also the SURDA system, an interactive data matrix preparation and editing system developed in the Computer Centre of Tampere University, offers flexible possibilities for handling data matrices. For example variable transformations can be carried our inside SURDA. There is a link between MATRIX and SUDRA, so that matrices prepared with the SURDA system can be handled with MATRIX and vice versa. The object of operations of the statistical data-processing system SURVO/71 (Pukkila 1980) is a data matrix created by SURDA. SURVO/71 is also developed in the Computer Centre of Tampere University. The programs we describe in the sequel can read SURDA files. There is thus a link between our programs and the SURDA, SURVO/71 and MATRIX systems.

4. MISSING DATA PROBLEM

Our GCM program for incomplete data is based on Kleinbaum's (1973) model. Kleinbaum generalizes the model (2.1) such that the n experimental units may be partitioned into u disjoint sets $S_1, S_2, \ldots, S_u$. On each individual in S_j, measurements are taken at the q_j ($g_j \leqq q$) time-points $t_{j1}, \ldots, t_{jq_j}$. Let K_j be a $q \times q_j$ incidence matrix of zeroes and ones which indicates the times with observations for subjects in S_j. No two different sets S_j and S_k can have measurements at the same q_j time points, although two sets can have the same number of measurements. Then the generalized growth curve model is as follows

$$E(Y_j) = X_j B T' K_j, \qquad (j = 1,2,\ldots,u)$$

where Y_j is an $n_j \times q_j$ matrix of observations and X_j is the $n_j \times m$ known across-individual design matrix. The parameter matrix B is estimated using Kleinbaum's approach.

Hypotheses regarding the parameters in B may be written as

$$H: CBD = \theta \tag{4.1}$$

for known matrices C of dimension $g \times m$ and rank g, and D of dimension $p \times v$ and rank v. To test the hypotheses (4.1) the Wald statistic W_n (Kleinbaum 1973) is computed. W_n is asymptotically distributed as χ^2_w, where $w = gv$. In the case where growth curves are associated with different groups of individuals (a) equality of curves, (b) parallelism of curves, and (c) $\underset{\sim}{\beta}_i = \underset{\sim}{0}$, $i = 1,2,\ldots,p$ are tested automatically, $\underset{\sim}{\beta}_i$ being the *ith* column of B. Any hypothesis can be tested by giving the matrices C and D. The program suggests polynomial growth curves, but arbitrary functions of time are allowed.

Use of Kleinbaum's methods requires a consistent estimator of Σ. As a first choice the program calculates the unbiased, consistent estimator S of $\Sigma = \{\sigma_{ij}\}$, which is a pooled estimator of each σ_{ij} from all those subjects for which the *ith* and *jth* observations are both obtained. However, this estimate of Σ often fails to be even positive semi-definite. In this case we can use a smoothed estimator suggested by Schwertman and Allen (1979) as an estimate of Σ. This is of the form $S^- = \sum_{\lambda_i > 0} \lambda_i^{-1} \underset{\sim}{d}_i \underset{\sim}{d}_i'$, where the λ_i and $\underset{\sim}{d}_i$ are the *ith* eigenvalue and the associated eigenvector of S. It is also possible to use only complete observation vectors to compute an estimate of Σ. Also a prior positive definite estimate of Σ can be used.

5. THE ESTIMATION OF MISSING VALUES

As one solution to the problem of missing data our GCM programs can also use missing data estimation techniques. Once missing values are estimated, one has a completed data set with which to work and standard analysis techniques can be applied. Our method of estimation of missing values is based on the Bayesian theory presented by Lindley and Smith (1972).

We consider now the case of several groups - in other words the one-way analysis of variance situation. For example, assume that we have a sample of bulls from different breeds, and we want to estimate growth curves for every breed.

According to the theory of hierarchical linear models put forward by Lindley and Smith (1972), we assume that each n individuals have a separate multiple regression model:

$$\underset{\sim}{y}_{ij} = X_{ij}\underset{\sim}{\beta}_{ij} + \underset{\sim}{\varepsilon}_{ij}, \qquad i = 1,2,\ldots,m \text{ and } j = 1,2,\ldots,n_i;$$

where $\underset{\sim}{y}_{ij}$ is a $q_{ij} \times 1$ vector, X_{ij} a $q_{ij} \times p$ matrix, $\underset{\sim}{\beta}_{ij}$ a $p \times 1$ vector and $\underset{\sim}{\varepsilon}_{ij}$ a $q_{ij} \times 1$ vector. Now $\underset{\sim}{y}_{ij}$ is the *jth* observation in the *ith* group, for example, the *jth* bull in the *ith* breed.

Assume now that we have some $q - k$ observations $\underset{\sim}{y}_{ij}$ on the *jth* individual in the *ith* group, and we wish to predict further k observations $\underset{\sim}{y}_{ij}^{(f)}$ when the observations $\underset{\sim}{y}_{i1}, \underset{\sim}{y}_{i2}, \ldots, \underset{\sim}{y}_{ij}, \ldots, \underset{\sim}{y}_{in_i}$ on the other individuals in the *ith* group are given. We assume that $\dot{y}_{ij} = (\underset{\sim}{y}_{ij}', \underset{\sim}{y}_{ij}^{(f)'})'$ follows a normal distribution. Further we denote $X' = (X_{ij}', X_{ij}^{(f)'})$ the design matrix corresponding to the *jth* individual in the *ith* group. Applying the linear model theory presented by Lindley and Smith (1972), we obtain estimates for missing values in the m group case as follows

$$\underset{\sim}{y}_{ij}^{(f)} = X_{ij}^{(f)}\underset{\sim}{\beta}_{ij}^{*}, \tag{5.1}$$

$$\underset{\sim}{\beta}_{ij}^{*} = W_{ij}\hat{\underset{\sim}{\beta}}_{ij} + (I - W_{ij})\Big(\sum_{k=1}^{n_i} W_{ij}\Big)^{-1}\Big(\sum_{k=1}^{n_i} W_{ik}\hat{\underset{\sim}{\beta}}_{ik}\Big), \tag{5.2}$$

where $\hat{\underset{\sim}{\beta}}_{ij} = (X'X)^{-1}X'(\underset{\sim}{y}_{ij}', \underset{\sim}{y}_{ij}^{(f)'})'$ and $W_{ij} = (s_i^{-2}X'X + \Sigma^{*-1})s_i^{-2}X'X$. The estimates s_i^2 and Σ^* need the estimates $\underset{\sim}{\beta}_{ij}^{*}$ in the same manner as given by Lindley and Smith. This is an iterative technique. It is also possible to use the computationally easier and faster Fearn method (Fearn 1975), which is not iterative.

6. DISCUSSION

Most existing growth curve fitting procedures assume that each sample member has a measurement on exactly each target occasion. When some observations are missing, we may try Kleinbaum's method for the analysis of growth curves. The advantage of this approach is its flexibility in expressing a variety of across-individual designs and tests of hypothesis.

If plenty of data are missing or the measurements are not taken on the same occasions, the Kleinbaum's approach is not applicable. In these cases the missing data estimation procedure which assumes a separate growth curve for each individual proves useful. This procedure is based on the theory of linear models by Lindley and Smith. One of the advantages of this approach is its flexibility in allowing each sample member to have measurements on arbitrary occasions. This kind of technique itself is also a growth curve estimation method. Finally, completed data afford the possibility of applying other existing methods developed for growth curve analysis.

REFERENCES

Baksalary, J.K., L.C.A. Corsten and R. Kala (1978). Reconciliation of two different views on estimation of growth curve parameters. *Biometrika* 65, 3, 662-665.

Bock, R.D. and D. Brandt (1980). Comparison of Some Computer Programs for Univariate and Multivariate Analysis of Variance. In: P.R. Krishnaiah, ed., *Handbook of Statistics*, Vol. I. North-Holland, Amsterdam.

Dempster, A.P., N.P. Laird and D.B. Rubin (1977). "Maximum likelihood from incomplete data via the EM Algorithm." *Journal of the Royal Statistical Society* B 39, 1-38.

Fearn, T. (1975). A Bayesian approach to growth curves. *Biometrika* 62, 1, 89-100.

Grizzle, E.G. and D. Allen (1969). Analysis of Growth and Dose Response Curves. *Biometrics*, 25, 357-381.

Khatri, C.G. (1966). A note on a MANOVA model applied to problems in growth curves. *Annals of the Institute of Statistical Mathematics*, 18, 75-86.

Kleinbaum, D.G. (1973). A generalization of the growth curve model which allows missing data. *Journal of Multivariate Analysis*, 3, 117-124.

Lindley, D.V. and A.F.M. Smith (1972). Bayes estimates for the linear model (with discussion). *J.R. Statist. Soc.*, B 34, 1-41.

Potthoff, R.F. and S.N. Roy (1964). A generalized multivariate analysis of variance model useful especially for growth curve problems. *Biometrika* 51, 313-326.

Pukkila, T. and S. Puntanen (1980). Computer usage on first statistics courses in the University of Tampere. Comstat 1980, *Proceedings in Computation Statistics*. Physica-Verlag, Wien, pp. 145-151.

Pukkila, T. and S. Puntanen (1982). The use of MATRIX, an interactive computer system for the analysis of matrices, in the teaching of statistics. Department of Mathematical Sciences, University of Tampere, Finland. Report A 79.

Rao, C.R. (1965). The theory of least squares when the parameters are stochastic and its application to the analysis of growth curves. *Biometrika* 52, 447-458.

Rao, C.R. (1966). Covariance adjustment and related problems in multivariate analysis. In: P.R. Krishnaiah, ed., *Multivariate analysis*. Academic Press, New York, 87-103.

Schwertman, N.C. and D.M. Allen (1979). Smoothing and Indefinite Variance-Covariance Matrix. *Journal of Statistical Computing and Simulation*, 9, 183-194.

Error Free and Formal Computations in Statistics

Error-free Computation in Design of Experiments

D. Collombier, Pau

SUMMARY : *This paper is devoted to the fractional factorial designs analysed by means of linear or log-linear models. The problems of connectedness, robustness against incomplete data and semi-separability of the irregular fractions are considered. Some computational solutions are described which use error-free procedures with fixed point and residue arithmetics.*

KEYWORDS : *Fractional Factorial Designs, Connectedness, Estimability, Robustness against missing data, Semi-separability, Error-free computation.*

1 - INTRODUCTION

The theory of fractional factorial designs - first introduced by Finney [1945] in an agricultural setting - has found increasing use in various fields of experimentation. However many problems in this area are still far from solved. Using error-free computation procedures it is possible to succeed in finding practical solutions to some of these problems.

If we consider the linear models, the problems either of estimability of given parametric functions or of connectedness of the design come first. Is the considered fraction an unbiased design ? Is it of resolution III ? (see Federer *et al.* [1981]). This type of questions may be asked *a priori,* that is before the experiment is started, or *a posteriori* if the experiment is going awry (missing or messy data). For some particular designs it is possible to answer these questions using analytical and combinatorial results. Consider for instance the regular (orthogonal) fractions constructed by the DSIGN method of Patterson [1976]. Using duality results in finite abelian groups Bailey [1977] succeeded in finding a theoretical solution to the problem of connectedness ; a practical solution is given by Kobilinsky *et al.* [1984].

It is well known that combinatorial arguments may be of great help when dealing with irregular fractions (see Birkes *et al.* [1976], Butz [1982]). These arguments solve the two-way and some three-way problems (see Bose [1947], Wynn [1977] and Collombier [1980, p. 147-154]). Answers to other problems are given either by automatic error-free computation of the rank of a design matrix X or by the automatic checking of the belonging of some vectors to $R[X']$ - the vector space spanned by the rows of X .

The fractional factorial designs analysed by means of log-linear or logistic-linear models give rise to similar problems. Thus the procedures presented in paragraph 2 are also available for this designs case.

COMPSTAT 1984 © Physica-Verlag, Vienna for IASC (International Association for Statistical Computing), 1984

The property of robustness against incomplete data as defined by Ghosh [1979] (that is to remain connected when there is a fixed number of missing observations) is considered in paragraph 3. Since it is closely related, the property of having no non-interactive experimental unit (or non-interactive cell see Bishop *et al.* [1975]) is also considered. Error-free automatic construction of a basis of $R[X]^{\perp}$ (the orghogonal complementary subspace to $R[X]$ in the space of observations) makes possible to check the above properties.

The construction of a basis of $R[X]^{\perp}$ gives also a practical answer to the question of the semi-separability of a fractional design (see Mantel [1970] and Collombier [1979]). If semi-separable the analysis of the design can be done by considering the components of the fraction independently.

For lake of space some related problems will not be considered here. For instance results in exact computation of either a so-called least square general-inverse or of the pseudo-inverse of the design matrix, although very useful in case of ill conditioning, will not be mentionned. Some error-free computation procedures are given in the literature by Dodge and Majumdar [1979] for the first type of g-inverse in the framework of the designing of experiments, by Stallings and Boullion [1972], Rao *et al.* [1976] for the pseudo-inverse of any integer matrix.

Accordindly some existence problems of certain types of fractions, like the balanced arrays, which may tackled by looking for the existence of solutions to linear diophantine systems will not be considered here (see Hurt and Waid [1970] for theoretical results and Worm *et al.* [1981] for an algorithm).

2 - CONNECTEDNESS OF FRACTIONAL FACTORIAL DESIGNS

2.1. Example

As an example let us consider as four-way fraction the following incomplete graeco-latin square (GLsquare) of order 4 analysed through a linear model.

Example 1 :

	j
i	k,ℓ

	1	2	3	4
1	1,1	2,2	3,3	4,4
2	2,3	1,4	4,1	
3	3,4	4,3	1,2	
4	4,2	3,1	2,4	

In this example three data are missing. Hence we have to look for resolution III of this irregular fraction (that is for estimability of all the main effects when all the interactions are assumed to be negligible).

The design above being saturated is of resolution III if and only if the associated design matrix is of full row-rank. This matrix is the restriction of the matrix of a GLsquare which is obtained by deleting three rows. But different models and their associated design matrices could be considered. Denote i,j,k and ℓ the levels of the four factors, $Y_{ijk\ell}$ the random yields and $\mathbb{E}Y_{ijk\ell}$ their expected means. Among others we may consider the following three models (the residuals being assumed homoscedastics and non correlated) :

1/ The usual overparametrized model :

$$\mathbb{E}Y_{ijk\ell} = \mu + \alpha_i^1 + \alpha_j^2 + \alpha_k^3 + \alpha_\ell^4$$

taken here without restraints. For this model the matrix of a GLsquare of order p has coefficients 0 or 1 . It is of order $p^2 \times (1 + 4p)$ and of rank $4p - 3$. However the design matrix of the above fraction is of order $(p^2 - 3) \times (1 + 4p)$, with $p = 4$, and the question is then whether its rank equals $4p - 3$.

2/ A model currently used in statistical computing :

$$\mathbb{E}Y_{ijk\ell} = \mu + \alpha_i^1 + \alpha_j^2 + \alpha_k^3 + \alpha_\ell^4 \quad \text{with} \quad \alpha_1^m = 0 \quad \text{for} \quad m = 1,2,3,4 \ .$$

In this second model the matrix of a GLsquare is of full column-rank when deleting as usual the columns corresponding to the α_1^m . Its coefficients are still 0 or 1 . The fraction of example 1 being saturated has a square design matrix X . It is of resolution III if and only if $\det [X] \neq 0$. Therefore the problem is to compute exactly this determinant.

3/ A contrasts model with integer design matrices :

Let for instance

$$P_0' = [1,1,1,1] \ , \quad P_1' = \begin{bmatrix} 1 & -1 & 0 & 0 \\ 1 & 0 & -1 & 0 \\ 1 & 0 & 0 & -1 \end{bmatrix}$$

In such a model a GLsquare of order 4 has a design matrix of the form

$$[P_0 \otimes P_0 \mid P_0 \otimes P_1 \mid P_1 \otimes P_0 \mid S_3 P_0 \otimes P_1 \mid S_4 P_0 \otimes P_1]$$

where $\otimes$ is the Kronecker product, S_3 and S_4 are suitable permutation matrices which depend on the structure of the GLsquare. The matrix above is of full column rank. In this context the question is still whether the determinant of the design matrix of the example 1 fraction is null.

Other popular models like the orthogonal polynomials ones are not investigated here. Their design matrices have in general irrational coefficients and cannot be exactly computer coded.

We now compare the models for the example from the computational viewpoint. Since the computation of a determinant of a square matrix is easier than the determination of the rank of a rectangular matrix, the overparametrized model is not interesting for our purpose. As for models 2 and 3, the model with smaller absolute value of the design matrix determinant det [X] will be retained. Indeed when computing this value overflows may occur. Hence we recommend to compute before hand some majorizations of |det [X]| .

Such majorizations are given by the Gram determinant of the GLsquare matrix columns (that is the determinant of their squared norms and scalar products). It is proved elsewhere that the Gram determinant of a GLsquare of order p ($p = 4$ in example 1) equals $p^{4(p-2)+2}$ in model 2 and p^{4p+2} in model 3 (see Collombier [1980, p. 95]). It is well known that these Gram determinants equal the sum of the squared minors of maximal order of the designs matrices (see for example Gantmacher [1966]). In example 1 we obtain 2^{10} for model 2 and 2^{18} for model 3 as majorizations of |det [X]| . Then model 2 seems more suitable for our purpose since it has smaller majorization of |det [X]| .

2.2. Choice of a model in the general case

We now consider the general case. Any fractional factorial design being given the aim is to find the rank of the design matrix defined within the frame of a given model. The smaller the order of this matrix the smaller are the computations of the rank. Therefore it seems here preferable to consider only non-overparametrized models. Hence our problem reduces to that of checking whether the design matrix is of full column-rank, or equivalently whether there is a non-null minor of size the number of columns of X . Thus, as previously seen in the example, we look here for a majorization of such minors. Hence we consider the Gram determinant of the columns of X but we use now the well known Hadamard inequality : the Gram determinant is less or equal to the product of the squared norms of the columns of X (see Gantmacher [1966]). Then the absolute values of these minors are less or equal to the square root of this product. The columns of a design matrix having greater norms in type 3 models than in type 2, type 2 models will be preferably retained. (Here type 2 models denote the models deduced from the usual overparametrized ones by restraining given parameters to zero ; type 3 are still contrasts models).

2.3. Algorithms and arithmetics

Simplest procedures for rank determination of a matrix or for determinant computation are triangulation algorithms. By premultiplying X - or its transpose X'- by a suitable matrix, and by permuting the columns if necessary, we transform X - or X' - into

$$\left[\begin{array}{c|c} T & * \\ \hline 0 & 0 \end{array}\right]$$

where T is an upper triangular and non-singular block, * is some block. The rank of X is equal to the order of T . The computations having to be done error-free, we cannot proceed through classical orthogonalizations (e.g. Householder or Givens transformations) which give rise to irrational numbers. However the gaussian procedure of elimination with total pivoting is here suitable but requires in this context fractional arithmetic.

Since fractional arithmetic is cumbersome or adaptated languages like MACSYMA are not currently available, we prefer to resort to a residue arithmetic. We hereafter briefly describe how this setting of the gaussian procedure.

Let $H[X]$ be the majorization of the Gram determinant of the columns of X given by the Hadamard inequality or by some other result. Consider m prime numbers represented in single precision and such that $2\sqrt{H[X]} < p_1 \times \dots \times p_m$. Denote x_{ij} the coefficients of X and for $k = 1,\dots,m$ X^k the matrix with coefficients (in the Galois field $GF[p_k]$) equal to x_{ij} mod p_k . X being a square matrix the Chinese remainder theorem states that :

1/ det $[X^k]$ (computed in $GF[p_k]$) is congruent to det [X] mod p_k ;

2/ det [X] is a simple function of the det $[X^k]$;

3/ there is a one to one correspondence between the integers z smaller than $p_1 \times \dots \times p_m$ in absolute value and the sets $\{z \bmod p ,\dots,z \bmod p_m\}$.

Accordingly for any matrix X (square or not)

$$\text{rank } [X] = \text{Max } \{\text{rank } [X^k] \mid k = 1,\dots,m\}$$

(see Collombier [1980, p. 98-99,111]). Hence successive gaussian eliminations in fixed finite fields enable an error-free computation, give without overflow the rank of the design matrix X and in case of saturated fraction det [X] .

In example 1 we had $\sqrt{H[X]} = 2^{10} = 1024$ as majorization for the absolute value of det [X] . Then it can be seen that for the prime numbers 3 , 7 , 11 and 13 :

$$2\sqrt{H[X]} = 2 \times 1024 < 3 \times 7 \times 11 \times 13 .$$

Furthermore det $[X^k]$, $k = 1,2,3,4$, can be computed by gaussian eliminations and have values 1 , 1 , 9 and 12 respectively. Thus the chinese remainder theorem gives $|\det [X]| = 64$ (see Collombier [1980, p. 100-103] for further details). Note that only one triangulation shows that X is of full rank

$$(\text{since } \det [X] = 0 \iff (\det [X^k] = 0 , \; k = 1,\dots,m)) .$$

Residue arithmetics, although easy to use on computers of any precision, may be thought very cumbersome compared to the fixed point arithmetic. Therefore we now present briefly other triangulation procedures of integer matrices where the computations are error-free feasible in fixed point arithmetic. In these methods all divisions give exact results (all remainders are null).

Without loss of generality suppose that all principal minors of X' are non null up to rank [X] . We note $X^0 = X', X^1, \dots, X^k, \dots$ the successive transformations of X' and x_{ij}^k the coefficients of X^k. To determine rank [X] one could think of the one step elimination procedure where for $i > k$ and $j \geq k$

$$x_{ij}^k = (x_{kk}^{k-1} x_{ij}^{k-1} - x_{ik}^{k-1} x_{kj}^{k-1}) / x_{k-1,k-1}^{k-2} \quad \text{with} \quad x_{00}^{-1} = 1$$

and $x_{ij}^k = x_{ij}^{k-1}$ elsewhere. Bareiss [1972] stated that divisions give exact integer results and that the pivots are the principal minors of X'. An Algol program of this procedure is given by Boothroyd [1966].

Other procedures use some greatest common divisor (GCD) computations. For instance for $i > k$ and $j \geq k$

$$x_{ij}^k = \begin{cases} x_{ij}^{k-1} & \text{when } x_{ij}^{k-1} = 0 \\ (x_{kk}^{k-1} x_{ij}^{k-1} - x_{ik}^{k-1} x_{kj}^{k-1}) / \mathrm{GCD}\,(x_{kk}^{k-1}, x_{ik}^{k-1}) & \end{cases}$$

and $x_{ij}^k = x_{ij}^{k-1}$ elsewhere. Collombier [1980, p. 104-108] stated that the pivots here again are the principal minors of X'.

Unfortunately overflows may occur in exact division procedures with fixed point arithmetic. However in example 1 computation of det [X] using the two procedures above had intermediate results with absolute value less or equal to 64 (see Collombier [1980, p. 107-108] for further details).

Note. Other methods using residue arithmetics to compute the rank of an integer matrix and other results could be investigated. For instance the Gauss-Jordan diagonalisation (see Howell [1971]) and the transformation in Smith or Hermite normal form (see Kannan and Bachem [1979]). Since they seem to be time consumming they will not be considered. Lastly some authors recommend to use p-adic instead of residue arithmetics (see Gregory [1980]).

2.4. Unbiasedness of a design

We may want to know if the fractional factorial design is unbiased with respect to given set of integer linear parametric functions. Consider the integer block matrix $[X' \mid C']$, where X is the design matrix and C is the matrix associated with the parametric functions. We recall that the design is unbiased if and only if rand $[X' \mid C']$ = rank [X]. Hence error-free triangulations where pivots are choosen in the first block enable to check the result. Further details are given by Collombier [1980, p. 130].

3 - ROBUSTNESS AND SEMI-SEPARABILITY

3.1. Construction of a basis for $R[X]^{\perp}$

We assume that a basis of $R[X]$ is given by the columns of a block X_1 in X. Let $X_1' = [X_{11}' \mid X_{21}']$, where X_{11}' is a regular square block of order equal to the rank of X. Then the rows of $[-X_{21} X_{11}^{-1} \mid I]$ define a basis of $R[X]^{\perp}$. Fur-

thermore $X_{21} X_{11}^{-1}$ is the solution of the matrix equation $X_{21} = Z X_{11}$ (1) , where Z is the unknown, X_{21} and X_{11} are integer matrices. Hence it suffices to solve this equation to construct a basis of $R[X]^{\perp}$.
Actually, for error-free computation system to solve is rather $K X_{21} = Y X_{11}$ (2). K is here a diagonal matrix with integer coefficients fixed *a priori* (*e.g.* $K = \det [X_{11}]I$) or in the computational process to ensure that Y is an integer matrix. A basis of $R[X]^{\perp}$ is then given by the rows of $[Y|-K]$. Equation (2) can be solved by triangulation of X_1 , moreover if exact divisions procedures are used rank $[X]$ is also given as a by-product.

Unfortunately overflows may occur. To prevent such possible failure gaussian eliminations with a residue arithmetic could be used. In that case the computation of rank $[X]$ have to be done before-hand. Indeed either some of the X^k matrices ($X^k = [x_{ij} \bmod p_k]$) have rank less than rank $[X]$ or the successive gaussian eliminations use different pivotings. Thus we have to choose the X_1 block of X (in other terms to select a basis of $R[X]$) before starting to solve equation (2) above.

Block X_1 being choosen, the moduli of the residue arithmetic (*i.e.* : the prime numbers p_k) are selected such that $2\sqrt{H[X_1]} < p_1 \times \ldots \times p_m$ and the algorithm proceeds in m steps. Let $B^k = X_{11} \bmod p_k$, $C^k = X_{21} \bmod p_k$ and $d_k = \det [X_{11}] \bmod p_k$. Two cases have to be considered.

1/ Det $[X_{11}] \not\equiv 0 \bmod p_k$: the equation (with Y^k as unknown) $d_k C^k = Y^k B^k$ is solved in $GF(p_k)$ by gaussian elimination and back substitution.

2/ Det $[X_{11}] \equiv 0 \bmod p_k$:

•/ if rank $B^k <$ rank $[X] - 1$, $Y^k = 0$ for the adjoint of B^k is then null in $GF(p_k)$,

•/ if rank $[B^k] =$ rank $[X] - 1$, $\tilde{B}^{k'}$ the adjoint of B^k in $GF(p_k)$ can be computed for example by an algorithm introduced by Shapiro [1963] and $Y^k = C^k \tilde{B}^{k'}$.

The different Y^k being thus obtained, the chinese remainder theorem gives the solution, Y , of equation (2) and $R[X]^{\perp}$ is spanned by the rows of

$$[Y \mid -\det[X_{11}]I] \quad .$$

3.2. Non-interactive units

We shall denote by z_i , $i = 1,\ldots,n$ the coordinates of a vector, z , in the observations space. Any experimental unit i such that $z_i = 0$ for all $z \in R[X]^{\perp}$ is said non-interactive (or a non-interactive cell for contingency tables, see Bishop *et al.* [1975]). A fractional factorial design is robust against missing of any one observation if and only if no unit is non-interactive. Indeed if there is a non-interactive unit necessarily rank $[X^*] =$ rank $[X] - 1$, where X^* is the design matrix of the reduced fraction (*i.e.* the fraction obtained by deleting that unit). Note that in a saturated and connected fraction all units are non-interactive.

Given a basis $\mathcal{B}$ of $R[X]^{\perp}$, unit i is non-interactive if and only if $z_i = 0$ for all $z \in \mathcal{B}$. Hence in order to look for the non-interactive units it suffices to construct a basis of $R[X]^{\perp}$ as above.

3.3. Semi-separable fractions

Suppose that $R[X]^{\perp}$ is a direct sum of q subspaces E_k such that, for all $k \neq m$ and $(y,z) \in E_k \times E_m$, $y_i \neq 0$ implies $z_i = 0$ then the design is said semi-separable (a terminology introduced by Mantel [1970] in the setting of two-way contingency tables).

Let $E_k = \{i \mid i = 1,\dots,n : z_i \neq 0 \ \forall z \in E_k\}$ be the kth component of the fraction and E_0 the set of all non-interactive units. In the so-called cell means models computation of all the parameters estimators (BLUES for linear models, maximum likelihood estimators for log linear models) is greatly facilitated since the different components E_k, $k = 0,1,\dots,q$, are considered independently. Moreover the estimations are equal to the observations in E (see Collombier [1980, p. 117]).

Given a basis of $R[X]^{\perp}$ a simple algorithm gives a decomposition of a fractional factorial design (through the semi-separability), if it exists (see Collombier [1979]). Moreover for two-way fractions analyzed without interaction terms this decomposition is unique and the different components are not semi-separable.

3.4. Further examples

Example 2 :

Consider the following two-way fraction analyzed without any interaction term.

1	1		
1	1	2	2
		2	2

(Here the experimental units are noted 1 or 2). $R[X]^{\perp}$ with dimension equal to 2 is spanned by the two following vectors.

1	-1		
-1	1	0	0
		0	0

0	0		
0	0	1	-1
		-1	1

No unit is non-interactive. The fraction is semi-separable in two components E_1 and E_2 with units which have been noted 1 and 2 respectively. The components are both 2×2 and complete. Hence, if considering a log-linear model, the maximum likelihood estimators of the expected cell means are given in closed form.

Example 3 :

Consider now a $2\times3\times3$ fractional design which is saturated in a model without interaction terms of any order.

0		

	1	1
0		

	1	1

(Here the experimental units are noted 0 or 1). By computing the rank of the design matrix it is proved that this fraction is not connected. Rank $[X] = 5$ and there is a square block X_{11} of order 5 with $|\det[X_{11}]| = 1$. Error-free solving a type (1) equation (see §.3.1) gives the following basis of $R[X]^{\perp}$.

0		

	1	-1
0		

	-1	1

Thus in this example there are two non-interactive units, those noted 0 . Moreover due to the structure of the confounding the set of the remaining units may viewed as a complete 2×2 fraction. Hence if considering a log-linear model maximum likelihood estimators of the cell means are in closed form (see Collombier [1980, p. 185]).

REFERENCES

BAILEY, R.A. (1977) : Patterns of confounding in factorial designs. *Biometrika*, 64, 597-603.

BAREISS, E.M. (1972) : Computational solutions of matrix problems over an integral domain. *J. Inst. Math. Appls*, 10, 68-104.

BIRKES, D., DODGE, Y. and SEELY, J. (1976) : Spanning sets for estimable contrasts in classification models. *Ann. Statist.*, 4, 86-107.

BISHOP, Y.M.M., FIENBERG, S.E. and HOLLAND, P.W. (1975) : *Discrete Multivariate Analysis : Theory and Practice*. The MIT Press, Cambridge (Mass.).

BOOTHROYD, J. (1966) : Algorithm 290 : linear equations, exact solutions. *Comm. A.C.M.*, 9, 683-684.

BOSE, R.C. (1947) : The design of experiments. *Proceedings of the 34th Indian Science Congress*, 1-25.

BUTZ, L. (1982) : *Connectivity in Multi-factor Designs : a combinatorial approach*. Heldermann Verlag, Berlin.

COLLOMBIER, D. (1979) : La semi-séparabilité dans les tables de contingence tronquées. *Data Analysis and Informatics*, E. *Diday et al.* (*eds*)., North-Holland Publishing Company, 169-179.

COLLOMBIER, D. (1980) : *Recherches sur l'analyse des tables de contingence : Etude et mise en oeuvre du modèle log-linéaire*. Université Paul Sabatier, Laboratoire de Statistique, Toulouse, FRANCE.

DODGE, Y. and MAJUMDAR, D. (1979) : An algorithm for finding least square generalized inverse for classification models with arbitrary patterns. *J. Statist. Comput. Simul.*, 9, 1-17.

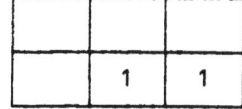

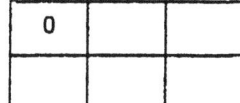

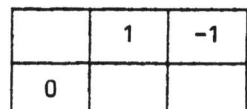

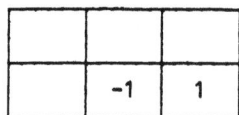

FEDERER, W.T., HEDAYAT, A. and RAKTOE, B.L. (1981) : *Factorial designs*. Wiley, New-York.

FINNEY, D.J. (1945) : The fractional replication of factorial arrangements. *Ann. Eugen.*, 12, 291-301.

GANTMACHER, F.R. (1966) : *Théorie des matrices*. Dunod, Paris.

GHOSH, S. (1979) : On robustness of designs against incomplete data. *Sankhyā*, B, 40, 204-208.

GREGORY, R.T. (1980) : *Error-Free Computation*. R. Krieger Publishing Company, New-York.

HOWELL, J.A. (1971) : Algorithm 406 : exact solution of linéar equations using residue arithmetic. *Comm. A.C.M.*, 14, 180-184.

HURT, M.F. and WAID, C. (1970) : A generalized inverse which gives all the integral solutions to a system of linear equations. *SIAM J. Appl. Math.*, 19, 547-550.

KANNAN, R. and BACHEM, A. (1979) : Polynomial algorithms for computing the Smith and Hermite normal forms of an integer matrix. *SIAM J. Comput.*, 8, 499-507.

KOBILINSKY, A., EL MOSSADEQ, A. et COLLOMBIER, D. (1984) : Orthogonalité dans les groupes abéliens finis et construction de plans factoriels. *Submitted for publication*.

MANTEL, N. (1970) : Incomplete contingency tables. *Biometrics*, 26, 291-304.

PATTERSON, H.D. (1976) : Generation of factorial designs. *J. Roy. Statist. Soc.*, B., 38, 175-179.

RAO, T.M., SUBRAMANIAN, K., KRISHNAMURTHY, E.V. (1976) : Residue arithmetic, algorithms for exact computation of g-inverses of matrices. *SIAM J. Num. Anal.*, 13, 155-171.

SHAPIRO, G. (1963) : Gauss elimination for singular matrices. *Math. Comp.*, 17, 441-445.

STALLINGS, W.T. and BOULLION, T.L. (1972) : Computation of pseudo-inverse matrices using residue arithmetic. *SIAM Review*, 14, 152-163.

WORM, G.H., FISKEAUX, C.D. and WLUDYKA, P.S. (1981) : A computational form for finding the integral solutions to a system of linear equations. *Comp. & Maths with Appls*, 7, 405-406.

WYNN, H.P. (1977) : The combinatorial characterization of certain connected $2 \times J \times K$ three way layouts. *Comm. Statist. Theor. Meth.*, A6, 945-953.

Formal Computation of Regression Parameters

E.-M. Tiit, Tartu

The possibility of formal calculation of regression coefficients and multiple correlation coefficient is demonstrated for a case when the correlation matrix of regressors and the correlation vector between regressors and regressand have a fixed structure. The cases of constant, cyclic and constant-block correlation matrices as well as those with given eigenvalues are considered. The applications of these theoretical results are: the best subset regression problem; the estimation of model bias; the testing of algorithms and their realizations; the studying of the properties of regression models.

Keywords: regression, multiple correlation coefficient, the best subset regression, correlation matrix, eigenvalues, testing of regression algorithms.

1. Introduction

An exact computation of the parameters of multivariate statistical methods, such as regression coefficients and multiple correlation coefficient, is useful for several purposes, for instance:

1) The prior estimation of the best model when a large set of models (regressors' subsets) is available.

2) The investigation of several theoretical properties of regression models and their interpretation in terms of data analysis.

3) Testing of algorithms and their realizations by means of statistical modelling or constructing of random vectors with given regression parameters.

The purpose of the paper is to give some simple possibilities of an exact calculation of regression parameters (regression coefficients b_i and multiple correlation coefficient R) for a suitable number of regressors. Some useful conclusions for data analysis will be made. Some results will be illustrated with computational examples.

We shall consider several types of initial data. Assuming that the regressors X_i ($i = 1, \ldots, m$) and regressand Y are standardized, instead of the covariation matrix, the correlation matrix will be used. The type of independence between the variables is fixed by means of the correlation matrix R of regressors and the correlation vector r, $r = (r_1, \ldots, r_m)$ between Y and X_i.

2. Assumptions

We shall make six different (alternative) assumptions, and, for deriving conclusions, choose a pair out of them.

Our assumptions are the following:

1° The correlation matrix R is constant:

$$R = R(\alpha) = \begin{pmatrix} 1 & \alpha & \alpha & \dots & \alpha \\ \alpha & 1 & \alpha & \dots & \alpha \\ \dots & \dots & \dots & \dots & \dots \\ \alpha & \alpha & \alpha & \dots & 1 \end{pmatrix}.$$

2° The correlation matrix is cyclic:

$$R = \begin{pmatrix} 1 & \alpha_1 & \alpha_2 & \dots & \alpha_{m-1} \\ \alpha_{m-1} & 1 & \alpha_1 & \dots & \alpha_{m-2} \\ \dots & \dots & \dots & \dots & \dots \\ \alpha_1 & \alpha_2 & \alpha_3 & \dots & 1 \end{pmatrix},$$

where $\alpha_i = \alpha_{m-i}$.

3° The correlation matrix consists of constant blocks:

$$R = \begin{pmatrix} R(\alpha_1) : R(\gamma_{12}) : & \dots & : R(\gamma_{1k}) \\ R(\gamma_{21}) : R(\alpha_2) : & \dots & : R(\gamma_{2k}) \\ \dots & \dots & \dots \\ R(\gamma_{k1}) : R(\gamma_{k2}) : & \dots & : R(\alpha_k) \end{pmatrix},$$

where the nondiagonal block consists of equal elements γ_{pq}, the size of blocks $R(\alpha_h)$ and $R(\gamma_{pq})$ are respectively $m_h \times m_h$ and $m_p \times m_q$.

4° The correlation matrix is given with its eigenvalues' matrix Λ, $\Lambda = (\lambda_1, \lambda_2, \dots, \lambda_m)$, and with specially chosen eigenvectors H, $H = (h_1, h_2, \dots, h_m)'$.

5° The correlation vector r is constant, $r = (r, r, \dots, r)'$.

6° The correlation vector consists of constant terms r (their number equals s, $r \neq 0$) and of zeros (their number is q, $q = m-s$).

3. Regression parameters of a constant correlation matrix

If assumption 1° is fulfilled, then the inverse correlation matrix R^{-1} is defined by its elements r^{ij} (Rao 1965):

$$r^{ij} = \begin{cases} (1+(m-2)\alpha)/((1+(m-1)\alpha)(1-\alpha)), & \text{if } i = j, \\ -\alpha/((1+(m-1)\alpha)(1-\alpha)), & \text{if } i \neq j. \end{cases}$$

If assumption 5° holds, too, then all the regression coefficients are equal:

$$b_i = r/(1+(m-1)\alpha), \quad i = 1, \dots, m.$$

Then it is also easy to find the squared multiple correlation coefficient R^2:

$$R^2 = mr^2/(1+(m-1)\alpha). \qquad (1)$$

The regression model exists if and only if $R \in [0, 1]$, that is, when the condition

$$r^2 \leqslant (1-\alpha)/m + \alpha$$

is fulfilled.

When assumption 5° is not fulfilled, then the regression coefficients are, as a rule, different. The multiple correlation coefficient's expression includes a term that depends on the 'variances' of the correlations between the regressand and regressors:

$$R^2 = m\bar{r}^2/(1+(m-1)\alpha) + mD/(1-\alpha), \qquad (2)$$

where $\bar{r}$ is the arithmetical mean of the components of the correlation vector r, and D is the 'variance' of the correlation vector components (Tiit 1981a):

$$\bar{r} = (\sum_i r_i)/m, \quad D = (\sum_i (r_i-\bar{r})^2)/m.$$

When assumption 1° is fulfilled, $\lambda_1 = 1+(m-1)\alpha$ is the largest eigenvalue of the matrix R. All other eigenvalues are equal among themselves and equal $1 - \alpha$. Thus, it is possible to express formulae (1) and (2) through the eigenvalues:

$$R^2 = m\bar{r}^2/\lambda_1, \qquad (1')$$
$$R^2 = m\bar{r}^2/\lambda_1 + mD/\lambda_2. \qquad (2')$$

It follows from the formula (2) that the best models include besides the highly correlated regressors the weakly correlated or uncorrelated ones, too.

The most illustrative proof of that fact becomes evident when we assume 6° to be fulfilled. In this case we get for $\bar{r}$ and D simple expressions:

$$\begin{cases} \bar{r} = rs/m \quad \text{and} \quad D = sr^2(m-s)/m^2, \\ R^2 = (sr^2(1+(m-s-1)\alpha))/((1+(m-1)\alpha)(1-\alpha)). \end{cases} \qquad (3)$$

The best model is defined by means of numbers of non-zero-regressors. The number of zero-regressors is then m - s. The extremum condition gives the formula for s (Tiit 1983):

$$\frac{\partial}{\partial s}R^2 = 0 \quad \Longrightarrow \quad s = (1+(m-1)\alpha)/(2\alpha).$$

This result is of interest from the viewpoint of data analysis, where the determination of best regression model is often a crucial problem.

<u>Example 1</u>

Let assumptions 1° and 6° be fulfilled, whereby $\alpha = 0.2$, $r = 0.4$

and s = m - s = 20. The task is to choose the best model that includes 10 regressors. All the possible models are described in table 1, where the R^2 is computed with the help of formula (3).

Table 1

m-s	0	1	2	3	4	5	6	7	8	9	10
R^2	0.57	0.64	0.69	0.70	0.69	0.64	0.57	0.47	0.40	0.21	0.00

The best model includes 3 zero-regressors.

4. Cyclic correlation matrix

Let us assume that the correlation matrix between regressors is cyclic, that means assumption 2° is fulfilled.

Such correlation matrices appear, for instance, when regressors form a time series. In this case the inverse correlation matrix R^{-1} is cyclic, too, and if assumption 5° is fulfilled, all regression coefficients are equal (Tiit 1983):

$$b_i = r/(1+(m-1)\bar{\alpha})$$

and the expression of R^2 is similar to that of the case of constant correlations:

$$R^2 = mr^2/(1+(m-1)\bar{\alpha}),$$

where $\bar{\alpha}$ is the arithmetical mean of correlations in one row of the matrix R.

5. Constant-block correlation matrix

Let us assume that assumption 3° is fulfilled, that means the correlation matrix R consists of the blocks $R(\alpha_i)$ (diagonal blocks) and $R(\gamma_{ij})$ (out-of-diagonal blocks).

If regression coefficients and the multiple correlation coefficients need to be computed from such block matrices, it is possible to replace every block of the initial matrix by an element of the new matrix and in such a way to get a new matrix A of size $k \times k$, where k is the number of blocks (Tiit 1981b).

The new matrix A is defined with the following expressions:

$$a_{pq} = \begin{cases} \sqrt{m_p m_q}\,\gamma_{pq}, & \text{if } p \neq q, \\ 1 + \alpha_p(m_p - 1), & \text{if } p = q. \end{cases}$$

In the similar way the new correlation vector c, $c = (c_1, \ldots, c_k)$ will be defined:

$$c_p = \sqrt{m_p}\, r,$$

and the formula of the multiple correlation coefficient will take the form:

$$R^2 = c'A^{-1}c.$$

In case assumption 5° is not fulfilled, the expression of R^2 includes several additional terms similar to formula (2). Then we must compute for every block number p of the vector r its arithmetical mean c_p and 'variance' D_p and at last we get

$$R^2 = c'A^{-1}c + \sum_p (m_p D_p/(1-\alpha_p)). \qquad (4)$$

From the formula obtained we can conclude that for getting a good regression model it is useful to include also the uncorrelated ones in the set of regressors, if it is possible.

Example 2

Let us regard the matrix R of size 60 × 60, that consists of 4 blocks, $m_1 = m_2 = 30$, $\alpha_1 = 0.3$, $\alpha_2 = 0.2$, $\gamma_{12} = 0.1$, $r = 0.4$, the number of 0-regressors is in each block 15. The problem is to find the best model that includes 10 regressors (Tiit 1982).

To solve such problems formula (4) is used with the help of the microcomputer APPLE II, and the correlation coefficients R for all possible models are computed. It became evident that all in there exist $7.54 \cdot 10^{10}$ different models, but the number of different multiple correlation coefficients is 286. The best model is characterized by parameters: $m_1 = 7$, $s_1 = 4$ (that means that the number of 0-regressors is 3) and $m_2 = s_2 = 3$. Then the value of R is 0.888.

If we regard the given R as a theoretical one and generate the sample random variable with given parameters and normal distribution, the calculations show that only in the case when the sample size is more than 1000, it is probable to get the best model by means of some step-regression-procedure. Nevertheless, the expectation of the 'best empirical model' overestimates the best theoretical model, the bias for N = 1000 is about 0.08 (Tiit 1982).

6. Correlation matrix with given eigenvectors

Assume that the matrix of the eigenvalues of the correlation matrix R is given, $\Lambda = \mathrm{diag}(\lambda_1, \lambda_2, \ldots, \lambda_m)$, $\min \lambda_i \neq 0$. It is evident that there exists an infinite set of correlation matrices with the eigenvalues' vector Λ. Every correlation matrix is identified by means of the matrix H of orthonormal eigenvectors, $H = (h_1 : \ldots : h_m)$, $h_i' = (h_{1i}, \ldots, h_{mi})$, $H'H = I$, $R = H\Lambda H'$. The inverse correlation matrix R^{-1} is also easily computable by means of H and Λ: $R^{-1} = H\Lambda^{-1}H'$.

If assumption 5° is also fulfilled, it is easy to see that $b' = r'H\Lambda^{-1}H'$ and $R^2 = r'H\Lambda^{-1}H'r$.

Let us assume that the matrix H is biorthogonal and uniformly normed, that means

$$HH' = I,$$
$$|h_{ij}| = 1/\sqrt{m}, \quad i,j = 1, \ldots, m. \qquad (5)$$

Then the matrices R and R^{-1} have only m different elements, in every row and every column the sum of elements is constant. Let us denote the eigenvalue that corresponds to the eigenvector with all positive terms, λ_1. It is always possible to choose the positive eigenvector when conditions (5) are fulfilled.

Then R^2 equals to the expression (1'), got for the constant matrix:

$$R^2 = mr^2/\lambda_1.$$

The eigenstructure H, satisfying conditions (5), exists always when $m = 2^h$, $h = 1, 2, \ldots$.

Example 3

Let us assume $m = 8$ and the eigenvalues of correlation matrix R form an arithmetical progression with $d = 0.25$ and $a_1 = 0.125$. Then one of correlation matrices corresponding to given eigenvector has the following form:

$$\begin{pmatrix} 1 & a & b & c & d & e & f & g \\ a & 1 & c & b & e & d & g & f \\ b & c & 1 & a & f & g & d & e \\ c & b & a & 1 & g & f & e & d \\ d & e & f & g & 1 & a & b & c \\ e & d & g & f & a & 1 & c & b \\ f & g & d & e & b & c & 1 & a \\ g & f & e & d & c & b & a & 1 \end{pmatrix}$$

with $a = 0.4375$, $b = 0.3125$, $c = e = f = 0$, $d = 0.1875$, $g = -0.125$. By means of mixtures of vectors with given (symmetrical) marginal distribution it is easy to generate a random vector with given R.

References

Rao C. R. (1965), Linear statistical inference and its applications, Wiley, New York.

Тийт Э.-М. (I98Ia), Выбор моделей в линейном регрессионном анализе, Труды ВЦ ТГУ, 46, 60-84.

Тийт Э.-М. (I98Iб), Класс допустимых эмпирических моделей в линейном регрессионном анализе, Вычислительные системы, 88, 99-I06.

Тийт Э -М. (I982), О стабильности модели линейного регрессионного анализа при большом числе регрессоров, Труды ВЦ ТГУ, 49, 76-95.

Тийт Э. А. (I983), Об априорном выборе регрессоров в линейной регрессионной модели, Заводская лаборатория, 49, № 7, 44-49.

Comparison of Some Algorithms for Discrete D-Optimal Design Generation

I.N. Vuchkow, D.L. Damgaliew, and A.N. Donev, Sofia

1. Introduction

In this paper the problem of generating D-optimal designs for the model

$$y_u = \sum_{i=1}^{k} \beta_i f_{iu} + \varepsilon_u = \underline{\beta}^T \underline{f}_u + \varepsilon_u$$

is considered, where by f_{iu} denote known functions of the $x_1, x_2, \ldots, x_m$ factors, β_i being unknown coefficient, ε_u - random disturbance with zero mean and σ^2 variance, y_u is the u-th responce, $u = 1,2,\ldots,N$. The problem aims to obtain a (NxK) design matrix F_N, formed by the values of f_{iu} in the different observations, seeks to a maximal determinant $|F_N^T F_N|$..

A number of algorithms to this and where developed over the recent 15 years and we shall refer to those described by Wynn (1971), Van-Shalkwyk (1971), Fedorov (1972), Mitchell (1974), Galil and Kiefer (1980). A comparison of some of them for a convex design space has been made by Cook and Nachtsheim (1980). Nevertheless a lot of problems relating to their applications still remain obscure. For example the search of the designs in a convex space gives rise to serious computational difficulties and only makes it possible with a comparatively small number of K in the f_{iu} functions.

This paper considers the algorithms when discrete factor spaces are used. The comparison of the algorithms is based on two counteracting requirements - efficiency of designs and speed of computations, the later being quite important, because the computational difficulties make some of the algorithms unuseable with large values of K.

As an efficiency measure we use the D-efficiency criterion:

$$D_{eff} = (|M(\xi_N)|/|M(\xi_*)|)^{1/k}$$

where $M(\ _N) = N^{-1} F_N^T F_N$ is the normalized information matrix of a design with N point and $|M(\xi_*)|$ is the corresponding continuous D-optimal design's determinant. The computational losses will be measured by the CPU-time of our IBM 370/145 computer.

2. Description of the algorithms compared.

We shall use the following notations: N is the finite number of points used for the design generation; $G_N = F_N^T F_N$ is the information matrix of an N-observations design, $G_N = C_N^{-1}$. Supposing (with no loss of generality) that $\sigma^2 = 1$, the covariance between the predicted va-

COMPSTAT 1984 © Physica-Verlag, Vienna for IASC (International Association for Statistical Computing), 1984

lues in points $\underline{x}_i$ and $\underline{x}_j$, is $d_{ij} = \underline{f}_i^T C_N \underline{f}_j$ and the variance in point $\underline{x}_i$ is d_{ii}, we use subscript m for a point added to the design and a subscript p for a point deleted from it. Provided the points are added one by one we can use the well known relations:

$$C_{N+1} = C_N - \frac{C_N \underline{f}_m \underline{f}_m^T C_N}{1 + d_{mm}(N)} \quad (2.1)$$

$$|G_{N+1}| = |G_N|[1+d_{mm}(N)] \quad (2.2)$$

$$d_{ii}(N+1) = d_{ii}(N) - \frac{d_{im}(N) d_{mi}(N)}{1 + d_{mm}(N)} \quad (2.3)$$

The same formulae can be used when a point is deleted by substituting p for subscript m and altering all the signs.

We call an itteration (step) the adding (or deleting) of one point. Following Mitchell (1974) we call an "excursion" of L lenght the sequential adding of L points to the design while deleting L others. Galil and Kiefer (1980) proposed to search the design repeatedly under different initial conditions. Each of these many starts is called a try.

We consider 3 algorithms for the important problem of an initial design choice. The first one which we denote by GKI was proposed by Galil and Kiefer (1980). It proposes to choose first $n_r < K$ points and to add subsequently to them new points one by one in a way that will maximize an information matrix determinant of n_r+j size, where j is the iteration number. Different designs can be obtained with different initial points and 100 tries are usually used for obtaining a better design.

The starting procedure I1 begins with an initial information matrix $G_o = rI_k$, where r being a small positive number ($r \approx 10^{-4} \div 10^{-6}$). A point which maximizes the $d_{ii}(N)$ value is added to the design and used as an initial design for the **next** step. The computation terminate after a given number of steps.

The procedure I2 includes all the stages of I1, but that it recalculates the matrix G_N in order to elliminate the influance of r and then makes an L long excursion.

We divide the procedures for D-optimal designs generation to non-sequential (NS) and sequential (S). When a NS procedure is used then the number of points N is preliminarily given and the program is carried out again from the begining to the end when N changes. With the S-procedures a design with N points is used as an initial one for generating an N+1-points design witbout repeating the procedure from the begining.

The Fedorov's (1972) procedure NSF is an example of NS-procedure. At each step it adds a point $\underline{x}_m$ and deletes simultaneously $\underline{x}_e$ so that

$|G_N|$ increases as much as possible. The Wynn-Mitchell's algorithm NSWM first adds the point which maximizes $|G_N|$ and then deletes those points which decrease its value less. The Van-Shalkwyk algorithm NSVS first deletes a point which decreases $|G_N|$ less and then adds a point maximizes it. An improved Mitchell's (1974) algorithm DETMAX uses excursions of L lenght. These algorithms are described by Cook and Nachtscheim (1981). The MD-algorithm of Galil and Kiefer (1980) is an improvement of DETMAX. It uses the starting procedure GKI and formulae (2.1)÷(2.3) and makes numerous tries. The best result of 100 tries is used.

We use an S-procedure which we call S1. By using I2 it generates an initial design with N=K points. It is improved by NSF algorithm and then formulae (2.1)÷(2.3) are used for sequential addition of points, each one of them ensuring the maximal increase of $|G_N|$.

3. Main results.

3.1. Comparison of the starting procedures.

Second order designs were generated supported by a 3^m set of points. Several tries were made with different number of points N. The procedures I1 and I2 were started only once and GKI -10- tries. The average values of D_{eff} for different N and given number of factors are shown in table 1.

Table 1

m	N	I1	I2	GKI
3	10,12,14,16,17,18,20	0.94632	0.94894	0.95109
4	15,17,18,25,26,27,28	0.91957	0.92507	0.92776
5	22,23,25,26	0.82557	0.86281	0.92699

It can be seen that with respect to the efficiency the best results are obtained by GKI procedure, followed by I2 and I1. The results for GKI and I2 for m=3 and 4 follow closely each other. The three procedures are commensurable with respect to speed of computation, but the time required for obtaining the results in Table 1 by GKI procedure proved 10 times as longer as that used for I1 and I2 because 10 tries had to be performed by it. Galil and Kiefer used GKI as a starting procedure for MD-algorithm making 100 tries, so the time is by far longer in theircase.

3.2. Comparison of the main procedures.

Second order designs with m = 3, 4 and 5 were generated on the basis 3^m support points. As a rule the efficiency decreases with m increasing. The results of comparison of 5 algorithms with m = 5; K = 21 and different values of N are given in Table 2 as typical for these algorithms.

Table 2

N	NSWM	NSVS	DETMAX	MD	S1
21	-	-	0.90021	0.90548	0.90548
22	0.88329	0.86012	0.92980	0.93023	0.91501
23	0.91457	0.88358	0.93173	0.93833	0.92398
24	0.92601	0.90142	-	-	0.93001
25	0.93113	0.94139	0.94129	0.94956	0.93523
26	0.93602	0.94176	0.93710	0.95046	0.93801
27	0.93901	0.94616	0.95163	0.95391	0.93989
28	0.94559	0.94776	0.94935	0.95499	0.94131
29	0.94494	0.94845	0.94775	0.95657	0.94085
AVERAGE	0.92757	0.92133	0.93611	0.94244	0.92997

It can be seen that with respect to the efficiency the algorithms can be arranged as follows: MD, DETMAX, S1, NSWM, NSVS. With m = 3 and 4 the results are similar, with m = 3 the procedure provides better results than DETMAX. The results of DETMAX and MD in Table 2 are taken from Mitchell and Bayne (1978) and Galil and Kiefer (1980).

As can be seen in Table 2 the differences between the Deff values grow larger for smaller values of ν = N-K and they decrease when ν increases. Fig. 1 shows the value of $D = D^k_{eff} - D^s_{eff}$ as a function of ν for m = 4 and 5. D^k_{eff} denotes the MD-procedure efficiency and D^s_{eff} refers to S1-procedure.

Consiquently using procedures of high computation speed is preperable when generating designs with comparetively large number of points ($\nu > 10$).

Comparing the results of Table 1 and 2 we can see that adequate results can often be obtained by the use of starting procedure alone. For example the results obtained with m = 5 for the starting procedure GKI are better than for the NSWM and NSVS.

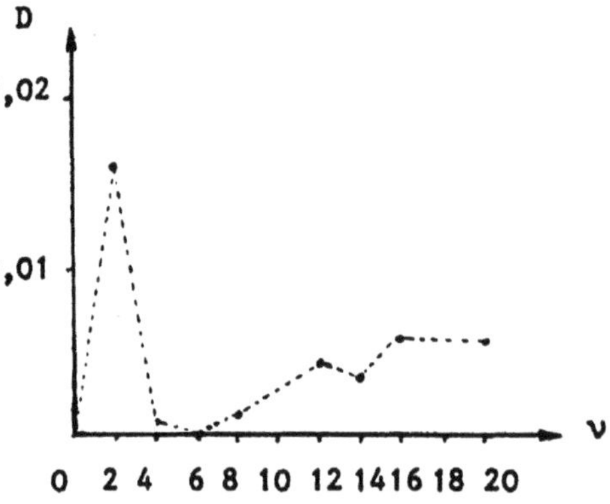

Fig. 1

Concerning the speed of computation the differences between the procedures are very substantial and they be the decisive factor in the choice of a procedure. As pointed out Galil and Kiefer (1980) their MD-procedure is usually 15 to 50 times faster than Mitchell's DETMAX. However it is too slow if compared with S1.

Table 3 compares the efficiencies for m = 3; k = 10 with S1 and MD-procedures as well as the CPU-time for obtaining a design with a given N by use of MD-procedure. The CPU times for N = 16,17,18,20 are not given because the corresponding results are taken from Galil and Kie-

Table 3

N	D^k_{eff}	D^s_{eff}	$D=D^k_{eff}-D^s_{eff}$	CPU time for MD
10	0.8631	0.8631	0.	232.79 s
12	0.9479	0.9312	0.0166	232.36 s
14	0,9757	0.9752	0.0005	293.90 s
16	0.9660	0.9660	0.	-
17	0.9668	0.9665	0.0003	-
18	0.9703	0.9696	0.0007	-
20	0.9779	0.9779	0.	-
22	0.9812	0.9771	0.0041	551.24 s
24	0.9847	0.9811	0.0036	480.22 s
26	0.9903	0.9840	0.0063	716.51 s
30	0.9995	0.9948	0.0047	395.11 s

fer's paper. As can be seen when N increases the time for computation strongly increases too. The total CPU-time for generating designs for 7 values of N (10,12,14,22,24,26,30) the MD-procedure is 2902,13 s. In order to generate designs with m = 3 and 21 values of N (from 10 to 30) the S1 procedure needed CPU time lapse equal to 5.65 s. It is clear that the sequental procedures are much faster than the nonsequental ones. When m increases the results are even more expressive. For example for one try following the MD-procedure with m = 5; K = 21; N = 29 we needed 132.38 s. That means that following Galil and Kiefer's recommendations we have to use about 3 hours and 40 minutes for 100 tries with the MD-procedure, while S1 procedure will only take about 20 s for generating 10 times as more points (N = 21 to 31).

Therefore the sequential procedures can be used with much larger dimensions and for large scale computation than the nonsequential ones. For example the S1 procedure was used for development of a catalogue which contains 123 sequentially generated designs (see Vuchkov et al. (1978)). The largest dimension of the model coefficients vector was K = 60. At the same time the MD-procedure is by far more difficult to be applied even with K = 21 because of the low computational speed. Futhermore as can be seen from the above tables the improved efficiency obtained by MD-procedure compared with S1 could scercely justify the computational losses.

4. Conclusions.

- The use of factor spaces with finite number of points makes it possilbe to substantially cut short the time for D-optimal designs generation which means solving large dimension problems.

- Best efficiency can be obtained by using the MD-procedure, but at the expense of large computational losses. In some cases the S1 procedure is comparable with the MD-procedure.

- With respect to the speed of computations best results can be obtained by S1-procedure.

- When the requierment to the efficiency are not too high sufficient results can be obtained by the starting procedures I2 and GKI.

5. Acknowledgments.

The authors wish to express their appreciation to the late J. Kiefer who kindly placed his MD-program at their disposal for use in the computations with GKI and MD-procedures that were performed.

REFERENCES

1. Cook R.D., C.J. Nachtsheim (1980). A comparison of algorithms for construction of exact D-optimal designs. Technometrics, 22, 315-324.
2. Fedorov V.V. (1972). Theory of optimal experiments. Academic Press, N.Y.
3. Galil Z. and J. Kiefer (1980). Time and space-saving computer methods, related to Mitchell's DETMAX, for finding D-optimum designs. Technometrics, 22, 301-313.
4. Mitchell T.J. (1974). An algorithm for construction of "D-optimal" experimental designs. Technometrics, 16, 203-210.
5. Mitchell T.J. and Bayne, C.K. (1978). D-optimal fractions of three-level factorial designs, Technometrics, 20, 369-379.
6. Van Schalkwyk D.J. (1971). On the design of mixture experiments. Ph.D. thesis, University of London.
7. Vuchkov I.N., Ch.A. Yontchev, D.L. Damgaliev, V.K. Tsochev, T.D. Dikova (1978). Catalogue of sequentially generated designs. Ministry of Education Publishing House, Sofia.
8. Wynn H.P. (1970). The sequential generation of D-optimal experimental designs. Ann. Math. Statist., 41, 1655-1664.

Computational Problems in Spatial Data Analysis

Means for Analysis of Space – Time Relationship

K. Košmelj, Ljubljana

In this paper simple means for analysis of space time relationship between units varying in space and time are introduced. The terms of distance, velocity and acceleration are defined, their characteristics listed. These terms are of great use in the analysis of space-time relationship between units and can be easily incorporated in other statistical methods. An example is presented.

Key words: time, space, distance, velocity, acceleration.

1. INTRODUCTION

In the set of instants of time

(1) $\mathcal{T} = \{t,\ t = t_1,\ t_2, \dots t_T\}$

a set of N units is observed. Let us denote this set $\mathcal{E}$ observed in $\mathcal{T}$ with

(2) $\mathcal{E}_{\mathcal{T}} = \{E^i(t),\ i = 1, 2, \dots N,\ t = t_1,\ t_2, \dots t_T\}$

where $E^i(t)$ represents the i-th unit at the instant t.
Further on, we denote with

(3) $\mathcal{E}_t = \{E^i(t),\quad i = 1, 2, \dots N\}$

the observed set at the instant $t \in \mathcal{T}$.

Each unit $E^i(t)$ is described by a vector of M predicates or variables

(4) $\underline{e}^i(t) = (e_1^i(t),\ e_2^i(t), \dots\ e_M^i(t))$

where $e_j^i(t)$ determines the value of the j-th variable on the i-th unit at the instant t, $e_j^i(t)$, $t = t_1,\ t_2, \dots t_T$ the corresponding time series.

The data can be represented as a three-dimensional matrix: units x variables x time of dimension N x M x T.

The following situation is very often encountered in everyday's life. Some examples: several variables are monthly (annually) observed for several countries in a definite time period, in meteorology many characteristics are daily measured in several stations etc.

The aim of this paper is to introduce a simple mathematical tool which is of great help in the analysis of space-time relationship between units varying in space and time. For that purpose the terms of distance, velocity and acceleration based on physical notions are introduced. In the second section the definitons of these terms are stated, their properties listed. These terms could be incorporated in several statistical methods. In the third section a simple example is presented: under some assumptions these terms are used in extrapolation. In the last section conclusions are presented.

2. DISTANCE, VELOCITY, ACCELERATION

To study the relationship between units observed in time we introduce the following three terms: distance, velocity and acceleration.

2.1. Distance

For each $t \in \mathcal{T}$ a functional d_t: $\mathcal{E}_t \times \mathcal{E}_t \longrightarrow \mathcal{R}^+ \cup \{0\}$ with the following characteristics is defined

D1. $d_t(E^i(t), E^j(t)) \geqq 0$

D2. $d_t(E^i(t), E^i(t)) = 0$

D3. $d_t(E^i(t), E^j(t)) = d_t(E^j(t), E^i(t))$

D4. $d_t(E^i(t), E^j(t)) = 0 \Longrightarrow E^i(t) = E^j(t)$

D5. $d_t(E^i(t), E^j(t)) \leqq d_t(E^i(t), E^k(t)) + d_t(E^k(t), E^j(t))$

D6. $d_t(E^i(t), E^j(t)) = d_t(E^i(t) + \underline{C}(t), E^j(t) + \underline{C}(t))$

$$\underline{C}(t) = (C_1(t), C_2(t), \ldots C_M(t)), \quad C_i(t) \in \mathcal{R}$$

D1. - D6. are valid for each $E^i(t)$, $E^j(t)$, $E^k(t) \in \mathcal{E}_t$.

The functional d_t is called the "distance". It fulfilles the metric axioms D1. - D5. and generates a metric space ($\mathcal{E}_t$, d_t) which is because of D6. called homogenous. Each unit $E^i(t)$ can be represented as a point in this space. A set of T homogenous metric spaces is obtained.

To be able to compare the relationship between units in time the set of homogenous metric spaces should be generated with the same metric $d = d_t$, for each $t \in \mathcal{T}$. The following homogenous metric spaces are obtained $(\mathcal{E}_{t_1}, d)$, $(\mathcal{E}_{t_2} d)$, ...$(\mathcal{E}_{t_T}, d)$.

Let us denote

$$(5) \quad d_t^{ij} = d(E^i(t), E^j(t))$$

d_t^{ij} measures the dissimilarity between i-th and j-th unit at the instant t, or, the distance between two points representing $E^i(t)$ and $E^j(t)$ in $(\mathcal{E}_t, d)$. The definition of d depends on the nature of the data.

For each $t \in \mathcal{T}$ a symetrical matrix of distances D_t with zeroes on the diagonal can be obtained. For further analysis only the upper (lower) triangles of the matrices D_{t_1}, D_{t_2},... D_{t_T} have to be calculated and stored in the computer memory.

2.2. Velocity

The set of distance matrices embodies all the information about the distances between points in the metric spaces. Further on, we are interested to know how the distances between points change from one metric space to the other. Therefore we introduce the term of the "velocity" in the following way

$$(6) \quad v_{t_{l,l+1}}^{ij} = (d_{t_{l+1}}^{ij} - d_{t_l}^{ij}) / (t_{l+1} - t_l) \qquad l=1, 2, \ldots T-1$$

The term $v_{t_{l,l+1}}^{ij}$ measures the velocity between the i-th and j-th unit within the time interval $[t_l, t_{l+1}]$.

The following properties of the velocity can be obtained from the definition (6).

$V_1.$ $v: (\mathbb{R}^+ \cup \{0\}) \times (\mathbb{R}^+ \cup \{0\})$

$V_2.$ $v^{ij}_{t_{l,l+1}} = v^{ji}_{t_{l,l+1}}$

$V_3.$ $v^{ii}_{t_{l,l+1}} = 0$

$V_4.$ $v^{ij}_{t_{l,l+1}} = 0 \rightarrow d^{ij}_{t_{l+1}} = d^{ij}_{t_l}$

$v^{ij}_{t_{l,l+1}} \gtrless 0 \rightarrow d^{ij}_{t_{l+1}} \gtrless d^{ij}_{t_l}$

The values of the velocity show the change of the distances between points in the observed time interval: the positivness of the velocity means that the distance between two points in the observed time interval is increasing, in case of zero velocity the distance is constant, in case of negative velocity the distance is diminishing.

For each couple (t_l, t_{l+1}), $l=1, 2, \dots T-1$ a symetrical velocity matrix $V_{t_{l,l+1}}$ with zeroes on the diagonal can be obtained. A set of T-1 upper (lower) triangles of velocity matrices $V_{t_{1,2}}, V_{t_{2,3}}, \dots V_{t_{T-1,T}}$ has to be calculated.

2.3. Acceleration

We proceed in the same way and introduce the term

$$a^{ij}_{t_{l,l+2}} = (v^{ij}_{t_{l+1,l+2}} - v^{ij}_{t_{l},t_{l+1}}) / (t_{l+2} - t_l), \quad l=1,2,\dots T-2 \tag{7}$$

which we call "the acceleration". Its properties can be easily derived, its meaning found out. A set of T-2 upper (lower) triangles of acceleration matrices $A_{t_{1,3}}, A_{t_{2,4}}, \dots A_{t_{T-2,T}}$ has to be calculated.

3. EXAMPLE

Suppose we are interested in the analysis of $(\mathcal{E}_{t_{T+1}}, d)$, $(\mathcal{E}_{t_{T+2}}, d)$... As there are no data available for t_{T+1}, t_{T+2}, ... an estimation of $\hat{D}_{t_{T+1}}$, $\hat{D}_{t_{T+2}}$, ... based on an acceptable model should be done.

In case of a sufficient length of the observed time interval each time series $e^i_j(t)$, $t=t_1, t_2, \ldots t_T$, for each i and j, $i=1,2,\ldots N$, $j=1,2,\ldots M$ can be modelled using an appropriate metodology. From the models the values $\hat{e}^i_j(t)$, $t=T+1, T+2, \ldots$, for each i and j are obtained, on their basis matrices $\hat{D}_t$, $t=T+1, T+2, \ldots$ can be calculated using the distance d.

The tedious work of modelling N x M time series can be avoided in case of short time intervals $h=t_{T+1}-t_T$, $h_1=t_{T+2}-t_{T+1}$ for the analysis of $(\mathcal{E}_{t_{T+1}}, d)$ and $(\mathcal{E}_{t_{T+2}}, d)$. We assume that the distance is a continous function of time with existing first and second derivatives e.g. velocity and acceleration. Using the Taylor's expansion the following approximations are obtained

$$(8) \quad \hat{D}_{t_{T+1}} = D_{t_T} + h\, V_{t_{T-1,T}}$$

$$(9) \quad \hat{D}_{t_{T+2}} = D_{t_T} + (h+h_1) V_{t_{T-1,T}} + h \cdot h_1\, A_{t_{T-2,T}}$$

The approximations (8) and (9) should be used strictly under the stated assumptions, otherwise the nonnegativness of the distance might be violated.

4. CONCLUSIONS

In this paper simple mathematical tools are developed which facilitate the analysis of time varying data. The definitions of the distance, velocity and acceleration based on physical notions are introduced, their characteristics presented. These terms help us to perceive the local changes in time and space.

The definition of the distance actually roots in the cluster analysis. For the case of clustering time varying data it is natural to introduce the term of the velocity and to incorporate it in the clustering procedure. As we are dealing with metric spaces it seems possible to modificate for example Wishart's Mode Analysis to cluster time varying data with same additional conditions on the velocity of the points. This approach may be restricted by the length of the time period and by the number of units. T distance matrices with N x (N-1)/2 elements have to be stored in the computer memory for further analysis which may be in case of "long" time series impossible and unresonable. For that case a clustering procedure is already developed and presented in Košmelj 1983.

REFERENCES

Bellacicco A.(1976): Clustering time varying data, Recent Developments in Statistics, Proc. of European Meeting of Statisticians, Grenoble, pp. 739-747.

Bohte, Čepar, Košmelj, (1980): Clustering of time series, COMPSTAT 80, Proc. in Computational Statistics, Edinburgh, pp 587-593.

Košmelj, K., Lachet B.(1983): Aspect temporel des variables hydriques du Haut Rhône français, Revue de Statistique Appliquée, XXXI, n°2, pp. 5-18.

Košmelj, K.(1983): Cluster Analysis Taking in Account the Dimension of Time, European Meeting of Classification Societies, Jouy en Josas, p. 127.

Wishart, D.(1969): Mode Analysis. V: A.J.Cole (Ed.): Numerical Taxonomy, Academic Press, New York.

Address-List of Authors

Anderson, A.J.B., Department of Statistics, King's College, Aberdeen AB9 2UB, U.K.
Ansley, C.F., Graduate School of Business, University of Chicago, 1101 E. 58th St. Chicago, Il. 60637, U.S.A.
Antoch, J., MFF UK, KPMS, Sokolovská 86, 186 00 Praha 8, Czechoslovakia
Banet Aluja, T., Dept. of Statistics, Facultad d' Informatica (UPC) C. Jordi Girona Salgado 31 Barcelona 34, Spain
Bethlehem, J., Central Bureau of Statistics, P.O. Box 959, 2270 AZ Voorburg, The Netherlands
Borzemski, L., Technical University of Wroclaw, Wyb. Wyspianskiego 27, 50–370 Wroclaw, Poland
Broesma, H. – see *Molenaar, J.W.*
Chytil, M.K., Physiological Institute, Czechoslovak Academy of Sciences, Vídeňská 1083, 142 20 Praha, Czechoslovakia
Collomb, G., Paul Sabatier University, Toulouse, France
Collombier, D., Université de Pau, Dépt. de Mathématiques et d' Informatique, Avenue Philippon, 64000 Pau, France
Damgaliev, D.L., see *Vuchkov, I.N.*
Dekker, A.L., United Nations, SIAP, Akasaka, P. O. Box 13, Tokyo–107, Japan
Dickson, Janet, AFRC Unit of Statistics, J.C.M.B., The King's Buildings, Mayfield Road, Edinburgh, U.K.
Diday, E., I.N.R.I.A., Domaine de Voluceau, B.P. 105, 78153 Le Chesnay Cedex, France
Dixon, J.T., see *Payne, R.W.*
Donev, A.N., see *Vuchkov, I.N.*
Edwards, D., RECKU, Vermundsgade 5, DK–2100 Copenhagen, Denmark
Gale, W., Bell Laboratories, Room 2C 278, 600 Mountain Ave., Murray Hill, N.J. 07974, U.S.A.
Gilchrist, R., Polytechnic of North London, Dept. Maths., Stats. and Computing, Holloway Road, London N78 DB, U.K.
Glasbey, C.A., University of Edinburgh, JCMB, The King's Buildings, Edinburgh EH9 3JZ, U.K.
Gomes, M.I., Dept. Statistics, Faculty of Science, University of Lisbon, 58 Rua Escola Politecnica, 1294 Lisboa Codex, Portugal
Gordesch, J., Freie Universität Berlin, Babelsbergerstr. 14–16, D-1000 Berlin 30, F.R.G.
Hájek, P., Mathemat. Institute, Czechoslovak Academy of Sciences, Žitná 25. 115 67, Czechoslovakia
Hassani, S., see *Collomb*
Heiser, W.J., see *Meulman, J.*
Jöckel, K.-H., Bremer Institut f. Prävention, Forschung u. Sozialmedizin, BIPS, Präsident-Kennedy-Platz, D-2800 Bremen 1, F.R.G.

Jomier, Geneviève, I.S.E.M.–Bat. 508, Université de Paris-Sud, 91405 Orsay Cedex, France

Jong, V.J. de, Institute of Econometrics, P.O. Box 800, The Netherlands

Kaishev, V.K., Laboratory of Computer Stochastics, Institute of Mathematics, Bulgarian Academy of Sciences, P.O. Box 373, Sofia, Bulgaria

Kezouit, O., see *Jomier, G.*

Kleffe, J., Institute of Mathematics, Academy of Sciences, Mohrenstr. 39, 1086 Berlin, G.D.R.

Kohn, R., see *C.F. Ansley*

Korhonen, P., Helsinki School of Economics, Runeberginkatu 14–16, 00100 Helsinki, Finland

Koszałka, L., Technical University of Wroclaw, Institute of Control and Systems Engineering, ul. Wyb. Wyspanskiego 27, 50–370 Wroclaw, Poland

Košmelj, Katarina, University of Ljubljana, Jamnikarieva 101, 61000 Ljubljana, Yugoslavia

Kowalski, A., Inst. of Computer Science, Polish Academy of Science, PKIN, P.O. Box 22, 00-901 Warsaw, Poland

Krzyśko, M., Mickiewicz University, Institute of Mathematics, 60-769 Poznań, Poland

Křivánek, M., VÚMS, Hviezdoslavova 33, 101 00 Praha 10, Czechoslovakia

Kubáček, L., Mathematical Institute, Slovak Academy of Sciences, Obrancov mieru 49, 814 73 Bratislava, Czechoslovakia

Kuik, D.J., Free Reformed University, Van der Boechorststraat 7, Amsterdam, The Netherlands

Kurzyński, M.W., Technical University of Wroclaw, Institute of Control and Systems Engineering, Wyb. Wyspiańskiego 27, 50-370 Wrocław, Poland

Kutas, T., Comp. and Aut. Institute of HAS, XI. Kende u. 13–17, H-1111 Hungary

Lafay de Micheaux, D., Université de Nice, Lab. Informatique, Parc Valrose, 06034 Nice Cedex, France

Lane, P.W., Northhamsted Experimental Station, Harpenden, Herts., U.K.

Läuter, J., Academy of Sciences of the G.D.R., Institute of Mathematics, Mohrenstr. 39, 1086 Berlin, G.D.R.

Lebart, L., Credoc, 142 rue du Chevaleret, 75013 Paris, France

Lebiediewa, S., Technical University of Wroclaw, Institute of Control and Systems Engineering, Wyb. Wyspanskiego 27, 53-503 Wroclaw, Poland

Lejeune, M., Université de Lausanne, Faculté des Lettres/BC, 1015 Lausanne, Switzerland

Liski, E., University of Tampere, P.O. Box 607, SF-33101 Tampere 10, Finland

Lužar, Vesna, University Computing Centre, Engelsova BB, 41000 Zagreb, Yugoslavia

Meester, L., Fuutlaan 132, 2623 MS Delft, The Netherlands

Mélard, G., Université Libre de Bruxelles, Institut de Statistique, CP 210, Campus Plaine ULB, Bd du Triomphe, B-1050 Bruxelles, Belgium

Messean, A., Laboratoire de Biométrie, C.N.R.Z., 78350 Jouyen-Josas, France
Meulman, Jacqueline, Dept. of Data Theory, University of Leiden, Middelstegracht 4, 2312 TW Leiden, The Netherlands
Michálek, Jaroslav, Dept. of Applied Mathematics, University of Brno, Janáčkovo nám. 2, 662 95, Brno, Czechoslovakia
Michálek, Jiří, Institute of Information Theory and Automation of the Czechoslovak Academy of Sciences, Pod vodárenskou věží 4, 182 08, Praha 8, Czechoslovakia
Molenaar, W., University of Groningen, Vakgroep S&M FSW, Oude Boteringestraat 23, 9712 GC Groningen, Netherlands
Morávek, J., Mathematical Institute, Czechoslovak Academy of Sciences, Žitná 25, 115 67 Praha 1, Czechoslovakia
Moreau, J. V., see *Diday, E.*
Morineau, A., CEPREMAP, 142 Rue du Chevaleret, 75013 Paris, France
Müller, H.-G., Institut für Angewandte Mathematik, Im Neuenheimer Feld 294, D-6900 Heidelberg, F.R.G.
Murphy, B., Faculty of Medicine, University of Western Australia, Nedlands 6009, Australia
Murtagh, F., Dept. A, Bldg. A 36, Joint Research Centre, 21020 Ispra (VA) Italy
Narula, S., School of Business, VCU, 1015 Floyd Ave. Richmond, VA 23284, U.S.A.
Nedoma, P., Institute of Information Theory and Automation, Pod vodárenskou věží 4, 182 08 Praha 8, Czechoslovakia
Němcová, M., see *Michálek Jaroslav*
Neumann, K., State Central Statistic Office, Heinz-Beimler-Str. 70/76, 1026 Berlin, G.D.R.
Nicole, O., see *Messean, A.*
Ososkov, G., Joint Institute for Nuclear Research, JINR, P.O. Box 79, 101000 Moscow, U.S.S.R.
Payne, R. W., Statistics Department, Rothamsted Experimental Station, Harpenden, Herts., AL5 2JQ
Perez, A., Institute of Information Theory and Automation of the Czechoslovak Academy of Sciences, Pod vodárenskou věží 4, 182 08 Praha 8, Czechoslovakia
Pokorný, D., Psychiatric Research Institute, Dept. of Applied Mathematics, 181 03 Praha 8 – Bohnice, Czechoslovakia
Popelinský, L., see *Michálek, Jaroslav*
Prékopa, A., Computer and Automation Institute, Hungarian Academy of Sciences, XI. Kende utca 13–17, H 1502 Budapest, Hungary
Pregibon, D., Bell Laboratories, 2C–272, 600 Mountain Ave., Murray Hill, N.J. 07974, U.S.A.
Pukkila, T., University of Tampere, Dept. of Mathematical Sciences, P.P. Box 607, SF. 33101 Tampere 10, Finland
Puntanen, S., University of Tampere, Dept. of Mathematical Sciences, P.O. Box 607, SF–33101 Tampere 10, Finland

Puranen, J., University of Helsinki, Dept. of Statistics, Aleksanterinkatu 7, 00100 Helsinki 10, Finland
Rafanelli, M., I.A.S.I.–C.N.R., Viale Manzoni 30, 00185 Roma, Italy
Ralambondrainy, H., I.N.R.I.A., B.P. 105, Domaine du Voluceau, 78150 Rocquencourt, France
Rauch, L., State Central Statistical Office, Hans Beimler-Str. 70-72, 1020 Berlin, G.D.R.
Ricci, F.L., I.S.R.D.S.–C.N.R., V.C. de Lollis 12, 00185 Roma, Italy
Ross, G.J.S., Rothamsted Experimental Station, Harpenden, Herts. AL5 2JQ, U.K.
Rudas, T., Institute of Sociology, Eotvos University, Pesti B.u.1., Budapest, H–1052 Hungary
Řebíčková, M., see *Michálek Jaroslav*
Scallan, A., University of Loughborough, Dept. of Mathematics, Loughborough, Leics., U.K.
Schafsma, W., see *Van der Sluis, D.M.*
Scherb, H., Inst. MEDIS GSF, Ingoldstädter Landstr. 1, 8042 München-Neuherberg, F.R.G.
Schütt, A., Universität Osnabrück, Rechenzentrum, 45 Osnabrück, Postfach 4469, F.R.G.
Sint, P., Institut für Sozio-ökonomische Entwicklungsforschung, Fleischmarkt 20, A–1010 Wien, Austria
Sluis, D.M. Van der, Rekencentrum der Rijksuniversiteit, Landleven 1, Paddepoel, 9747 AD Groningen, Netherlands
Smoczynski, D., see *Krzyśko, M.*
Stenman, O., University of Tampere, Ippisenkatu 3, 33300 Tampere 30, Finland
Still, A.W., see *White, A.P.*
Tajuddin, I.H., see *Tracy, D.S.*
Tiit, E., Tartu State University, J. Liivi str. 2, 202400 Tartu, Estonian SSR, U.S.S.R.
Tracy, D.S., Dept. of Mathematics, University of Windsor, Windsor, Ont. Canada
Vajda, I., Institute of Information Theory and Automation of the Czechoslovak Academy of Sciences, Pod vodárenskou věží 4, 182 08 Praha 8, Czechoslovakia
Vajteršic, M., Institute of Technical Cybernetics. Slovak Academy of Sciences, Bratislava, Czechoslovakia
Vila, J.-P., Laboratoire de Biométrie, C.N.R.Z., 78350 Jouy-en-Josas, France
Vuchkov, I., Higher Institute of Chemical Technology, Dept. of Automation, 1156 Sofia, Bulgaria
Wellington, J.F., Cannon University, Erie, Pennsylvania, U.S.A.
Welzl, G., see *Scherb, H.*
White, A.P., The Centre for Computing and Computer Science, The University of Birmingham, P.O. Box 363, Birmingham B15 2TT, U.K.
Whittaker, J., Dept. of Mathematics, University of Lancaster, Lancaster, U.K.
Zagoruiko, N., Institute of Mathematics, Sibirian Division of ANUSSSR, Novosibirsk, U.S.S.R.

MIX
Papier aus verantwortungsvollen Quellen
Paper from responsible sources
FSC® C105338

If you have any concerns about our products,
you can contact us on
ProductSafety@springernature.com

In case Publisher is established outside the EU,
the EU authorized representative is:
Springer Nature Customer Service Center GmbH
Europaplatz 3, 69115 Heidelberg, Germany

Printed by Libri Plureos GmbH
in Hamburg, Germany